T0310137

THE JPEG 2000 SUITE

Wiley-IS&T Series in Imaging Science and Technology

Series Editor:
Michael A. Kriss

Consultant Editors:
Anthony C. Lowe
Lindsay W. MacDonald
Yoichi Miyake

Reproduction of Colour (6th Edition)
R. W. G. Hunt

Colour Appearance Models (2nd Edition)
Mark D. Fairchild

Colorimetry: Fundamentals and Applications
Noburu Ohta and Alan R. Robertson

Color Constancy
Marc Ebner

Color Gamut Mapping
Ján Morovič

Panoramic Imaging: Sensor-Line Cameras and Laser Range-Finders
Fay Huang, Reinhard Klette, and Karsten Scheibe

Digital Color Management (2nd Edition)
Edward J. Giorgianni and Thomas E. Madden

The JPEG 2000 Suite
Peter Schelkens, Athanassios Skodras, and Touradj Ebrahimi (Eds.)

Published in Association with the Society for Imaging Science
and Technology

imaging.org

THE JPEG 2000 SUITE

Edited by

Peter Schelkens
Vrije Universiteit Brussel, Belgium

Athanassios Skodras
Hellenic Open University, Greece

Touradj Ebrahimi
EPFL, Switzerland

A John Wiley and Sons, Ltd., Publication

This edition first published 2009

© 2009, John Wiley & Sons Ltd

Registered office

John Wiley & Sons Ltd, The Atrium, Southern Gate, Chichester, West Sussex, PO19 8SQ, United Kingdom

For details of our global editorial offices, for customer services and for information about how to apply for permission to reuse the copyright material in this book please see our website at www.wiley.com.

Library of Congress Cataloging-in-Publication Data

The JPEG 2000 suite / [edited] by Peter Schelkens, Athanassios Skodras, Touradj Ebrahimi.
    p. cm.
  Includes bibliographical references and index.
  ISBN 978-0-470-72147-6 (cloth)
  1. JPEG (Image coding standard) 2. Image compression. I. Schelkens, Peter. II. Skodras, Athanassios.
III. Ebrahimi, Touradj.
  TA1638.J656 2009
  006.6 – dc22

                2009015214

A catalogue record for this book is available from the British Library.

ISBN: 978-0-470-72147-6 (H/B)

Typeset in 10/12 Times by Laserwords Private Limited, Chennai, India

Printed and bound in Great Britain by CPI Antony Rowe, Chippenham, Wiltshire

Contents

10   Security Applications for JPEG 2000 Imagery   **273**
John Apostolopoulos, Frédéric Dufaux, and Qibin Sun

11   Video Surveillance and Defense Imaging   **291**
Touradj Ebrahimi and Frédéric Dufaux

Contributor Biographies

Yiannis Andreopoulos obtained an Electrical Engineering Diploma and an MSc degree from the University of Patras, Patras, Greece, and a PhD in Applied Sciences from the Vrije Universiteit Brussel (Belgium) in May 2005. During his postdoctoral work at the University of California Los Angeles (US) he performed research on cross-layer optimization of wireless media systems, video streaming, and theoretical aspects of rate-distortion complexity modeling for multimedia systems. From October 2006 to December 2007, he was Lecturer at the Electronic Engineering Department of Queen Mary University of London. Since December 2007, he has been Lecturer at the Electronic and Electrical Engineering Department of UCL. During 2002–2003, Dr Andreopoulos made several decisive contributions to the ISO/IEC JTC1/SC29/WG11 (Motion Picture Experts Group – MPEG) committee in the early exploration of scalable video coding. In 2007, he won the Most-Cited Paper Award from the Elsevier EURASIP *Journal of Signal Processing: Image Communication*. His overall research interests are in the fields of signal transforms, fast algorithms, video coding, and video transmission through unreliable media, e.g. wireless networks and the Internet.

John Apostolopoulos is part of Hewlett-Packard Laboratories, Palo Alto, CA, where he is currently a Distinguished Technologist and Lab Director for the Multimedia Communications and Networking Lab. He is also a Consulting Associate Professor of EE at Stanford University. While at graduate school he contributed to the US Digital TV standard. In collaboration with Susie Wee, he developed an approach for media transcoding in the middle of a network while preserving end-to-end security (secure transcoding) which was adopted by the JPEG 2000 Security (JPSEC) standard. His research interests include improving the reliability, fidelity, scalability, and security of multimedia communications over wired and wireless packet networks. He received his BS, MS, and PhD degrees in EECS from MIT.

Joeri Barbarien obtained the degree of Master in Electrical Engineering in 2000 and the degree of Dr in Engineering Sciences in 2006, both from Vrije Universiteit Brussel, Belgium. Since October 2000, he has been a member of the Department of Electronics and Informatics, Vrije Universiteit Brussel, Belgium, where he is currently working as a postdoctoral researcher and part-time professor. He is also actively involved as a project coordinator in the Interdisciplinary Institute for Broadband Technology (IBBT). His research interests include scalable video and still-image coding and implementation aspects thereof. In 2007, he was the co-recipient of the Most Cited Paper Award

2007 (period 2004–2006) given by Elsevier for the paper 'In-band motion compensated temporal filtering,' published in *Signal Processing: Image Communication*.

Bernard Brower is ITT Technology Fellow of the ITT Space Systems Division. Mr Brower has over 20 years of experience in the development, optimization, and operation of remote sensing systems. He was the lead engineer in the development, optimization, and implementation of the downlink compression for ITT's commercial remote sensing systems (IKONOS, QuickBird, WorldView-1, OrbView-1), and is currently serving as the Head of the US Delegation to the ISO/IEC SC 29/JTC 1 WG 1 (JPEG) committee and the Chairman of the International Committee for Technology Standards (INCITS) L3.2 (Still Image Coding). Mr Brower has a Master of Science/Bachelor of Science from the Rochester Institute of Technology, Center for Imaging Science (1993).

Tim Bruylants graduated as a Master of Science in 2001 at the University of Antwerp. He started out working for a small private company as a systems designer and programmer, creating document publishing software. In 2005, he participated as a member of the Forms Working Group (W3C). In 2006, Tim Bruylants became a PhD student at the Vrije Universiteit Brussel, Belgium. The main topic of his research is the compression of volumetric data sets, using wavelet and geometric transforms. Since 2005, Tim Bruylants has also been an active member of the JPEG committee. He is co-editor of the JPEG 2000 Part 10 (JP3D) specification.

Robert Buckley is a Research Fellow with the Xerox Research Center Webster in Webster, NY. He has been with Xerox since 1981, when he joined the Xerox Palo Alto Research Center after receiving a PhD in Electrical Engineering from MIT. He also has an MA in Psychology and Physiology from the University of Oxford, which he attended as a Rhodes Scholar, and a BSc in Electrical Engineering from the University of New Brunswick. During his career at Xerox, he has held research management and project leadership positions in color imaging and systems and has worked on color printing, image processing, enterprise coherence, and standards for color documents and images. He is the Xerox representative on the US JPEG 2000 committee and was the Project Editor for Part 6 of the JPEG 2000 standard, which defines the JPEG 2000 file format for compound and document images. He currently chairs the CIE Technical Committee on Archival Color Imaging and was founding co-chair of the IS&T Archiving Conference. He is a Fellow of the Society for Imaging Science anad Technology (IS&T) and Past President of the Inter-Society Color Council.

Greg Colyer graduated in Physics & Theoretical Physics from the University of Cambridge and went on to research electrodynamics and quantum mechanics at the University of Sussex. From 1996 to 2004 he worked for Elysium Ltd in Crowborough, UK. He helped to create the Pandora JPEG 2000 demonstrator produced by the MIGRATOR 2000 project and the open-source C implementation of a JPIP proxy server produced by the 2KAN project. As a UK delegate to the JPEG committee, Greg worked on the design of the JPEG 2000 file format and the JPIP network protocol, co-editing ISO/IEC 15444-5/Amd.1 and ISO/IEC 15444-9. With Richard Clark he wrote PD 6777, *Guide to the practical implementation of JPEG 2000*, published by the British Standards Institution. From 2005 to

2007 he lectured in computing and physics, and since 2006 has been a systems analyst, at the University of Guelph, Ontario, Canada.

Eric Delfosse received the MS degree of the Civil Electrotechnical Engineer in Telecommunications from the Free University of Brussels (VUB), Brussels, Belgium, in 1999. From 1999 to 2001, he was a Researcher at the VUB's Telecommunication Research Group. In 2001, he joined the Multimedia Research group of the IMEC Research Center, Leuven, Belgium. He is currently Multimedia Activity Leader at IMEC. His current research interests include wavelet-based image coding algorithms, 3-D graphics, and quality of service. Since December 2001, he has been an active contributor to the ISO/IEC MPEG Standardization Committee, focusing on MPEG-4 Animated Framework eXtension (ISO/IEC 14 496-16) and MPEG-21 Digital Item Adaptation (ISO/IEC 21000-7).

Frédéric Dufaux received his MSc in Physics and PhD in Electrical Engineering from the Swiss Federal Institute of Technology (EPFL), Lausanne, Switzerland, in 1990 and 1994 respectively. From 1990 to 1994, he was a research assistant at the Signal Processing Laboratory at EPFL. In 1994 and 1995, he was a Postdoctoral Fellow at the Media Laboratory of the Massachusetts Institute of Technology. From 1995 till 2001, he was a senior member of research staff at the Cambridge Research Laboratory of Compaq Computer Corporation. In 2001, he joined Genimedia SA as a principal solutions architect. He is currently on the research staff at EPFL. His research interests include image and video coding, motion estimation, image and video analysis, media search and retrieval, archival of media content, media security, media transmission over wireless, and computer vision. He has been involved in the standardization of digital video and imaging technologies in the MPEG and JPEG committees. He is the author or co-author of more than 70 research publications and holds 10 patents in the field of media technologies.

Touradj Ebrahimi received his MSc and PhD, both in Electrical Engineering, from the Swiss Federal Institute of Technology (EPFL), Lausanne, Switzerland, in 1989 and 1992 respectively. He is currently Professor at EPFL heading its Multimedia Signal Processing Group. He is also adjunct Professor with the Center of Quantifiable Quality of Service at Norwegian University of Science and Technology (NTNU). Professor Ebrahimi has been the recipient of various distinctions and awards, such as the IEEE and Swiss national ASE award, the SNF-PROFILE grant for advanced researchers, Four ISO-Certificates for key contributions to MPEG-4 and JPEG 2000, and the best paper award of *IEEE Transactions on Consumer Electronics*. He became a Fellow of the International Society for Optical Engineering (SPIE) in 2003. Professor Ebrahimi has initiated more than two dozen National, European, and International cooperation projects with leading companies and research institutes around the world. He is also the head of the Swiss delegation to MPEG, JPEG, and SC29, and acts as the Chairman of Advisory Group on Management in SC29. He is a co-founder of Genista SA, a high-tech start-up company in the field of multimedia quality metrics. In 2002, he founded Emitall SA, a start-up active in the area of media security and surveillance. In 2005, he founded EMITALL Surveillance SA, a start-up active in the field of privacy and protection. He is or has been associate Editor with various IEEE, SPIE, and EURASIP journals, such as *IEEE Signal Processing Magazine, IEEE Transactions on Image Processing, IEEE Transactions on Multimedia,*

EURASIP Image Communication Journal, EURASIP Journal of Applied Signal Processing, and *SPIE Optical Engineering Magazine*. His research interests include still, moving, and 3-D image processing and coding, visual information security (rights protection, water-marking, authentication, data integrity, steganography), new media, and human computer interfaces (smart vision, brain computer interface). He is the author or the co-author of more than 200 research publications and holds 14 patents. Professor Ebrahimi is a member of IEEE, SPIE, ACM, and IS&T. Email: Touradj.Ebrahimi@epfl.ch.

Robert Fiete is Chief Technologist at the ITT Space Systems Division. Dr Fiete received his BS in Physics and Math from Iowa State University and his MS and PhD in Optical Sciences from the University of Arizona. In 1987 he joined Eastman Kodak's Federal Systems Division as a senior project engineer and later became manager of the Imaging Systems Analysis group. During this time he developed a digital image simulation process for designing remote sensing systems by mathematically modeling the image formation process of the entire imaging chain and generating image simulations to assess the resulting image quality. This process has been used to assess and develop many of the compression algorithms in use today. He has authored over 30 technical papers, received 9 patents, and was awarded the Rudolf Kingslake Medal by SPIE, the International Society of Optical Engineering.

Siegfried Fößel, born in 1964, received his Diploma degree (MS) in Electronic Engineering from the University of Erlangen, Germany, in 1989. He started his professional career as a scientist at the Fraunhofer Institute IIS in Erlangen. He was team leader and project manager for several projects in the field process automation, image processing systems, and digital camera design. In 2000 he received his PhD degree. Since 2001 he has been head of the digital cinema group within the Fraunhofer IIS. He was responsible for projects like ARRI D20, DCI certification plan, JPEG 2000 standardization for Digital Cinema or field recorder Megacine, and coordinated several funded projects like WorldScreen. Siegfried is a member of the ISO SC29/JPEG and TC36 group. Within the JPEG group he is chairing the ad hoc group for Digital Cinema and Motion JPEG 2000. In addition he is a member of SMPTE, FKTG, ISDCF, and DIN. Within the EDCF (European Digital Cinema Forum) he is a member of the technical board.

Ron Gut received a BS in Electrical Engineering from the Massachusetts Institute of Technology in 1992. Until 2007 he was employed at Aware, Inc. in Bedford, MA, where he participated in and led the development of several image compression software products, including spending eight years leading Aware's JPEG 2000 software development team. He was a member of the JPEG standards body, and authored or co-authored several papers and technical reports on image compression.

Hans Hoffman is currently program manager at the European Broadcast Union (EBU). Previously, he was member of the research staff at the Institut für Rundfunktechnik (IRT) in Germany. In 2007, Hans Hoffman obtained his PhD degree at Brunel University (UK) on the subject 'Image quality considerations for HDTV formats in the flat panel display environments.' He is SMPTE Governor for Europe, the Middle East, Africa, Central and South America Region for a Two-Year Term, 2008–2009, and Co-chairman of the Joint

EBU-SMPTE Task Force on Time Labelling and Synchronization. In the past he has been SMPTE Engineering Director of Television.

J. Scott Houchin is an Engineering Specialist at The Aerospace Corporation and was previously a Research Associate at The Eastman Kodak Company. Throughout the JPEG 2000 standardization process, he chaired the file format ad hoc group and was responsible for leading the development of the JP2 and JPX file formats and the metadata transfer capability of the JPEG 2000 internet protocol. Before JPEG 2000, Mr Houchin served as the chief architect of the Flashpix™ digital image format and was a major contributor to the Professional Extensions to the Photo CD system. He has written several papers on the JPEG 2000 file formats and other digital image file format technologies. Mr Houchin is a recipient of a 2006 INCITS Technical Excellence Award for his work on the JPEG 2000 family of standards.

Paul W. Jones is currently Vice President and Director of Research and Development of Certifi Media Inc. and has 24 years of technical experience in the areas of digital image and video processing, image compression, image security, and image quality. He has a proven record of innovation and currently holds 29 US patents and has eight US patent applications. As a Senior Principal Scientist at the Eastman Kodak Company, he made significant contributions to a diverse range of digital imaging applications, including consumer and professional photography, document imaging, medical imaging, and motion picture imaging. Paul was also a co-recipient of the 2005 CEK Mees Award (Kodak's highest research honor) for his work on digital image watermarking. He holds BS and MS degrees in Imaging Science from Rochester Institute of Technology and an MS degree in Electrical Engineering from Rensselaer Polytechnic Institute. He is co-author of the textbook *Digital Image Compression Techniques*, currently in its seventh printing, and has co-authored three book chapters on image compression. He has taught numerous seminars on image compression and has presented new technical contributions at a variety of conferences and forums and in journal articles. He was a co-recipient of the 2007 Journal Award for the best paper of the Society of Motion Picture and Television Engineers (SMPTE).

Rajan L. Joshi received his Bachelor of Technology degree in Electrical Engineering and Masters of Technology degree in Communications Engineering from the Indian Institute of Technology, Mumbai, in 1988 and 1990, respectively. He received his PhD degree in Electrical Engineering from Washington State University, Pullman, in 1996. Presently, he is a Senior Member of Technical Staff at Thomson Corporate Research, Burbank, CA. From 1996 to 2006 he was a Principal Scientist at Kodak Research Laboratories, Rochester, NY. His research interests include image and video compression, information theory, wavelet analysis, and image processing. He has co-authored a book chapter and tutorial in the area of wavelet image compression and JPEG 2000. He holds 12 US patents.

James Kasner received a BS degree in Electrical Engineering from the University of Akron, Akron, OH, in 1987, an MS degree in Electro-optics from the University of Dayton, Dayton, OH, in 1990, and a PhD degree in Electrical Engineering from the

University of Arizona, Tucson, AZ, in 1995. From 1995 to 1999, he was with Optivision, Inc., Palo Alto, CA, where he developed JPEG and wavelet-based image compression algorithms. Thereafter, he was with The Aerospace Corporation, Chantilly, VA, where he conducted research on multi- and hyperspectral image compression, and with Eastman Kodak Corporation. Currently, he is employed at ITT Corporation. Since 1995 he has been active in the ISO JPEG group and has participated in several standards efforts, including the development of the emerging JPEG 2000 standard. He is editor of ISO/IEC 15444-2 and editor of the Multiple Component Transform (MCT) framework Annex J.

Adi Kouadio obtained his MSc degree in communication systems in 2007 from the Ecole Polytechique Fédérale de Lausanne, Switzerland. Since 2007, he has been working as project manager at the European Broadcast Union (EBU) in Geneva, Switzerland. He is chairman of the EBU Correspondence Group on 1080P Picture Format for HDTV (D/1080P). This group reports on the state-of-the-art and industry developments concerning 1080p, monitors development in the standardization bodies, and recommends actions when necessary to ensure open and interoperable systems.

Gauthier Lafruit was a Research Scientist with the Belgian National Foundation for Scientific Research from 1989 to 1994, being mainly active in the area of wavelet image compression implementations. Subsequently, he was a Research Assistant with the Vrije Universiteit Brussel (VUB), Brussels, Belgium. In 1996, he became the recipient of the Scientific Barco Award and joined IMEC, where he was involved as a Senior Scientist with the design of low-power VLSI for combined JPEG/wavelet compression engines. He is currently the Principal Scientist in the Digital Components Unit of the Department on Smart Systems and Energy Technology at IMEC. His main interests include progressive transmission in still image, video, and 3-D object coding, as well as scalability and resource monitoring for advanced, scalable video, and 3-D coding applications. He is the author or co-author of a significant amount of scientific publications, MPEG standardization contributions, and patents (applications) and has participated (and been appointed as evaluator) in several national and international projects. Since 2008, he is also associate editor of *IEEE Transactions on Circuits and Systems for Video Technology*.

Daniel T. Lee is a seasoned technology executive with over 25 years of management experience in the high-tech industry. He is the General Manager of eBay Global Development Centers, with development centers in Shanghai, China, and Chennai, India. Prior to eBay, he was the CTO of Yahoo! Asia, and before that he was with HP, where he held a number of management positions at HP Labs as Manager of Imaging Technology and at HP Internet Imaging Operation where he led the development of OpenPix imaging software and a photo hosting service. He also worked at IBM Research where he worked on image compression. In standardization, he serves as the Convener of ISO/IEC JTC1 SC29/WG1 – JPEG Standards Committee, a position he has held since 1996. As Convener, he works with over 100 experts from 16 countries in developing the highly successful JPEG 2000 family of imaging standards that has been adopted by many groups in the imaging industry. Dr Lee received a BS degree from Cornell University and MS and PhD degrees from Stanford University. He completed the Executive Development Program at Kellogg School of Management, Northwestern University. He has published over 70 publications and serves on the Advisory Committee of the Chinese University of Hong Kong.

Margaret Lepley is a principal scientist at the MITRE Corporation with 25 years of experience and related educational background in mathematical analysis, image compression, spatial and spectral feature analysis, and scientific programming. As an active member of the JPEG committee, she helped develop Parts 1 and 2 of the JPEG 2000 image compression standard. Margaret has applied her knowledge of JPEG 2000 and wavelet compression in general in the fields of remote sensing and fingerprint imaging.

Simon McPartlin works as a software engineer for a think-cell at Software GmbH in Berlin, Germany. He studied Computer Science at the University of Edinburgh and his interests include digital image processing and still image compression. He is co-editor of the JPEG 2000 Part 6 ISO standard and developed the first commercial JPM implementation for LuraTech GmbH.

Michael Marcellin holds the title of Regents Professor of Electrical and Computer Engineering at the University of Arizona. Dr Marcellin was a major contributor to JPEG 2000. Throughout the standardization process, he chaired the JPEG 2000 Verification Model ad hoc group. He is co-author of the book, *JPEG 2000: Image Compression Fundamentals, Standards and Practice*, Kluwer Academic Publishers. Dr Marcellin served as a consultant to Digital Cinema Initiatives (DCI) on the development of the JPEG 2000 profiles for digital cinema. Professor Marcellin is a Fellow of the IEEE and is a member of Tau Beta Pi, Eta Kappa Nu, and Phi Kappa Phi. He is a 1992 recipient of the National Science Foundation Young Investigator Award, and a co-recipient of the 1993 IEEE Signal Processing Society Senior (Best Paper) Award.

Joerg Mohr studied Electrotechnical Engineering at the Friedrich-Alexander University in Erlangen, Germany, where he graduated in 1999. He started working as research engineer at the Fraunhofer Institute for Integrated Circuits IIS, Germany. After his participation in several projects of camera and imaging device development, Joerg Mohr focused on video compression and Digital Cinema technology. He was responsible for JPEG 2000 hardware and software development and participated at various public and private founded research projects. As a member of the JPEG 2000 standardization committee he contributed mainly to Motion JPEG 2000 and Digital Cinema activities. Since 2008 he is head of R&D at KERN electronic GmbH, a Germany-based manufacturer of professional wireless video transmission equipment.

Luk Overmeire obtained his MSc degree of Electrotechnical Engineering at the University of Ghent, Belgium. After his studies he joined Alcatel Bell Antwerp where he designed software-based services and digital filters for ADSL. In 2002, he joined the R&D department of VRT, where he did research on metadata-based video coding and bit-rate control. Currently, he is leading a competence group on media production and processing of audiovisual data. He is a member of the EBU working group on P/TVFile (MXF), AMWA (Advanced Media Workflow Association) and SMPTE's W25. Luk Overmeire is the co-author of several scientific publications and patents.

Robert Prandolini graduated in Electrical and Electronics Engineering from the Queensland University of Technology (QUT) in 1986. While on the academic staff of QUT (1987–1992), he was a consultant in electronic forensics. This led to his PhD studies (QUT) in nonlinear spread-spectrum, which he graduated from in 1996. In 1995 he joined the Australian Federal Police (AFP) as an expert in forensic electronic recording. He moved to the Defence Science and Technology Organisation in 1999 to work on image coding and management, including JPEG 2000 for military applications. He was the chair and co-editor for Part 9 of the JPEG 2000 standard. Over the past few years Dr Prandolini has been working on future imagery and geospatial information systems, was the lead for the ground environment technical implementation team in the Australian 2006 North-West Shelf UAV Trial, and is presently advising Defence on agile dissemination technology for UAVs.

Majid Rabbani received his MS and PhD degrees in electrical engineering from the University of Wisconsin in Madison in 1980 and 1983, and joined Eastman Kodak in the same year. He is a Kodak Distinguished Research Fellow and project manager of video processing at Eastman Kodak. Rabbani is the recipient of the 1988 Kodak C. E. K. Mees Award, and co-recipient of two Engineering Emmy Awards in 1990 and 1997, respectively. He has been actively involved in organizing technical conferences and panels and teaching short courses, both internal and external to Kodak. His current research interests span the various aspects of digital imaging, where he has published many technical articles, four book chapters, and holds more than 30 issued patents. Majid Rabbani was actively involved in the JPEG 2000 standardization process. He is a Fellow of IEEE, a Fellow of SPIE, and a Kodak Distinguished Inventor. He is the co-author of the book *Digital Image Compression Techniques* published by SPIE Press in 1991 and the Editor of the SPIE Milestone Series on *Image Coding and Compression*, published in 1992.

Peter Schelkens received his Electronic Engineering degree in VLSI Design from the Industriële Hogeschool Antwerpen-Mechelen (IHAM), Campus Mechelen, in 1991. Thereafter, he obtained an Electrical Engineering degree (MSc) in Applied Physics in 1994, a Biomedical Engineering degree (Medical Physics) in 1995, and a PhD degree in Applied Sciences in 2001 from the Vrije Universiteit Brussel (VUB). Peter Schelkens currently holds a professorship at the Department of Electronics and Informatics (ETRO) at the Vrije Universiteit Brussel (VUB) and in addition a postdoctoral fellowship with the Fund for Scientific Research – Flanders (FWO), Belgium. Peter Schelkens is a member of the management board of the Interdisciplinary Institute for Broadband Technology (www.IBBT.be), Belgium. Additionally, he has been, since 1995, also affiliated as scientific collaborator to the Interuniversity Microelectronics Institute (www.IMEC.be), Belgium. Peter Schelkens coordinates a research team in the field of multimedia coding, communication, and security. He has published over 100 papers in journals and conference proceedings and holds several patents. His team is participating in ISO/IEC JTC1/SC29/WG1 (JPEG), WG11 (MPEG), and ITU-T standardization activities. Peter Schelkens is the Belgian head of delegation for the ISO/IEC JPEG standardization committee, editor/chair of Part 10 of JPEG 2000: Extensions for Three-Dimensional Data. He is a member of IEEE, SPIE, ACM, and is currently the Belgian EURASIP Liaison Officer. Email: Peter.Schelkens@vub.ac.be.

Louis Sharpe is President of Picture Elements Inc., which develops algorithms, boards, chips, sensors, and software for document scanners. He has been actively involved in JPEG 2000 standardization and is co-editor of Part 6 of the standard and of the color amendment to JBIG2. He has been involved in document imaging and scanner design since the early 1980s. He was a co-author of the SCSI scanner command set, which is still used in USB scanners. He has consulted with the Library of Congress, the Federal Reserve Banks, and the US Patent Office to develop high-quality approaches to imaging documents. He holds a degree in Physics from the University of Colorado Boulder.

Roddy Shuler is Senior Staff Engineer of Image Science Products at the ITT Space Systems Division. Mr Shuler has extensive experience in applying and optimizing JPEG 2000 and JPIP technology to GIS and remote sensing applications. He has been a key software developer for several JPEG 2000 projects, including ITT's Image Access Solutions (IAS) product line, and was the lead image scientist for the development of ITT's JPEG 2000 compression hardware. He has also contributed to the success of the ITT downlink compression algorithm used on the WorldView-1 and GeoEye-1 satellites. He is currently the lead engineer for ITT's airborne compression and dissemination solutions based on JPEG 2000 and JPIP. Mr Shuler holds a Master of Science Degree in Electrical Engineering from Stanford University (1993).

Athanassios Skodras studied Physics at the Aristotle University of Thessaloniki, Greece, and Computer Engineering and Informatics at the University of Patras, Greece. He holds a PhD degree in Electronics from the University of Patras. Since 1986 he has held teaching and research positions at the Departments of Physics and Computer Engineering and Informatics of the University of Patras and the Research Academic Computer Technology Institute, Patras, Greece. From October 2002 he has been Professor of Digital Systems and Head of Computer Science, School of Science and Technology, Hellenic Open University, Patras, Greece. During the academic years 1988–1989 and 1996–1997 he has been a Visiting Research Scientist with the DEEE, Imperial College, London, UK. His research interests include image and video coding, digital watermarking for IPR protection, and video analysis. He has published over 100 technical papers in journals and conference proceedings, authored or co-authored four books, two book chapters, and holds two international patents on compressed domain image processing. He is the co-recipient of the first place Chester Sall Award for the best paper in the *2000 IEEE Transactions on Consumer Electronics* for the work on JPEG 2000 standards. He is a Chartered Engineer, Senior Member of the IEEE, Chair of the Greek Association of Image Processing and Digital Media, and Technical Coordinator of the WG6 on image and video coding of the Greek Organization for Standardization. Email: skodras@eap.gr.

Qibin Sun received a PhD degree in Electrical Engineering from the University of Science and Technology of China (USTC), China, in 1997. Prior to joining Hewlett-Packard China, he worked in the Institute for Infocomm Research, Singapore, and Columbia University, USA. Dr Sun actively participates in professional activities including IEEE ICME, IEEE ISCAS, IEEE ICASSP, and ACM MM. He serves as a member of the Editorial Board of *IEEE Multimedia Magazine*, a member of the Editorial Board of *LNCS Transactions on Data Hiding and Multimedia Security*, and is the Associate Editor of *IEEE Transactions on*

Circuits and Systems on Video Technology and the Associate Editor of *IEEE Transactions on Multimedia*.

Klaas Tack received a BSc degree from Katholieke Hogeschool Brugge Oostende, Belgium, in 1998 and an MEng degree in Electrical Engineering from the Katholieke Universiteit Leuven in 2001. He obtained his PhD degree in Electrical Engineering at the Katholieke Universiteit Leuven in 2007. His research interests focus on terminal QoS and resource management for three-dimensional scalable coding and rendering applications.

Frederik Temmermans studied theoretical Computer Science at the University of Brussels (Vrije Universiteit Brussel) and specialized in artificial intelligence. He graduated in 2006. Thereafter, he started a PhD at the Department of Electronics and Informatics (ETRO) of the same university. His research is mainly situated in the area of medical image retrieval. In the context of his PhD research, Frederick has been following the activities of JPSearch, a part of the JPEG committee, focusing on interoperability between image retrieval systems.

Alexis Tzannes received a BS in Electrical Engineering from the University of Central Florida in 1990, an MS in Electrical Engineering from Brown University in 1992, and a PhD in Electrical Engineering from Northeastern University in 1998. Since 1999 he has been employed at Aware, Inc. in Bedford, MA, where he has led their JPEG 2000 software development team. For the past 10 years he has also been an active participant in the JPEG standards body. He was co-editor of JPEG 2000 Part 10 (JP3D) and is Chair of the JPIP sub-group and the Medical Imaging ad hoc group. He has presented various tutorials and papers on JPEG 2000 and its applications at scientific conferences and symposiums. He was also the author of DICOM Supplement 105, which incorporated the Multi-Component Transformations of JPEG 2000 Part 2 into the DICOM standard. He is the author of several papers in the areas of image compression and point target detection and tracking.

Wolfgang Van Raemdonck received a degree in Industrial Engineering in Electronics from the Karel de Grote Hogeschool, Antwerpen, Belgium, in 1999 and received thereafter in 2003 an MS degree in Artificial Intelligence at the Catholic University of Leuven, Leuven, Belgium. In 1999, he joined the Interuniversity Micro-electronics Center (IMEC), Leuven, as a multimedia developer on wavelet-based image compression algorithms, where he was working on resource constrained three-dimensional (3-D) graphics coding and rendering systems. His main interests include game programming, scalable 3-D modeling, and augmented reality. Since 2008, he has been working for Alcatel-Lucent as a research engineer.

Susie Wee obtain her BS, MS, and PhD degrees from the Massachusetts Institute of Technology in 1990, 1991, and 1996, respectively. Susie Wee is the Vice President of the Experience Software Business in HP's Personal Systems Group. The goal of ESB is to create compelling, easy-to-use experiences and services on HP's personal computing products. She is building a team of experienced designers and software engineers in Cupertino, California, and Shanghai, China. Prior to this, Susie was the lab director of the

HP Labs Mobile and Media Systems Lab, which included research in experience design, streaming media, networking, computer vision and graphics, media security, semantic data management, and next-generation mobile multimedia systems. MMSL had activities in the US, Japan, and England, and included collaborations with partners around the world. Susie was the co-editor of the JPSEC standard for the security of JPEG-2000 images and the editor of the JPSEC amendment on File Format Security. She was formerly an associate editor for *IEEE Transactions on Circuits, Systems, and Video Technology* and for *IEEE Transactions on Image Processing*. In addition to working at HP Labs, Susie has been a consulting assistant professor at Stanford University since 1999. At Stanford she co-taught a graduate-level course on digital video processing. Susie received a Technology Review's Top 100 Young Innovators Award in 2002, the ComputerWorld Top 40 Innovators under 40 Award in 2007, and the INCITs Technical Excellence Award in 2007. She has over 50 international publications and over 50 granted or pending patents. She authors a blog entitled 'Research, Technology, and Teamwork.'

Zhishou Zhang is Principal Investigator at the Institute for Infocomm Research, Singapore. He received his BSc degree in Computer Engineering from Nanyang Technological University in 1998, and his MSc degree in Computer Science and his PhD in Electrical and Computer Engineering from the National Unversity of Singapore in 2002 and 2007, respectively. He has significant research and development experience in MNCs (JVC, Siemens Germany, and HP USA) and research institutes. His expertise encompasses image/video coding, video streaming, media security, networking protocols, network management, and cryptography. He received the Paper Award at ICME 2006 (selected out of 550 papers) and was Best Student Paper Finalist at ICASSP 2007 (selected out of over 1000 papers published). Zhishou Zhang actively participated in the JPSEC standardization process and served as the co-editor for JPEG 2000 Part 8, Amendment 1.

Foreword

Image coding is one of the most popular fields of research since the advent of digital communication in modern times. Researchers are attracted to image coding research for many reasons. It is a very stimulating field requiring firm groundings in image science, digital signal processing, information theory, and systems concepts. A larger reason is the satisfaction of being able to directly see the visual results of the images, often manifested in an interactive way where not only can one appreciate the beauty but also the innovations applied to the images in the research.

The industry also sees the attractiveness of imaging as a significant market potential. As more researchers are attracted to the field, technology coming from research has led to developments of new imaging products and services. Having attractive products, however, is not good enough to ensure market success. As the image itself is often an integral part of the product, market success of the imaging products will depend on the adoption of well-defined imaging standards to support the interoperability of features and functions among the imaging products.

A good example is the success of the consumer digital camera market, which has made the use of digital photos pervasive in recent years. The joy of sharing digital photos is a new phenomenon in human social interactions enjoyed by millions of Internet users today. A key element of this success is due to the availability of a well-adopted standard, in this case the JPEG standard, a highly successful standard that was published in 1988.

With the ubiquity of broadband networks (wired and wireless), a growing number of new applications such as high-resolution imagery, digital libraries, cultural archives, high-fidelity color imaging, Internet applications, wireless, medical imaging, digital cinemas, etc., requires additional, enhanced functionalities from a compression standard, which JPEG cannot satisfy due to some of its inherent shortcomings – design points that were beyond the scope of JPEG when it was developed in the previous decade. The shortcomings of JPEG can be seen in a number of areas: distortion and artifacts, ineffective handling of high-quality images, poor compression for lossless images, lack of effective colorspace support, and lack of resolution scaling. In the mid-1990s, the JPEG committee had an opportunity to start a *new work item* to address these issues and the result is the JPEG 2000 family of standards.

JPEG 2000 makes use of several advances in compression technology to deliver superior compression performance and provides many advanced features in scalability, flexibility, and systems functionalities that outperform its predecessor. In particular, JPEG 2000 uses the discrete wavelet transform (DWT) in place of the discrete cosine transform (DCT) of JPEG. It uses a more sophisticated coding mechanism that supports more flexible, finely embedded representation of the image so that many desirable features are provided in one single bit-stream. JPEG 2000 places a strong emphasis on scalability to the extent that virtually all JPEG 2000 bit-streams are highly scalable.

After the publications of the first six parts of JPEG 2000 standards, which included the core, extensions, motion, conformance, reference software, and multilayer compound image file format, the JPEG committee began investigation of four application areas that are considered important applications for JPEG 2000. Four new parts of JPEG 2000 were established to address these applications: JPSEC, JPIP, JP3D, and JPWL, which deal with security, interactive protocols, multidimensional data sets and wireless applications, respectively. These new parts were designed to address standardization needs of specific application areas to which the rich set of technology from JPEG 2000 apply.

Since its publication, JPEG 2000 has seen successful adoptions in many areas – digital cinemas, security applications, video surveillance, defense imaging, remote sensing, medical imaging, digital culture imaging, broadcast applications, 3-D graphics, etc. The work of JPEG 2000 is continuing with technology maintenance and a new part, Part 14: JPXML – XML Interface to JPEG 2000 Objects. JPEG 2000 is indeed a triumph of innovations and teamwork in the formal standardization process, where the best minds in image coding technology gathered from all over the world to work on a consensus driven process and develop the specification of the most comprehensive image coding system, which will last through many years to come. *The JPEG 2000 Suite* is a remarkable collection of work that documents the development of the JPEG 2000 standards and the related applications that make use of this powerful standard.

The following observations are offered as the success factors learned in the development of future standards based on the experience of managing a complex standardization project such as JPEG 2000.

Importance of the Standardization Process

The foremost objective for developing standards in technology is to ensure compliance in interoperability of the technology among different implementations so that applications/users will benefit from uniformity in functionalities and features when the technology is applied. In the standardization process there are essentially two approaches: formal standards managed by organizations such as ISO/IEC and ITU-T and industry standards managed by an industry consortium. Both approaches have their relative merits – the formal standard approach tends to be more encompassing, inclusive, while the industry consortium based standard approach tends to be more selective and narrowly focused. Either approach will need well-established procedures and rules to manage the standard development process. More importantly, the product life cycle management of the standard once it is published is key to the success of the longevity of the standard. In the formal standard approach, organizations like ISO/IEC and ITU-T, a consensus based approach to the decision making process is adopted, which requires accepting opinions and contributions from minority representatives. This approach will help to lessen the danger of anticompetitive behavior in any settings that involve major forces in the industry. In the case of JPEG 2000, the ISO standardization process has proven to be the right channel to develop such a standard.

When to Develop a Standard?

Timing, maturity of the technology, and market requirements are all key decision factors on when to develop a standard. If the technology is too far ahead of its adoption curve or key technical elements are not ready, then pushing for standardization too early may create nonrealistic requirements and immature solutions that will lead to unsuccessful standards. On the other hand, if the technology is ready but the market window has gone by where alternative technology has taken root and become pervasive such that switching costs may become too high, then the chance for adopting the technology may be lost. In the case of JPEG, the timing for standardization of Motion JPEG was a bit late and hence different versions of Motion JPEG came to the market without a standardized version. In the case of JPEG 2000, the committee recognized the need for a Motion JPEG 2000 standard and made the right decision to open up JPEG 2000 Part 3 as a new standard, which led to the adoption by the Digital Cinema Group.

Value of Standards and Standards Participation

Without the support of a large number of organizations, the JPEG 2000 program would never have developed into such a success in the form of a very powerful image coding system with superior performance meeting all the goals set in the requirements objectives. The support is not only measured in financial terms, the travel and meeting costs for the participants, but in terms of innovations contributed by members of the different organizations, which are the true value of the standard. Further, the credibility that the organizations bring to the working committee lends to the branding and trust of the JPEG 2000 standard. For the participants, working in a standard's body opens one's professional and personal relationships with one's peers and creates a nonstop learning opportunity in one's professional growth.

Daniel Lee
Convener, ISO/IEC JTC1 SC29/WG1
(commonly known as the JPEG committee –
Joint Photographic Experts Group)

Series Editor's Preface

The 8th offering of the Wiley-IS&T Series in Imaging Science and Technology, **The JPEG 2000 Suite**, edited by Peter Schelkens, Athanassios Skodras, and Touradj Ebrahimi, marks two milestones in the series. This is the first edited offering of the series where a group of outstanding scientists and engineers have come together under the umbrella of three editors to write a comprehensive text on an evolving topic in imaging science. Secondly, this is the first text in the series to delve into the physical nature of digital images, the intricate relationships between compression and image quality, and the practical implications of compression on a wide range of image applications.

When most images were recorded and stored on visually addressable media like film, there was no need for compression, and the "shoe box" or photo album was a suitable storage device. Compression algorithms focused on how to store text, or more accurately ASCII code, in a file to save space on magnetic tape or disk drives. While these encoding techniques, like differential pulse code modulation, entropy encoding, run-length coding, Limpel-Ziv-Welch (LZW) compression, etc, were not aimed specifically at images, they would become very important as the digitization of images became commonplace with the introduction of scanned images and digital photography.

It is important to put JPEG 2000 compression into some historical context. While JPEG 2000 represents a significant advance over the series of JPEG compression algorithms, it is still evolving and finding its proper and cost effective place in the world of imaging and telecommunications. This text provides an outstanding snapshot of the current status of JPEG 2000.

One of the earliest applications of image compression was the NTSC (PAL, SECAM) method for encoding color television systems. Prior to color television, about 5 megahertz of bandwidth had been allocated to each television channel with about 5 megahertz separation. The monotone images were of sufficient quality to ensure the rapid development of broadcast television in the industrial countries of the world. When color television became a technical reality, the spectrum allocation developed for monochrome television represented a serious problem. In the absence of compression, three times the bandwidth would be required, which would result in a serious loss of viable broadcasters and a significant loss of service and revenue. To solve this problem the first example of combining the knowledge of color vision and compression was developed. It was clear that a conventional Red-Green-Blue image could be transformed into a Luminance-Chrominance-Chrominance image where the two Chrominance channels demonstrated very little image detail, which was concentrated in the Luminance Channel. Since the two Chrominance Channels contained little detail, their respective bandwidths were greatly reduced, hence the first example of a compressed video signal. The two

Chrominance Channels were further adapted to the 5 megahertz bandwidth limitation by placing them in quadrature (using a sine and cosine wave carrier for the two channels, respectively) and interlacing the quadrature signal between the Luminance Channel signal peaks. While this made color television possible, as home color receivers became better and better, the poor image quality (particularly in the reds) became evident. The above method also introduced very annoying artifacts such as color banding when a very strong Luminance signal "bleeds" into the two Chrominance Channels, such as was found in a "busy" jacket of a newscaster or the striped shirts of referees. This was the first significant example of compression image artifacts. As I write this preface, the United States is just five days away from having only digital TV broadcasts and most of these old NTSC artifacts will be eliminated (but replaced by some different digital artifacts).

While the earliest "electronic" still cameras used NTSC-like formats to record images (from image sensors rather than film), still electronic cameras quickly adopted a digital storage format. So just how much "space" is required to store a digital image in terms of a standard 8-bit byte for each color of each pixel? The answer, of course, depends on the size of the image sensor or the resolution of the scanning device (used to scan a photographic negative, slide or print). The earliest days of digital photography were driven by the common monitor resolutions available at a modest price. These included 320×240 and 640×480 resolutions. As monitor resolution increased, the image sizes increased to 720×480, 1280×720, 1440×1080, and today to HDTV resolution of 1920×1080. Consider a 640×480 image, which contains 307,200 pixels, each of which must have three (red-green-blue) 8-bit values to form the final digital image. Thus, each image has 921,600 bytes of information, or nearly a mega-byte of data. At the same time (in the mid 1990's), internal solid-state storage rarely exceeded a few megabytes. Since a typical roll of film would "store" 24 or 36 images, either much higher solid-state storage capacity devices would be required or some form of image compression would be needed. Since memory was expensive, this led to the development of JPEG (Joint Picture Expert Group) compression where, upon much experimentation and competition, a 8×8 block compression using DCT (Discrete Cosine Transforms) preceded by a Luminance-Chrominance transformation (with various sub-sampling) was used to decrease the size of the image data required for a good image. The image quality-compression tradeoff was then introduced by means of a variable Quantization Matrix (8×8), which further reduced the complexity of the resulting DCT coefficients, which were in turn encoded by means of RLE and entropy or Huffman encoding, resulting in much reduced (up to 10 fold for "poorer" images) file sizes. As image sensors became larger, JPEG compression allowed for the storage of many images per storage media element. IC chips were developed to perform this compression within the camera digital pipelines and this is the current status at the time of writing.

Today, JPEG compression on large images provides outstanding images. So why is there a need for a better compression suite? To answer this question we need to look to the ever-growing internet. A single frame of HDTV would require more than 6 megabytes of uncompressed information, and in an hour show there are 108,000 frames, making for a total of over 650 megabytes of information, or about the storage capacity of current CD-ROM. Today, digital cameras often have more than 12-mega pixels per sensor, meaning 36 megabytes of data are required to store a single color frame. Hence to store 100 images, over 3.6 gigabytes of storage would be required. At the same time, one can purchase storage cards that have over 32 gigabytes of storage, making it possible (at some expense) to store about 900 images without compression. So it is clear that we can compress and store images easily, but can we move them around on the internet or even within devices like printers easily, quickly and without loss of quality? Transmission of image data is one of the most motivating reasons for better compression. Printers and copiers use compression between the host computer and printer/copier hardware to keep the transmission rates, and hence cost, down. Everyday, images are downloaded over the Internet; mostly over connections that allow rates less than one megabit per second. So one frame from a HDTV signal, without compression, would take 48 seconds, which cannot support on-demand movies let alone normal broadcast cable HDTV. It is estimated that with compression a six-megabit bandwidth (MPEG2) would be required to support HDTV. Thus the compression ratio about eight is required to make this possible. The compressed frames must be free of a wide range of image artifacts.

JPEG 2000, with the use of wavelets, provides a means to obtain the critical high compression ratio and still retain high image quality, thus making better use of existing and future telecommunication bandwidths. When one considers that some images, like medical images, high-dynamic range still and motion images, three dimensional graphics, etc may require up to 16 bits per pixel per color, the need of high compression without artifacts is paramount. Given the ever growing demand to "send" images across communication networks (even within equipment like copiers and printers) the value of a highly compressed, artifact free image will make JPEG 2000 a viable alternative to the current JPEG algorithms serving us to date.

The JPEG 2000 Suite provides scientists and engineers with a text that will allow them to understand and implement compression algorithms based on JPEG 2000. The basic principles of JPEG 2000 are given in the first two chapters, while chapters three and four provide in-depth understanding of file formats. Chapters five through eight deal with securing and extending JPEG 2000 into interactivity and three-dimensional data. Chapters nine through sixteen provide a series of important applications of JPEG 2000, ranging from medical imaging to digital cinema to broadcast application. All in all, *The JPEG 2000 Suite* provides a unique source of basic understanding and applications of compression in a digital world. It belongs on the desk of every serious student of imaging science and technology, as well as every active participant in digital still camera development and all forms of telecommunications products and applications.

Michael A. Kriss
Formerly of Eastman
Kodak Research Laboratories
and the University of Rochester

Preface

The Joint Photographic Experts Group (JPEG)[1] committee has a long tradition in the development of still image coding standards. JPEG is a joint working group of the International Standardization Organization (ISO)[2] and the International Electrotechnical Commission (IEC).[3] It resides under JTC1, which is the ISO/IEC Joint Technical Committee for Information Technology. More specifically, the JPEG committee is Working Group 1 (WG1), Coding of Still Pictures, of JTC 1's subcommittee 29 (SC29),[4] Coding of Audio, Picture, Multimedia and Hypermedia Information. Another working group within SC29 is WG11 – Coding of Moving Pictures and Audio – well known as the Moving Picture Experts Group (MPEG).

The word 'Joint' in JPEG, however, does not refer to the joint efforts of ISO and IEC, but to the fact that the JPEG activities are the result of an additional collaboration with the International Telecommunication Union ITU.[5] Hence, the JPEG standard has the following 'dual' numbering: ISO/IEC IS 10918−1| ITU-T Recommendation T.81 (the T appended to ITU refers to the fact that it concerns a telecommunication standard). This specification was published in 1994 as the result of a process that started in 1986. Though this standard is generally considered as a single specification, in reality it is composed of four separate parts and an amalgam of coding modes. Part 1 of JPEG (IS 10918) specifies the core coding technology and incorporates many options for encoding photographic images. Part 2 defines the compliance testing. Part 3 defines a set of extensions to the coding technologies of Part 1, and via an amendment the SPIFF file format. Finally, Part 4 focuses on the registration of JPEG profiles, SPIFF profiles, SPIFF tags, SPIFF color spaces, and SPIFF compression types, and defines the Registration Authorities. Without any doubt, it can be stated that JPEG has been one of the most successful multimedia standards defined so far.

In parallel, the WG1 committee also launched other initiatives in the context of bilevel image coding and lossless coding of photographic images. The Joint Bilevel Image experts Group (JBIG) of WG1 delivered the JBIG-1 (ISO/IEC 11544:1993| ITU-T Rec. T.82:1993) and JBIG-2 standards (ISO/IEC 14492:2001| ITU-T Rec. T.88: 2000) that offer solutions for the progressive, lossy/lossless encoding of bilevel and halftone images. JPEG-LS, on the other hand, was defined to address the need for effective lossless and near-lossless compression of continuous-tone still images. This standard can be broken

---

[1] www.jpeg.org.
[2] www.iso.org.
[3] www.iec.ch.
[4] www.itscj.ipsj.or.jp/sc29.
[5] www.itu.int.

into two parts: ISO/IEC 14495-1:1999| ITU-T Rec. T.87 (1998), defining the core technology, and ISO/IEC 14495-2:2003| ITU-T Rec. T.870 (03/2002), containing the extension. This specification is an underestimated part of the JPEG family of standards, though it delivers excellent coding performance while having a limited complexity requirement.

In the last five years of the previous century, the WG1 community realized that though the JPEG standard was perfectly hitting the consumer market – and the real boost had still to come with the advent of the digital consumer camera – a more efficient technology was needed in terms of rate-distortion performance and functionality. JPEG delivered a performance that fulfilled the needs of the market at that time, but in parallel a new promising technology was seeing the daylight. Wavelet technology had been around for some years, but during that period the underlying theory had really matured and people had also succeeded in building codecs that exploited maximally the promises theory had brought in terms of rate-distortion output and other additional functionalities. In particular, scalability functionality could now be exploited without any loss in compression efficiency, which had not been the case in earlier solutions (also supported in some of the coding modes of JPEG, although rarely implemented in commercial solutions). Hence, inspired by these evolutions, in 1997, WG1 launched a new standardization process that would result in 2000 in a new standard: JPEG 2000 (ISO/IEC 15444−1| ITU-T Rec. T.800). As in the case of the original JPEG compression algorithm, the JPEG 2000 standard is the result of a consensus process in which many different technologies were competing. Finally, a wavelet-based image coding solution was selected that included the embedded block coding by optimized truncation (EBCOT) algorithm. Chapter 1 of this book elaborately outlines the technology and features that this standard is offering. In addition to excellent rate-distortion performance, JPEG 2000 is bringing functionalities such as lossy-to-lossless progressive coding, quality scalability, resolution scalability, region-of-interest functionality, and random code-stream access – to mention a few. As with JPEG, the JPEG 2000 is a collection of a number of standard parts, which together shape the complete standard; hence the title of this book is *The JPEG 2000 Suite*. At the time of writing this book, the following parts were defined or in progress (the corresponding chapter number is given in parentheses):

- Part 1 – Core Coding System (*Chapter 1*)
- Part 2 – Extensions (*Chapter 2*)
- Part 3 – Motion JPEG 2000 (*Chapter 3*)
- Part 4 – Conformance Testing (*Chapter 17*)
- Part 5 – Reference Software (*Chapter 17*)
- Part 6 – JPM: Compound Image File Format (*Chapter 4*)
- Part 8 – JPSEC: Secure JPEG 2000 (*Chapter 5*)
- Part 9 – JPIP: Interactivity Tools, APIs, and Protocols (*Chapter 6*)
- Part 10 – JP3D: Extensions for Three-Dimensional Data (*Chapter 7*)
- Part 11 – JPWL: Wireless (*Chapter 8*)
- Part 12 – ISO Base Media File Format (*Chapter 3*)
- Part 13 – An Entry Level JPEG 2000 Encoder
- Part 14 – JPXML: XML Structural Representation and Reference

Part 7 was originally intended to deliver reference hardware implementations but this part was abandoned because of lack of interest. Consequently, WG1 additionally launched

Part 13, which was intended to contain a precise specification of the JPEG 2000 encoder, in order to provide implementers with an explicit example. Since this specification is very close to the technologies described in Part 1, Part 13 does not bring any special details and therefore this book does not cover it. Finally, Part 14 – JPXML, an XML structural representation and reference, is an activity only launched very recently by WG1, and hence was not yet complete to be covered in this book.

Meanwhile, JPEG 2000 has known vast adoption across a wide range of applications. One of the most successful adoptions is probably the acceptance by the Digital Cinema Initiatives (DCI).[6] JPEG 2000 was selected as the encoding technology for motion pictures for use in movie theaters. In this book, we discuss several professional markets where the JPEG 2000 family of standards is being used, or has the potential to have a significant impact:

- Digital cinema *(Chapter 9)*
- Security applications *(Chapter 10)*
- Video surveillance and defense imaging *(Chapter 11)*
- Remote sensing *(Chapter 12)*
- Medical imaging *(Chapter 13)*
- Digital culture imaging *(Chapter 14)*
- Broadcast applications *(Chapter 15)*
- 3-D graphics *(Chapter 16)*

Because the algorithmic complexity of JPEG 2000 is significantly higher than its predecessor JPEG, it has been perceived for a long time as less suitable for consumer market applications, despite offering superior coding performance and a set of novel features. While the early chapters in this book demonstrate the suitability of JPEG 2000 for digital photography, Chapter 17 of this book tackles the implementation complexity and proposes a few commercial implementations without attempting to give a complete market survey.

Finally, Chapter 18 focuses on current activities of WG1, which are JPEG XR (ISO/IEC FCD 29199-2) for high dynamic range (HDR) imaging, JPSearch (ISO/IEC TR 24800-1:2007), aiming at support for image search applications, and advance image coding and evaluation methodologies (AIC, ISO/IEC 29170).

This book has been produced by contributions from core people involved in the standardization process of JPEG 2000 and gives the reader a clear insight into the technologies and motivations that drove its standardization process. *The JPEG 2000 Suite* is of great appeal to researchers, practicing electronics engineers, and hardware and software developers using and developing image coding techniques. Graduate students taking courses on image compression, digital archiving, and data storage techniques will also find the book useful, as will graphic designers, artists, and decision makers in industries developing digital applications.

Peter Schelkens, Athanassios Skodras
and Touradj Ebrahimi
The Editors

---

[6] www.dcimovies.com.

Acknowledgments

This book is the result of over two years of cumulative effort of many researchers, academics, and developers from different institutions, universities, and companies in the world, the common component of all of them being their intense involvement in the development and the use of the JPEG 2000 standard. The editors would like to express their gratitude to all the authors who have contributed to this book.

The editors would also like to thank all of the WG1 colleagues who have contributed to the standardization process of JPEG 2000 that led to this standard suite.

Finally, the support of the Wiley editorial staff (Emily Bone, Simone Taylor, Laura Bell, Nicky Skinner, Alexandra King, Liz Benson, and Georgia Pinteau) is gratefully acknowledged.

The Editors
**Peter Schelkens, Athanassios Skodras
and Touradj Ebrahimi**

List of Acronyms

AAF	*Advanced Authoring Format*
AC	*Alternating Current*
ACE	*Association des Cinémathèques Européennes*
ACR	*American College of Radiology*
ADS	*Arbitrary Decomposition Styles marker segment*
AES	*Advanced Encryption Standard*
AES	*Audio Engineering Society*
AIC	*Advanced Image Coding and evaluation methodologies*
ARQ	*Automatic Repeat reQuest*
ASC	*American Society of Cinematographers*
ATK	*Arbitrary Transform Kernel marker segment*
AVC	*Advanced Video Coding*
BER	*Bit Error Rate*
BiM	*MPEG–7 BInary forMat*
bpp	*Bits Per Pixel*
CAP	*extended CAPabilities marker segment*
CBAP	*Code-Block Anchor Point*
CBD	*Component Bi- depth Definition marker segment*
CBR	*Constant Bit-Rate*
CCITT	*Consultative Committee for International Telephone and Telegraph*
CDL	*Color Decision List*
CIF	*Common Interchange Format*
COC	*COding style Component marker segment*
COD	*CODing style default marker segment*
COM	*COMment marker segment*
CON	*complexity CONstant*
CPL	*Composition PLaylist*
CRC	*Cyclic Redundancy Check*
CRG	*Component ReGistration marker segment*
CSF	*Contrast Sensitivity Function*
CT	*Computed Tomography*
DC	*Direct Current*
DCDM	*Digital Cinema Distribution Master*
DCI	*Digital Cinema Initiatives LLC*
DCO	*variable DC Offset*
DCP	*Digital Cinema Package*

DCT	*Discrete Cosine Transform*
DFS	*Downsampling Factor Styles marker segment*
DICOM	*Digital Imaging and COmmunications in Medicine*
DPCM	*Differential Pulse Code Modulation*
DRM	*Digital Rights Management*
DSA	*Digital Signature Algorithm*
DSC	*Digital Still Camera*
DSL	*Digital Subscriber Line*
DSM	*Digital Source Master*
DVB	*Digital Video Broadcasting*
DVB-C	*Digital Video Broadcasting – Cable*
DVB-H	*Digital Video Broadcasting – Handheld*
DVB-S	*Digital Video Broadcasting – Satellite*
DVB-T	*Digital Video Broadcasting – Terrestrial*
DVP	*Digital Video Package*
DWT	*Discrete Wavelet Transform*
EBCOT	*Embedded Block Coding by Optimized Truncation*
EBU	*European Broadcast Union*
ECC	*Error Correcting Code*
ECTCQ	*Entropy-Coded TCQ*
EDL	*Editing Decision List*
EOC	*End Of Code-stream marker segment*
EPB	*Error Protection Block marker segment*
EPC	*Error Protection Capability marker segment*
EPH	*End of Packet Header marker segment*
ESB	*Enterprise Service Bus*
ESD	*Error Sensitivity Descriptor marker segment*
ETS	*Executable Test Set*
EZW	*Embedded Zero-tree Wavelet coding*
FEC	*Forward Error Correcting*
FFA	*FilmFörderAnstalt*
FIAF	*Fédération Internationale des Archives du Film*
FIR	*Finite Impulse Response*
FLUTE	*File Delivery over Unidirectional Transport*
GIS	*Geographical Information System*
GWG	*Geospatial-intelligence Working Group*
HDR	*High Dynamic Range*
HDTV	*High Definition Television*
HMAC	*Hash Message Authentication Code*
HP	*High Pass*
HS	*Half-sample Symmetric*
HVS	*Human Visual System*
IAP	*Intermediate Access Package*
ICC	*International Color Consortium*
ICT	*Irreversible Color Transform*
IEC	*International Electrotechnical Commission*
IJG	*Independent JPEG Group*
IP	*Internet Protocol*
IPR	*Intellectual Property Rights*
IROI	*Interactive ROI*

ISO	*International Standardisation Organisation*
ITU	*International Telecommunication Union*
ITU-T	*ITU's Telecommunication Standardization Sector*
IUT	*Implementation Under Test*
IWT	*Inverse Wavelet Transform*
JBIG	*Joint Bilevel Image experts Group*
JP2	*JPEG 2000 file format*
JPEG	*Joint Photographic Experts Group*
JPEG-LS	*JPEG Lossless*
JPIP	*JPEG 2000 Interactive Protocol*
JPM	*JPEG 2000 Multilayer file format*
JPSEC	*JPEG 2000 Security*
JPWL	*JPEG 2000 Wireless*
JPX	*JPEG 2000 eXtended file format (Part 2)*
KDM	*Key Delivery Message*
KLT	*Karhunen–Loévé Transform*
LABR	*Lowest Authentication Bit-Rate*
LBT	*Lapped Biorthogonal Transform*
LIDAR	*LIght Detection And Ranging of Laser Imaging Detection And Ranging*
LP	*Low Pass*
LSB	*Least Significant Bit*
LUT	*Look-Up Table*
MAC	*Message Authentication Code*
MAM	*Media Asset Management*
MAP	*Master Archive Package*
MCC	*Multiple Component transform Collection marker segment*
MCO	*Multiple Component transform Ordering marker segment*
MCT	*Multiple Component Transform marker segment*
MIME	*Multipurpose Internet Mail Extensions*
MJ2	*Motion JPEG 2000 file format*
MPEG	*Moving Picture Experts Group*
MRC	*Mixed Raster Content*
MRI	*Magnetic Resonance Imaging*
MSB	*Most Significant Bit*
MSE	*Mean Squared Error*
MSSIM	*Mean Structural SIMilarity index*
MXF	*Material eXchange Format*
NDNP	*National Digital Newspaper Program*
NEMA	*National Electrical Manufacturers Association*
NLT	*NonLinear point Transform marker segment*
NSI	*additional dimension image and tile size*
OOD	*Object Oriented Design*
PCRD-opt	*PostCompression Rate Distortion Optimization*
PCS	*Profile Connection Space*
PCT	*Photo Core Transform*
PDF	*Portable Document Format*
PET	*Positron Emission Tomography*
PKL	*Packaging List*
PLM	*Packet Length, Main header marker segment*
PLR	*Packet Loss Rate*

PLT	*Packet Length, Tile-part header marker segment*
POC	*Progression Order Change marker segment*
POT	*Photo Overlap Transform*
PPM	*Packed Packet headers, Main header marker segment*
PPT	*Packed Packet headers, Tile-part header marker segment*
PR	*Perfect Reconstruction*
PRNG	*Pseudo-Random Noise Generator*
PSNR	*Peak Signal-to-Noise Ratio*
QCC	*Quantization Component marker segment*
QCD	*Quantization Default marker segment*
QCIF	*Quarter CIF*
QP	*Quantization Parameters*
QPC	*Quantization Precinct Component marker segment*
QPD	*Quantization Precinct Default marker segment*
RA	*Registration Authority*
RCT	*Reversible Color Transform*
R-D	*Rate Distortion*
RED	*Residual Error Descriptor marker segment*
RGB	*Red Green Blue*
RGN	*ReGioN of interest marker segment*
RLE	*Run-Length Encoding*
ROI	*Region of Interest*
RTP	*Real-time Transport Protocol*
RVLC	*Reversible Variable Length Code*
SCLA	*Spatial Combinatorial Lifting Algorithm*
SDI	*Serial Digital Interface*
SDTV	*Standard Definition TV*
SEC	*SECurity marker segment*
SHA	*Secure Hash Algorithm*
SIZ	*image and tile SIZe marker segment*
SMPTE	*Society of Motion Pictures and Television Engineers*
SNR	*Signal-to-Noise Ratio*
SOA	*Service Oriented Architecture*
SOC	*Start Of Code-stream marker segment*
SOD	*Start Of Data marker segment*
SOP	*Start Of Packet marker segment*
SOT	*Start Of Tile-part marker segment*
SPIHT	*Set Partitioning In Hierarchical Trees*
SSO	*Single-Sample Overlap*
StEM	*Standard Evaluation Material*
TCQ	*Trellis Coded Quantization*
TIFF	*Tagged Image File Format*
TLM	*Tile-part Lengths Marker segment*
TSSO	*Tile Single-Sample Overlap*
UEP	*Unequal Error Protection*
UML	*Unified Modeling Language*
UUID	*Universal Unique IDentifier*
VBR	*Variable Bit Rate*
VDCO	*Variable DC Offset*
VIF-P	*Visual Information Fidelity measure in Pixel domain*

VLC	*Variable Length Code*
VM	*Verification Model*
VMS	*Visual Masking marker Segment*
VSQ	*Variable Scalar Quantization*
VTR	*Video Tape Recorder*
WAN	*Wide Area Network*
WLAN	*Wireless Local Area Network*
WS	*Whole-sample Symmetric*
XML	*eXtensible Markup Language*
ZOI	*Zone Of Influence*

Part A

1

JPEG 2000 Core Coding System (Part 1)

Majid Rabbani, Rajan L. Joshi, and Paul W. Jones

1.1 Introduction

The Joint Photographic Experts Group (JPEG) committee was formed in 1986 under the joint auspices of ISO and ITU-T[1] and was chartered with the 'digital compression and coding of continuous-tone still images.' In 1993, the committee published an International Standard known as JPEG Part 1 (ISO/IEC, 1993; Pennebaker and Mitchell, 1993), which provides a toolkit of compression techniques from which applications can select various elements to satisfy particular requirements. This toolkit includes: (i) the JPEG baseline system, which is a discrete cosine transform (DCT)-based lossy compression algorithm that uses Huffman coding and is restricted to 8 bits/pixel input; (ii) an extended system, which provides enhancements to the baseline algorithm, such as 12 bits/pixel input and arithmetic coding, to satisfy a broader set of applications; and (iii) a lossless mode, which uses a predictive coding method that is independent of the DCT. Of these various components, only the JPEG baseline system has seen widespread adoption. While the JPEG baseline system offers good compression efficiency with modest computational complexity, its success as an image compression standard can really be attributed to the availability of a free software implementation that was released by the Independent JPEG Group (IJG).[2] The IJG code allowed developers to build efficient and robust JPEG applications quickly.

Despite the phenomenal success of the JPEG baseline system, it has various limitations that hinder its use in applications such as medical diagnostic imaging, mobile communications, digital cinema, enhanced internet browsing, and multimedia. The extended JPEG

---

[1] Formerly known as the Consultative Committee for International Telephone and Telegraph (CCITT).

[2] Independent JPEG Group, JPEG Library (Version 6b), available from http://www.ijg.org/ or ftp://ftp.uu.net/graphics/jpeg/ (tar.gz format archive).

The JPEG 2000 Suite   Edited by Peter Schelkens, Athanassios Skodras and Touradj Ebrahimi
© 2009 John Wiley & Sons, Ltd

system addresses only some of these limitations, while also being subject to intellectual property rights (IPR) issues for certain technology components. The desire to provide a broad range of features for numerous applications in a single compressed bit-stream prompted the JPEG committee in 1996 to investigate possibilities for a new compression standard that was subsequently named JPEG 2000. In March 1997, a call for proposals was issued (ISO/IEC, (1997a, 1997b)), seeking to produce a standard to 'address areas where current standards failed to produce the best quality or performance,' 'provide capabilities to markets that currently do not use compression,' and 'provide an open system approach to imaging applications.' After evaluating more than 20 algorithms and performing hundreds of technical studies known as 'core experiments,' the JPEG committee issued JPEG 2000 Part 1 as an International Standard in December 2000 (ISO/IEC International Standard 15444-1, ITU Recommendation T.800).

In the same vein as the JPEG baseline system, Part 1 of JPEG 2000 defines a core coding system that provides high coding efficiency with minimal complexity, and is aimed at satisfying 80% of potential applications (ISO/IEC International Standard 15444-1, ITU Recommendation T.800). In addition, it defines an optional file format that includes essential information for the proper rendering of the image. To help drive the adoption of Part 1, the JPEG 2000 committee worked diligently to ensure that it could be made available on a royalty- and fee-free basis. Most of the technologies that were excluded from Part 1 of the JPEG 2000 standard because of IPR or complexity issues were included in Part 2 (ISO/IEC International Standard 15444-2, ITU Recommendation T.801). The division of the JPEG 2000 standard into various parts allows an application to minimize complexity by using only those parts that are needed to satisfy its requirements.

The incentive behind the development of the JPEG 2000 system was not just to provide higher compression efficiency than the baseline JPEG system. Instead, it was to provide a new image representation with a rich set of features, all supported within the same compressed bit-stream, that can address a variety of existing and emerging compression applications. In particular, Part 1 of the JPEG 2000 standard addresses some of the shortcomings of baseline JPEG by providing the following features:

- Improved compression efficiency
- Lossy to lossless compression
- Multiple resolution representation
- Embedded bit-stream, including progressive decoding and signal-to-noise ratio (SNR) scalability
- Tiling
- Region-of-interest (ROI) coding
- Error resilience
- Random code-stream access and processing
- Improved performance to multiple compression/decompression cycles
- Flexible file format

The JPEG 2000 standard makes use of several advances in compression technology in order to achieve these features. The block-based DCT of JPEG has been replaced by the full-frame discrete wavelet transform (DWT). The DWT inherently provides a multiresolution image representation, and it also improves compression efficiency because of good energy compaction and the ability to decorrelate the image across a larger scale.

Furthermore, integer DWT filters can be used to provide both lossless and lossy compression within a single compressed bit-stream. Embedded coding is achieved by using a uniform quantizer with a central dead zone (with twice the step size). When the output index of this quantizer is represented as a series of binary symbols, a partial decoding of the index is equivalent to coarser quantization, where the effective quantizer step size is equivalent to the scaling of the original step size by a power of two. To encode the binary bit-planes of the quantizer index, JPEG 2000 has replaced the Huffman encoder of baseline JPEG with a context-based adaptive binary arithmetic encoder that is known as the MQ-coder. The embedded bit-stream that results from bit-plane coding provides SNR scalability in addition to the capability of compressing to a target file size. Furthermore, the bit-planes in each subband are encoded in independent rectangular blocks and in three fractional bit-plane passes to provide an optimal embedded bit-stream, improved error resilience, partial spatial random access, ease of certain geometric manipulations, and an extremely flexible code-stream syntax. Finally, the introduction of a canvas coordinate system facilitates certain operations in the compressed domain such as cropping, rotations by multiples of 90°, flipping, etc., in addition to improved performance to multiple compression/decompression cycles.

Several excellent review papers on JPEG 2000 Part 1 have appeared in the literature (Adams *et al.*, 2000; Christopoulos, Skodras, and Ebrahimi, 2000; Ebrahimi *et al.*, 2000; Gormish, Lee, and Marcellin, 2000; Marcellin *et al.*, 2000; Rabbani and Joshi, 2000; Santa-Cruz and Ebrahimi, 2000; Taubman and Marcellin, 2002). Two comprehensive books describing the technical aspects of the standard have been published (Acharya and Tsai, 2005; Taubman and Marcellin, 2001) and a Special Journal Issue was dedicated to JPEG 2000 (SPIC, 2002). The goal of this chapter is to provide a technical description of the fundamental building blocks of JPEG 2000 Part 1 and to explain the rationale behind the selected technologies. Emphasis is placed on general encoder and decoder technology issues to provide a better understanding of the standard in various applications, and many specific implementation issues have been omitted. As a result, readers who plan on implementing the standard are encouraged to refer to the actual standard (ISO/IEC International Standard 15444-1, ITU Recommendation T.800). It is worth noting that the standard document is written from the standpoint of the decoder, while the focus of this chapter is mainly from the standpoint of the encoder.

This chapter is organized as follows. In Section 1.2, the fundamental building blocks of the JPEG 2000 Part 1 standard, such as preprocessing, DWT, quantization, and entropy coding, are described. In Section 1.3, the syntax and organization of the compressed bit-stream is explained. In Section 1.4, various rate control strategies that can be used by the JPEG 2000 encoder for achieving an optimal SNR or visual quality for a given bit-rate are discussed. In Section 1.5, the tradeoffs between the various choices of encoder parameters are illustrated through an extensive set of examples. Finally, Section 1.6 contains a brief description of some additional JPEG 2000 Part 1 features such as ROI, error resilience, and file format.

1.2 JPEG 2000 Fundamental Building Blocks

The fundamental building blocks of a JPEG 2000 encoder are shown in Figure 1.1. These components include preprocessing, DWT, quantization, arithmetic coding (tier-1 coding),

Figure 1.1   JPEG 2000 fundamental building blocks

and bit-stream organization (tier-2 coding). In the following, each of these components is discussed in more detail.

The input image to JPEG 2000 may contain one or more components. Although a typical color image would have three components (e.g. RGB or YC_bC_r), up to 16 384 (2^{14}) components can be specified for an input image to accommodate multispectral or other types of imagery. The sample values for each component can be either signed or unsigned integers with a bit-depth in the range of 1 to 38 bits. Given a sample with a bit-depth of B bits, the unsigned representation would correspond to the range $[0, 2^B - 1]$, while the signed representation would correspond to the range $[-2^{B-1}, 2^{B-1} - 1]$. The bit-depth, resolution, and signed versus unsigned specification can vary for each component. If the components have different bit-depths, the most significant bits of the components should be aligned when estimating the distortion at the encoder.

1.2.1   Preprocessing

The first step in preprocessing is to partition the input image into rectangular and nonoverlapping tiles of equal size (except possibly for those tiles at the image borders). The tile size is arbitrary and can be as large as the original image itself (i.e. only one tile) or as small as a single pixel. Each tile is compressed independently using its own set of specified compression parameters. Tiling is particularly useful for applications where the amount of available memory is limited compared to the image size.

Next, unsigned sample values in each component are level shifted (DC offset) by subtracting a fixed value of 2^{B-1} from each sample to make its value symmetric around zero. Signed sample values are not level shifted. Similar to the level shifting performed in the JPEG standard, this operation simplifies certain implementation issues (e.g. numerical overflow, arithmetic coding context specification, etc.), but has no effect on the coding efficiency. Part 2 of the JPEG 2000 standard allows for a generalized DC offset, where a user-defined offset value can be signaled in a marker segment.

Finally, the level-shifted values can be subjected to a forward point-wise intercomponent transformation to decorrelate the color data. One restriction on applying the intercomponent transformation is that the components must have identical bit-depths and dimensions. Two transform choices are allowed in Part 1, where both transforms operate on the first three components of an image tile with the implicit assumption that these components correspond to RGB. One transform is the *irreversible color transform* (ICT), which is identical to the traditional RGB to YC_bC_r color transformation and can only be used for lossy coding. The forward ICT is defined as

$$\begin{pmatrix} Y \\ C_b \\ C_r \end{pmatrix} = \begin{pmatrix} 0.299 & 0.587 & 0.114 \\ -0.16875 & -0.33126 & 0.500 \\ 0.500 & -0.41869 & -0.08131 \end{pmatrix} \times \begin{pmatrix} R \\ G \\ B \end{pmatrix}. \tag{1.1}$$

This can alternatively be written as

$$Y = 0.299(R - G) + G + 0.114(B - G), \quad C_b = 0.564(B - Y), \quad C_r = 0.713(R - Y),$$

while the inverse ICT is given by

$$\begin{pmatrix} R \\ G \\ B \end{pmatrix} = \begin{pmatrix} 1.0 & 0 & 1.402 \\ 1.0 & -0.34413 & -0.71414 \\ 1.0 & 1.772 & 0 \end{pmatrix} \times \begin{pmatrix} Y \\ C_b \\ C_r \end{pmatrix}. \tag{1.2}$$

The other transform is the *reversible color transform* (RCT), which is a reversible integer-to-integer transform that approximates the ICT for color decorrelation and can be used for both lossless and lossy coding. The forward RCT is defined as

$$\tilde{Y} = \left\lfloor \frac{R + 2G + B}{4} \right\rfloor, \quad D_b = B - G, \quad D_r = R - G, \tag{1.3}$$

where $\lfloor w \rfloor$ denotes the largest integer that is smaller than or equal to w. The $\tilde{Y}$ component has the same bit-depth as the RGB components while the color difference components D_b and D_r have one extra bit of precision. The inverse RCT, which is capable of exactly recovering the original RGB data, is given by

$$G = \tilde{Y} - \left\lfloor \frac{D_b + D_r}{4} \right\rfloor, \quad B = D_b + G, \quad R = D_r + G. \tag{1.4}$$

At the decoder, the decompressed image is subjected to the corresponding inverse color transform if necessary, followed by the removal of the DC level shift.

Because each component of each tile is treated independently, the basic compression engine for JPEG 2000 will only be discussed with reference to a single tile component, after the application of an intercomponent transform if one is used.

1.2.2 The Discrete Wavelet Transform (DWT)

The block DCT transformation in baseline JPEG has been replaced with the full-frame DWT in JPEG 2000. The DWT has several characteristics that make it suitable for fulfilling some of the requirements set forth by the JPEG 2000 committee. For example, a multiresolution image representation is inherent to the DWT. Furthermore, the full-frame nature of the transform decorrelates the image across a larger scale and eliminates blocking artifacts at high compression ratios. Finally, the use of integer DWT filters allows for both lossless and lossy compression within a single compressed bit-stream. In the following, we first consider a one-dimensional (1-D) DWT for simplicity, and then extend the concepts to two dimensions.

1.2.2.1 The 1-D DWT

The forward 1-D DWT at the encoder is best understood as successive applications of a pair of lowpass and highpass filters, followed by downsampling by a factor of two

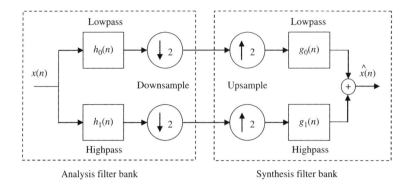

Figure 1.2 1-D, two-band wavelet analysis and synthesis filter banks

(e.g. discarding odd indexed samples) after each filtering operation, as shown in Figure 1.2. The lowpass and highpass filter pair is known as an *analysis filter bank*. The lowpass filter preserves the low frequencies of a signal while attenuating or eliminating the high frequencies, and the result is a blurred version of the original signal. Conversely, the highpass filter preserves the high frequencies in a signal such as edges, texture, and detail, while removing or attenuating the low frequencies.

Consider a 1-D signal $x(n)$ (such as the pixel values in a row of an image) and a pair of lowpass and highpass filters designated by $h_0(n)$ and $h_1(n)$, respectively. An example of a lowpass filter is $h_0(n) = (-1\ 2\ 6\ 2\ -1)/8$, which is symmetric and has five integer coefficients (or taps). An example of a highpass filter is $h_1(n) = (-1\ 2\ -1)/2$, which is symmetric and has three integer taps. The analysis filter bank used in this example was first proposed in LeGall and Tabatabai (1988) and is often referred to as the (5, 3) filter bank, indicating a lowpass filter of length five and a highpass filter of length three. To ensure that the filtering operation is defined at the signal boundaries, the 1-D signal must be extended in both directions. When using odd-tap filters, the signal is symmetrically and periodically extended as shown in Figure 1.3. The extension for even-tap filters (allowed in Part 2 of the standard) is more complicated and is explained in JPEG 2000 Part 2 (ISO International Standard 15444-2, ITU Recommendation T.801).

The filtered samples that are outputted from the forward DWT are referred to as *wavelet coefficients*. Because of the downsampling process, the total number of wavelet coefficients is the same as the number of original signal samples. When the DWT decomposition is applied to sequences with an odd number of samples, either the lowpass or the highpass sequence will have one additional sample to maintain the same number of coefficients as original samples. In JPEG 2000, this choice is dictated by the positioning of the

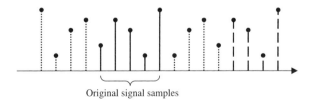

Original signal samples

Figure 1.3 Symmetric and periodic extension of the input signal at boundaries

input signal with respect to the canvas coordinate system, which will be discussed in Section 1.3.1. The (h_0, h_1) filter pair is designed in such a manner that after downsampling the output of each filter by a factor of two, the original signal can still be completely recovered from the remaining samples in the absence of any quantization errors. This is referred to as the *perfect reconstruction* (PR) property.

Reconstruction of a signal from the wavelet coefficients at the decoder is performed with another pair of lowpass and highpass filters (g_0, g_1), known as the *synthesis filter bank*. Referring to Figure 1.2, the downsampled output of the lowpass filter $h_0(n)$ is first upsampled by a factor of two by inserting zeroes in between every two samples. The result is then filtered with the synthesis lowpass filter $g_0(n)$. The downsampled output of the highpass filter $h_1(n)$ is also upsampled and filtered with the synthesis highpass filter $g_1(n)$. The results are added together to produce a reconstructed signal $\hat{x}(n)$, which, assuming sufficient precision, will be identical to $x(n)$ because of the PR property.

For perfect reconstruction, the analysis and synthesis filters have to satisfy the following two conditions:

$$H_0(z)G_0(z) + H_1(z)G_1(z) = 2, \tag{1.5}$$

$$H_0(-z)G_0(z) + H_1(-z)G_1(z) = 0, \tag{1.6}$$

where $H_0(z)$ is the Z-transform of $h_0(n)$, $G_0(z)$ is the Z-transform of $g_0(n)$, etc. The condition in Equation (1.6) can be satisfied by choosing

$$G_0(z) = -cz^{-l}H_1(-z) \quad \text{and} \quad G_1(z) = cz^{-l}H_0(-z), \tag{1.7}$$

where l is an integer constant and c is a scaling factor. Combining this result with Equation (1.5) indicates that the analysis filter pair (h_0, h_1) has to be chosen to satisfy

$$-cz^{-l}H_0(z)H_1(-z) + cz^{-l}H_1(z)H_0(-z) = 2. \tag{1.8}$$

The constant l represents a delay term that imposes a restriction on the spatial alignment of the analysis and synthesis filters, while the constant c affects the filter normalization. The filter bank that satisfies these conditions is known as the *biorthogonal* filter bank. This name stems from the fact that h_0 and g_1 are orthogonal to each other and h_1 and g_0 are orthogonal to each other. A particular class of biorthogonal filters is one where the analysis and synthesis filters are finite impulse response (FIR) and linear phase (i.e. they satisfy certain symmetry conditions) (Vetterli and Kovacevic, 1995). Then, it can be shown that in order to satisfy Equation (1.8), the analysis filters h_0 and h_1 have to be of unequal lengths. If the filters have an odd number of taps, their length can differ only by an odd multiple of two, while for even-tap filters the length difference can only be an even multiple of two.

While the (5, 3) filter bank is a prime example of a biorthogonal filter bank with integer taps, the filter banks that result in the highest compression efficiency often have floating-point taps (Villasenor, Belzer, and Liao, 1995). The most well-known filter bank in this category is the Daubechies (9, 7) filter bank, introduced in Antonini *et al.* (1992) and characterized by the filter taps given in Table 1.1. For comparison, the analysis and synthesis filter taps for the integer (5, 3) filter bank are specified in Table 1.2. It can be easily verified that these filters satisfy Equations (1.7) and (1.8) with $l = 1$ and $c = 1.0$. As

Table 1.1 Analysis and synthesis filter taps for the floating-point Daubechies (9, 7) filter bank

n	Lowpass, $h_0(n)$	Lowpass, $g_0(n)$
0	+0.602949018236360	+1.115087052457000
±1	+0.266864118442875	+0.591271763114250
±2	−0.078223266528990	−0.057543526228500
±3	−0.016864118442875	−0.091271763114250
±4	+0.026748757410810	

n	Highpass, $h_1(n)$	n	Highpass, $g_1(n)$
−1	+1.115087052457000	1	+0.602949018236360
−2, 0	−0.591271763114250	0, 2	−0.266864118442875
−3, 1	−0.057543526228500	−1, 3	−0.078223266528990
−4, 2	+0.091271763114250	−2, 4	+0.016864118442875
		−3, 5	+0.026748757410810

Table 1.2 Analysis and synthesis filter taps for the integer (5, 3) filter bank

n	$h_0(n)$	$g_0(n)$	n	$h_1(n)$	n	$g_1(n)$
0	3/4	+1	−1	+1	1	+3/4
±1	1/4	+1/2	−2, 0	−1/2	0, 2	−1/4
±2	−1/8				−1, 3	−1/8

is evident from Tables 1.1 and 1.2, the filter h_0 is centered at zero while h_1 is centered at −1. As a result, the downsampling operation effectively retains the even-indexed samples from the lowpass output and the odd-indexed samples from the highpass output sequence, where the indices are defined relative to the reference grid (Section 1.3.1).

The frequency responses of the (9, 7) and (5, 3) analysis filter pairs are shown in Figure 1.4. For convenience, the filter amplitudes have been normalized to approximately

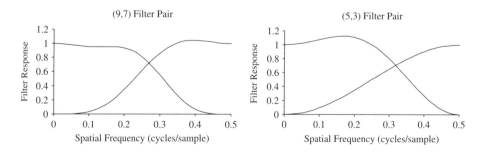

Figure 1.4 Frequency responses of (9, 7) and (5, 3) filter pairs with (1, 1) normalization

the same range (refer to Section 1.2.2.4 for the actual normalizations). It is easy to see the lowpass and highpass nature of the filter pairs, with the (9, 7) filter pair having better frequency discrimination than the (5, 3) filter pair. However, even the (9, 7) filter pair has substantial overlap between the lowpass and highpass filters. For either filter pair, the frequency content that is above a normalized frequency of 0.25 cycles/sample for the lowpass filter and below 0.25 cycles/sample for the highpass filter will alias when the filtered outputs are decimated by a factor of two. While it may seem surprising, the perfect reconstruction property guarantees that the aliased content will be canceled during reconstruction in the absence of any errors in the wavelet coefficients.

After the 1-D signal has been decomposed into two frequency bands, the lowpass output is still highly correlated and can be subjected to another stage of two-band decomposition to achieve additional decorrelation. In comparison, there is generally little to be gained by further decomposing the highpass output. In most DWT decompositions, only the lowpass output is further decomposed to produce what is known as a *dyadic* or *octave* decomposition. Part 1 of the JPEG 2000 standard supports only dyadic decompositions, while Part 2 also allows for the further splitting of the high-frequency bands. Figure 1.5 shows an example of the effective filter responses that are produced by recursive filtering of the lowpass output in a dyadic decomposition. This example uses the (9, 7) filter pair and illustrates a five-level decomposition, which produces six frequency bands (i.e. resolution levels). The frequency discrimination of the frequency bands becomes increasingly tighter as one moves from the highest frequency band to the lowest frequency band.

1.2.2.2   The 2-D DWT

The 1-D DWT can be easily extended to two dimensions (2-D) by applying the filter bank in a separable manner. At each level of the wavelet decomposition, each row of a 2-D image is first transformed using a 1-D horizontal analysis filter bank (h_0, h_1). The same filter bank is then applied vertically to each column of the filtered and subsampled data. Given the linear nature of the filtering process, the order in which the horizontal and the vertical filters are applied does not affect the final values of the 2-D subbands in an ideal implementation.

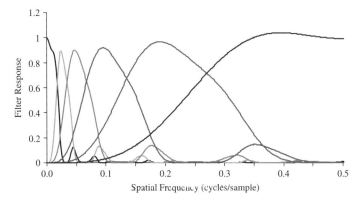

Figure 1.5   Frequency responses for a five-level dyadic decomposition (six frequency bands) using the (9, 7) filter pair (see Plate 1)

This separable filtering with a lowpass/highpass pair creates four frequency bands, and correspondingly there are four sets of wavelet coefficients. The sets of wavelet coefficients represent filtered and subsampled versions of the input image, which are referred to as subband images, or simply *subbands*. Thus, one-level, 2-D wavelet decomposition produces four subbands. In a 2-D dyadic decomposition, the lowest frequency subband (denoted as the LL band to indicate lowpass filtering in both directions) is further decomposed into four smaller subbands, and this process may be repeated as desired to achieve gains in compression efficiency and/or for easy access to lower resolution versions of the image. Figure 1.6 shows a three-level, 2-D dyadic decomposition and the corresponding labeling for each subband. For example, the subband label kHL indicates that a horizontal highpass (H) filter has been applied to the rows, followed by a vertical lowpass (L) filter applied to the columns during the kth level of the DWT decomposition. As a convention, the subband 0LL refers to the original image (or image tile). Figure 1.7 shows a three-level, 2-D DWT decomposition of the 'Lena' image using the (9, 7) filter bank as specified in Table 1.1, and it clearly demonstrates the energy compaction property of the DWT (i.e. most of the image energy is found in the lower frequency subbands). To visualize the subband energies better, the AC subbands (i.e. all the subbands except for LL) have been scaled up by a factor of four. However, as will be explained in Section 1.2.2.4, in order to show the actual contribution of each subband to the overall image energy, the wavelet coefficients in each subband should be scaled by the weights given in the last column of Table 1.3.

The DWT decomposition provides a natural solution for the multiresolution requirement of the JPEG 2000 standard. The lowest resolution at which the image can be reconstructed is referred to as resolution zero. For example, referring to Figure 1.6, the 3LL subband would correspond to resolution zero for a three-level decomposition. For an N_L-level[3] DWT decomposition, the image can be reconstructed at $N_L + 1$ resolutions. In general, to reconstruct an image at resolution $r(r > 0)$, subbands $(N_L - r + 1)$HL, $(N_L - r + 1)$LH, and $(N_L - r + 1)$HH need to be combined with the image at resolution

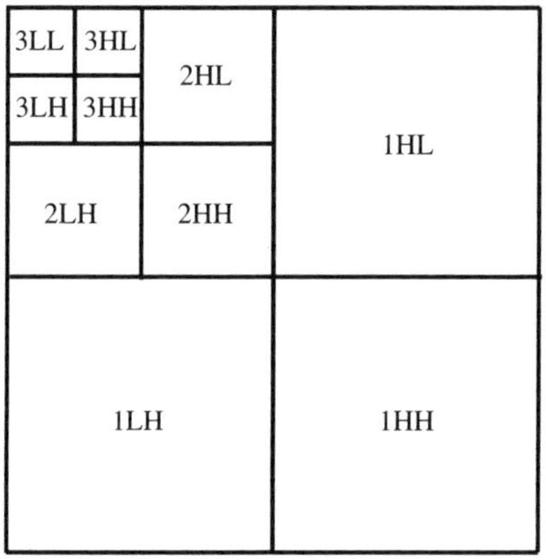

Figure 1.6   Labeling of subbands produced by a three-level, 2-D wavelet decomposition

---

[3] N_L is the notation that is used in the JPEG 2000 document to indicate the number of resolution levels, although the subscript L might be somewhat confusing, as it would seem to indicate a variable.

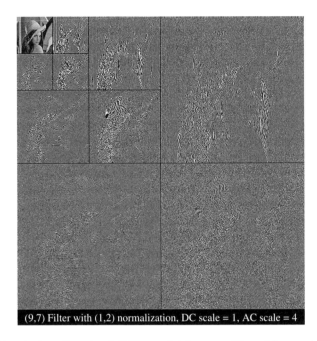

(9,7) Filter with (1,2) normalization, DC scale = 1, AC scale = 4

Figure 1.7   Subbands for a three-level, 2-D wavelet decomposition of Lena using the (9, 7) filter bank

Table 1.3 L_2-norms of the DWT subbands after a 2-D, three-level wavelet decomposition

	($\sqrt{2}$, $\sqrt{2}$) Normalization		(1, 2) Normalization	
Subband	(5, 3) filter	(9, 7) filter	(5, 3) filter	(9, 7) filter
3LL	0.67188	1.05209	5.37500	8.41675
3HL	0.72992	1.04584	2.91966	4.18337
3LH	0.72992	1.04584	2.91966	4.18337
3HH	0.79297	1.03963	1.58594	2.07926
2HL	0.79611	0.99841	1.59222	1.99681
2LH	0.79611	0.99841	1.59222	1.99681
2HH	0.92188	0.96722	0.92188	0.96722
1HL	1.03833	1.01129	1.03833	1.01129
1LH	1.03833	1.01129	1.03833	1.01129
1HH	1.43750	1.04044	0.71875	0.52022

$(r - 1)$. These subbands (excluding the lower-resolution image) are referred to as belonging to resolution r. Resolution zero consists of only the N_LLL band. Because the subbands are encoded independently, the image can be reconstructed at any resolution level by simply decoding those portions of the code-stream that contain the subbands corresponding to that resolution and all the previous resolutions. For example, referring to Figure 1.6, the image can be reconstructed at resolution two by combining the resolution one image and the three subbands labeled 2HL, 2LH, and 2HH.

1.2.2.3   The DWT as a Basis Function Decomposition

The DWT can also be viewed as a basis function decomposition. The basis functions are simply the impulse responses of the spatial filters that comprise the wavelet filter bank. In a dyadic decomposition, the low-frequency filters have large basis functions (corresponding to a narrow range of frequencies) while the high-frequency filters have small basis functions (corresponding to a wide range of frequencies). Because the analysis and synthesis filter banks are different, the corresponding analysis and synthesis basis functions will also be different.

The basis function viewpoint is most useful when considering the reconstruction process that occurs with the inverse DWT during decompression. The basis functions are the fundamental building blocks that are used to reconstruct an image, and the wavelet coefficients indicate how much of each basis function is needed at a given point in the image. Wavelet coefficient errors that are introduced during compression will result in artifacts in the decompressed image that have the appearance of either isolated basis functions or combinations of basis functions. Figure 1.8 shows the 2-D synthesis basis functions that are used to reconstruct images with the (9, 7) filter pair and a three-level dyadic decomposition. It can be seen that the basis functions vary in size (corresponding to the various resolution levels) and also in orientation (corresponding to the various lowpass/highpass filtering combinations in the horizontal and vertical directions).

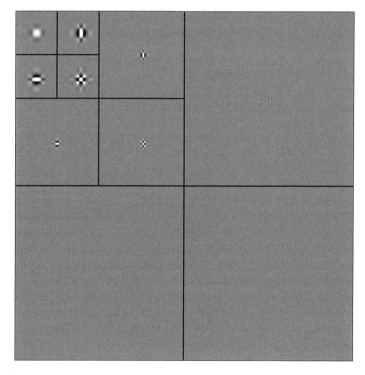

Figure 1.8   Wavelet basis functions for reconstructing an image from a three-level decomposition using the (9, 7) synthesis filters

A key aspect of the DWT is that adjacent basis functions overlap each other during image reconstruction. This overlap results in compression artifacts that are much smoother than the blocky artifacts that are produced by the nonoverlapping basis functions of the block-based DCT that is used in JPEG and MPEG compression. This is one reason why wavelet-based compression techniques can often yield superior quality at low bit rates as compared to DCT-based techniques. Figure 1.9 illustrates the smooth nature of

(a)

(b)

Figure 1.9 Reconstruction of the Lena image using wavelet basis functions from a three-level decomposition: (a) single lowest frequency basis function and (b) four lowest frequency basis functions

images that are reconstructed with wavelet basis functions. In Figure 1.9(a), the Lena image is reconstructed with only the lowest frequency basis function of the three-level decomposition; in Figure 1.9(b), the four lowest frequency basis functions are used for the reconstruction.

1.2.2.4 Wavelet Filter Normalization

The output of an invertible forward transform can generally have any arbitrary normalization (scaling) as long as it is undone by the inverse transform. In the case of DWT filters, the analysis filters h_0 and h_1 can be normalized arbitrarily. Referring to Equation (1.8), the normalization chosen for the analysis filters will influence the value of c, which in turn determines the normalization of the synthesis filters, g_0 and g_1. The normalization of the DWT filters is often expressed in terms of the DC gain of the lowpass analysis filter h_0 and the Nyquist gain of the highpass analysis filter h_1. The DC gain and the Nyquist gain of a filter $h(n)$, denoted by G_{DC} and $G_{Nyquist}$, respectively, are defined as

$$G_{DC} = \left| \sum_n h(n) \right|, \quad G_{Nyquist} = \left| \sum_n (-1)^n h(n) \right|. \tag{1.9}$$

The (9, 7) and the (5, 3) analysis filter banks as defined in Tables 1.1 and 1.2 have been normalized so that the lowpass filter has a DC gain of 1 and the highpass filter has a Nyquist gain of 2. This is referred to as the (1, 2) normalization and is the one adopted by Part 1 of the JPEG 2000 standard. Other common normalizations that have appeared in the literature are $(\sqrt{2}, \sqrt{2})$ and (1, 1). Once the normalization of the analysis filter bank has been specified, the normalization of the synthesis filter bank is automatically determined by reversing the order and multiplying by the scalar constant c of Equation (1.8).

In the baseline JPEG standard, the scaling of the forward DCT is defined to create an *orthonormal* transform, which has the property that the sum of the squares of the image samples is equal to the sum of the squares of the transform coefficients (Parseval's theorem). Furthermore, the orthonormal normalization of the DCT has the useful property of the mean squared error (MSE) of the quantized DCT coefficients being equal to the MSE of the reconstructed image. This provides a simple means for quantifying the impact of coefficient quantization on the reconstructed image MSE. Unfortunately, this property does not hold for a DWT decomposition.

If we consider a 1-D DWT, the reconstructed signal $\hat{x}(n)$ can be expressed as a weighted sum of the 1-D basis functions, where the weights are the wavelet coefficients (either quantized or unquantized). Let $\psi_m^b(n)$ denote the basis function corresponding to a coefficient $y_b(m)$, the mth wavelet coefficient from subband b. Then,

$$\hat{x}(n) = \sum_b \sum_m y_b(m) \psi_m^b(n). \tag{1.10}$$

In general, the basis functions of a DWT decomposition are not orthogonal; hence, Parseval's theorem does not apply. Woods and Naveen (1992) have shown that for quantized wavelet coefficients under certain assumptions on the quantization noise, the MSE of the reconstructed image can be approximately expressed as a weighted sum of the MSE of the wavelet coefficients, where the weight for subband b is

$$\alpha_b^2 = \sum_n |\psi^b(n)|^2. \tag{1.11}$$

The coefficient α_b is referred to as the L_2-norm[4] for subband b. For an orthonormal transform, all the α_b values would be unity. The knowledge of the L_2-norms is essential for the encoder, because they represent the contribution of the quantization noise of each subband to the overall MSE and are a key factor in designing quantizers or prioritizing the quantized data for coding.

The DWT filter normalization impacts both the L_2-norm and the dynamic range of each subband. Given the normalization of the 1-D analysis filter bank, the *nominal* dynamic range of the 2-D subbands can be easily determined in terms of the bit-depth of the tile component R_I (after application of an intercomponent transform, if one is used). In particular, for the (1, 2) normalization, the kLL subband will have a nominal dynamic range of R_I bits. However, the *actual* dynamic range might be slightly larger. In JPEG 2000, this situation is handled by using *guard bits* to avoid the overflow of the subband value. For the (1, 2) normalization, the nominal dynamic ranges of the kLH and kHL subbands are $R_I + 1$, while that of the kHH subband is $R_I + 2$.

Table 1.3 shows the L_2-norms of the DWT subbands after a three-level decomposition with either the (9, 7) or the (5, 3) filter bank and using either the $(\sqrt{2}, \sqrt{2})$ or the (1, 2) filter normalization. Clearly, the $(\sqrt{2}, \sqrt{2})$ normalization results in a DWT that is closer to an orthonormal transform (especially for the (9, 7) filter bank), while the (1, 2) normalization avoids the dynamic range expansion at each level of the decomposition.

1.2.2.5 DWT Implementation Issues and the Lifting Scheme

In the development of the existing DCT-based JPEG standard, great emphasis was placed on the implementation complexity of the encoder and decoder, which included such issues as memory requirements, number of operations per sample, and amenability to hardware or software implementation, e.g. transform precision, parallel processing, etc. The choice of the 8×8 block size for the DCT was greatly influenced by these considerations.

In contrast to the limited buffering required for the 8×8 DCT, a straightforward implementation of the 2-D DWT decomposition requires the storage of the entire image in memory. The use of small tiles reduces the memory requirements without significantly affecting the compression efficiency (see Section 1.5.1.1). In addition, some clever designs for line-based processing of the DWT have been published that substantially reduce the memory requirements depending on the size of the filter kernels (Chrysafis and Ortega, 2000). An alternative implementation of the DWT has also been developed, known as the *lifting scheme* (Daubechies and Sweldens, 1998; Sweldens, 1995, 1996, 1998). In addition to providing a significant reduction in the memory and the computational complexity of the

---

[4] We have ignored the fact that, in general, the L_2-norm for the coefficients near the subband boundaries are slightly different from the rest of the coefficients in the subband.

DWT, lifting provides in-place computation of the wavelet coefficients by overwriting the memory locations that contain the input sample values. The wavelet coefficients computed with lifting are identical to those computed by a direct filter bank convolution, in much the same manner as a fast Fourier transform results in the same DFT coefficients as a brute force approach. Because of these advantages, the specification of the DWT kernels in JPEG 2000 is only provided in terms of the lifting coefficients and not the convolutional filters.

The lifting operation consists of several steps. The basic idea is to first compute a trivial wavelet transform, also referred to as the *lazy* wavelet transform, by splitting the original 1-D signal into odd and even indexed subsequences, and then modifying these values using alternating prediction and updating steps. Figure 1.10 depicts an example of the lifting steps corresponding to the integer (5, 3) filter bank. The sequences $\{s_i^0\}$ and $\{d_i^0\}$ denote the even and odd sequences, respectively, resulting from the application of the lazy wavelet transform to the input sequence.

In JPEG 2000, a *prediction* step consists of predicting each odd sample as a linear combination of the even samples and subtracting it from the odd sample to form the prediction error $\{d_i^1\}$. Referring to Figure 1.10, for the (5, 3) filter bank, the prediction step consists of averaging the two neighboring even sequence pixels and subtracting the average from the odd sample value, i.e.

$$d_i^1 = d_i^0 - \frac{1}{2}\left(s_i^0 + s_{i+1}^0\right). \qquad (1.12)$$

Because of the simple structure of the (5, 3) filter bank, the output of this stage, $\{d_i^1\}$, is actually the highpass output of the DWT filter. In general, the number of even pixels employed in the prediction and the actual weights applied to the samples depend on the specific DWT filter bank.

An *update* step consists of updating the even samples by adding to them a linear combination of the already modified odd samples, $\{d_i^1\}$, to form the updated sequence $\{s_i^1\}$. Referring to Figure 1.10, for the (5, 3) filter bank, the update step consists of the following:

$$s_i^1 = s_i^0 + \frac{1}{4}\left(d_{i-1}^1 + d_i^1\right). \qquad (1.13)$$

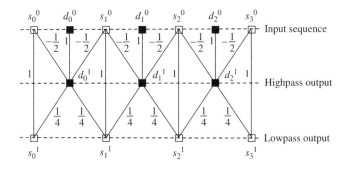

Figure 1.10 Lifting prediction/update steps for the (5, 3) filter bank

For the (5, 3) filter bank, the output of this stage, $\{s_i^1\}$, is actually the lowpass output of the DWT filter. Again, the number of odd pixels employed in the update and the actual weights applied to each sample depend on the specific DWT filter bank. The prediction and update steps are generally iterated N times, with different weights used at each iteration. This can be summarized as

$$d_i^n = d_i^{n-1} + \sum_k P_n(k)s_k^{n-1}, \quad n \in [1, 2, \ldots, N], \tag{1.14}$$

$$s_i^n = s_i^{n-1} + \sum_k U_n(k)d_k^n, \quad n \in [1, 2, \ldots, N], \tag{1.15}$$

where $P_n(k)$ and $U_n(k)$ are, respectively, the prediction and update weights at the nth iteration. For the (5, 3) filter bank, $N = 1$, while for the Daubechies (9, 7) filter bank, $N = 2$. The output of the final prediction step will be the highpass coefficients up to a scaling factor K_1, while the output of the final update step will be the lowpass coefficients up to a scaling constant K_0. For the (5, 3) filter bank, $K_0 = K_1 = 1$. The lifting steps corresponding to the (9, 7) filter bank (as specified in Table 1.1) are shown in Figure 1.11. The general block diagram of the lifting process is shown in Figure 1.12.

A nice feature of the lifting scheme is that it makes the construction of the inverse transform straightforward. Referring to Figure 1.12 and working from right to left, first the lowpass and highpass wavelet coefficients are scaled by $1/K_0$ and $1/K_1$ to produce $\{s_i^N\}$ and $\{d_i^N\}$. Next, $\{d_i^N\}$ is taken through the update stage $U_N(z)$ and subtracted from $\{s_i^N\}$ to produce $\{s_i^{N-1}\}$. This process continues, where each stage of the prediction and

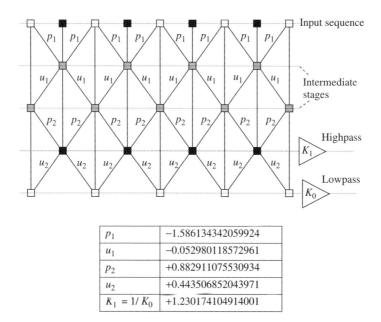

p_1	-1.586134342059924
u_1	-0.052980118572961
p_2	$+0.882911075530934$
u_2	$+0.443506852043971$
$K_1 = 1/K_0$	$+1.230174104914001$

Figure 1.11   Lifting steps for the (9, 7) filter bank

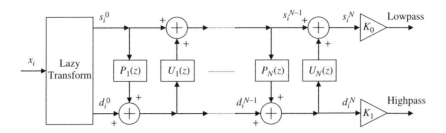

Figure 1.12   General block diagram of the lifting process

update is undone in the reverse order that it was constructed at the encoder until the image samples have been reconstructed.

1.2.2.6   Integer-to-Integer Transforms

Although the input image samples to JPEG 2000 are integers, the output wavelet coefficients are floating point when using floating-point DWT filters. Even when dealing with integer filters such as the (5, 3) filter bank, the precision required for achieving mathematically lossless performance increases significantly with every level of the wavelet decomposition and can quickly become unmanageable. An important advantage of the lifting approach is that it can provide a convenient framework for constructing integer-to-integer DWT filters from any general filter specification (Adams and Kossentini, 2000; Calderbank *et al.*, 1998).

This can be best understood by referring to Figure 1.13, where quantizers are inserted immediately after the calculation of the prediction and the update terms but before modifying the odd or the even sample value. The quantizer typically performs an operation such as truncation or rounding to the nearest integer, thus creating an integer-valued output. If the values of K_0 and K_1 are approximated by rational numbers, it is easy to verify that the resulting system is mathematically invertible despite the inclusion of the quantizer. If the underlying floating-point filter uses the (1, 2) normalization and $K_0 = K_1 = 1$, as is the case for the (5, 3) filter bank, the final lowpass output will have roughly the same bit precision as that of the input sample, while the highpass output will have an extra bit of precision. This is because, for input samples with a large enough dynamic range

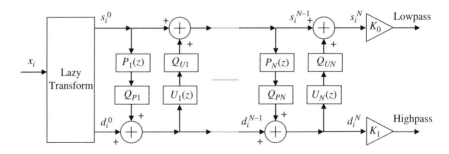

Figure 1.13   General block diagram of a forward integer-to-integer transform using lifting

(e.g. 8 bits or higher), rounding at each lifting step has a negligible effect on the nominal dynamic range of the output.

As described in the previous section, the inverse transformation is simply performed by undoing all the prediction and update steps in the reverse order that they were performed at the encoder. However, the resulting integer-to-integer transform is nonlinear and hence, when extended to two dimensions, the order in which the transformation is applied to the rows or the columns will impact the final output. To recover the original sample values losslessly, the inverse transform must be applied in exactly the reverse row–column order of the forward transform. In a JPEG 2000 encoder, the columns are processed first with the integer-to-integer wavelet transform, followed by the rows. Hence, in a JPEG 2000 decoder, the order of the 2-D reversible wavelet transform is rows, followed by columns. An extensive performance evaluation and analysis of reversible integer-to-integer DWT for image compression has been published in Adams and Kossentini (2000).

As an example, consider the conversion of the $(5, 3)$ filter bank into an integer-to-integer transform by adding the two quantizers $Q_{P1}(w) = -\lfloor -w \rfloor$ and $Q_{U1}(w) = \lfloor w + 1/2 \rfloor$ to the prediction and update steps, respectively, in the lifting diagram of Figure 1.10. The resulting forward transform is given by

$$
\begin{cases}
y(2n + 1) = x(2n + 1) - \left\lfloor \dfrac{x(2n) + x(2n + 2)}{2} \right\rfloor \\[2em]
y(2n) = x(2n) + \left\lfloor \dfrac{y(2n - 1) + y(2n + 1) + 2}{4} \right\rfloor
\end{cases} . \tag{1.16}
$$

The required precision of the lowpass band stays roughly the same as the original sample, while the precision of the highpass band grows by one bit. The inverse transform, which losslessly recovers the original sample values, is given by

$$
\begin{cases}
x(2n) = y(2n) - \left\lfloor \dfrac{y(2n - 1) + y(2n + 1) + 2}{4} \right\rfloor \\[2em]
x(2n + 1) = y(2n + 1) - \left\lfloor \dfrac{x(2n) + x(2n + 2)}{2} \right\rfloor
\end{cases} . \tag{1.17}
$$

1.2.2.7 DWT Filter Choices in JPEG 2000 Part 1

Part 1 of the JPEG 2000 standard has adopted only two choices for the DWT filters. One is the Daubechies $(9, 7)$ floating-point filter bank (as specified in Table 1.1), which has been chosen for its superior lossy compression performance. The other is the lifted integer-to-integer $(5, 3)$ filter bank, also referred to as the *reversible $(5, 3)$ filter bank*, as specified in Equations (1.16) and (1.17). This choice was driven by requirements for low implementation complexity and lossless capability. The performance of these filters is compared in Section 1.5.1.3. The mathematical properties of the $(5, 3)$ and $(9, 7)$ filters, such as Riesz bounds, order of approximation, and regularity, have been studied in great detail (Unser and Blu, 2003), but are beyond the scope of this chapter. Part 2 of the JPEG 2000 standard allows for arbitrary filter specifications in the code-stream, including filters with an even number of taps.

1.2.3   Quantization

The JPEG baseline system employs a uniform quantizer and an inverse quantization process that reconstructs the quantized coefficient to the midpoint of the quantization interval. A different step size is allowed for each DCT coefficient to take advantage of the sensitivity of the human visual system (HVS), and these step sizes are conveyed to the decoder via an 8×8 quantization table (q-table) using one byte per element. The quantization strategy employed in JPEG 2000 Part 1 is similar in principle to that of JPEG, but it has a few important differences to satisfy some of the JPEG 2000 requirements.

One difference is in the incorporation of a central dead zone in the quantizer. It was shown in Sullivan (1996) that the rate-distortion (R-D) optimal quantizer for a continuous signal with Laplacian probability density (such as DCT or wavelet coefficients) is a uniform quantizer with a central dead zone. The size of the optimal dead zone as a fraction of the step size increases as the variance of the Laplacian distribution decreases; however, it always stays less than two and is typically closer to one. In Part 1, the dead zone has twice the quantizer step size, as depicted in Figure 1.14, while in Part 2, the size of the dead zone can be parameterized to have a different value for each subband.

Part 1 adopted the dead zone with twice the step size due to its optimal embedded structure (Marcellin *et al.*, 2002). Briefly, this means that if an M_b-bit quantizer index resulting from a step size of Δ_b is transmitted progressively, starting with the most significant bit (MSB) and proceeding to the least significant bit (LSB), the resulting index after decoding only N_b bits is identical to that obtained by using a similar quantizer with a step size of $\Delta_b 2^{M_b - N_b}$. This property allows for *SNR scalability*, which in its optimal sense means that the decoder can cease decoding at any truncation point in the code-stream and still produce exactly the same image that would have been encoded at the bit-rate corresponding to the truncated code-stream. This property also allows a target bit-rate or a target distortion to be achieved exactly, while the baseline JPEG standard generally requires multiple encoding cycles to achieve the same goal. This allows an original image to be compressed with JPEG 2000 to the highest quality required by a given set of clients (through the proper choice of the quantization step sizes) and then disseminated to each client according to the specific image quality (or target file size) requirement without the need to decompress and recompress the existing code-stream. Importantly, the code-stream can also be reorganized in other ways to meet the various requirements of the JPEG 2000 standard, as will be described in Section 1.3.

Another difference is that the inverse quantization of JPEG 2000 explicitly allows for a reconstruction bias from the quantizer midpoint for nonzero indices to accommodate the

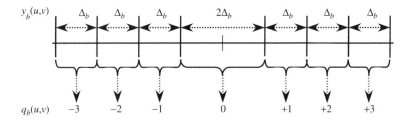

Figure 1.14   Uniform quantizer with a central dead zone with step size Δ_b

skewed probability distribution of the wavelet coefficients. In the JPEG baseline, a simple biased reconstruction strategy has been shown to improve the decoded image PSNR by approximately 0.25 dB (Price and Rabbani, 1999). Similar gains can be expected with the biased reconstruction of wavelet coefficients in JPEG 2000. The exact operation of the quantization and inverse quantization is explained in more detail in the following sections.

1.2.3.1 Quantization at the Encoder

For each subband b, a quantizer step size Δ_b is selected by the user and is used to quantize all the coefficients in that subband. The choice of Δ_b can be driven by the perceptual importance of each subband based on HVS data (Albanesi and Bertoluzza, 1995; Jones, 2007; Jones et al., 1995; O'Rourke and Stevenson, 1995; Watson et al., 1997) or it can be driven by other considerations such as rate control. The quantizer maps a wavelet coefficient $y_b(u, v)$ in subband b to a quantized index value $q_b(u, v)$, as shown in Figure 1.14. The quantization operation is an encoder issue and can be implemented in any desired manner. However, it is most efficiently performed according to

$$q_b(u, v) = sign(y_b(u, v)) \left\lfloor \frac{|y_b(u, v)|}{\Delta_b} \right\rfloor. \tag{1.18}$$

The step size Δ_b is represented with a total of two bytes, an 11-bit mantissa μ_b and a 5-bit exponent ε_b, according to the relationship:

$$\Delta_b = 2^{R_b - \varepsilon_b} \left(1 + \frac{\mu_b}{2^{11}}\right), \tag{1.19}$$

where R_b is the number of bits representing the nominal dynamic range of the subband b, which is explained in Section 1.2.2.4. This limits the largest possible step size to approximately twice the dynamic range of the input sample (when μ_b has its maximum value and $\varepsilon_b = 0$), which is sufficient for all practical cases of interest. When the reversible (5, 3) filter bank is used, Δ_b is set to one by choosing $\mu_b = 0$ and $\varepsilon_b = R_b$. The quantizer index $q_b(u, v)$ will have M_b bits if fully decoded, where $M_b = G + \varepsilon_b - 1$. The parameter G is the number of guard bits signaled to the decoder and is typically one or two.

For the irreversible (9, 7) wavelet transform, two modes of signaling the value of Δ_b to the decoder are possible. In one mode, which is similar to the q-table specification used in the current JPEG, the (ε_b, μ_b) value for every subband is explicitly transmitted. This is referred to as *expounded quantization*. The values can be chosen to take into account the HVS properties (Zheng, Daly, and Lei, 2002) and/or the L_2-norm of each subband in order to align the bit-planes of the quantizer indices according to their true contribution to the MSE. In another mode, referred to as *derived quantization*, a single value (ε_0, μ_0) is sent for the LL subband and the (ε_b, μ_b) values for each subband are derived by scaling the Δ_0 value by some power of two, depending on the level of decomposition associated with that subband. In particular,

$$(\varepsilon_b, \mu_b) = (c_0 - N_L + n_b, \mu_0), \tag{1.20}$$

where N_L is the total number of decomposition levels and n_b is the decomposition level corresponding to subband b. It is easy to show that Equation (20) scales the step sizes for each subband according to a power of two that best approximates the L_2-norm of a subband relative to the LL band (refer to Table 1.3). This procedure approximately aligns the quantized subband bit-planes according to their proper MSE contribution.

1.2.3.2 Inverse Quantization at the Decoder

When the irreversible (9, 7) filter bank is used, the reconstructed transform coefficient, $Rq_b(u, v)$, for a quantizer step size of Δ_b is given by

$$Rq_b(u, v) = \begin{cases} (q_b(u, v) + \gamma)\Delta_b, & \text{if } q_b(u, v) > 0, \\ (q_b(u, v) - \gamma)\Delta_b, & \text{if } q_b(u, v) < 0, \\ 0, & \text{otherwise,} \end{cases} \qquad (1.21)$$

where $0 \leq \gamma < 1$ is a reconstruction parameter arbitrarily chosen by the decoder. A value of $\gamma = 0.50$ results in midpoint reconstruction as in the existing JPEG standard. A value of $\gamma < 0.50$ creates a reconstruction bias toward zero, which can result in improved reconstruction PSNR (peak SNR) when the probability distribution of the wavelet coefficients falls off rapidly away from zero (e.g. a Laplacian distribution). A popular choice for biased reconstruction is $\gamma = 0.375$. If all of the M_b bits for a quantizer index are fully decoded, the step size is equal to Δ_b. However, when only N_b bits are decoded, the step size in Equation (1.21) is equivalent to $\Delta_b 2^{M_b - N_b}$. The reversible (5, 3) filter bank is treated in the same way (with $\Delta_b = 1$), except when the index is fully decoded to achieve lossless reconstruction, in which case $Rq_b(u, v) = q_b(u, v)$.

1.2.4 Entropy Coding

The quantizer indices corresponding to the quantized wavelet coefficients in each subband are entropy encoded to create the compressed bit-stream. The choice of the entropy encoder in JPEG 2000 is motivated by several factors. One is the requirement to create an embedded bit-stream, which is made possible by bit-plane encoding of the quantizer indices. Bit-plane encoding of wavelet coefficients has been used by several well-known embedded wavelet encoders such as EZW (Shapiro, 1993) and SPIHT (Said and Pearlman, 1996). However, these encoders use coding models that exploit the correlation between subbands to improve coding efficiency. This adversely impacts error resilience and severely limits the flexibility of an encoder to arrange the bit-stream in an arbitrary progression order. In JPEG 2000, each subband is encoded independently of the other subbands. In addition, JPEG 2000 uses a block coding paradigm in the wavelet domain as in the embedded block coding with optimized truncation (EBCOT) algorithm (Taubman and Marcellin, 2001), where each subband is partitioned into small rectangular blocks, referred to as *code-blocks*, and each code-block is independently encoded. The nominal dimensions of a code-block are free parameters specified by the encoder, but are subject to the following constraints: they must be an integer power of two; the total number of coefficients in a code-block cannot exceed 4096; and neither the height nor the width of the code-block can be less than four.

The independent encoding of the code-blocks has many advantages including localized random access into the image, parallelization, improved cropping and rotation functionality, improved error resilience, efficient rate control, and maximum flexibility in arranging progression orders (see Section 1.3.6). It may seem that failing to exploit inter-subband redundancies would have a sizable adverse effect on coding efficiency. However, this is more than compensated for by the finer scalability that results from multiple-pass encoding of the code-block bit-planes. By using an efficient rate control strategy that independently optimizes the contribution of each code-block to the final bit-stream (see Section 1.4.2), the JPEG 2000 Part 1 encoder achieves a compression efficiency that is superior to other existing approaches (Taubman *et al.*, 2002).

Figure 1.15 shows a schematic of the multiple bit-planes that are associated with the quantized wavelet coefficients. The symbols that represent the quantized coefficients are encoded one bit at a time, starting with the MSB and proceeding to the LSB. During this progressive bit-plane encoding, a quantized wavelet coefficient is called *insignificant* if the quantizer index is still zero (e.g. the example coefficient in Figure 1.15 is still insignificant after encoding its first two MSBs). Once the first nonzero bit is encoded, the coefficient becomes *significant* and its sign is encoded. Once a coefficient becomes significant, all subsequent bits are referred to as *refinement* bits. Because the DWT packs most of the energy in the low-frequency subbands, the majority of the wavelet coefficients will have low amplitudes. Consequently, many quantized indices will be insignificant in the earlier bit-planes, leading to a very low information content for those bit-planes. JPEG 2000 uses an efficient coding method for exploiting the redundancy of the bit-planes known as context-based adaptive binary arithmetic coding.

1.2.4.1   Arithmetic Coding and the MQ-Coder

Arithmetic coding uses a fundamentally different approach from Huffman coding in that the entire sequence of source symbols is mapped into a single codeword (albeit a very long codeword). This codeword is developed by recursive interval partitioning using the symbol probabilities and the final codeword represents a binary fraction that points to the subinterval determined by the sequence.

An adaptive binary arithmetic encoder can be viewed as an encoding device that accepts the binary symbols in a source sequence, along with their corresponding probability estimates, and produces a code-stream with a length at most two bits greater than the combined ideal code-lengths of the input symbols (Pennebaker *et al.*, 1988). Adaptivity

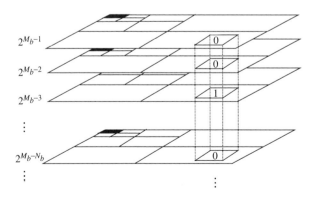

Figure 1.15   Bit-plane coding of quantized wavelet coefficients

is provided by updating the probability estimate of a symbol based upon its present value and history. In essence, arithmetic coding provides the compression efficiency that comes with Huffman coding of large blocks, but only a single symbol is encoded at a time. This single-symbol encoding structure greatly simplifies probability estimation because only individual symbol probabilities are needed at each subinterval iteration (not the joint probability estimates that are necessary in block coding). Furthermore, unlike Huffman coding, arithmetic coding does not require the development of new codewords each time the symbol probabilities change. This makes it easy to adapt to the changing symbol probabilities within a code-block of quantized wavelet coefficient bit-planes.

Practical implementations of arithmetic coding are always less efficient than an ideal one. Finite-length registers limit the smallest probability that can be maintained, and computational speed requires approximations such as replacing multiplies with adds and shifts. Moreover, symbol probabilities are typically chosen from a finite set of allowed values, so the true symbol probabilities must often be approximated. Overall, these restrictions result in a coding inefficiency of approximately 6% compared to the ideal code-length of the symbols encoded (Pennebaker and Mitchell, 1993). It should be noted that even the most computationally efficient implementations of arithmetic coding are significantly more complex than Huffman coding in both software and hardware.

One of the early practical implementations of adaptive binary arithmetic coding was the Q-coder developed by IBM (Pennebaker *et al.*, 1988). Later, a modified version of the Q-coder, known as the QM-coder, was chosen as the entropy encoder for the JBIG standard and the extended JPEG mode (Pennebaker and Mitchell, 1993). However, IPR issues have hindered the use of the QM-coder in the JPEG standard. As a result, the JPEG 2000 committee adopted another modification of the Q-coder, named the MQ-coder. The MQ-coder was also adopted for use in the JBIG2 standard (ISO/IEC, 2000). The companies that own the IPR on the MQ-coder have made it available on a license-fee-free and royalty-free basis for use in the JPEG 2000 standard. Differences between the MQ- and the QM-coders include 'bit stuffing' versus 'byte stuffing,' decoder versus encoder carry resolution, hardware versus software coding convention, and the number of probability states. The specific details of these coders are beyond the scope of this chapter and the reader is referred to (Slattery and Mitchell, 1998) and the MQ-coder flowcharts in the standard document (ISO/IEC International Standard 15444-1, ITU Recommendation T.800). We mention in passing that the specific realization of the 'bit stuffing' procedure in the MQ-coder (which costs approximately 0.5% in coding efficiency) creates a redundancy such that any two consecutive bytes of encoded data are always forced to lie in the range of hexadecimal '0000' through 'FF8F' (Taubman and Marcellin, 2001). This leaves the range of 'FF 90' through 'FFFF' unattainable by encoded data, and the JPEG 2000 syntax uses this range to represent unique marker codes that facilitate the organization and parsing of the bit-stream as well as improve error resilience.

In general, the probability distribution of each binary symbol in a quantized wavelet coefficient is influenced by all the previously encoded bits corresponding to that coefficient, as well as by the value of its immediate neighbors. In JPEG 2000, the probability of a binary symbol is estimated from a *context* formed from its current significance as well as the significance information of its immediate eight neighbors as determined from the previous bit-plane and the current bit-plane, based on encoded information up to that point. In context-based arithmetic coding, separate probability estimates are maintained for each context, which is updated according to a finite-state machine every time a symbol is encoded in that context.[5] For each context, the MQ-coder can choose from a total of 46 probability states (estimates), where states $0-13$ correspond to start-up states (also referred to as *fast-attack*) and are used for rapid convergence to a stable probability estimate. States $14-45$ correspond to steady-state probability estimates, and once this range of states has been entered from a start-up state, it can never be left by the finite-state machine. There is also an additional nonadaptive state (state 46), which is used to encode symbols with equal probability distribution, and can neither be entered nor exited from any other probability state.

1.2.4.2 Bit-Plane Coding Passes

The quantized coefficients in a code-block are bit-plane-encoded independently from all other code-blocks when creating an embedded bit-stream. Instead of encoding the entire bit-plane in one coding pass, each bit-plane is encoded in three sub-bit-plane passes with the provision of truncating the bit-stream at the end of each coding pass. A main advantage of this approach is near-optimal embedding, where the information that results in the largest reduction in distortion for the smallest increase in file size is encoded first. Moreover, the large number of potential truncation points facilitates the implementation of an optimal rate control strategy where a target bit-rate is achieved by including those coding passes that minimize the total distortion.

Referring to Figure 1.16, consider the encoding of a single bit-plane from a code-block in three coding passes (labeled A, B, and C), where a fraction of the bits are encoded at each pass. Let the distortion and bit-rate associated with the reconstructed image prior and subsequent to the encoding of the entire bit-plane be given by (D_1, R_1) and (D_2, R_2), respectively. The two coding paths ABC and CBA correspond to coding the same data in a different order, and they both start and end at the same rate-distortion points. However, their embedded performances are significantly different. In particular, if the encoded bit-stream is truncated at any intermediate point during the encoding of the bit-plane, the path ABC would have less distortion for the same rate, and hence would possess a superior embedding property. In optimal embedding, the data with the highest distortion

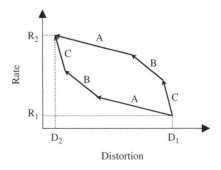

Figure 1.16 R-D path for optimal embedding

---

[5] In the MQ-coder implementation, a symbol's probability estimate is actually updated only when at least one bit of coded output is generated.

reduction per average bit of compressed representation should be encoded first (Li and Lei, 1999).

For a coefficient that is still insignificant, it can be shown that given reasonable assumptions regarding its probability distribution, the distortion reduction per average bit of compressed representation increases with increasing probability of becoming significant, p_s (Li and Lei, 1999; Ordentlich, Weinberger, and Seroussi, 1998). For a coefficient that is being refined, the distortion reduction per average bit is smaller than an insignificant coefficient, unless p_s for that coefficient is less than 1%. As a result, optimal embedding can theoretically be achieved by first encoding the insignificant coefficients starting with the highest p_s until that probability reaches about 1%. At that point, all the refinement bits should be encoded, followed by all the remaining coefficients in the order of their decreasing p_s. However, the calculation of the p_s values for each coefficient is a tedious and approximate task, so the JPEG 2000 encoder instead divides the bit-plane data into three groups and encodes each group during a fractional bit-plane pass. Each coefficient in a block is assigned a binary state variable called its *significance state*, which is initialized to zero (insignificant) at the start of the encoding. The significance state changes from zero to one (significant) when the first nonzero magnitude bit is found. The context vector for a given coefficient is the binary vector consisting of the significance states of its eight immediate neighbor coefficients, as shown in Figure 1.17. During the first pass, referred to as the *significance propagation* pass, the insignificant coefficients that have the highest probability of becoming significant, as determined by their immediate eight neighbors, are encoded. In the second pass, known as the *refinement* pass, the significant coefficients are refined by their bit representation in the current bit-plane. Finally, during the *cleanup* pass, all the remaining coefficients in the bit-plane, which have the lowest probability of becoming significant, are encoded. The order in which the coefficients in each pass are visited is data dependent and follows a deterministic stripe-scan order with a height of four pixels, as shown in Figure 1.18. This stripe-based scan has been shown to facilitate software and hardware implementations (Marcellin *et al.*, 1999). In the following, each coding pass is described in more detail.

Significance Propagation Pass

During this pass, the insignificant coefficients that have the highest probability of becoming significant in the current bit-plane are encoded. The data are scanned in the stripe order shown in Figure 1.18. Every sample that is currently insignificant but has at least one significant immediate neighbor, based on encoded information up to that point, is encoded. As soon as a coefficient is encoded, its significance state is updated so that it can effect the inclusion of subsequent coefficients in that coding pass. The significance state of the

Figure 1.17   Neighboring pixels used in context selection

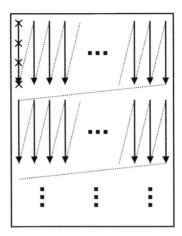

Figure 1.18 Scan order within a code-block

coefficient is arithmetic encoded using contexts that are based on the significance states of its immediate neighbors. In general, the significance states of the eight neighbors can create 256 different contexts.[6] However, many of these contexts have similar probability estimates and can be merged together. A context-reduction mapping reduces the total number of contexts to only nine to improve the efficiency of the MQ-coder probability estimation for each context. Because the code-blocks are encoded independently, if a sample is located at the code-block boundary, only its immediate neighbors that belong to the current code-block are considered and the significance states of the missing neighbors are assumed to be zero. Finally, if a coefficient becomes significant in this coding pass, its sign needs to be encoded. The sign value is also arithmetic encoded using five contexts that are determined from the significance and the sign of the coefficient's four horizontal and vertical neighbors.

Refinement Pass
During this pass, the magnitude bit of a coefficient that has already become significant in a previous bit-plane is arithmetic encoded using three contexts. In general, the refinement bits have an even distribution unless the coefficient has just become significant in the previous bit-plane (i.e. the magnitude bit to be encoded is the first refinement bit). This condition is first tested and if it is satisfied, the magnitude bit is encoded using two coding contexts based on the significance of the eight immediate neighbors. Otherwise, it is encoded with a single context regardless of the neighboring values.

Cleanup Pass
All the remaining coefficients in the code-block are encoded during the cleanup pass. Generally, the coefficients that are encoded in this pass have a very small p_s value and are expected to remain insignificant. As a result, a special mode, referred to as the run

---

[6] Technically, the combination where all the neighbors are insignificant cannot happen in this pass. However, this combination is given its own context (labeled zero) and is used during the cleanup pass.

mode, is used to aggregate the coefficients that have the highest probability of remaining insignificant. More specifically, a run mode is entered if all of the four samples in a vertical column of the stripe have insignificant neighbors. In the run mode, a binary symbol is arithmetic encoded in a single context to specify whether all of the four samples in the vertical column remain insignificant. An encoded value of zero implies insignificance for all four samples, while an encoded value of one implies that at least one of the four samples becomes significant in the current bit-plane. An encoded value of one is followed by two additional arithmetic-encoded bits that specify the location of the first nonzero coefficient in the vertical column. Because the probabilities of these additional two bits are nearly evenly distributed, they are encoded with a uniform context, which uses state 46 of the MQ-coder as its probability estimate. It should be noted that the run mode has a negligible impact on the coding efficiency and is primarily used to improve the throughput of the arithmetic encoder through symbol aggregation.

After the position of the first nonzero coefficient in the run is specified, the remaining samples in the vertical column are encoded in the same manner as in the significance propagation pass and use the same nine coding contexts. Similarly, if at least one of the four coefficients in the vertical column has a significant neighbor, the run mode is disabled and all the coefficients in that column are encoded according to the procedure employed for the significance propagation pass.

For each code-block, the MSB planes that are entirely zero are skipped, and the number of such planes is signaled in the bit-stream. Because the significance state of all the coefficients in the first nonzero MSB is zero, only the cleanup pass is applied to the first nonzero bit-plane. Subsequent bit-planes employ all three coding passes (significance propagation, magnitude refinement, and cleanup).

1.2.4.3 Entropy Coding Options

The coding models used by the JPEG 2000 entropy coder employ 18 coding contexts, in addition to a uniform context, according to the following assignment. Contexts 0–8 are used for significance coding during the significance propagation and cleanup passes, contexts 9–13 are used for sign coding, contexts 14–16 are used during the refinement pass, and an additional context is used for run coding during the cleanup pass. Each code-block employs its own MQ-coder to generate an arithmetic code-stream for the entire code-block. In the default mode, the coding contexts for each code-block are initialized at the start of the coding process and are not reset at any time during the encoding process. Furthermore, the resulting codeword can only be truncated at the coding pass boundaries to include a different number of coding passes from each code-block in the final code-stream. All contexts are initialized to uniform probabilities except for the zero context (all insignificant neighbors) and the run context, where the initial less probable symbol (LPS) probabilities are set to 0.030053 and 0.063012, respectively.

In order to facilitate the parallel encoding or decoding of the sub-bit-plane passes of a single code-block, it is necessary to decouple the arithmetic encoding of the sub-bit-plane passes from one another. Hence, JPEG 2000 allows for the termination of the arithmetic-encoded bit-stream as well as the reinitialization of the context probabilities at each coding pass boundary. If any of these two options is flagged in the code-stream, it must be executed at every coding pass boundary. The JPEG 2000 also provides for another coding option known as *vertically stripe-causal* contexts. This option is aimed at enabling the parallel decoding of the coding passes as well as reducing the external memory utilization. In this mode, during the encoding of a certain stripe of a code-block, the significances of the samples in future stripes within that code-block are ignored. Because the height of the vertical columns is four pixels, this mode only affects the pixels in the last row of each stripe. The combination of these three options, namely arithmetic encoder termination, reinitialization at each coding pass boundary, and the vertically stripe-causal context, is often referred to as the *parallel* mode.

Another entropy coding option, aimed at reducing computational complexity, is the *selective arithmetic coding bypass* mode, where the arithmetic encoder is entirely bypassed in certain coding passes. It is common to refer to this as the 'lazy' coding mode. More specifically, after the encoding of the fourth most significant bit-plane of a code-block, the arithmetic encoder is bypassed during the encoding of the first and second sub-bit-plane coding passes (i.e. significance propagation and refinement) of subsequent bit-planes. Instead, their content is included in the code-stream as raw data. In order to implement this mode, it is necessary to terminate the arithmetic encoder at the end of the cleanup pass preceding each raw coding pass and to pad the raw coding pass data to align it with the byte boundary. However, it is not necessary to reinitialize the MQ-coder context models. The lazy mode can also be combined with the parallel mode in the *lazy–parallel* mode. The impact of the lazy, parallel, and lazy–parallel modes on the coding efficiency is studied in Section 1.5.1.5.

1.2.4.4   Tier-1 and Tier-2 Coding

The arithmetic coding of the bit-plane data is referred to as *tier-1* coding. Figure 1.19 illustrates a simple example of the compressed data generated at the end of tier-1 encoding.

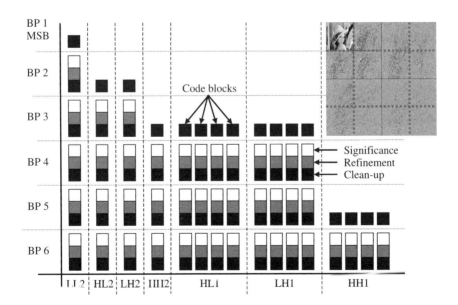

Figure 1.19   Example of compressed data associated with various sub-bit-plane coding passes

The example image (whose wavelet-transformed subbands are shown at the top right of Figure 1.19) is of size 256×256 with two levels of decomposition, and the code-block size is 64×64. Each square box in the figure represents the compressed data associated with a single coding pass of a single code-block. Because the code-blocks are independently encoded, the compressed data corresponding to the various coding passes can be arranged in different configurations to create a rich set of progression orders to serve different applications. The only restriction is that the sub-bit-plane coding passes for a given code-block must appear in a causal order starting from the most significant bit-plane. The compressed sub-bit-plane coding passes can be aggregated into larger units called *packets*. This process of packetization along with its supporting syntax, as will be explained in Section 1.3, is often referred to as *tier-2* coding.

1.3   JPEG 2000 Bit-Stream Organization

JPEG 2000 offers significant flexibility in the organization of the compressed bit-stream to enable such features as random access, region of interest coding, and scalability. This flexibility is achieved partly through the various structures of components, tiles, subbands, resolution levels, and code-blocks that are discussed in Section 1.2. These structures partition the image data into: (1) color channels (through components); (2) spatial regions (through tiles); (3) frequency regions (through subbands and resolution levels); and (4) space-frequency regions (through code-blocks). Tiling provides access to the image data over large spatial regions, while the independent coding of the code-blocks provides access to smaller units. Code-blocks can be viewed as a tiling of the coefficients in the wavelet domain. JPEG 2000 also provides an intermediate space-frequency structure known as a *precinct*. A precinct is a collection of spatially contiguous code-blocks from all subbands at a particular resolution level.

In addition to these structures, JPEG 2000 organizes the compressed data from the code-blocks into units known as *packets* and *layers* during the tier-2 coding step. For each precinct, the compressed data for the code-blocks is first organized into one or more packets. A packet is simply a continuous segment in the compressed code-stream that consists of a number of bit-plane coding passes for each code-block in the precinct. The number of coding passes can vary from code-block to code-block (including zero coding passes). Packets from each precinct at all resolution levels in a tile are then combined to form layers. In order to discuss packetization of the compressed data, it is first necessary to introduce the concepts of *resolution grids* and *precinct partitions*. Throughout the following discussion, it will be assumed that the image has a single tile and a single component. The extension to multiple tiles and components (which are possibly sub-sampled) is straightforward, but tedious, and it is not necessary for understanding the basic concepts. Section B.4 of the JPEG 2000 Part 1 standard (ISO/IEC International Standard 15444-1, ITU Recommendation T.800) provides a detailed description and examples for the more general case.

1.3.1   Canvas Coordinate System

During the application of the DWT to the input image, successively lower resolution versions of the input image are created. The input image can be thought of as the highest resolution version. The pixels of the input image are referenced with respect to a high-resolution grid, known as the *reference grid*. The reference grid is a rectangular grid of points with indices from (0, 0) to ($Xsiz$-1, $Ysiz$−1).[7] If the image has only one component, each image pixel corresponds to a high-resolution grid. In case of multiple components with differing sampling rates, the samples of each component are at integer multiples of the sampling factor on the high-resolution grid. An *image area* is defined by the parameters ($XOsiz$, $YOsiz$), which specify the upper left corner of the image, and extends to ($Xsize$−1, $Ysiz$−1), as shown in Figure 1.20.

The spatial positioning of each resolution level, as well as each subband, is specified with respect to its own coordinate system. We will refer to each coordinate system as a *resolution grid*. The collection of these coordinate systems is known as the *canvas coordinate system*. The relative positioning of the different coordinate systems corresponding to the resolution levels and subbands is defined in Section B.5 of the JPEG 2000 standard (ISO/IEC International Standard 15444-1, ITU Recommendation T.800), and is also specified later in this section. The advantage of the canvas coordinate system is that it facilitates the compressed domain implementation of certain spatial operations, such as cropping and rotation by multiples of 90°. As will be described in Section 1.5.1.6, proper use of the canvas coordinate system improves the performance of the JPEG 2000 encoder in the case of multiple compression cycles when the image is being cropped between compression cycles.

1.3.2 Resolution Grids

Consider a single component image that is wavelet transformed with N_L decomposition levels, creating $N_L + 1$ distinct resolution levels. An image at resolution level r ($0 \leq r \leq N_L$) is represented by the subband ($N_L - r$)LL. Recall from Section 1.2.2.2 that the image at resolution r ($r > 0$) is formed by combining the image at resolution ($r - 1$) with the subbands at resolution r, i.e. subbands ($N_L - r + 1$)HL, ($N_L - r + 1$)LH, and ($N_L - r + 1$)HH. The image area on the high-resolution reference grid as specified by ($Xsiz$, $Ysiz$) and ($XOsiz$, $YOsiz$) is propagated to lower resolution levels as follows. For

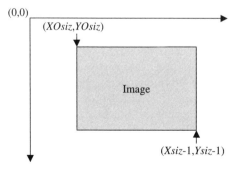

Figure 1.20 The reference grid

---

[7] The coordinates are specified as (x, y), where x refers to the column index and y refers to the row index.

the image area at resolution level r ($0 \le r \le N_L$), the upper left-hand corner is (xr_0, yr_0) and the lower right-hand corner is ($xr_1 - 1$, $yr_1 - 1$), where

$$xr_0 = \left\lceil \frac{XOsiz}{2^{N_L-r}} \right\rceil, \; yr_0 = \left\lceil \frac{YOsiz}{2^{N_L-r}} \right\rceil, \; xr_1 = \left\lceil \frac{Xsiz}{2^{N_L-r}} \right\rceil \text{ and } yr_1 = \left\lceil \frac{Ysiz}{2^{N_L-r}} \right\rceil, \quad (1.22)$$

and $\lfloor w \rfloor$ denotes the smallest integer that is greater than or equal to w.

The high-resolution reference grid is also propagated to each subband as follows. The positioning of the subband n_bLL is the same as that of the image at a resolution of ($N_L - n_b$). The positioning of subbands n_bHL, n_bLH, and n_bHH is specified as

$$(xb_0, yb_0) = \begin{cases} \left(\left\lceil \frac{XOsiz - 2^{n_b-1}}{2^{n_b}} \right\rceil, \left\lceil \frac{YOsiz}{2^{n_b}} \right\rceil \right) & \text{for } n_b \text{ HL band,} \\[2ex] \left(\left\lceil \frac{XOsiz}{2^{n_b}} \right\rceil, \left\lceil \frac{YOsiz - 2^{n_b-1}}{2^{n_b}} \right\rceil \right) & \text{for } n_b \text{ LH band,} \quad (1.23) \\[2ex] \left(\left\lceil \frac{XOsiz - 2^{n_b-1}}{2^{n_b}} \right\rceil, \left\lceil \frac{YOsiz - 2^{n_b-1}}{2^{n_b}} \right\rceil \right) & \text{for } n_b \text{ HH band.} \end{cases}$$

The coordinates (xb_1, yb_1) can be obtained from Equation (1.23) by substituting $XOsiz$ by $Xsiz$ and $YOsiz$ by $Ysiz$. The extent of subband b is from (xb_0, yb_0) to ($xb_1 - 1$, $yb_1 - 1$). These concepts are best illustrated by a simple example. Consider a three-level wavelet decomposition of an original image of size 768 (columns) $\times$ 512 (rows). Let the upper left reference grid point ($XOsiz$, $YOsiz$) be (7, 9) for the image area. Then ($Xsiz$, $Ysiz$) is (775, 521). Resolution one extends from (2, 3) to (193, 130) while subband 3HL, which belongs to resolution one, extends from (1, 2) to (96, 65).

1.3.3 Precinct and Code-Block Partitioning

Precincts represent a coarser partition of the wavelet coefficients than code-blocks, and they provide an efficient way to access spatial regions of an image. A precinct is defined as a group of code-blocks from all of the subbands at a specific resolution, such that the group nominally corresponds to the same spatial region in the original image. Figure 1.21 shows an example of precinct and code-block partitions for a three-level decomposition of a 768 $\times$ 512 image (single tile-component). In this example, there are six precincts at each resolution level and each precinct corresponds to a 256 $\times$ 256 region in the original image.

In Figure 1.21, the highlighted precincts in resolutions 0–3 correspond roughly to the same 256 $\times$ 256 region in the original image.[8] For the subbands at resolution 1, i.e. subbands 3HL, 3LH, and 3HH, a 32 $\times$ 32 region nominally corresponds to a 256 $\times$ 256 region in the original image. Hence, the precinct size in the subbands 3HL, 3LH, and 3HH is 32 $\times$ 32. However, in JPEG 2000, the size of the precinct partition is actually defined at the resolution level, instead of the subband level. With this convention, the precinct size at resolution 1 is 64 $\times$ 64, which nominally corresponds to a 256 $\times$ 256 region in the original image. Note that the precinct size in the subbands at resolution 1 is half (32 $\times$ 32) that of the precinct size in the image at resolution 1 (64 $\times$ 64). In general,

---

[8] Here we have neglected expansion of the region due to the support of the wavelet filters.

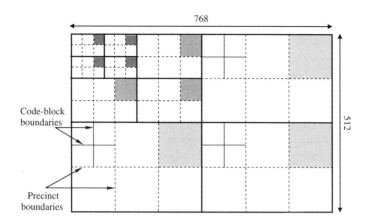

768

Code-block
boundaries

512

Precinct
boundaries

Figure 1.21   Examples of precincts and code-blocks

the precinct partition at a resolution $r(r > 0)$ induces an effective precinct partitioning of the subbands at *the same resolution level*. Except for resolution 0, the size of the induced precinct partition in the subbands is half (in each dimension) of the precinct size at the corresponding resolution. For resolution 0, the size of the induced precinct partition in the subband matches the precinct size at image resolution 0 because the image resolution 0 is the same as subband N_LLL. For the example in Figure 1.21, the precinct sizes at resolutions 0–3 are 32×32, 64×64, 128×128, and 256×256, respectively, and the induced subband precinct sizes for resolutions 0–3 are 32×32, 32×32, 64×64, and 128×128, respectively.

While the precinct size can vary from resolution to resolution, it is always restricted to be a power of two. Recall that each subband is also divided into rectangular code-blocks with dimensions that are a power of two. The precinct and code-block partitions are both anchored at (0, 0). Each precinct boundary coincides with a code-block boundary, but the reverse is not necessarily true because a precinct may consist of multiple code-blocks. The induced subband precinct size also imposes some constraints on the code-block size at a given resolution level. Code-blocks from all resolution levels are constrained to have the same size, except when constrained by the induced subband precinct size. For example, in Figure 1.21, the nominal code-block size is chosen to be 64×64, but the induced subband precinct size for subbands 3LL, 3HL, 3LH, and 3HH (i.e. the subbands that correspond to resolutions 0 and 1) is 32×32. This restricts the code-block size for resolutions 0 and 1 to be 32×32.

The precinct size can be chosen so that an entire subband belongs to a single precinct or the induced precinct size matches the code-block size. In the former case, the amount of overhead due to the introduction of precincts is very small, but the spatial accessibility is very poor. In the latter case, the amount of overhead is high due to the large number of code-blocks, but spatial accessibility is very precise.

1.3.4   Layers and Packets

The compressed bit-stream for each code-block is distributed across one or more layers in the code-stream. All of the code-blocks from all subbands and components of a tile contribute compressed data to each layer. For each code-block, a number of consecutive coding passes (possibly including zero) are included in a layer. Each layer represents a quality increment. The number of coding passes included in a specific layer can vary from one code-block to another and is typically determined by the encoder as a result of postcompression rate-distortion optimization, as will be explained in Section 1.4.2. This feature offers great flexibility in ordering the code-stream. It also enables spatially adaptive quantization. Recall that all the code-blocks in a subband must use the same quantizer step size. However, the layers can be formed in such a manner that certain code-blocks, which are deemed perceptually more significant, contribute a greater number of coding passes to a given layer. As discussed in Section 1.2.3, this reduces the effective quantizer step size for those code-blocks by a power of two compared to other code-blocks with less coding passes in that layer.

The compressed data belonging to a specific tile, component, resolution, layer, and precinct is aggregated into a packet. The compressed data in a packet needs to be contiguous in the code-stream. If a precinct contains data from more than one subband, it appears in the order HL, LH, and HH. Within each subband, the contributions from code-blocks appear in the raster order. Figure 1.21 shows an example of code-blocks belonging to a precinct. The numbering of the code-blocks represents the order in which the encoded data from the code-blocks will appear in a packet.

1.3.5   Packet Header

A packet is the fundamental building block in a JPEG 2000 code-stream. Each packet starts with a *packet header*. The packet header contains information regarding the number of coding passes for each code-block in the packet. It also contains the length of the compressed data for each code-block. The first bit of a packet header indicates whether the packet contains data or is empty. If the packet is nonempty, code-block inclusion information is signaled for each code-block in the packet. This information indicates whether any compressed data from a code-block is included in the packet. If compressed code-block data has already been included in a previous packet, a single bit is used to signal this information. Otherwise, it is signaled with a separate *tag tree* for each subband of the corresponding precinct. The tag tree is a hierarchical data structure that is capable of exploiting spatial redundancy. If code-block data are being included for the first time, the number of most significant bit-planes that are entirely zero is also signaled with another set of tag trees for the precinct. After this, the number of coding passes for the code-block and the length of the corresponding compressed data are signaled.

1.3.5.1   Tag Trees

The concept of tag tree encoding, which is a particular type of quadtree structure, is best illustrated by an example. Consider a 2-D array of nonnegative integers consisting of two rows and three columns, as shown at the top of Figure 1.22. The array values may represent the layer number in which a code-block is included for the first time. Alternatively, the array values may represent the number of most significant bit-planes that are entirely zero. The values of the original array are the leaf nodes of the tag tree structure. The node value at the next level in the tag tree is the minimum of the values of the child nodes. Nominally, there are four child nodes, but there may be less than

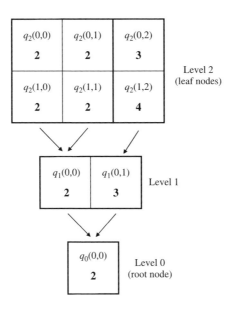

Figure 1.22   Example of a tag tree

four at the right and bottom edges, as shown in Figure 1.22. This process of forming a parent node from the minimum of the child nodes is repeated until only a single root node remains. The value of a particular tag tree node is denoted by $q_z(i, j)$, where z is the level within the quadtree, with leaf nodes at the highest level and the root of the quadtree at level 0, and the row and column indices are represented by i and j, respectively. We will use $q_z(i, j)$ to refer to the tag tree node as well as the actual value of the tag tree node. The exact meaning should be clear from the context.

To encode a specific leaf node from the tag tree, all of the ancestors of the leaf node are encoded, starting with root node at level 0 and proceeding to the higher levels. Any ancestor that has previously been encoded is skipped. As a first example of tag tree encoding, consider the encoding of the 2-D array values of Figure 1.22 in a raster scan order, starting with $q_2(0, 0)$. The ancestor nodes of $q_2(0, 0)$ are $q_1(0, 0)$ and $q_0(0, 0)$. The root node, $q_0(0, 0)$, is encoded first, using a very simple code where a node value of $B \geq 0$ is represented by B zeros followed by a one. Thus, $q_0(0, 0) = 2$ is encoded as '**001**.' The next node to be encoded is $q_1(0, 0)$. For nodes at level 1 or higher, only the difference between the current node and the parent node is encoded using the code described above. This difference is always nonnegative because of the way the tag tree is constructed using the minimum value for the parent node. The difference between $q_1(0, 0)$ and $q_0(0, 0)$ is 0, which is encoded as '**1**.' Similarly, $q_2(0, 0)$ is encoded as '**1**.' For $q_2(0, 1)$, its ancestors, $q_0(0, 0)$ and $q_1(0, 0)$, have already been encoded. Thus, $q_2(0, 1)$ is encoded as '**1**.' To encode $q_2(0, 2)$, it is necessary to encode $q_1(0, 1)$ first. The nodes $q_1(0, 1)$ and $q_2(0, 2)$ are encoded as '**011**.' The remaining nodes, $q_2(1, 0)$, $q_2(1, 1)$, and $q_2(1, 2)$, are encoded as '**1101**.' Thus the entire array is encoded as '**0011110111101**.'

As a second example of tag tree encoding, suppose that each entry in the 2-D array represents the number of zero MSB bit-planes for each code-block belonging to a particular subband and precinct. As described previously, this information is included in the packet header immediately after the first time a code-block makes a contribution to a layer. As a result, the order of tag tree encoding may not follow a raster scan order. As an example, suppose that code-blocks corresponding to $q_2(0, 2)$ and $q_2(1, 2)$ are included for the first time in layer 0 and the remaining code-blocks are included for the first time in layer 1. Then, the order in which the leaf nodes get encoded is $q_2(0, 2), q_2(1, 2), q_2(0, 0),$ $q_2(0, 1), q_2(1, 0),$ and $q_2(1, 1)$. The encoding order including the ancestor nodes is $q_0(0, 0), q_1(0, 1), q_2(0, 2), q_2(1, 2), q_1(0, 0), q_2(0, 0), q_2(0, 1), q_2(1, 0),$ and $q_2(1, 1)$. Following the method described in the previous paragraph, the tag tree encoding is '**0010110111111**.' This encoding has exactly 13 bits as before. Thus, the number of bits produced by a tag tree encoding does not depend on the order in which the leaf nodes are encoded.

The method that was just described is suitable for encoding the tag tree containing zero MSB bit-planes information. However, it is not suitable for encoding the code-block inclusion tag tree. This is because it is only necessary to indicate whether a code-block contributes to the current layer if it has not contributed to any previous layer. It is not necessary to specify the exact layer number in which it gets included for the first time. Following the encoding method described above would mean that inclusion information for all the layers is aggregated upfront, which leads to a suboptimal embedding. To rectify this shortcoming, the tag tree encoding method can be modified to make it hierarchical.

To illustrate hierarchical encoding of a tag tree, assume that the values of the 2-D array in Figure 1.22 represent the layers in which the code-blocks are included for the first time. Instead of coding the exact layer number in which the code-block is included for the first time, the only information included for packet w (corresponding to layer w) is whether the code-block is included in layer w for the first time. This is accomplished by hierarchical encoding of the tag tree. It is necessary to keep track of two more variables for each node in the tag tree, the current value cv and the state S. A state value of 1 indicates that the exact value for a tag tree node can be deduced from the information already encoded. The tag tree decoder mimics the encoding steps.

At the beginning of the tag tree encoding procedure, the current value and state for each node is initialized to zero. The interpretation of the current value for a tag tree node is as follows. If the state S is 0, $q_z(i, j) \geq cv_z(i, j)$. Let the total number of layers be W and let the leaf nodes in the tag tree be at level Z. A leaf node $q_Z(i, j)$ has ancestors $q_z(i_z, j_z), 0 \leq z < Z$, where $i_z = \lfloor i/2^{(Z-z)} \rfloor$ and $j_z = \lfloor j/2^{(Z-z)} \rfloor$.

Encode:

for each layer $w, w = 0 : (W - 1)$
    for each leaf node $q_Z(i, j)$
        for each $q_z(i_z, j_z), 0 \leq z \leq Z$
            if $(S_z(i_z, j_z) == 0)$
                if $(z > 0)$ and $cv_z(i_z, j_z) < cv_{(z-1)}(i_{z-1}, j_{z-1})$
                        $cv_z(i_z, j_z) = cv_{(z-1)}(i_{z-1}, j_{z-1})$ // Due to tag tree
                construction
                if $(cv_z(i_z, j_z) \leq w)$
                    if $(q_z(i_z, j_z) \leq w)$ Emit '**1**,' set $S_z(i_z, j_z) = 1$
                    else emit '**0**,' increment $cv_z(i_z, j_z)$ by 1.

Decode:

for each layer $w, w = 0 : (W - 1)$

    for each leaf node $q_Z(i, j)$

        for each $q_z(i_z, j_z), 0 \leq z \leq Z$

            if $(S_z(i_z, j_z) = 0)$

                if $(z > 0)$ and $cv_z(i_z, j_z) < cv_{(z-1)}(i_{z-1}, j_{z-1})$

                    $cv_z(i_z, j_z) = cv_{(z-1)}(i_{z-1}, j_{z-1})$ // Due to tag tree

                construction

                if $(cv_z(i_z, j_z) \leq w)$

                    if next bit is '**1**,' set $S_z(i_z, j_z) = 1$ and

                $q_z(i_z, j_z) = cv_z(i_z, j_z)$

                    else increment $cv_z(i_z, j_z)$ by 1.

Referring to the tag tree of Figure 1.22, we will now illustrate the hierarchical encoding process. All the state and current value variables are initialized to 0. For $w = 0, cv_0(0, 0) \leq w$ and $q_0(0, 0) > w$. Hence, a '0' bit is emitted and $cv_0(0, 0)$ is incremented to 1. From the interpretation of cv, this implies that $q_0(0, 0) \geq cv_0(0, 0) = 1$. As a result of the manner in which the tag tree is constructed, $q_z(i_z, j_z) \geq 1, 0 \leq z < Z$. Thus, all the remaining states are incremented by 1 and no further bits are emitted for layer 0. Then for $w = 1, cv_0(0, 0) \leq w$ and $q_0(0, 0) > w$. Hence a '0' bit is emitted and $cv_0(0, 0)$ is incremented to 2. As before, all the remaining states are incremented to 2 and no further bits are emitted for layer 1. For $w = 2, cv_0(0, 0) \leq w$ and $q_0(0, 0) \leq w$. Hence a '1' is emitted and $S_0(0, 0)$ is set to 1. Now proceeding to $q_1(0, 0), q_2(0, 0)$, and $q_2(0, 1)$, a '1' is emitted for each, and the corresponding states are set to 1. Proceeding to $q_1(0, 1), cv_1(0, 1) \leq w$ and $q_1(0, 1) > w$. Hence a '0' bit is emitted and $cv_1(0, 1)$ is incremented to 3. Then $cv_2(0, 1)$ is set to 3 and no further bits are emitted. For $q_2(1, 0)$ and $q_2(1, 1)$, a '1' is emitted for each and their states are set to 1. Proceeding to $q_2(1, 2), cv_2(1, 2)$ is incremented to 3 to equal $cv_1(0, 1)$ and no bits are emitted. For $w = 3$, a '1' is emitted for $q_1(0, 1)$ as well as $q_2(0, 2)$, and their states are set to 1. Proceeding to $q_2(1, 2)$, a '0' bit is emitted and $cv_2(1, 2)$ is incremented to 4. Finally, for $w = 4$, a '1' is emitted for $q_2(1, 2)$ and its state is set to 1. Thus, the encoded tag tree bit-stream is '**0 0 1111011 110 1**.' The extra spaces indicate layer separation.

Note that the encoded tag tree also consists of 13 bits for this hierarchical encoding method. Furthermore, the number of 1's and the number of 0's are also exactly the same as in the case of nonhierarchical encoding. Because of its general applicability, the hierarchical encoding method is used for encoding all tag trees in JPEG 2000.

1.3.6 *Progression Order*

The arithmetic encoding of the bit-planes is referred to as tier-1 coding, whereas the packetization of the compressed data and encoding of the packet header information is known as tier-2 coding. In order to change the sequence in which the packets appear in the code-stream, it is necessary to decode the packet header information, but it is not necessary to perform arithmetic decoding. This allows the code-stream to be reorganized with minimal computational complexity.

The order in which packets appear in the code-stream is called the progression order and is controlled by specific markers. Regardless of the ordering, it is necessary that coding passes for each code-block appear in the code-stream in causal order from the most significant bit to the least significant bit. For a given tile, four parameters are needed to uniquely identify a packet. These are component, resolution, layer, and position (precinct). The packets for a particular component, resolution, and layer are generated by scanning the precincts in a raster order. All the packets for a tile can be ordered by using nested 'for loops' where each 'for loop' varies one parameter from the above list. By changing the nesting order of the 'for loops,' a number of different progression orders can be generated. JPEG 2000 Part 1 allows only five progression orders, which have been chosen to address specific applications. They are: (i) layer–resolution–component–position progression, (ii) resolution–layer–component–position progression, (iii) resolution–position–component–layer progression, (iv) position–component–resolution–layer progression, and (v) component–position–resolution–layer progression. These progression orders share some similarities with the different modes of the extended DCT-based JPEG standard, as will be pointed out in the subsequent subsections.

To illustrate these different orderings, consider a three-component color image of size 768×512 with two layers and three decomposition levels (corresponding to four resolution levels). The precinct partition is as shown in Figure 1.21. The component, resolution, layer, and position are indexed by c, r, l, and k, respectively. It is possible that the components of an image have different numbers of resolution levels. In that case, the LL subbands of different components are aligned.

1.3.6.1 Layer–Resolution–Component–Position Progression (LRCP)

This type of progression is obtained by arranging the packets in the following order:

for each $l = 0, 1$
        for each $r = 0, 1, 2, 3$
                for each $c = 0, 1, 2$
                        for each $k = 0, 1, 2, 3, 4, 5$
                                packet for component c, resolution r, layer l, and position k.

This type of progression order is useful in an image database-browsing application, where progressively refining the quality of an image may be desirable. This mode has no exact counterpart in the existing JPEG. However, the 'sequential progressive' mode of extended JPEG (component noninterleaved format) provides similar functionality for a single resolution image.

1.3.6.2 Resolution–Layer–Component–Position Progression (RLCP)

This type of progression order is obtained by interleaving the 'for loops' in the order r, l, c, and k, starting with the outermost 'for loop.' It is useful in a client–server application, where different clients might demand images at different resolutions. This progression order is similar to the 'hierarchical progressive' mode of extended JPEG where each resolution is further encoded with the 'sequential progressive' mode (component noninterleaved format).

1.3.6.3 Resolution–Position–Component–Layer Progression (RPCL)

This type of progression order is obtained by interleaving the 'for loops' in the order r, k, c, and l, starting with the outermost 'for loop.' It can be used when resolution scalability is needed, but within each resolution it is desirable that all packets corresponding to a precinct appear contiguously in the compressed bit-stream. For JPEG systems, the 'resolution–position–component' order for a single layer can be obtained using the hierarchical progressive mode of extended JPEG with each resolution encoded with baseline JPEG (component interleaved format).

1.3.6.4 Position–Component–Resolution–Layer Progression (PCRL)

This type of progression order is obtained by arranging the 'for loops' in the order k, c, r, and l, starting with the outermost 'for loop.' It should be used if it is desirable to refine the image quality at a particular spatial location. The 'position–component' order is similar to the JPEG baseline where the image is sequentially compressed by compressing the component interleaved 8×8 blocks in a raster order fashion.

1.3.6.5 Component–Position–Resolution–Layer Progression (CPRL)

This type of progression order is obtained by arranging the 'for loops' in the order c, k, r, and l, starting with the outermost 'for loop.' It should be used if it is desirable to obtain the highest quality image for a particular spatial location only for a specific image component. The 'component–position' order is similar to the JPEG baseline where the image is sequentially compressed by compressing each color component separately in a raster order fashion.

In the last three progression orders, the 'for loop' corresponding to the variable k, which determines the order in which the precincts appear in the code-stream, can become complicated if different components have different precinct sizes, as explained in the standard document (ISO/IEC International Standard 15444-1, ITU Recommendation T.800). The JPEG 2000 syntax offers the flexibility of changing from one progression order to another in the middle of the codestream. For example, a digital camera image might start out in the RLCP order to provide a thumbnail. The order then may be switched to LRCP to facilitate rate control and truncation after the image has been captured.

Figures 1.23 to 1.25 illustrate some of these progression orders for the 'Boy' image (768×512, monochrome). In these examples, the DWT has three decomposition levels, the (9, 7) filter bank is used, and the precinct sizes at resolutions 0, 1, 2, and 3 are 32×32, 64×64, 128×128, and 256×256, respectively. The code-block size is 64×64, except for resolutions 0 and 1, where the code-block size is constrained to the subband precinct size of 32×32. Thus, there are four resolutions, six precincts per resolution, and two layers, resulting in 48 packets. Figure 1.23 shows the LRCP progression order (Section 1.3.6.1). The image has been reconstructed at the two quality levels of 0.125 bits/pixel and 0.5 bits/pixel by decoding 24 and 48 packets, respectively. Figure 1.24 illustrates the RLCP ordering (Section 1.3.6.2). The figure shows images reconstructed after decoding resolutions 0, 1, 2, and 3 (12, 24, 36, and 48 packets), respectively. Figure 1.25 illustrates the PCRL ordering (Section 1.3.6.4). The image has been reconstructed after decoding 32 packets corresponding to the first four precincts.

Figure 1.23 Example of layer progressive bit-stream ordering: (left) 0.125 bpp; (right) 0.50 bpp

Figure 1.24 Example of resolution progressive bit-stream ordering

Figure 1.25 Example of spatially progressive bit-stream ordering (four precincts decoded)

It should be noted that because of the prediction step in the hierarchical progressive mode of the extended JPEG, before decoding any data at a given resolution it is necessary to fully decode all the data corresponding to the lower resolution versions of the image. This interresolution dependency makes it impossible to achieve certain progression orders, e.g. LRCP. Also, rearranging the JPEG compressed data from one progression mode to another generally requires an inverse DCT, e.g. when converting from the hierarchical progressive to the sequential progressive mode. With JPEG 2000, a given progression order can be converted into another without the need for arithmetic decoding or inverse wavelet transform by simply rearranging the packets. This only requires decoding of the packet headers to determine the length of each packet. However, the decoding of packet headers can be avoided by inclusion of PLT (or PLM) marker segments, which are described in the next subsection.

1.3.7   Code-Stream Organization and Syntax

A JPEG 2000 code-stream consists of two fundamental types of data: (1) compressed data in the form of packets and (2) syntactical data in the form of markers and marker segments that define the characteristics of the image and delimit the code-stream. Certain markers and marker segments are combined to form headers, and the JPEG 2000 standard defines two kinds of headers, a main header and a tile-part header. At the highest structural level, a JPEG 2000 code-stream consists of a main header, one or more tile parts, and an end of code-stream (EOC) marker. Multiple tile parts are formed, if desired, by breaking the compressed data for a tile at any packet boundary. Each tile part consists of a tile-part header and compressed data for the tile part as a sequence of packets. Tile parts from different tiles can be interleaved in the code-stream. Figure 1.26 shows an example of a JPEG 2000 code-stream consisting of two tiles, where tile 0 has a single tile part and tile 1 has two tile parts.

1.3.7.1   Markers and Marker Segments

A marker is always two bytes, and the first byte is always 0xFF.[9] The second byte denotes the specific marker, with a value in the range 0x01 to 0xFE. The use of the 0xFF byte as a prefix allows the markers to be located easily when a code-stream is parsed. Typically, a marker is followed by a parameter list, and together the marker and the parameter list form a marker segment. The marker occupies the first two bytes of a marker segment. The marker is followed by a two-byte length parameter, which is an unsigned integer in the big-endian format that specifies the length of the marker segment. The length of the marker segment includes the two bytes for the length parameter itself but does not include the two bytes for the marker. The marker segments are always multiples of 8 bits and all multibyte parameter values in a marker segment are big endian. There are six types of marker and marker segments:

- delimiting markers and marker segments: start of code-stream (SOC), start of tile part (SOT), start of data (SOD), and end of code-stream (EOC);
- fixed information marker segments: image and tile size (SIZ);

---

[9] The prefix 0x indicates that the number following the prefix is in hexadecimal notation.

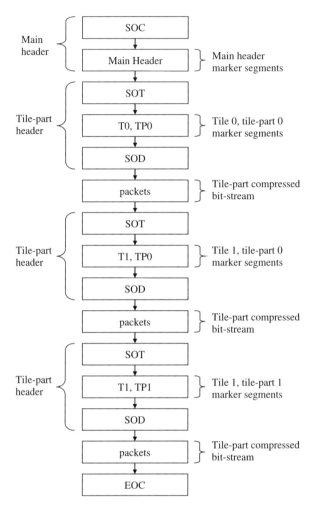

Figure 1.26   JPEG 2000 code-stream organization

- functional marker segments: coding style default (COD), coding style component (COC), region of interest (RGN), quantization default (QCD), quantization component (QCC), and progression order change (POC);
- pointer marker segments: tile-part lengths (TLM), packet length, main header (PLM), packet length, tile-part header (PLT), packed packet headers, main header (PPM), and packed packet headers, tile-part header (PPT);
- in bit-stream markers and marker segments: start of packet (SOP) and end of packet header (EPH);
- informational marker segments: component registration (CRG) and comment (COM).

The delimiting and fixed information markers and marker segments must be present in specific locations in all JPEG 2000 compliant code-streams. The code-stream must begin with the SOC marker, followed by the SIZ marker in the main header. Each tile-part header must contain the SOT marker as its first marker segment. The SOD marker must be present after each tile-part header. Finally, a valid JPEG 2000 code-stream must end with the EOC marker.

The SIZ marker segment as well as the functional marker segments contain information concerning the image and the code-stream. The SIZ marker segment contains information regarding the image and tile sizes, and their positioning with respect to the reference grid, as well as component subsampling factors and bit-depths. The COD marker segment contains information related to coding parameters such as precinct and code-block sizes, entropy-coding mode, number of layers, progression order, etc. The COC marker segment contains the same type of information as the COD marker, but it applies to a specific component. The QCD and QCC marker segments contain quantizer step size information for all components and for a specific component, respectively. The RGN marker segment can be used to specify a region of interest and the POC marker is used to signify a change in the progression order.

The pointer marker segments provide length information or pointers into the code-stream. They are useful for providing random access to certain parts of the code-stream without having to parse the whole code-stream. The SOP marker segment may appear before each packet and the EPH marker may appear immediately after a packet header. The SOP and EPH markers are useful for error detection and synchronization. Finally, CRG and COM are informational marker segments. They are not necessary for correct decoding of the JPEG 2000 code-stream. The CRG marker specifies the specific registration of different components, which may be helpful for rendering purposes. The COM marker segment allows storage of unstructured comment information.

1.3.7.2 Allowable Marker Segments and Scope

The markers and marker segments that can be present in the main header are SOC, SIZ, COD, COC, QCD, QCC, RGN, POC, PPM, TLM, PLM, CRG, and COM. The SOC, SIZ, COD, and QCD are required in the main header, while the rest of the markers are optional. The SOC and SIZ markers must be the first two markers in the main header. The rest of the marker segments can appear in an arbitrary order within the main header.

For each tile-part header, SOT must be the first marker segment and SOD must be the last marker. Each tile-part header may optionally contain POC, PPT, PLT, or COM marker segments. If the tile-part header is the first for a given tile, it can optionally contain COD, COC, QCD, QCC, or RGN marker segments.

The scope of a marker segment depends on its type. The scope of a marker segment in the main header extends to the whole image, while the scope of a marker segment in a tile-part header extends only to the corresponding tile part. If a marker segment is component specific, its scope is restricted accordingly. For example, if a COC marker segment is present in a tile-part header, its scope extends to the specific component in that tile part. When there are multiple marker segments with overlapping scopes, the following precedence rules apply. Within a tile-part header, a component-specific marker segment takes precedence over a general tile-part header marker segment for the specific component. Similarly, within the main header, a component-specific segment takes precedence over a general main header marker segment for that specific component. Furthermore, a marker segment appearing in a tile-part header takes precedence over a marker segment appearing in the main header for the specific tile. As an example, consider a code-stream with the main header containing COD and COC marker segments and a tile-part header also containing COD and COC marker segments. For this example, the marker segments in decreasing order of precedence are: tile-part COC, tile-part COD, main COC, and main COD.

1.4 JPEG 2000 Rate Control

Rate control refers to the process of allocating bits to an image and is strictly an encoder issue. In constant bit-rate (CBR) systems, the goal is to achieve a target file size with the highest possible image quality. In variable bit-rate (VBR) systems, the goal is to achieve a target image quality with the smallest possible file size. Depending upon the application, either CBR or VBR may be preferable. The metric that is used to assess image quality is typically the mean squared error between the original and reconstructed image. However, the use of MSE is primarily motivated by mathematical convenience, and it is well known that MSE does not always correlate well with perceived quality. As a result, visually weighted MSE or more sophisticated visual distortion metrics can also be used (Liu, Karam, and Watson, 2006: Marziliano *et al.*, 2004).

The structure of the JPEG 2000 encoding process provides new opportunities for rate control that are not available in conventional JPEG encoding. In the existing JPEG standard, the user only has control over the quantization and Huffman tables, and there is no easy mechanism for compressing an image to a desired bit-rate. A typical JPEG rate control algorithm for CBR encoding starts with a basic q-table and iteratively modifies the q-table elements (e.g. by a scale factor) until the desired file size (bit-rate) is achieved. In contrast, the embedded block encoding scheme of the JPEG 2000 and its flexible code-stream syntax allow for the noniterative generation of an R-D optimized code-stream for a given file size or a given image quality level. Each JPEG 2000 encoder can perform its own optimization (based on the distortion metric used) to generate a code-stream that conforms to the standardized syntax. In the following, a brief discussion of several possible approaches to JPEG 2000 rate control is provided.

1.4.1 Rate Control Using an Explicit q-Table

One approach to rate control is to use an explicit q-table (or q-tables when encoding color images) in a manner similar to JPEG, where a quantizer step size is specified for each subband and signaled explicitly as header information. However, for CBR encoding, this approach suffers from the same drawback as JPEG in that the q-table needs to be modified (e.g. scaled or otherwise adjusted) iteratively to achieve the desired bit-rate. Although there is no need to perform the wavelet transform at each iteration, the quantization and encoding processes still need to be performed repeatedly.

The use of explicit q-tables is more applicable to VBR systems, where the goal is constant image quality and the bit-rate will fluctuate with the image content. Because human observers are the ultimate judges of image quality in most applications, it is necessary to consider the properties of the HVS when designing a q-table (Albanesi and Bertoluzza, 1995; Jones, 2007; Jones *et al.*, 1995; O'Rourke and Stevenson, 1995; Watson *et al.*, 1997; Zeng, Daly, and Lei, 2002). The general approach to perceptually based q-table design is to take advantage of the sensitivity variations of the HVS to different spatial frequencies. The DWT in JPEG 2000 conveniently provides a frequency decomposition of the input image, and the HVS response can be mapped on to the wavelet subbands. Although visually based q-tables can be designed through actual observer experiments, as was done in developing the example q-tables specified in the existing JPEG standard, such experiments are laborious and must be repeated each time the viewing conditions are changed. A more efficient approach is to use a computational model of the contrast sensitivity function (CSF) as described in Jones *et al.* (1995). The CSF quantifies the detection threshold of an observer to different spatial frequencies, and the goal in applying the CSF to image compression is to ensure that all compression errors are kept below the detection threshold for specified viewing conditions. If this goal is achieved, the compressed image quality is referred to as *visually lossless* for the specified viewing conditions. The viewing conditions include such parameters as the viewing distance, displayed pixel size, display noise, and light adaptation level. The type of analysis that is needed to determine visually lossless q-tables for JPEG 2000 is described in Jones (2007). It is noted that the use of explicit q-tables can also be combined with the rate-distortion optimization method of the next section if the visually based VBR encoding is subject to a maximum file size constraint.

1.4.2 Rate Control Using the EBCOT Algorithm (PCRD-opt)

In 2000, Taubman proposed an efficient rate control method for the EBCOT compression algorithm that achieves a desired rate in a single iteration with minimum distortion. This method can also be used by a JPEG 2000 encoder, with several possible variations, including achieving a desired distortion with the minimum bit-rate.

In the basic approach, each subband is first quantized using a very fine step size, and the bit-planes of the resulting code-blocks are entropy encoded. This typically generates more coding passes for each code-block than will be eventually included in the final code-stream. This situation is known as *overcoding*, and it represents a computational inefficiency that is necessary to achieve the desired rate control benefits with this method. Heuristics can be used to reduce the amount of overcoding, but they require stable image statistics to perform well. If the quantizer step size is chosen to be small enough, the R-D performance of the algorithm is independent of the initial choice of the step size. Next, a Lagrangian R-D optimization is performed to determine the number of coding passes from each code-block that should be included in the final compressed bit-stream to achieve the desired bit-rate. If more than a single layer is desired, this process can be repeated at the end of each layer to determine the additional number of coding passes from each code-block that need to be included in the next layer.

The Lagrangian R-D optimization works in the following manner. The compressed bit-stream from each code-block contains a large number of potential truncation points that can occur at the end of each sub-bit-plane pass. The wavelet coefficients $y(u, v)$ contained in a code-block of subband b are initially quantized with a step size of Δ_b, resulting in an M_b-bit quantizer index for each coefficient. If the code-block bit-stream is truncated so that only N_b bits are decoded, the effective quantizer step size for the coefficients is $\Delta_b 2^{M_b - N_b}$. The inclusion of each additional bit-plane in the compressed bit-stream will decrease the effective quantizer step size by a factor of two. However, the effective quantizer step size might not be the same for every coefficient in a given code-block due to the inclusion of some coefficients in the sub-bit-plane at which the truncation occurs. For each sub-bit-plane, the increase in bit-rate and the reduction in distortion resulting from the inclusion of that sub-bit-plane in the bit-stream are calculated. The distortion measure selected is usually MSE or visually weighted MSE, although any general distortion measure that is additive across code-blocks can be used. Let the total number of code-blocks for the entire image be P and let the code-blocks in the image be denoted by B_i, $1 \leq i \leq P$. For a given truncation point t in code-block B_i, the associated weighted MSE distortion D_i^t is given by

$$D_i^t = \alpha_b^2 \sum_{u,v} w_i(u, v) \left[y_i(u, v) - y_i^t(u, v) \right]^2, \tag{1.24}$$

where u and v represent the coefficient row and column indices within the code-block B_i, $y_i(u, v)$ is the original coefficient value, $y_i^t(u, v)$ is the quantized coefficient value for truncation point t, $w_i(u, v)$ is a weighting factor for coefficient $y_i(u, v)$, and α_b is the L_2-norm for subband b. Under certain assumptions for the quantization noise, this distortion is additive across code-blocks. At the given truncation point t, the size of the associated compressed bit-stream (i.e. the rate) for the code-block B_i is determined and denoted by R_i^t.

Given a total bit budget of R bytes for the compressed bit-stream, the EBCOT rate control algorithm finds the truncation point for each code-block that minimizes the total distortion D. This is equivalent to finding the optimal bit allocation for all of the code-blocks, R_i^*, $1 \leq i \leq P$, such that

$$D = \sum_i D_i^* \text{ is minimized subject to } \sum_i R_i^* \leq R. \tag{1.25}$$

In the JPEG 2000 literature, this rate control algorithm is also referred to as the *postcompression R-D optimization (PCRD-opt)* algorithm. If the weighting factor $w_i(u, v)$ is set to unity for all subband coefficients, the distortion metric reduces to the MSE. A visual weighting strategy can also be used in conjunction with the EBCOT rate control algorithm, as will be discussed next.

1.4.2.1 Fixed Visual Weighting

The CSF model used to design the q-tables for explicit quantization of the wavelet coefficients can also be used to derive the weighting factors $w_i(u, v)$. For example, once the CSF-based quantization step sizes have been computed for a given viewing condition (Section 1.4.1), the weighting factor for all the coefficients in a subband can be set equal to the square of the reciprocal of these step sizes. Table J-24 from Part 1 of the JPEG 2000 standard (ISO/IEC International Standard 15444-1, ITU Recommendation T.800) lists recommended frequency weightings for three different viewing conditions. This approach is known as *fixed visual weighting*.

1.4.2.2 Bit-Stream Truncation and Layer Construction

In the EBCOT rate control algorithm, an image is compressed in such a way that the minimum distortion is achieved at the desired bit-rate. However, it is sometimes desirable to truncate an existing JPEG 2000 code-stream to achieve a smaller bit-rate. For example, this scenario could take place in a digital camera where the already captured and compressed images have to be truncated to enable storage of a newly captured image. The question that arises is whether the truncated bit-stream also achieves the minimum distortion for the smaller bit-rate. In other words, we want the visual quality of the image from the truncated code-stream to be as close as possible to the visual quality of the image that would be produced by compressing directly to that bit-rate.

This property can be achieved only if the image was initially encoded with a number of layers using the LRCP ordering of the packets as described in Section 1.3.6.1. The layers can be designed using R-D optimization so that the minimum distortion for the resulting bit-rate is achieved at each layer boundary. However, the quality of the resulting truncated image might not be optimal if the truncation point for the desired bit-rate does not fall on a layer boundary. This is because the nonboundary truncation of a layer in LCRP ordering will result in a number of packets being discarded. If the desired bit-rates or quality levels are known in advance for a given application, it is recommended that the layers be constructed accordingly. If the exact target bit-rates are not known *a priori*, it is recommended that a large number of layers (e.g. 50) be formed. This provides the ability to approximate a desired bit-rate while still truncating at a layer boundary. As demonstrated in Section 1.5.2.2, the impact of the resulting overhead on PSNR is quite small.

1.4.2.3 Progressive Visual Weighting

In fixed visual weighting, the visual weights are chosen according to a single viewing condition. However, if the bit-stream is truncated, this viewing condition may be inappropriate for the reduced quality image. Consider a case where the compressed bit-stream has a number of layers, each corresponding to a potential truncation point. If the bit-stream is truncated at a layer boundary with a very low bit-rate, the resulting image quality would be poor and the image might be viewed at a larger viewing distance than the one intended for the original compressed bit-stream. As a result, in some applications it might be desirable to have each layer correspond to a different viewing condition. In an embedded encoder such as JPEG 2000, it is not possible to change the subband quantization step sizes for each layer. However, if a nominal viewing condition can be associated with each layer, a corresponding set of visual weighting factors $w_i(u, v)$ can be used during the R-D optimization process for that layer (Li, 1999). This is known as *progressive visual weighting*.

1.5 Performance Comparison of the JPEG 2000 Encoder Options

The JPEG 2000 standard offers a number of encoder options that directly affect the coding efficiency, speed, and implementation complexity. In this section, we primarily compare the effects of various coding options on the coding efficiency for lossless compression and on the rate-distortion performance for lossy compression. It is more difficult to compare the speed and implementation complexity of different coding options accurately, so we only point out the relative speed/complexity advantages of certain options.

To obtain the reported results, the three test images shown in Figure 1.27, 'Bike,' 'Café,' and 'Woman' of size 2048 (columns) × 2560 (rows) were chosen from the JPEG

Figure 1.27   Test images (from left to right) Bike, Café, Woman

2000 test set. All three images are grayscale and have a bit-depth of 8 bits/sample. For lossy compression, distortion was characterized by the peak signal-to-noise ratio (PSNR), which for an 8-bit decompressed image is defined as

$$\text{PSNR} = 10\log_{10}\left(\frac{255^2}{\text{MSE}}\right), \tag{1.26}$$

where MSE is the mean squared error between the original image and the reconstructed image. In most cases, the results are presented as the average PSNR of the three images. We use the average PSNR instead of the PSNR corresponding to the average MSE in accordance to the practice of JPEG 2000 core experiments.

During the development of the JPEG 2000 standard, the committee maintained a software implementation of an encoder and decoder that contained all the technologies considered for the inclusion in the standard as of that time. This was also accompanied by a textual description of the technologies. Both the software and the textual description were referred to as the Verification Model (VM). After each meeting of the JPEG committee, the VM was updated to reflect any approved modifications. For the results in this section, we used the JPEG 2000 Verification Model Version 8.6 (VM8.6) (ISO/IEC, 2000). During each simulation, only one parameter was varied while the others were kept constant in order to study the effect of a single parameter on compression performance.

As a reference implementation, we used the following set of compression parameters: single tile; five levels of wavelet decomposition; 64×64 code-blocks; and a single layer. The subbands at each resolution were treated as a single precinct. In the case of irreversible (9, 7) lossy compression, the reciprocal of the L_2-norm was used as the fundamental quantization step size for each subband. In the case of reversible (5, 3) lossless and lossy compression, the quantizer step size was set to unity for all subbands as required by the JPEG 2000 standard. Hence, when using the reversible (5, 3) filter bank for lossy compression, rate control is possible only by discarding bits from the integer representation of the index for the quantized wavelet coefficient.

The results for lossy coding are reported for bit-rates of 0.0625, 0.125, 0.25, 0.5, 1.0, and 2.0 bits/pixel (bpp). To achieve a target bit-rate, the compressed code-block bit-streams were truncated to form a single layer. The truncation points are determined using the EBCOT postcompression R-D optimization procedure as described in Section 1.4.2. We chose this alternative instead of varying the quantizer step size for the following reason. Suppose that a particular step size is used to achieve a target bit-rate without any truncation of the compressed bit-stream. Then, varying a single coding parameter, while keeping the step size the same, results in a different distortion as well as a different rate. In that case, the only meaningful way to compare results is by plotting the rate-distortion curves (as opposed to single R-D points). Hence, it is more effective to present the comparisons in a tabular form by comparing the PSNRs of different encoding options for fixed bit-rates by using the EBCOT rate control algorithm.

1.5.1 Lossy Results

1.5.1.1 Tile Size

JPEG 2000 allows spatial partitioning of the image into tiles. Each tile is wavelet transformed and encoded independently. In fact, a different number of decomposition levels can be specified for each component of each tile. Smaller tile sizes are particularly desirable in memory-constrained applications or when access to only a small portion of the image is desired (e.g. remotely roaming over a large image) without the extra complexity of small precincts. Table 1.4 compares the R-D performance of the JPEG 2000 encoder for various tile sizes with the (9, 7) filter. It is evident that the compression performance decreases with decreasing tile size, particularly at low bit-rates. Furthermore, at low bit-rates where the tile boundaries are visible in the reconstructed image, the perceived quality of the image might be lower than that indicated by the PSNR. The impact of the boundary artifacts can be reduced by using post-processing techniques, such as those employed in reducing the blocking artifacts in low bit-rate DCT-based JPEG and MPEG images. Part 2 of the JPEG 2000 standard offers the option of using a single-sample overlap DWT (SSO-DWT), which reduces edge artifacts at the tile boundaries.

1.5.1.2 Code-Block Size

The code-blocks in JPEG 2000 are rectangular with user-defined dimensions that are identical for all subbands. Each dimension has to be a power of two and the total number of samples in a code-block cannot exceed 4096. Furthermore, when the induced subband

Table 1.4 Comparison of R-D performance for different tile sizes with the (9, 7) filter bank

Rate (bits/pixel)	Average PSNR in dB				
	No tiling	512×512	256×256	192×192	128×128
0.0625	22.82	22.73 (−0.09)	22.50 (−0.32)	22.22 (−0.60)	21.79 (−1.03)
0.125	24.84	24.77 (−0.07)	24.59 (−0.25)	24.38 (−0.46)	24.06 (−0.78)
0.25	27.61	27.55 (−0.06)	27.41 (−0.20)	27.20 (−0.41)	26.96 (−0.65)
0.5	31.35	31.30 (−0.05)	31.19 (−0.16)	30.99 (−0.36)	30.82 (−0.53)
1.0	36.22	36.19 (−0.03)	36.11 (−0.11)	35.96 (−0.26)	35.85 (−0.37)
2.0	42.42	42.40 (−0.02)	42.34 (−0.08)	42.22 (−0.20)	42.16 (−0.26)

Table 1.5 Comparison of R-D performance for various code-block sizes with the (9, 7) filter bank

Rate (bits/pixel)	Average PSNR in dB			
	64×64	32×32	16×16	8×8
0.0625	22.82	22.78 (−0.04)	22.62 (−0.20)	22.27 (−0.55)
0.125	24.84	24.78 (−0.06)	24.57 (−0.27)	24.13 (−0.71)
0.25	27.61	27.52 (−0.09)	27.23 (−0.38)	26.63 (−0.98)
0.5	31.35	31.22 (−0.13)	30.84 (−0.51)	30.04 (−1.31)
1.0	36.22	36.09 (−0.13)	35.68 (−0.54)	34.70 (−1.52)
2.0	42.42	42.28 (−0.14)	41.83 (−0.59)	40.70 (−1.72)

precinct size for a particular subband is less than the code-block size, the code-block size is set equal to the induced subband precinct size.

Table 1.5 compares the effect of varying the code-block size on R-D performance with the (9, 7) filter. There is very little loss of PSNR (maximum of 0.14 dB) in going from a code-block size of 64×64 to a code-block size of 32×32. However, code-block sizes smaller than 32×32 result in a significant drop in PSNR. There are several factors that contribute to this phenomenon. One factor is the overhead information contained in the packet header. The packet header contains information regarding the number of coding passes and the length of compressed data for each code-block, so the total size of the header information increases with an increasing number of code-blocks. Another factor is the independent encoding of each code-block that requires the reinitialization of the arithmetic encoding models. As the code-block size becomes smaller, the number of samples required to adapt to the underlying probability models constitutes a greater portion of the total number of samples encoded. In addition, the pixels that lie on the boundary of a code-block have an incomplete context because pixels from neighboring code-blocks cannot be used in forming the coding contexts. As the code-block size decreases, the percentage of boundary pixels with incomplete contexts increases.

It can also be concluded from Table 1.5 that the loss in compression performance with decreasing code-block size is more pronounced at higher bit-rates. This can be explained as follows. When the bit-rate is high, more coding passes are encoded, and the inefficiencies that are attributable to model mismatch and incomplete contexts add up. In comparison, at low bit-rates, many code-blocks from the higher frequency subbands contribute no compressed data to the compressed bit-stream. The JPEG 2000 bit-stream syntax has provisions to signal this information very efficiently, so for these code-blocks, a smaller size has almost no impact on the coding efficiency. Moreover, these high frequency subbands represent a large percentage of the total number of code-blocks.

1.5.1.3 DWT Filters

Part 1 of JPEG 2000 offers a choice of either the (9, 7) or the (5, 3) filter bank for lossy compression. Figure 1.28 compares the energy compaction of the (9, 7) and the (5, 3) filter banks graphically. Each subband has been scaled with its L_2-norm to reflect its proper contribution to the overall energy. Moreover, for better visualization of the

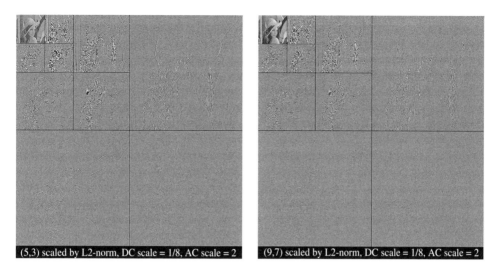

(5,3) scaled by L2-norm, DC scale = 1/8, AC scale = 2   (9,7) scaled by L2-norm, DC scale = 1/8, AC scale = 2

Figure 1.28 Energy compaction comparison between the irreversible (5, 3) and (9, 7) filter banks

subband energies, the AC subbands of both images have been scaled up by a factor of two, while the LL subbands have been scaled down by a factor of eight. It can be seen that the LL subband of the (9, 7) filter bank has a higher contrast, which implies superior energy compaction. However, it is worth noting that the (5, 3) filter has much less computational complexity than the (9, 7) filter.

Table 1.6 compares the R-D performance of the two filter banks in the lossy compression mode. The (9, 7) filter bank consistently outperforms the (5, 3) filter bank with the performance gap increasing with increasing bit-rate. However, it should be noted that the (5, 3) filter bank is also capable of performing lossless compression. When the target lossy bit-rate equals the lossless bit-rate for a particular image, the (5, 3) filter bank can produce zero MSE (or infinite PSNR) in the lossless mode, whereas the (9, 7) filter bank always produces a nonzero MSE. Lossless compression ratios are typically around 2:1 and thus, for bit-rates in the range of 4.0 bits/pixel or higher (with 8-bit images), lossless compression with the (5, 3) filter bank will outperform lossy compression with the (9, 7)

Table 1.6 Comparison of R-D performance of the irreversible (9, 7) and the reversible (5, 3) filter banks

Rate (bits/pixel)	Average PSNR in dB	
	Irreversible (9, 7)	Reversible (5, 3)
0.0625	22.82	22.37 (−0.45)
0.125	24.84	24.37 (−0.47)
0.25	27.61	27.04 (−0.57)
0.5	31.35	30.74 (−0.61)
1.0	36.22	35.48 (−0.74)
2.0	42.42	41.33 (−1.09)

filter bank. Specific lossless compression results for the (5, 3) filter bank are presented in Section 1.5.2.

1.5.1.4 Wavelet Decomposition Levels

The number of decomposition levels affects the coding efficiency of a JPEG 2000 encoder as well as the number of resolutions at which an image can be decompressed. In general, the number of decomposition levels does not impact the computational complexity significantly because only the LL band is further split at each level. Table 1.7 compares the R-D performance of the JPEG 2000 encoder for different numbers of decomposition levels with the (9, 7) filter. Our simulations show that the PSNRs resulting from five and eight levels of decomposition are practically indistinguishable. The use of fewer than five levels results in a loss in coding efficiency, with increasing loss as the number of levels is reduced. The loss is greatest at lower bit-rates and tapers off with increasing bit-rate. We can conclude that a five-level decomposition is adequate in terms of coding efficiency, although it still may be desirable to use more than five levels to provide easy access to lower resolution versions of high-resolution images.

1.5.1.5 Lazy, Parallel, and Lazy–Parallel Modes

As mentioned in Section 1.2.4, JPEG 2000 provides several entropy-coding options that facilitate the parallel processing of the quantized coefficient bit-planes. The collection of these encoding options is termed the parallel mode. Another option that reduces the computational complexity of the entropy encoder (especially at high bit-rates) is the lazy (i.e. selective arithmetic coding bypass) mode, where only the cleanup pass is arithmetic encoded after the fourth most significant bit-plane. Table 1.8 shows the R-D performance of the parallel, lazy, and lazy–parallel modes relative to the reference implementation. It can be seen that the loss in PSNR is generally small (0.01–0.3 dB) and increases with increasing bit-rate.

1.5.1.6 Effect of Multiple Compression Cycles

Table 1.9 examines the effect of multiple compression cycles on PSNR where an image is compressed and reconstructed multiple times to the same bit-rate. Our reference

Table 1.7 Comparison of R-D performance for various levels of decomposition with the (9, 7) filter bank

Rate (bits/pixel)	Average PSNR in dB				
	5 levels	4 levels	3 levels	2 levels	1 level
0.0625	22.82	22.77 (−0.05)	22.47 (−0.35)	21.50 (−1.30)	17.68 (−5.12)
0.125	24.84	24.80 (−0.04)	24.62 (−0.22)	23.91 (−0.93)	21.67 (−3.17)
0.25	27.61	27.57 (−0.04)	27.45 (−0.16)	26.94 (−0.67)	25.54 (−2.07)
0.5	31.35	31.33 (−0.02)	31.24 (−0.11)	30.87 (−0.48)	29.71 (−1.64)
1.0	36.22	36.21 (−0.01)	36.15 (−0.07)	35.91 (−0.31)	35.15 (−1.07)
2.0	42.42	42.42 (−0.01)	42.37 (−0.05)	42.26 (−0.16)	41.71 (−0.71)

Table 1.8 R-D performance of lazy, parallel, and lazy–parallel modes with the (9, 7) filter bank

Rate (bits/pixel)	Average PSNR in dB			
	Reference	Lazy	Parallel	Lazy–parallel
0.0625	22.82	22.81 (−0.01)	22.76 (−0.06)	22.75 (−0.07)
0.125	24.84	24.82 (−0.02)	24.76 (−0.08)	24.74 (−0.10)
0.25	27.61	27.57 (−0.04)	27.49 (−0.12)	27.46 (−0.15)
0.5	31.35	31.28 (−0.07)	31.19 (−0.16)	31.14 (−0.21)
1.0	36.22	36.10 (−0.12)	36.03 (−0.19)	35.94 (−0.28)
2.0	42.42	42.28 (−0.14)	42.22 (−0.20)	42.12 (−0.30)

Table 1.9 R-D performance of multiple compression cycles with the (9, 7) filter bank

Rate (bits/pixel)	Average PSNR in dB			
	1 iteration	4 iterations	8 iterations	16 iterations
0.0625	22.82	22.78 (−0.04)	22.77 (−0.05)	22.76 (−0.06)
0.125	24.84	24.80 (−0.04)	24.78 (−0.06)	24.76 (−0.08)
0.25	27.61	27.57 (−0.04)	27.56 (−0.05)	27.54 (−0.07)
0.5	31.35	31.32 (−0.03)	31.30 (−0.05)	31.28 (−0.07)
1.0	36.22	36.19 (−0.03)	36.17 (−0.05)	36.16 (−0.06)
2.0	42.42	42.39 (−0.03)	42.37 (−0.05)	42.36 (−0.06)

Table 1.10 R-D performance of multiple compression cycles with cropping with the (9, 7) filter bank

Rate (bits/pixel)	Average PSNR in dB				
	Reference	No canvas coordinate system		Canvas coordinate system	
		4 iterations	16 iterations	4 iterations	16 iterations
0.0625	22.82	21.14 (−1.68)	18.58 (−4.24)	22.78 (−0.04)	22.76 (−0.06)
0.125	24.84	22.74 (−2.10)	20.30 (−4.54)	24.80 (−0.04)	24.76 (−0.08)
0.25	27.61	25.16 (−2.45)	22.75 (−4.86)	27.57 (−0.04)	27.54 (−0.07)
0.5	31.35	28.61 (−2.74)	26.40 (−4.95)	31.32 (−0.03)	31.28 (−0.07)
1.0	36.22	33.30 (−2.92)	31.29 (−4.93)	36.19 (−0.03)	36.16 (−0.06)
2.0	42.42	39.26 (−3.16)	37.08 (−5.34)	42.39 (−0.03)	42.36 (−0.06)

implementation with the (9, 7) filter was used in all cases. The postcompression R-D optimization engine is used to achieve the desired bit-rate at each iteration. It can be seen from Table 1.9 that multiple compression cycles cause very little degradation (0.03–0.08 dB) in compression performance when the compression parameters are held constant.

Table 1.10 examines the effect of multiple compression cycles when one image column is cropped from the left side in between compression cycles. Two scenarios are explored.

In one case, the image is always anchored at (0, 0) so that the canvas coordinate system shifts by one column as the image is cropped in between compression cycles. This changes the alignment of the code-blocks. Furthermore, the column index for the samples changes from odd to even and even to odd, which results in a completely different set of wavelet coefficients. In the other case, the anchoring point is shifted to preserve the code-block alignment using the canvas coordinate system. In this case, only the wavelet coefficients near the boundary of the image are affected by cropping. From Table 1.10, it can be seen that maintaining the code-block alignment leads to superior compression performance. More performance comparisons can be found in Joshi, Rabbani, and Lepley (2000).

1.5.1.7   JPEG 2000 versus JPEG Baseline

Table 1.11 compares the R-D performance of JPEG 2000 with JPEG baseline at equivalent bit-rates for the reference test set. Our reference implementation with the (9, 7) filter bank was used. The JPEG baseline PNSR results were generated by iteratively compressing with JPEG baseline to within 1% of the file size of the JPEG 2000 compressed image (including the file headers). The IJG code with the example luminance q-table and a local Huffman table was used for this purpose.[10] For at least one image from our test set, rates of 0.0625 and 0.125 bits/pixel were not achievable even when using a q-table with all the entries set to the highest possible value of 255; hence JPEG baseline results for those rates are not listed in Table 1.11. It can be seen that the use of JPEG 2000 results in approximately 2–4 dB higher PSNR than JPEG baseline depending on the bit-rate.

1.5.2   Lossless Results

1.5.2.1   Reversible Color Transform (RCT)

It is well known that decorrelating the components of an image by applying a color transform improves the coding efficiency. For example, RGB images are routinely transformed into YC_bC_r before applying JPEG compression. In a similar fashion, a lossless component transform can be beneficial when used in conjunction with lossless coding. Table 1.12

Table 1.11   R-D performance of JPEG 2000 and JPEG baseline for the Lena image

Rate (bits/pixel)	Average PSNR in dB	
	JPEG 2000	JPEG baseline
0.0625	22.82	–
0.125	24.84	–
0.25	27.61	25.65
0.5	31.35	28.65
1.0	36.22	32.56
2.0	42.42	38.24

---

[10] Independent JPEG Group, JPEG Library (Version 6b), available from http://www.ijg.org/ or ftp://ftp.uu.net/ graphics/jpeg/ (tar.gz format archive).

Table 1.12 Comparison of lossless bit-rates for color images with and without RCT

	Bit-rate in bits/pixel	
Image	No RCT	RCT
Lena	13.789	13.622
Baboon	18.759	18.103
Bike	13.937	11.962
Woman	13.892	11.502

compares the performance of the JPEG 2000 algorithm for lossless coding, with and without applying the RCT transform. The results are based on using the reversible (5, 3) filter bank with the reference set of compression parameters. Instead of using our reference 8-bit test images, we used the 24-bit color (i.e. 8 bits per component) version of Lena and 'Baboon' images (of size 512×512), in addition to 24-bit versions of the Bike and Woman images. From the table, it can be seen that applying the RCT transform prior to lossless compression results in savings of 0.16–2.39 bpp, which is quite significant in the context of lossless coding.

1.5.2.2 Lossless Encoder Options

Tables 1.13 to 1.16 summarize the lossless compression performance of Part 1 of the JPEG 2000 standard as a function of tile size, number of decomposition levels, code-block size, and lazy–parallel modes. The bit-rates have been averaged over the three test images (Café, Bike, and Woman) and the reversible (5, 3) filter bank has been used. A rather

Table 1.13 Comparison of average lossless bit-rates (bits/pixel) for different tile sizes

No tiling	512×512	256×256	128×128	64×64	32×32
4.797	4.801	4.811	4.850	5.015	5.551

Table 1.14 Comparison of average lossless bit-rates (bits/pixel) for different numbers of decomposition levels

5 levels	4 levels	3 levels	2 levels	1 level	0 levels
4.797	4.798	4.802	4.818	4.887	5.350

Table 1.15 Comparison of average lossless bit-rates (bits/pixel) for different code-block sizes

64×64	32×32	16×16	8×8
4.797	4.846	5.005	5.442

Table 1.16   Comparison of average lossless bit-rates
(bits/pixel) for 'lazy,' 'parallel,' and 'lazy–parallel' modes

Reference	Lazy	Parallel	Lazy–parallel
4.797	4.799	4.863	4.844

surprising finding is that the average lossless performance difference between the one-level and five-level decompositions is very small (<0.1 bpp). This suggests that the three-pass bit-plane entropy-coding scheme and the associated contexts efficiently exploit the redundancy of correlated samples. There is a small (although significant) performance penalty when using a code-block size of 16×16 or smaller, or a tile size of 64×64 or smaller. Finally, there is only a slight decrease in coding efficiency when using the lazy, parallel, or lazy–parallel modes.

Table 1.17 compares the effect of multiple layers on the lossless coding efficiency. As mentioned in Section 1.4.2.2, in order to facilitate bit-stream truncation, it is desirable to construct as many layers as possible. However, the number of packets increases linearly with the number of layers, which also increases the overhead associated with the packet headers. As can be seen from the table, the performance penalty for using 50 layers is small for lossless compression. However, this penalty is expected to increase at lower bit-rates (Marziliano *et al.*, 2004), whereas increasing the number of layers from 7 to 50 does not linearly increase the lossless bit-rate because the header information for the increased number of packets is encoded more efficiently. In particular, the percentage of code-blocks that do not contribute to a given packet increases with the number of layers, and the packet header syntax allows this information to be encoded very efficiently using a single bit.

1.5.2.3   Lossless JPEG 2000 versus JPEG-LS

Table 1.18 compares the lossless performance of JPEG 2000 with JPEG-LS (ISO/IEC, 1999). Although the JPEG-LS has only a small performance advantage (3.4%) over JPEG 2000 for the images considered in this study, it has been shown that for certain classes

Table 1.17   Comparison of average lossless
bit-rates (bits/pixel) for different numbers of layers

1 layer	7 layers	50 layers
4.797	4.809	4.829

Table 1.18   Comparison of average lossless
bit-rates (bits/pixel) for JPEG 2000 and JPEG-LS

JPEG 2000	JPEG-LS
4.797	4.633

of imagery (e.g. the 'cmpnd 1' compound document from the JPEG 2000 test set), the JPEG-LS bit-rate is only 60% of that of JPEG 2000 (Marziliano *et al.*, 2004).

1.5.3   Bit-Plane Entropy Coding Results

In this section, we examine the redundancy contained in the various bit-planes of the quantized wavelet coefficients. These results were obtained by quantizing the wavelet coefficients of the Lena image with the default quantization step size for VM8.6 ('step 1/128.0'). Because Lena is an 8-bit image, the actual step size used for each band was 2.0 divided by the L_2-norm of that band. This had the effect that equal quantization errors in each subband had roughly the same contribution to the reconstructed image MSE. Hence, the bit-planes in different subbands were aligned by their LSBs. Eleven of the resulting bit-planes were encoded, starting with the most significant bit-plane.

One way to characterize the redundancy is to count the number of bytes that are generated by each sub-bit-plane coding pass. The number of bytes generated from each sub-bit-plane coding pass is not readily available unless each coding pass is terminated. However, during postcompression R-D optimization, VM8.6 computes the number of additional bytes needed to uniquely decode each coding pass using a 'near optimal length calculation' algorithm (ISO/IEC, 2000). It is not guaranteed that the near optimal length calculation algorithm will determine the minimum number of bytes needed for unique decoding. This means that the estimated bytes for a coding pass contain some data from the next coding pass, which can lead to some unexpected results. With this caveat in mind, Table 1.19 contains the number of bytes generated from each sub-bit-plane coding pass. The estimated bytes for each coding pass were summed across all the code-blocks in the image to generate these entries.

During the encoding of the first bit-plane, there is only a cleanup pass and 36 coefficients turn significant. All of these significant coefficients belong to the 5LL subband. In the refinement pass of the next bit-plane, only these 36 coefficients are refined. Surprisingly, the first refinement bit for all of these 36 coefficients are zero. Due to the fast model adaptation of the MQ-coder, very few refinement bits are generated for the

Table 1.19   Encoded bytes resulting for sub-bit-plane passes of the Lena image

Bit-plane Number	Significance bytes	Refinement bytes	Cleanup bytes	Total for current BP	Total for all BPs
1	0	0	21	21	21
2	18	0	24	42	63
3	38	13	57	108	171
4	78	37	156	271	442
5	224	73	383	680	1122
6	551	180	748	1479	2601
7	1243	418	1349	3010	5611
8	2315	932	2570	5817	11428
9	4593	1925	5465	11983	23411
10	10720	3917	12779	27416	50827
11	25421	8808	5438	39667	90494

second bit-plane. This, in conjunction with the possibility of overestimating the number of bytes in the cleanup pass of the first bit-plane, leads to the rather strange result that the refinement pass for the second bit-plane requires zero bytes. It is also interesting that the number of bytes needed to encode a given bit-plane is usually greater than the total number of bytes used to encode all of the bit-planes prior to it (except for bit-plane 11).

Figure 1.29 shows images reconstructed from the first nine bit-planes and Table 1.20 provides the corresponding PSNRs. Table 1.20 also shows the percentage of the coefficients that are refined at each bit-plane, the percentage of the coefficients that are found to be significant at each bit-plane, and the percentage of the coefficients that remain insignificant after completion of the encoding of a bit-plane. It is interesting to note that approximately 72% of the coefficients still remain insignificant after encoding the tenth bit-plane.

Figure 1.29   Reconstructed Lena image after decoding bit-planes 1−9 (from left to right and top to bottom)

Table 1.20   Coding statistics resulting from encoding the first eleven wavelet coefficient
bit-planes of the Lena image

BP	Compression ratio	Rate (bits/pixel)	PSNR (dB)	Percent refined	Percent significant	Percent insignificant
1	12483	0.000641	16.16	0.00	0.01	99.99
2	4161	0.00192	18.85	0.01	0.04	99.95
3	1533	0.00522	21.45	0.05	0.06	99.89
4	593	0.0135	23.74	0.11	0.12	99.77
5	233	0.0343	26.47	0.23	0.32	99.43
6	101	0.0792	29.39	0.57	0.75	98.68
7	47	0.170	32.54	1.32	1.59	97.09
8	23	0.348	35.70	2.91	3.10	93.99
9	11.2	0.714	38.87	6.01	6.33	87.66
10	5.16	1.55	43.12	12.34	15.78	71.88
11	2.90	2.76	49.00	28.12	25.08	46.80

Figure 1.30 provides a graphic representation of the data that is contained in Table 1.20
for bit-planes 4, 6, 8, and 11. For each bit-plane, the green pixels denote the location
of the wavelet coefficients that become significant during the significance propagation
pass for that bit-plane, the red pixels denote the location of those coefficients that turn
significant during the cleanup pass, the black pixels denote the location of the coefficients
that get refined, and the white pixels denote coefficients that still remain insignificant after
the encoding of that bit-plane.

1.6   Additional Features of JPEG 2000 Part 1

1.6.1   Region-of-Interest (ROI) Coding

In some applications, it might be desirable to encode certain portions of the image (called
the *region of interest*, or ROI) at a higher level of quality relative to the rest of the
image (called the background). Alternatively, one might want to prioritize the compressed
data corresponding to the ROI relative to the background so that it appears earlier in
the code-stream. This feature is desirable in progressive transmission in case of early
termination of the code-stream.

ROI coding can be accomplished by encoding the quantized wavelet coefficients corresponding to the ROI with a higher precision relative to the background, e.g. by scaling
up the ROI coefficients or scaling down the background coefficients. A scaling-based
ROI encoding method would generally proceed as follows (Atsumi and Farvardin, 1998).
First, the ROI(s) are identified in the image domain. Next, a binary mask in the wavelet
domain, known as the *ROI mask*, is generated. The ROI mask has a value of one at
those coefficients that contribute to the reconstruction of the ROI and has a value of
zero elsewhere. The shape of the ROI mask is determined by the image domain ROI as
well as the wavelet filter bank, and it can be computed in an efficient manner for most
regular ROI shapes (Marziliano *et al.*, 2004). Prior to entropy coding, the bit-planes of
the coefficients belonging to the ROI mask are shifted up (or the background bit-planes

Figure 1.30   From left to right and top to bottom are shown bit-planes 4, 6, 8, and 11 of the JPEG 2000 encoded Lena image as described in Table 1.20. Green pixels denote the location of the wavelet coefficients that become significant during the significance propagation pass in that bit-plane, red pixels denote the coefficients that turn significant during the cleanup pass in that bit-plane, black pixels denote coefficients that are refined, and white pixels denote coefficients that remain insignificant after completion of the three coding passes (see Plate 2)

are shifted down[11] by a desired amount that can vary from one ROI to another within the same image. The ROI shape information (in the image domain) and the scaling factor used for each ROI is also encoded and included in the code-stream. In general, the overhead associated with the encoding of an arbitrary-shaped ROI might be large unless the ROI has a regular shape, e.g. a rectangle or a circle, which can be described with a

---

[11] The main idea is to store the magnitude bits of the quantized coefficients in the most significant part of the implementation register so that any potential precision overflow would only impact the LSB of the background coefficients.

small set of parameters. At the decoder, the ROI shape and scaling factors are decoded and the quantized wavelet coefficients within each ROI (or background) coefficient are scaled to their original values.

The procedure described above requires the generation of an ROI mask at both the encoder and decoder, as well as the encoding and decoding of the ROI shape information. This increased complexity is balanced by the flexibility to encode ROIs with multiple qualities and to control the quality differential between the ROI and the background. To minimize decoder complexity while still providing ROI capability, JPEG 2000 Part 1 has adopted a specific implementation of the scaling-based ROI approach known as the *Max-shift* method (Christopoulos, Askelof, and Larsson, 2000; Nister and Christopoulos, 1999).

In the Max-shift method, the ROI mask is generated in the wavelet domain, and all wavelet coefficients that belong to the background are examined and the coefficient with the largest magnitude is identified. Next, a value s is determined such that 2^s is larger than the largest magnitude background coefficient. To ensure that the smallest nonzero ROI coefficient is still larger than the largest background coefficient, s LSBs are added to each wavelet coefficient and all bit-planes of the background coefficients are shifted down by s bits, as shown in Figure 1.31. The presence of ROI is signaled to the decoder by a marker segment and the value of s is included in the code-stream. At the decoder, those wavelet coefficients whose values are more than 2^s belong to the foreground and are scaled down to their original value. In the Max-shift method, the decoder is not required to generate an ROI mask or to decode any ROI shape information. Furthermore, the encoder can encode any arbitrary shape ROI within each subband, and it does not need to encode the ROI shape information (although it may still need to generate an ROI mask). The main disadvantage of the Max-shift method is that ROIs with multiple quality differentials cannot be encoded.

In the Max-shift method, the ROI coefficients are prioritized in the code-stream so that they are received (decoded) before the background. However, if the entire code-stream is decoded, the background pixels will eventually be reconstructed to the same level of quality as that of the ROI. In certain applications, it may be desirable to encode the ROI to a higher level of quality than the background, even after the entire code-stream has been

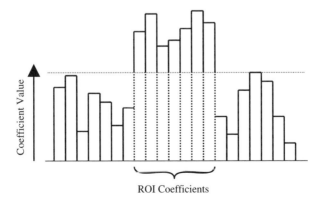

Figure 1.31   Max-shift method of ROI coding in JPEG 2000 Part 1

decoded. The complete separation of the ROI and background bit-planes in the Max-shift method can be used to achieve this purpose. For example, all the wavelet coefficients are quantized to the precision desired for the ROI. The ROI coefficients are encoded first, followed by the encoding of the background coefficients in one or more layers. By discarding a number of layers corresponding to the background coefficients, any desired level of quality can be achieved for the background.

Given that the encoding of the ROI and the background coefficients in the Max-shift method are completely disjoint processes, it might seem that the ROI needs to be completely decoded before any background information is reconstructed. However, this limitation can be circumvented to some extent. For example, if the data are organized in the resolution progressive mode, the ROI data are decoded first followed by the background data for each resolution. As a result, at the start of decoding for each resolution, the reconstructed image will contain all the background data corresponding to the lower resolutions. Alternatively, because of the flexibility in defining the ROI shape for each subband, the ROI mask at each resolution or subband can be modified to include some background information. For example, the entire LL subband can be included in the ROI mask to provide low-resolution information regarding the background in the reconstructed image.

Experiments show that for the lossless coding of images with ROIs, the Max-shift method increases the bit rate by 1–8% (depending on image size and ROI size and shape) compared to the lossless coding of the images without ROI (Christopoulos, Askelof, and Larsson, 2000). This is a relatively small cost for achieving the ROI functionality.

1.6.2 Error Resilience

Many emerging applications of the JPEG 2000 standard require the delivery of the compressed data over communications channels with different error characteristics. For example, wireless communication channels are susceptible to random and burst channel errors, while Internet communication is prone to data loss due to traffic congestion. To improve the transmission performance of JPEG 2000 in error-prone environments, Part 1 of the standard provides several options for error resilience. The error resilience tools are based on different approaches such as compressed data partitioning and resynchronization, error detection, and quality of service (QoS) transmission based on priority. The error resilience bit-stream syntax and tools are provided both at the entropy-coding level and the packet level (Liang and Talluri, 1999; Moccagata et al., 2000).

As discussed before, one of the main differences between the JPEG 2000 coder and previous embedded wavelet coders is in the independent encoding of the code-blocks. Among the many advantages of this approach is improved error resilience, because any errors in the bit-stream, corresponding to a code-block, will be contained within that code-block. In addition, certain entropy-coding options described in Section 1.2.4.3 can be used to improve error resilience. For example, the arithmetic coder can be terminated at the end of each coding pass and the context probability models can be reset. The optional lazy mode allows the bypassing of the arithmetic coder for the first two coding passes of each bit-plane after the fourth most significant bit-plane and can help protect against catastrophic error propagation that is characteristic of all variable-length coding schemes. Finally, JPEG 2000 provides for the insertion of error resilience *segmentation symbols* at the end of the cleanup pass of each bit-plane that can serve in error detection. The segmentation symbol is a binary ' 1010' symbol whose presence is signaled in the marker segments. It is encoded with the uniform arithmetic coding context, and its correct decoding at the end of each bit-plane confirms the correctness of the decompressed data corresponding to that bit-plane. If the segmentation symbol is not decoded correctly, the data for that bit-plane and all the subsequent bit-planes corresponding to that code-block should be discarded. This is because the data encoded in the subsequent coding passes of that code-block depend on the previously encoded data.

Error resilience at the packet level can be achieved by using the SOP marker, which provides for spatial partitioning and resynchronization. This marker is placed in front of each packet in a tile and numbers the packets sequentially starting at zero. Also, the packet headers can be moved to either the main header (for all tiles, using PPM markers) or the tile header (using PPT markers), to create what is known as *short packets*. In a QoS transmission environment, these headers can be protected more heavily than the rest of the data. If there are errors present in the packet compressed data, the packet headers can still be associated with the correct packet by using the sequence number included in the resynchronization marker. The combination of these error resilience tools can often provide adequate protection in some of the most demanding error-prone environments.

1.6.3   File Format

Most digital imaging standards provide a file format structure to encapsulate the encoded-image data. While the code-stream specifies the compressed image, the file format serves to provide useful information regarding the characteristics of the image and its proper use and display. Sometimes the file format includes redundant information that is also included in the code-stream, but such information is useful in that it allows trivial manipulation of the file without any knowledge of the code-stream syntax. A minimal file format, such as the one used in the JPEG baseline system, includes general information regarding the number of image components, their corresponding resolutions and bit depths, etc. However, two important components of a more comprehensive file format are *color space* and *metadata*. Without this information, an application might not know how to use or display an image properly. The color space defines how the decoded component values relate to real-world spectral information (e.g. sRGB or YC_bC_r), while the metadata provides additional information regarding the image. For example, metadata can be used to describe how the image was created (e.g. the camera type or photographer's name) as well as to describe how the image should be used (e.g. IPRs related to the image, default display resolution, etc.). It also provides the opportunity to extract information regarding an image without the need to decode it, which enables a fast text-based search in databases. The SPIFF file format defined in Part 3 extensions of the existing JPEG standard (ISO/IEC, 1995) was targeted at 8-bit-per-component sRGB and YC_bC_r images, and there was limited capability for metadata. The file format defined by the JPEG 2000 standard is much more flexible with respect to both the color space specification and the metadata embedding.

Part 1 of the JPEG 2000 standard defines a file format referred to as *JP2*. Although this file format is an optional part of the standard, it is expected to be used by many applications. It provides a flexible, but restricted, set of data structures to describe the encoded-image data. In order to balance flexibility with interoperability, the JP2 format defines two methods of color space specification. One method (known as the *enumerated* method) limits flexibility, but provides a high degree of interoperability by directly specifying only three color spaces, sRGB, gray scale, and sYCC. Another method known as the *restricted ICC* (International Color Consortium) (ICC, 1998) method, allows for the specification of a color space using a subset of standard ICC profiles, referred to in the ICC specification as 'three-channel matrix-based and monochrome input profiles.' These profiles, which specify a transformation from the reconstructed code values to the profile connection space (PCS), contain at most three 1-D look-up tables followed by a 3×3 matrix. These profile types were chosen because of their simplicity. The restricted ICC method can simply be thought of as a data structure that specifies a set of color space transformation equations. Finally, the JP2 file format also allows for displaying palletized images, i.e. single component images where the value of the single component represents an index into a palette of colors.

The JP2 file format also provides two mechanisms for defining and embedding metadata in a compressed file. The first method uses a universal unique identifier (UUID) while the second method uses XML (W3C, 2006). For both methods, the individual blocks of metadata can be embedded almost anywhere in the file. Although very few metadata fields have been defined in the JP2 file format, its basic architecture provides a strong foundation for extension.

Part 2 of the standard defines extensions to the JP2 file format, encapsulated in an extended file format called *JPX*. These extensions increase the color space flexibility by providing more enumerated color spaces (and also allow vendors to register additional values for color spaces) as well as providing support for all ICC profiles. They also add the capability for specifying a combination of multiple images using composition or animation, and add a large number of metadata fields to specify image history, content, characterization, and IPR.

Acknowledgments

The authors would like to thank Brian Banister for generating the bit-plane results in Section 1.5.3 and Roddy Shuler for reviewing this chapter for technical accuracy.

References

Acharya, T. and Tsai, P.-S. (2005) *JPEG 2000 Standard for Image Compression – Concepts, Algorithms and VLSI Architectures*, John Wiley & Sons, Inc., Hoboken, NJ.

Adams, M. D. and Kossentini, F. (2000) Reversible integer-to-integer wavelet transforms for image compression: performance evaluation and analysis, *IEEE Transactions on Image Processing*, **9**(6), June, 1010–1024.

Adams, M. D., Man, H., Kossentini, F. and Ebrahimi, T. (2000) JPEG 2000: The Next Generation Still Image Compression Standard, ISO/IEC JTC1/SC29/WG1N1734, June 2000.

Albanesi, M. and Bertoluzza, S. (1995) Human vision model and wavelets for high-quality image compression, in *Proceedings of the 5th International Conference on Image Processing and Its Applications*, vol. 410, Edinburgh, UK, July 1995, pp. 311–315.

Antonini, M., Barlaud, M., Mathieu, P. and Daubechies, I. (1992) Image coding using wavelet transform, *IEEE Transactions on Image Processing*, **1**(2), April, 205–220.

Atsumi, E. and Farvardin, N. (1998) Lossy/lossless region-of-interest image coding based on set partitioning in hierarchical trees, in *Proceedings of the IEEE International Conference on Image Processing*, Chicago, IL, October 1998, pp. 87–91.

Calderbank, R. C., Daubechies, I., Sweldens, W. and Yeo, B.-L. (1998) Wavelet transforms that map integers to integers, *Applied and Computational Harmonic Analysis*, **5**(3), July, 332–369.

Christopoulos, C., Askelof, J. and Larsson, M. (2000) Efficient methods for encoding regions of interest in the upcoming JPEG 2000 Still Image Coding Standard, *IEEE Signal Processing Letters*, **7**(9), September, 247–249.

Christopoulos, C., Skodras, A. and Ebrahimi, T. (2000) The JPEG 2000 still image coding system: an overview, *IEEE Transactions on Consumer Electronics*, **46**(4), November, 1103–1127.

Chrysafis, C. and Ortega, A. (2000) Line-based, reduced memory, wavelet image compression, *IEEE Transactions on Image Processing*, **9**(3), March, 378–389.

Daubechies, I. and Sweldens, W. (1998) Factoring wavelet transforms into lifting steps, *Journal of Fourier Analysis Applications*, **4**(3), 247–269.

Ebrahimi, T., Santa-Cruz, D., Askelöf, J., Larsson, M. and Christopoulos, C. (2000) JPEG 2000 still image coding versus other standards, in *Proceedings of SPIE*, vol. 4115, San Diego, CA, July/August 2000, pp. 446–454.

Gormish, M. J., Lee, D. and Marcellin, M. W. (2000) JPEG 2000: overview, architecture, and applications, in *Proceedings of the IEEE International Conference on Image Processing*, Vancouver, Canada, September 2000.

ICC (International Color Consortium) (1998) ICC Profile Format Specification, ICC.1: 1998-09.

ISO/IEC (1993) Information Technology – Digital Compression and Coding of Continuous-Tone Still Images – Part 1: Requirements and Guidelines, ISO/IEC International Standard 10918-1, ITU-T Recommendation T.81.

ISO/IEC (1995) Information Technology – Digital Compression and Coding of Continuous-Tone Still Images – Part 3: Extensions, ISO/IEC International Standard 10918-3, ITU-T Recommendation T.84.

ISO/IEC (1997a) Call for Contributions for JPEG 2000 (JTC 1.29.14, 15444): Image Coding System, ISO/IEC JTC1/SC29/WG1N505, March 1997.

ISO/IEC (1997b) New Work Item: JPEG 2000 Image Coding System, ISO/IEC JTC1/SC29/WG1N390R, March 1997.

ISO/IEC (1999) Information Technology – Lossless and Near Lossless Compression of Continuous-Tone Still Images, ISO/IEC International Standard 14495-1, ITU Recommendation T.87.

ISO/IEC (2000) JPEG 2000 Verification Model 8.6 (Software), ISO/IEC JTC1/SC29/WG1N1894, December 2000.

Jones, P. W. (2007) Efficient JPEG 2000 VBR compression with true constant quality, *SMPTE Journal on Motion Imaging*, **7/8**, July/August.

Jones, P., Daly, S., Gaborski, R. and Rabbani, M. (1995) Comparative study of wavelet and DCT decompositions with equivalent quantization and encoding strategies for medical images, in *Proceedings of SPIE*, vol. 2431, San Diego, CA, February 1995, pp. 571–582.

Joshi, R. L., Rabbani, M. and Lepley, M. (2000) Comparison of multiple compression cycle performance for JPEG and JPEG 2000, in *Proceedings of SPIE*, vol. 4115, San Diego, CA, July/August 2000, pp. 492–501.

LeGall, D. and Tabatabai, A. (1988) Subband coding of digital images using symmetric kernel filters and arithmetic coding techniques, in *Proceedings of International Conference on Acoustic Speech and Signal Processing*, New York, April 1988, pp. 761–764.

Li, J. (1999) Visual progressive coding, in *Proceedings of SPIE*, vol. 3653, San Jose, CA, January 1999.

Li, J. and Lei, S. (1999) An embedded still image coder with rate-distortion optimization, *IEEE Transactions on Image Processing*, **8**(7), July, 913–924.

Liang, J. and Talluri, R. (1999) Tools for robust image and video coding in JPEG 2000 and MPEG-4 standards, in *Proceedings of SPIE*, vol. 3653, San Jose, CA, January 1999, pp. 40–51.

Liu, Z., Karam, L. J. and Watson, A. B. (2006) JPEG 2000 encoding with perceptual distortion control, *IEEE Transactions on Image Processing*, **15**(7), July, 1763–1778.

Marcellin, M., Flohr, T., Bilgin, A., Taubman, D., Ordentlich, E., Weinberger, M., Seroussi, G., Chrysafis, C., Fischer, T., Banister, B., Rabbani, M. and Joshi, R. (1999) Reduced Complexity Entropy Coding, ISO/IEC JTC1/SC29/WG1 Document N1312, June 1999.

Marcellin, M. W., Gormish, M. J., Bilgin, A. and Boliek, M. P. (2000) An overview of JPEG-2000, in *Proceedings of the Data Compression Conference*, Snowbird, UT, March 2000, pp. 523–541.

Marcellin, M. W., Lepley, M. A., Bilgin, A., Flohr, T. J., Chinen, T. T. and Kasner, J. H. (2002) An overview of quantization in JPEG 2000, *Signal Processing: Image Communications*, **17**(1), January, 73–84.

Marziliano, P., Dufaux, F., Winkler, S. and Ebrahimi, T. (2004) Perceptual blur and ringing metrics: application to JPEG 2000, *Signal Processing: Image Communication*, **19**(2), February, 163–172.

Moccagata., I., Sodagar, S., Liang, J. and Chen, H. (2000) Error resilient coding in JPEG-2000 and MPEG-4, *IEEE Journal of Selected Areas in Communications*, **18**(6), June, 899–914.

Nister, D. and Christopoulos, C. (1999) Lossless region of interest with embedded wavelet image coding, *Signal Processing*, **78**(1), October, 1–17.

Ordentlich, E., Weinberger, M. J. and Seroussi, G. (1998) A low complexity modeling approach for embedded coding of wavelet coefficients, in *Proceedings of the Data Compression Conference*, Snowbird, UT, March 1998, pp. 408–417.

O'Rourke, T. and Stevenson, R. (1995) Human visual system based wavelet decomposition for image compression, *Journal of Visual Communications and Image Representation*, **6**(2), June, 109–121.

Pennebaker, W. B. and Mitchell, J. L. (1993) *JPEG Still Image Data Compression Standard*, Van Nostrand Reinhold, New York.

Pennebaker, W. B., Mitchell, J. L., Langdon Jr, G. G. and Arps, R. B. (1988) An overview of the basic principles of the Q-coder adaptive binary arithmetic coder, *IBM Journal of Research Development*, **32**(6), November, 717–726.

Price, J. R. and Rabbani, M. (1999) Biased reconstruction for JPEG decoding, *Signal Processing Letters*, **6**(12), December, 297–299.

Rabbani, M. and Joshi, R. L. (2000) An overview of the JPEG 2000 still image compression standard, *Signal Processing: Image Communications*, **17**(1), January, 3–48.

Said, A. and Pearlman, W. A. (1996) A new fast and efficient image codec based on set partitioning in hierarchical trees, *IEEE Transactions on Circuits System Video Technology*, **6**(3), June, 243–250.

Santa-Cruz, D. and Ebrahimi, T. (2000) An analytical study of JPEG 2000 functionalities, in *Proceedings of IEEE International Conference on Image Processing*, Vancouver, Canada, September 2000.

Shapiro, J. M. (1993) Embedded image coding using zero trees of wavelet coefficients, *IEEE Transactions on Signal Processing*, **41**(12), December, 3445–3462.

Slattery, M. J. and Mitchell, J. L. (1998) The Qx-coder, *IBM Journal of Research Development*, **42**(6), November, 767–784.

SPIC (2002) Special Issue on JPEG 2000 still image compression standard, *Signal Processing: Image Communication*, **17**(1), January.

Sullivan, G. (1996) Efficient scalar quantization of exponential and Laplacian variables, *IEEE Transactions on Information Theory*, **42**(5), September, 1365–1374.

Sweldens, W. (1995) The lifting scheme: a new philosophy in biorthogonal wavelet constructions, in *Proceedings of SPIE*, vol. 2569, September 1995, pp. 68–79.

Sweldens, W. (1996) The lifting scheme: a custom-design construction of biorthogonal wavelets, *Applied and Computational Harmonic Analysis*, **3**(2), April, 186–200.

Sweldens, W. (1998) The lifting scheme: a construction of second generation wavelets, *Siam Journal of Mathematical Analysis*, **29**(2), March, 511–546.

Taubman, D. (2000) High performance scalable image compression with EBCOT, *IEEE Transactions on Image Processing*, **9**(7), July, 1158–1170.

Taubman, D. and Marcellin, M. W. (2001) *JPEG 2000: Image Compression Fundamentals, Practice and Standards*, Kluwer Academic Publishers, Boston, MA.

Taubman, D. S. and Marcellin, M. W. (2002) JPEG 2000: standard for interactive imaging, *Proceedings of the IEEE*, **90**(8), August, 1336–1357.

Taubman, D., Ordentlich, E., Weinberger, M. J. and Seroussi, G. (2002) Embedded block coding in JPEG 2000, *Signal Processing: Image Communications*, **17**(1), January, 49–72.

Unser, M. and Blu, T. (2003) Mathematical properties of the JPEG 2000 wavelet filters, *IEEE Transactions on Image Processing*, **12**(9), September, 1080–1090.

Vetterli, M. and Kovacevic, J. (1995) *Wavelet and Subband Coding*, Prentice Hall, Englewood Cliffs, NJ.

Villasenor, J. D., Belzer, B. and Liao, J. (1995) Wavelet filter evaluation for image compression, *IEEE Transactions on Image Processing*, **4**(8), August, 1053–1060.

W3C (2006) W3C, Extensible Markup Language (XML) 1.0, 4th edition, September 2006. Available at: http://www.w3.org/TR/Rec-xml.

Watson, A. B., Yang, G. Y., Solomon, J. A. and Villasenor, J. (1997) Visibility of wavelet quantization noise, *IEEE Transactions on Image Processing*, **6**(8), August, 1164–1175.

Woods, J. W. and Naveen, T. (1992) A filter based bit allocation scheme for subband compression of HDTV, *IEEE Transactions on Image Processing*, **1**(3), July, 436–440.

Zeng, W., Daly, S. and Lei, S. (2002) An overview of visual optimization tools in JPEG 2000, *Signal Processing: Image Communications*, **17**(1), January, 85–105.

2

JPEG 2000 Extensions (Part 2)

Margaret Lepley, J. Scott Houchin, James Kasner, and Michael Marcellin

2.1 Introduction

As JPEG 2000 developed, many ideas for value-added capabilities emerged. It was not feasible to include them in the Part 1 Core (ISO/IEC, 2004a) – originally published in 2000 – so additional parts were created. The Part 2 standard, published as ISO/IEC 15444-2 or ITU Recommendation T.801 (ISO/IEC, 2004b), contains multiple extensions of JPEG 2000 that were not large enough to merit entire documents of their own.

Unlike the JPEG 2000 Part 1 Core, where decoders are expected to handle all the code-stream functionality, Part 2 is a collection of options that may be implemented à la carte to meet specific market requirements. Moreover, sections within an extension annex can be implemented separately (e.g. subsets of the extended file format JPX). Hence, some extension features may appear across a wide spectrum of JPEG 2000 applications, while others will be less common in decoders.

The extensions in Part 2 cover a disparate set of topics that modify or add to the Part 1 JPEG 2000 processing chain. Some tools improve the compression efficiency and/or visual appearance of compressed images, while others modify or extend the functionality in other ways. To set the stage for the remainder of this chapter, we list the major topics with a pointer to their location within the Part 2 standard:

- Compression efficiency
    - Variable DC offset (VDCO)   –Annex B
    - Variable scalar quantization (VSQ)   –Annex C
    - Trellis coded quantization (TCQ)   –Annex D
    - Extended visual masking   –Annex E
    - Arbitrary wavelet decomposition   –Annex F
    - Arbitrary wavelet transform kernel   –Annexes G and H
    - Multiple component transform   –Annex J
    - Nonlinear point transform   –Annex K

The JPEG 2000 Suite   Edited by Peter Schelkens, Athanassios Skodras and Touradj Ebrahimi
© 2009 John Wiley & Sons, Ltd

- Functionality
    - Geometric manipulation   −Annex I
    - Single-sample overlap (SSO/TSSO)   −Annex I
    - Precinct-dependent quantization   −Amendment 1
    - Extended region of interest   −Annex L
    - Extended file format/metadata (JPX)   −Annexes M and N
    - Extended capabilities signaling   −Amendment 2

Figure 2.1 provides a graphic overview of how these extensions fit together within the more familiar Part 1 decoder. The shaded boxes correspond to Part 1 technologies and the remaining boxes represent Part 2 technologies. Figure 2.1 establishes the processing order that a JPEG 2000 decoder must follow to apply the technologies in Part 2 properly when decoding a Part 2 code-stream.

The syntax additions and modifications from Part 1 are described fully in ISO/IEC 15444-2:2004 Annex A and are not covered in this chapter. However, the chapter does refer to some key elements for transmitting code-stream control parameters, namely marker segments. Table 2.1 summarizes the marker segments that are extended or new in Part 2. In particular, note that the Rsiz parameter of the SIZ marker segment is modified in Part 2 to (1) flag Part 2 use and (2) indicate which extensions or technologies are present and whether they are 'required to decode' or 'useful to decode' a Part 2

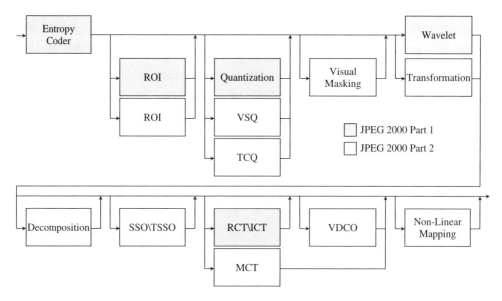

Figure 2.1   Decoder block diagram (based on ISO/IEC 15444-2, Figure 6.1). © ISO. This material is reproduced from ISO/IEC 15444−2:2004 with the permission of the American National Standards Institute (ANSI) on behalf of the International Organization for Standardization (ISO). No part of this material may be copied or reproduced in any form, electronic retrieval system or otherwise or made available on the Internet, a public network, by satellite or otherwise without the prior written consent of the ANSI. Copies of this standard may be purchased from the ANSI, 25 West 43rd Street, New York, NT 10036, (212) 642−4900, http://webstore.ansi.org

Table 2.1 Summary of extended and new marker segments. © ISO. This material is reproduced from ISO/IEC 15444–2:2004 with the permission of the American National Standards Institute (ANSI) on behalf of the International Organization for Standardization (ISO). No part of this material may be copied or reproduced in any form, electronic retrieval system or otherwise or made available on the Internet, a public network, by satellite or otherwise without the prior written consent of the ANSI. Copies of this standard may be purchased from the ANSI, 25 West 43rd Street, New York, NT 10036, (212) 642–4900, http://webstore.ansi.org

Extension	Extended marker segments	New marker segments
All extensions	SIZ (Rsiz)	–
Variable DC offset	–	DCO
Variable scalar quantizer	QCD, QCC	–
Trellis-coded quantization	QCD, QCC	–
Precinct-dependent quantization	SOT	QPD, QPC
Visual masking	–	VMS
Nonlinear point transform	–	NLT
Arbitrary wavelet decompositions	COD, COC	DFS, ADS
Arbitrary wavelet transforms	COD, COC	ATK
Multiple component transform	COD	CBD, MCT, MCC, MCO
Geometric manipulation	COD	–
Single sample overlap	COD, COC	–
Extended region of interest	RGN	–
Extended capabilities marker	SIZ optional	CAP

Based on ISO/IEC 15444-2, Table A.1.

code-stream. 'Required to decode' means that no useful image can be reconstructed without the extension. 'Useful to decode' means image quality will improve using the extension, but the image may be decoded without its use.

The extended file format JPX wraps a code-stream and associated metadata within a separate container, using a 'box' to contain specific elements (e.g. a complete code-stream or an XML document containing intellectual property information) instead of using markers to delimit different elements. JPX is described in Section 2.14.

2.2 Variable DC Offset

A common stage of many transforms and compression operations is a DC offset intended to center the data distribution near zero or remove the so-called DC component. Although not explicitly shown in Figure 2.1, JPEG 2000 Part 1 uses a default offset derived from the bit-depth and signedness of each component. These fast Part 1 offsets (0 for signed data and half the bit-depth for unsigned data) work well over a wide range of imagery.

However, in some situations the ability to customize the DC offset can yield benefits. For example, consider an image of relatively low dynamic range as a whole, but containing small regions with bright burnt-in text or other bright content. If the bulk of the image data lies well below the default DC offset, many of the lowpass wavelet coefficients will be large and negative and will not compress as well as values near zero. Assigning a DC offset of zero, or a value closer to the center of the dynamic range of the anomalous image data, can yield dramatic improvements in compression efficiency on such images.

A custom DC offset can also be used to achieve goals other than improved compression. For example, WSQ (CJIS, 1997), a wavelet compression format used for fingerprints, varies the DC offset for each image. Variable DC offset usage within JPEG 2000 facilitates highly accurate conversion of WSQ to JPEG 2000 (Lepley, 2001).

Mathematically the Part 2 variable DC offset is quite simple. The compressor subtracts an offset, O_c, from each sample of component c prior to applying any RCT/ICT or wavelet transform. The decompressor adds the offsets back into the component samples after the wavelet transform and any RCT/ICT.

Although variable DC offset increases overall compression efficiency, it typically causes a few of the wavelet coefficients to become even larger. The number of guard bits (signaled in QCD/QCC) should be altered to handle this increase. If the offsets fall within the dynamic range of the original data, then increasing the number of guard bits by one will suffice. As Figure 2.1 shows, the variable DC offset cannot be used in conjunction with the multiple component transform extension, since that extension includes its own techniques for allowing data offsets.

The least significant bit in Rsiz warns the decoder than variable DC offsets are required to decode the image. A DCO marker segment transmits offsets O_c for each component as either integers or floats. When used in conjunction with a reversible transformation, the offsets should be integer valued. Otherwise any value may be chosen, although good choices typically fall within the dynamic range of the original data.

JPEG 2000 decoders that implement this extension will be able to read the DCO marker segment and apply the appropriate offsets. Decoders without this extension, if they do not halt with an error, will skip the DCO marker and decode with default offsets, causing a wrap-around effect in the output image.

2.3 Variable Scalar Quantization

JPEG 2000 Part 1 uses a uniform dead-zone scalar quantizer. This quantizer has properties that make it practical from the standpoint of both simple implementation and naturally embedded accuracy (Marcellin *et al.*, 2002). When using this type of quantization, reducing the number of bit-planes has the same effect as multiplying the step size by two. This allows bit-plane reduction to act as a stand-in for increased step size and makes the actual step size recorded in the QCD/QCC less important.

However, the Part 1 quantizer has a dead zone (zero bin) that is twice as long as the other bins. Forcing so many small wavelet coefficients to zero can cause faint lines or low-level textures to disappear. This concern becomes particularly important if both very strong and very faint edges appear in an image, without a broad spectrum between the two. Radar reflectance data can fall in this category. Large step sizes work best to quantize large coefficients efficiently, while small step sizes perform better for maintaining small coefficients. Using a medium step size with a smaller dead zone is a useful compromise available in Part 2 via the variable scalar quantizer (VSQ). Other methods of addressing this problem include TCQ (Section 2.4) and the nonuniform quantization resulting from self-contrast masking (Section 2.5).

2.3.1 Theory

VSQ allows the default dead-zone size to be modified by setting an adjustment factor, τ, in the range $[-1, 1)$. The adjusted dead-zone size is $2(1 - \tau)\Delta$, where Δ is the quantizer step size. This allows the dead-zone size to range from almost zero up to four times the quantizer step size. When $\tau = 0$ behavior is identical to the Part 1 quantizer; see Figure 2.2(a) for a graphical representation.

The forward VSQ formula

$$q = \begin{cases} \text{sgn}(x) \left\lfloor \dfrac{|x| + \tau\Delta}{\Delta} \right\rfloor, & |x| \geq -\tau\Delta \\[2ex] 0, & \text{otherwise} \end{cases}$$

generates the quantized coefficient q from the wavelet coefficient x. If p bit-planes are missing from q during reconstruction, giving q_p, then the inverse VSQ formula is

$$\hat{x} = \begin{cases} \text{sgn}(q_p)(|q_p| + r2^p - \tau)\Delta, & q_p \neq 0 \\[2ex] 0, & q_p = 0 \end{cases},$$

where r is a decoder chosen reconstruction factor frequently set to 0.5.

The VSQ dead-zone size tends towards 2Δ as more bit-planes are omitted from the code-stream (see Figure 2.2(b)). Therefore the variable scalar quantizer has most impact when wavelet coefficients are fully decoded, and the user should choose initial step sizes with care.

2.3.2 Signaling

The second least significant bit in Rsiz signals that the use of VSQ is required to decode the code-stream, and extended QCD/QCC marker segments transmit the adjustment factors, τ. Each wavelet subband may have a different τ value or all subbands may use the same value. This extension can only be performed in conjunction with irreversible transforms.

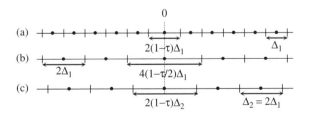

Figure 2.2 Embedded quantizer comparison. Upright lines separate the quantizer bins and circles indicate midpoint reconstruction: (a) variable scalar quantizer with step size Δ_1 and $\tau = 0.25$; (b) embedded results of (a) with one bit missing; (c) variable scalar quantizer with step size $\Delta_2 = 2\Delta_1$ and $\tau = 0.25$. Notice that (b) and (c) are different

2.4   Trellis-Coded Quantization

Trellis-coded quantization (TCQ) can be considered a special case of structured vector quantization. TCQ employs the trellises and sets partitioning ideas from trellis-coded modulation (Ungerboeck, 1982) to encode long data sequences. TCQ can achieve MSE performance very close to that promised by rate-distortion theory. Many variants of TCQ have been explored (Marcellin and Fischer, 1990; Fischer, Marcellin, and Wang, 1991; Fischer and Wang, 1992; Marcellin, 1994; Joshi, Crump, and Fischer, 1995; Kasner, Marcellin, and Hunt, 1999; Bilgin, Sementilli, and Marcellin, 1999). JPEG 2000 Part 2 allows for the use of entropy-coded TCQ (ECTCQ) as a replacement for scalar quantization. In JPEG 2000, the theoretical MSE advantage of ECTCQ is often seen only at high encoding rates (≥ 2 bits/sample). However, significant improvements in perceptual quality are usually present across the entire gamut of encoding rates. We describe here only the variant of ECTCQ supported in JPEG 2000.

TCQ takes a scalar codebook and partitions it into four subsets called D_0, D_1, D_2, and D_3. The codebook employed is composed of all integer multiples of a quantization step size Δ. Figure 2.3 illustrates the partition used for JPEG 2000; subsets obtained in this fashion are then associated with branches of a trellis as diagrammed in Figure 2.4. Each column of heavy dots (or nodes) in the trellis represents eight possible states at one point in time. The states are implicitly labeled 0, 1, 2, ..., 7 from top to bottom. Each branch in the trellis represents a transition from one state to another at the next point in time

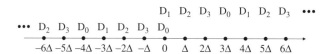

Figure 2.3   Codebook partition for TCQ. Reproduced by permission of Kluwer

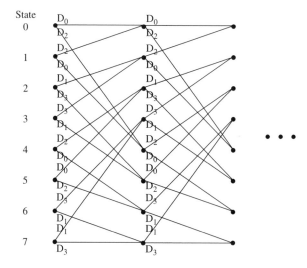

Figure 2.4   Eight-state trellis for JPEG 2000. Reproduced by permission of Kluwer

(next stage). Specifying a path through the trellis is equivalent to specifying a sequence of subsets.

The set partitioning and trellis structure introduce constraints on the codewords that the encoder can choose. For example, when beginning in state 0, the first codeword can only come from D_0 or from D_2. Specifically, if the encoder chooses the upper branch when leaving state 0, the first codeword must be chosen from $D_0 = \{\ldots, -8\Delta, -4\Delta, 0, 3\Delta, 7\Delta, \ldots\}$. Similarly, if the encoder chooses the lower branch when leaving state 0, the first codeword must come from $D_2 = \{\ldots, -6\Delta, -2\Delta, \Delta, 5\Delta, \ldots\}$.

For a given sequence of data samples (wavelet coefficients in JPEG 2000), the Viterbi algorithm is used to find the sequence of codewords (as allowed by the trellis structure) that minimizes the MSE between the data and the sequence of chosen codewords. Specifically, for a sequence of data samples x_i, $i = 0, 1, \ldots, m - 1$, a trellis of m stages is employed. Such a trellis has $m + 1$ columns of states, which we label $S_{i,l}$, $i = 0, 1, \ldots, m$, $l = 0, 1, \ldots, 7$. For each state $S_{i+1,l}$ let $S_{i,l'}$ and $S_{i,l''}$ be the two states having branches ending in $S_{i+1,l}$. Also, let $D^{l',l}$ and $D^{l'',l}$ be the subsets associated with those branches, respectively. Let $c_{l',l}$ and $c_{l'',l}$ be the codewords in $D^{l',l}$ and $D^{l'',l}$ that minimize $\rho(x_i, c) = (x_i - c)^2$, and let $d_{l',l} = (x_i - c_l')^2$ and $d_{l'',l} = (x_i - c_l'')^2$. Finally, let $s_{i+1,l}$ be the 'survivor distortion' associated with the survivor path at state $S_{i+1,l}$.

The ith step ($i = 0, \ldots, m - 1$) in the Viterbi algorithm then consists of setting $s_{i+1,l} = \min\{s_{i,l'} + d_{l',l}, s_{i,l''} + d_{l'',l}\}$, preserving the branch that achieves this minimum, while deleting the other branch from the trellis. If two values compared for minimum survivor distortion are equal, the 'tie' can be resolved arbitrarily with no impact on the MSE.

When the end of the data is reached ($i = m - 1$), the trellis is traced back from the final state having the lowest survivor distortion to find the sequence of chosen TCQ codewords. For long data sequences, the choice of initial state has a negligible impact on the MSE. Thus, we arbitrarily fix the initial state at 0. This is easily done by setting $s_{0,0} = 0$ and $s_{0,l} = \infty$, $l = 1, 2, \ldots, 7$.

The computational requirements and memory usage are both proportional to m. On a per sample basis, however, the computational requirements are independent of m. More specifically, the ith step (corresponding to the ith sample x_i) in the Viterbi algorithm requires essentially four scalar quantizer operations (to find the best codeword in each subset), followed by four add–multiply operations (to compute the distortion associated with each such codeword). Finally, to determine the survivor at each of the eight states, a total of 16 adds and 8 compares are required.

The trellis of Figure 2.4 has the property that the subsets associated with the two branches leaving any state are either D_0 and D_2, or D_1 and D_3. The 'union codebooks' $D_0 \cup D_2$ and $D_1 \cup D_3$ are shown in Figure 2.5, along with the ECTCQ index for each codeword. Note that the indices are signed. The compressed code-stream includes the ECTCQ indices corresponding to the sequence of codewords obtained via the Viterbi search. The ECTCQ decoder uses these indices to produce the sequence of codewords chosen by the ECTCQ encoder.

Note that, from the point of view of the decoder, there are two possible codewords for each index. However, the trellis structure resolves this ambiguity. For example, with an initial state of 0 in the trellis of Figure 2.4, the index sequence $(1, 2, -3, \ldots)$ would be decoded as $(\Delta, 3\Delta, -5\Delta, \ldots)$. This can be seen by noting that at state 0, $D_0 \cup D_2$ is the

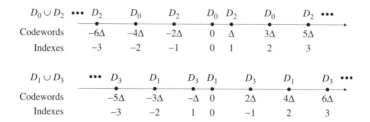

Figure 2.5   Union codebooks. Reproduced by permission of Kluwer

appropriate union codebook. Thus, the index 1 indicates that the codeword is Δ, which is in D_2, indicating (by examination of the trellis branch labels) that the next state is 4. At state 4, the appropriate union codebook is also $D_0 \cup D_2$. The index 2 then indicates that the codeword is 3Δ, which is in D_0, indicating that the next state is 6, at which the union codebook is $D_1 \cup D_3$. The index -3 indicates that the codeword is -5Δ, which is in D_3, indicating that the next state is 3, and so on.

At this point, we mention the seemingly strange sign flipping of the ± 1 indices in $D_1 \cup D_3$. In fact, flipping the signs in this fashion for all the indices in $D_1 \cup D_3$ is desirable. For symmetric PDFs, this would cause the indices to have identical distributions in both $D_0 \cup D_2$ and $D_1 \cup D_3$. This, in turn, would allow the trellis state to be omitted from any context model employed for entropy coding of ECTCQ indices. Unfortunately, flipping all the signs destroys any possibility of successive refinement, which is important for quality scalability.

A compromise solution is to flip only the sign of ± 1 and omit the trellis state from the context model. Simple analysis and experimental results show that the resulting distribution mismatch produces only a negligible loss in entropy coding efficiency. This allows the bit-plane coding of JPEG 2000 Part 1 to be used on the ECTCQ indices without any modification. Accordingly, ECTCQ processes each code-block independently, using the same scan order as the JPEG 2000 Part 1 bit-plane coder.

As with the dead-zone scalar quantization employed in JPEG 2000 Part 1, the bit-plane coder employs the sign magnitude representation of the ECTCQ indices. This allows an (approximate) embedding for ECTCQ. Although inverse ECTCQ requires exact knowledge of each index to track the state progression through the trellis, a partial inverse is still possible. Examination of Figure 2.5 shows that if the LSB of an ECTCQ index is unknown, the ambiguity is limited to the choice of four codewords. For example, if the binary representation (sign-magnitude form) of an index is known to be $+1$? (where '?' denotes the missing LSB), the correct index is guaranteed to be either 2 or 3, and the codeword can be any one of 3Δ, 4Δ, 5Δ, or 6Δ. One reasonable reconstruction policy in this case would be to choose $x = 4\Delta$. Extending this argument to the general case of p missing LSBs, a reasonable reconstruction policy is to set the missing LSBs to 0 to obtain an approximate index q and then setting $x = 2q\Delta$. This is in fact the policy recommended for use with JPEG 2000 (Bilgin, Sementilli, and Marcellin, 1999).

When there are missing LSBs, this form of ECTCQ provides MSE performance slightly worse than that of scalar quantization. However, when all bits are decoded, the full benefit of ECTCQ is realized and the MSE performance is superior to that of scalar quantization. To incorporate ECTCQ into JPEG 2000, each JPEG 2000 code-block is quantized independently using the same scan order as the JPEG 2000 Part 1 bit-plane coder. The step sizes employed by TCQ are signaled via QCD and/or QCC marker segments. It is worth noting that the values signaled in these marker segments are actually twice the TCQ step sizes employed by the encoder. Thus, TCQ decoders should divide these values by two before executing inverse TCQ processing. The reason for signaling twice the 'correct' step size is to allow 'non-TCQ' decompressors to decode TCQ code-streams in some reasonable fashion. From the discussion above, it is clear that inverse embedded scalar quantization using 2Δ is equivalent to approximate inverse embedded TCQ using Δ. Thus, a JPEG 2000 decompressor that blindly applies inverse scalar quantization using the step sizes signaled in QCD and/or QCC will produce a reasonable decoding.

The discussion of the previous paragraph is actually only applicable if there is at least one missing LSB. Clearly, if no LSBs are missing, the full inverse TCQ should be employed rather than the approximate inverse. If no inverse TCQ processor is available, it is actually preferable to discard the LSB prior to inverse scalar quantization. This is because the state-dependent sign flipping will make the LSBs appear corrupted if they are used in inverse scalar quantization to approximate inverse ECTCQ.

On a related note, it is worth commenting on layering in the presence of TCQ. Recall that each bit-plane is coded in three passes and that each code-block may contribute an arbitrary number of coding passes to each layer. In general, a nonconnected subset of LSBs is useless for inverse TCQ; all LSBs of a code-block must be available before full inverse TCQ processing can occur. For this reason, the three LSB coding passes of a given code-block should all be placed in the same packet. If postcompression rate-distortion optimization (PCRD-opt) is used to drive the formation of layers, the desired result may be obtained by setting the estimated distortion reduction to zero in the first two coding passes of the least significant bit-plane of each code-block.

2.5 Precinct-Dependent Quantization

JPEG 2000 Part 1 supports only scalar quantization. Rate control is generally performed using PCRD-opt. As a result, the step sizes used to quantize wavelet coefficients have little influence on the resulting bit-rate and/or image quality. This need not be the case. The standard allows the encoder to choose quantization step sizes as a method of rate/quality control and omit PCRD-opt. In fact, this is the preferred method when TCQ is used in JPEG 2000 Part 2 because, as mentioned in Section 2.4, the full benefits of TCQ are not obtained until all bit-planes are decoded.

Precincts allow low-memory implementation of JPEG 2000 without the need to break large images into tiles. This avoids the possibility of any block artifacts at tile boundaries. In JPEG 2000 Part 1, quantization step sizes can be chosen independently for each wavelet transform subband. However, within a subband, the step size cannot vary by precinct.

JPEG 2000 Part 2 allows the step size to be chosen independently for each region corresponding to a precinct within a subband. This extension allows for considerable flexibility when rate control is performed via step size selection. This is particularly useful for 'scan-based' coding (Flohr, Marcellin, and Rountree, 2000) of very large images. For such images, coding of precincts near the top of the image may need to be finalized before the entire image has even been acquired.

Step sizes for precinct-dependent quantization are signaled via QPD (quantization precinct default) and QPC (quantization precinct component) marker segments. The Rsiz field signals that precinct-dependent quantization is required to decode the code-stream.

2.6 Extended Visual Masking

While the normative portion of JPEG 2000 Part 1 does not explicitly mention the human visual system (HVS), JPEG 2000 includes a variety of optimization techniques that can use HVS modeling to improve perceived image quality. These tools make it possible to achieve visual frequency weighting and some limited visual masking within Part 1.

Visual frequency weighting is very effective for long viewing distances or high-resolution displays or prints. For close viewing, low-resolution monitors, and interactive zooming, however, its impact is greatly reduced. Visual masking, a phenomenon where local texture hides (or masks) image distortions to some degree, provides more leverage for improving visual quality in those cases. Image compression improves if fewer bits are encoded in well-masked regions and more bits are devoted to areas that have a low degree of masking.

In JPEG 2000 Part 1 the rate-control metric can be modified to take into account the visual masking effect. However, since all coefficients in a code-block must share the same truncation point, it is not possible to achieve point-by-point control of the masking operation in Part 1. This coarse-grain masking approach is referred to as block-based neighborhood masking. It tends to smooth out fine texture, but protects areas around high-contrast edges in larger images. However, it does not work as well on small images due to the block-based limitation.

In Part 2, the visual optimization options are expanded to include Zeng, Lei, and Daly's point-wise extended visual masking (Daly *et al.*, 2000; Zeng, Daly, and Lei, 2000a, 2000b, 2002). This technique self-adjusts at each coefficient, and includes both self-contrast masking and neighborhood masking to achieve very good visual quality. Self-masking protects fine texture well, making it especially suitable for high-quality photographic images that show human faces, but suffers near sharp edges, especially at low bit-rates. Neighborhood masking protects high-contrast edges well, but tends to smooth out fine texture. Part 2 extended visual masking combines the strengths of both approaches and minimizes the weaknesses. Experiments have shown that for some images the extended masking approach yields an improvement over no masking equivalent to a savings of up to 50% in bit-rate.

2.6.1 Theory

The self-contrast masking in Part 2 puts normalized wavelet coefficients through a power function with an exponent α of [0, 2). When combined with uniform quantization, this has the effect of nonuniform quantization on the original coefficients. The eye readily perceives small changes relative to low contrast but not high contrast, so $\alpha < 1$ can be expected to improve visual performance, since larger coefficients are more coarsely quantized. With wavelet coefficients normalized by wavelet filter gain, the exponent α is constant across all subbands. A typical choice of self-masking exponent is $\alpha = 0.7$.

The neighborhood masking in Part 2 next computes a masking factor with large values in busy areas and small values at edges with low-contrast background. The mask factor

x	x	x	x	x
x	x	x	x	x
x	x	o		

Figure 2.6 Causal neighborhood of 'o' when no restricting boundaries are nearby ($min\_width = 2$, $Size(N) = 12$); 'x' marks coefficients in the causal neighborhood

at position i, m_i, is computed from an average power of the surrounding coefficients and divided into the self-masked coefficients for a prequantization value z_i:

$$m_i = \begin{cases} 1 + a^\beta \dfrac{\sum_{k \in N(i)} |\hat{x}_k|^\beta}{Size(N(i))}, & Size(N(i)) > 0 \\ 1, & Size(N(i)) = 0 \end{cases} \quad ; \quad z_i = \text{sgn}(x_i) \dfrac{|x_i|^\alpha}{m_i} ,$$

where a is a fixed constant based on the image bit-depth, $N(i)$ is the neighborhood of coefficient x_i, and $\hat{x}_k$ are reduced quality estimates of neighboring values. Choosing β to be a very small fraction, such as $\beta = 0.2$, minimizes the impact of a few strong edges and maintains the impact in areas with many edges of intermediate contrast.

Since m_i must be available at the decoder as well as the encoder, it is computed using reconstructed coefficient values from only causal neighbors (see Figure 2.6). The extent of the causal neighborhood is at most $(min\_width + 1) \times (2 \ min\_width + 1)$. The neighborhood excludes the current coefficient and automatically shrinks to respect subband boundaries. If parallel code-block processing is desired, $respect\_block\_boundaries = 1$ will cause the neighborhood to respect code-block boundaries as well.

Subbands with resolution below $minlevel$ are omitted entirely from the masking process. Typically $minlevel$ is set to at least 1 so that the mask processing omits the low-resolution band.

The intermediate reconstructions $\hat{x}_k$ are computed from quantized values where only a limited number of bit-planes, $bits\_retained$, are available. This is achieved by locally quantizing, shifting, and dequantizing z using the Part 1 quantizer and a rounding reconstruction parameter of zero. The same m_i is used for all embedding levels. Although this causes a coarser masking granularity, experiments have shown that the performance is not very sensitive to the accuracy of the neighboring coefficients.

2.6.2 Signaling

The fourth least significant bit in Rsiz signals the presence of pointwise extended visual masking. The VMS marker segment transmits the values α, β, min_width, $bits\_retained$, $minlevel$, and $respect\_block\_boundaries$. Different components can have different VMS segments or one VMS may cover all components. This extension is considered 'useful for decoding,' but without it some images may appear distorted.

2.6.3 Interactions

Either self-masking or neighborhood masking may be used alone (self-masking only: $\beta = 0$ and $min\_width = 0$, neighborhood masking only: $\alpha = 1$), but the best performance typically comes from using both of them together.

If Part 2 visual masking is applied, then any encoder CSF frequency weighting that follows the quantization, e.g. in rate distortion, must modify the CSF weights to apply in the z domain, i.e. raise them to the power α.

Self-masking and TCQ produce similar results, so self-masking can be omitted when using TCQ, but neighborhood masking remains beneficial. VSQ should not be used with neighborhood masking as it interacts poorly with the local Part 1 quantization estimates. Extended visual masking is incompatible with reversible processing.

2.7 Arbitrary Decomposition

JPEG 2000 Part 1 allows only one wavelet decomposition structure: the well-known Mallat dyadic decomposition. Although this decomposition represents a good first choice across a broad spectrum of imagery, other decomposition styles can improve the image quality over specialized image classes and allow unequal size reductions in the horizontal and vertical dimensions of reduced resolution extracts.

Other decomposition styles described with some regularity in the wavelet literature include the full packet tree and its derivatives. The packet decomposition derivatives can outperform the dyadic decomposition at maintaining regular fine-grain texture and work well on synthetic aperture radar imagery. The US Federal Bureau of Investigation uses a 500 ppi fingerprint compression standard, WSQ (CJIS, 1997), with a decomposition specialized for the characteristics of fingerprint imagery at 500 dpi. Figure 2.7 shows some of these decompositions.

In addition to prespecified decomposition structures, wavelet packet analysis can be used to design custom decompositions for specific images or image types (Coifman and Wickerhauser, 1992; Ramchandan anad Vetterli, 1993; Meyer, Averbuch, and Stromberg, 2000). Such approaches start with a large decomposition tree and locate a good decomposition based upon a chosen optimization metric.

Another useful decomposition tool is the one-dimensional transform. This can be beneficial when processing very wide but shallow image tiles. Applying a dyadic decomposition to such a tile component will quickly create an LL subband that is only one row high.

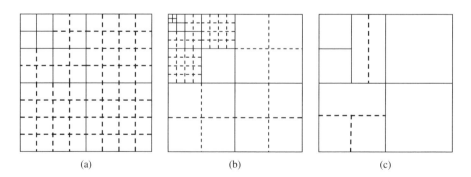

| (a) | (b) | (c) |

Figure 2.7 Part 2 decomposition examples. Solid lines are resolution decomposition. Dashed lines are extra sublevel decomposition. (a) Full packet decomposition at NL = 3: Ddfs = 111, Doads = 321, Dsads = all 1s. (b) FBI decomposition at NL = 5: Ddfs = 11111, Doads = 2321, Dsads = 11101111111111111. (c) Arbitrary example (see Figure 2.8 for decomposition parameters)

Performing row-only transforms at times helps maintain more rows for later stages in processing.

JPEG 2000 Part 2 arbitrary decomposition allows many of these other decomposition styles to be used. Signaling limitations restrict the maximum decomposition splitting within each resolution level. This can affect wavelet packet analysis but is not a serious concern in most cases. In addition to increasing compression performance and providing alternate resolution scaling alternatives, the wider range of decomposition structures allows for easy transcoding of compressed imagery from other formats using nondyadic decompositions.

2.7.1   Theory

A decomposition style can be considered in two separate parts: the underlying resolution decomposition and any extra sublevel decomposition of the highpass or detail subbands. The Part 2 signaling is separated in this fashion as well.

The underlying resolution reduction is formed by splitting the lowpass band at each level either in both directions or one direction only. The sequence of resolution reduction splits is signaled using an array of two-bit codes (2 = row transform only, 3 = column transform only, and 1 = both) in the DFS (downsampling factor styles) marker segment. If all resolution reduction splits maintain the horizontal/vertical aspect ratio, then no DFS is needed. Figure 2.7(c) shows a decomposition including some row-only and column-only transforms.

A marker segment called ADS (arbitrary decomposition styles) signals extra decomposition splits within the highpass subbands. The ADS syntax allows only two extra levels of splitting for the original highpass bands, which somewhat restricts wavelet packet analysis on the higher resolution subbands. ADS includes two arrays: the maximum number of split levels per resolution in DOads and the type of each extra split (1 = row and column, 2 = row only, 3 = column only, 4 = none) in DSads. If no extra highpass decompositions occur at a given resolution, then the DOads entry is 1. Extra highpass decomposition levels cause DOads entries to increase, up to a maximum of three. The DSads array is more complex and can be viewed as a depth-first traversal of the decomposition tree. The order in which the subbands are traversed for DSads is highest resolution to lowest, and within each level HH, LH, HL, LL or HX/XH, LX/XL.

Figure 2.8 demonstrates how the arrays in DFS and ADS are derived from the decomposition tree and Figure 2.7 gives the content of these arrays for the decompositions shown. Note that Part 2 coders that allow arbitrary decomposition may have other ways of inputting this information, but ultimately DFS and ADS marker segments signal all of it.

2.7.2   Implementation Hints

For the purposes of computing contexts and subband gains LX and XL are considered the same as LL, XH is considered the same as LH, and HX is considered the same as HL. When a highpass band adds extra decompositions, the subband gains at each level multiply. This affects the nominal dynamic range used in the quantizer step size formula. For calculating derived step sizes, the 'number' of decomposition levels, n_b, increases

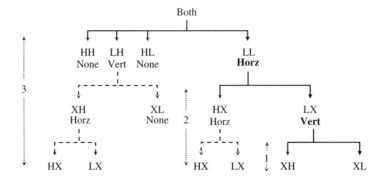

Figure 2.8 Sample decomposition tree for Figure 2.7(c). Resolution decomposition in solid lines (Ddfs = 123). Sublevel decompositions in dashed lines. Maximum number of sublevels per resolution level is between the double arrows (DOads = 321). Depth-first search of sublevels, left to right, skipping the initial resolution decomposition, gives DSads = 032002

by $^1/_2$ for every horizontal- or vertical-only decomposition. Formulas for sizing subband precincts alter in the expected fashion, with positioning following suit.

2.7.3 Signaling

The sixth least significant bit of Rsiz flags when an arbitrary decomposition is required to decode the code-stream, and the COD/COC, ADS, and DFS marker segments contain the information needed to create the decomposition structure. The COD/COC points to the ADS/DFS markers that apply for a given tile component. (DFS can only vary per component, not by tile. AFS can alter on a tile-by-tile basis.) The content of the DFS and ADS marker segments is described in Section 2.7.1.

The syntax allows the Ddfs, DOafs, and DSdfs arrays to terminate early if they end with a continuously repeated value. See Part 2 Annex F for details.

2.8 Arbitrary Wavelet Transforms

JPEG 2000 Part 1 includes only two wavelet transforms: the irreversible 9-7I and the reversible 5-3R, both specified with periodic symmetric preextension of the signal at the boundaries. While these filters work well for compressing a wide range of image types, certain image classes compress better with other wavelets. To allow this flexibility, Part 2 broadens the range of wavelet transforms that can be used to include not only the wider range of whole-sample symmetric (WS) filters, but also half-sample symmetric (HS) filters and generic nonsymmetric filters. This ability to handle generic filters makes JPEG 2000 powerful not only for niche compression applications but also as a research tool.

2.8.1 Transform via Lifting

Part 1 filters are implemented via lifting, so all Part 2 extended transforms are specified via lifting coefficients as well. Like their Part 1 counterparts, Part 2 filters must be normalized to achieve unit lowpass gain (Brislawn and Wohlberg, 2004), but otherwise a great deal of flexibility is allowed. Some of the choices include: whether the first decoding lifting step is highpass or lowpass m_0, offsets for each lifting step, and additive residues β for reversible lifting steps, as well as the actual lifting coefficients and scaling factors themselves. As an example, a generic reversible decomposition lifting step ignoring boundary handling has this form

$$
y(2n + m) = y(2n + m) + \left\lfloor \frac{\beta + \sum_{k=0}^{L-1} \alpha_k y(2n + 1 - m + 2(k + \mathit{off}))}{2^\varepsilon} \right\rfloor .
$$

The irreversible lifting is much the same, but does not need the floor function or additive residues, and the scaling takes place in a separate step. For exact arbitrary transform formulas see Part 2 Annex H.

Since WS filters are very common, some implementations may choose to handle only those filters or to streamline processing in this case. Part 2 Annex G gives formulas to use for such a simplification. It should be noted, however, that the generic formulas and the simplified formulas produce the same results on WS filters.

2.8.2 Boundary Extension

For WS filters the boundary extension defined in Part 1 works well. However, handling the HS filters was more difficult because the Part 1 preextension is not resolution scalable on even-length filters and the commonly used HS preextension does not interface well with the lifting implementation. Brislawn, Wohlberg, and Percus (2003) explain these issues in detail. To avoid these problems, Part 2 changes the placement of the boundary extension operation.

Instead of performing a long boundary extension prior to any lifting (preextension), in Part 2 each lifting step includes boundary extension. When WS extension is used on WS filters, the preextension policy is equivalent to the lifting step extension policy, so even the Part 1 transforms can be implemented in this way without any change in the expected results. Even better, lifting step extensions can be used without alteration on arbitrary filters as well as HS filters without resolution scalability problems.

In addition to moving the extension into the lifting steps, Part 2 allows two types of boundary extension. The typical periodic symmetric WS boundary extension can be used for any filter and is required for Part 1 transforms and any filter marked as WS. While conceptually simple, the WS extension is not always easy to compute, so a lower complexity constant (CON) extension policy is also available. The CON extension uses the two outermost coefficients within each subband (noninterleaved) as constant value extensions to the left and right, so only two values must be stored in memory at each lifting stage. Visually the CON and WS extensions perform similarly down to 0.5 bpp (Brislawn, Wohlberg, and Percus, 2003).

2.8.3 Signaling

The arbitrary transform kernel (ATK) marker segment contains all the information needed to describe a wavelet transform. Contents include the ATK index, type of transform (irreversible or reversible, WS or arbitrary), the type of boundary extension, whether the first lifting step is highpass or lowpass, the number of lifting steps, scaling factors, offsets, additive residues, and the lifting coefficients themselves. Part 2 transforms that are specifically labeled as WS filters must use the WS boundary extension, but transforms that are labeled as arbitrary may use either boundary extension. Although a code-stream may include several ATK markers, the wavelet described in a particular ATK marker segment is only in use when its index is referenced in a COD/COC or MCC marker.

The seventh least significant bit of Rsiz indicates when the arbitrary transform capability is required to decode the code-stream.

2.9   Multiple-Component Transform Extensions

Part 2 Annex J includes extensions to Part 1 that provide improved compression performance for imagery with more than one image band or component. These extensions are known collectively as the multiple-component transform (MCT) framework. The MCT framework extensions are implemented as a 'wrapper' around the Part 1 JPEG 2000 system. This allows a Part 1 decoder to partially decode a code-stream that employs the MCT framework, although the partially decoded file may be of limited or no use. More importantly, the wrapper implementation allows the MCT framework to be added on to an existing Part 1 codec while minimizing the amount of programming needed to do so. To facilitate better understanding of the MCT framework, this section considers both encoder and decoder. It is important, however, to remember that Part 2 normatively defines only how the MCT framework applies to a Part 1 decoder.

Figure 2.9 illustrates the wrapper implementation of the MCT framework from an encoder perspective. The top half of the diagram shows a Part 1 encoder; the bottom portion shows the addition of the MCT framework. From the encoder's point of view the MCT framework preprocesses the input image components prior to compression. The decoding process is the reverse, with the MCT framework serving as a postprocessing step to the Part 1 decoder. It is important to note that the MCT framework does not extend other JPEG 2000 encoding/decoding processes, most notably the entropy encoding processes. Thus, the MCT framework is not a full three-dimensional compression algorithm in that not all encoding/decoding processes extend into the third (component) dimension. Note that the number of decoded image components may not equal the number of components encoded in the code-stream.

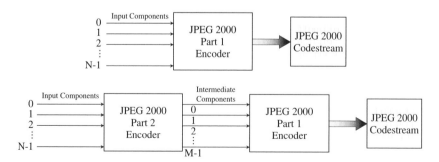

Figure 2.9   MCT wrapper (encoder)

2.9.1 Interactions

The MCT framework interacts with other sections of Part 2 as well as those from Part 1. Not all technologies within Part 2 (or Part 1) are compatible with the MCT framework; these incompatibilities are detailed below. The MCT framework also necessitates addition of several new code-stream marker segments (see Annex A of Part 2) as well as adoption of the JPX file format (see Annex M).

If a Part 2 decoder does not support the MCT framework then it is likely that no useful information will be obtained if it attempts to decode a code-stream that employs the MCT. Code-streams that make use of the MCT framework are prohibited from using the variable DC offset (VDCO), the default Part 1 DC offset, and the Part 1 RCT/ICT transforms (see Figure 2.1). The MCT framework also requires a subtle reinterpretation of the SIZ marker segment relative to its meaning in Part 1 of JPEG 2000. This will be discussed below.

The MCT framework may be applied in concert with any extensions from Part 2 other than the VDCO. In particular, other quantization (VSQ and TCQ) and wavelet techniques (arbitrary decomposition, arbitrary filters) are appropriate for use with the MCT framework and may improve coding performance.

2.9.2 MCT Framework

We now consider the MCT framework itself. The framework is composed of three main parts: transformation stages, component collections, and the transforms themselves. The specific multiple-component transform applied to a given image can be very simple or very complex, depending upon an application's needs. The particular multiple-component transform used within a code-stream can be changed on a per-tile basis in a similar fashion to other Part 1 coding constructs (QCC, QCD, COD, and COC).

2.9.2.1 Transformation Stages

A multiple-component transform consists of one or more transformation, or transform, stages. The number of transform stages used in a multiple-component code-stream depends on the application; for most applications a single-transform stage is often sufficient. Transform stages allow intermediate components within the code-stream to be processed through several successive transforms. Figure 2.10 illustrates this concept from the encoder's point of view. The input image components are passed through the first transform stage, creating a set of output intermediate components. These output intermediate components serve as input intermediate components to the next transform stage. This continues until after the final Kth transform stage, when the output intermediate components are passed off to the two-dimensional wavelet transform. These final output intermediate components will become the compressed code-stream components in the JPEG 2000 file.

At each transform stage the number of input intermediate components need not equal the number of output intermediate components. Since the number of compressed code-stream components need not equal the number of input image components, a subtle reinterpretation of the SIZ marker segment and the addition of a component bit-depth definition (CBD) marker segment are required to interpret the compressed data properly. The SIZ

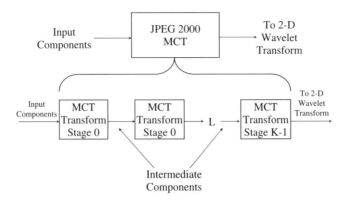

Figure 2.10   MCT transform stages

marker segment represents the number of *compressed code-stream* components as well as their bit-depths and whether they are signed or unsigned. This is a slightly different point of view from Part 1 where the SIZ marker segment described the *decoded image* components. When using the MCT framework the CBD marker segment describes the number of decoded image components, their bit-depths, and sign type.

Within the code-stream the multiple-component transform ordering (MCO) marker segment communicates the ordering of transform stages required by the decoder to implement the inverse multiple-component transform. The MCO marker segment contains a sequence of references to multiple-component transform collection (MCC) marker segments. The MCC marker segment encodes the component collections and transforms employed within a transform stage (see below).

2.9.2.2   Component Collections

Each transform stage is composed of one or more component collections. Associated with each component collection is a transform type that is applied to the input intermediate components of that collection. There is no requirement that all input intermediate components participate in a given component collection, but all input intermediate components of a transform stage must appear in at least one component collection. This requirement prevents an implementation from accidentally overlooking input intermediate components. Note that components may be discarded by utilizing the transforms (e.g. a nonsquare matrix transform) or by simply not passing them through on the output side.

Figure 2.11 illustrates the relationship between the component collections and transform stages. Each component collection is reflected in the MCC marker segment for this stage. This example shows Q collections; therefore the MCC marker segment must describe Q collections. The MCC marker segment is also responsible for permuting the input and output intermediate components associated with each transform.

Figure 2.12 (Wilkinson *et al.*, 2001) illustrates several possible component collections for a transform stage with six input intermediate components and six output intermediate components. As the figure shows, different transform types can be intermixed within a transform stage. Input components may be reused and the transforms need not be one-to-one. Furthermore, components may be permuted on the input and output.

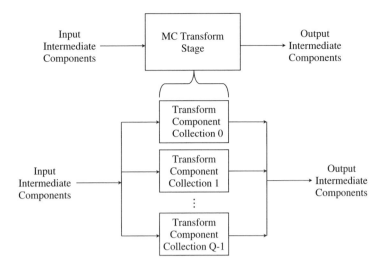

Figure 2.11   MCT component collections

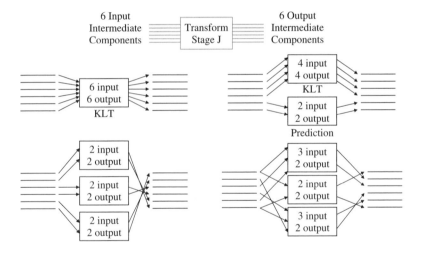

Figure 2.12   MCT component collection examples

There are many reasons why an application might use more than one component collection (or transform stage). For example, consider the correlation matrix for the hyperspectral scene shown in Figure 2.13 (Shen, Kasner, and Wilkinson, 2001). Brighter areas indicate regions of higher correlation between spectral components.

The white boxes in Figure 2.13 denote the component collections utilized by the transform stage. The four small boxes around components – [0, 2], [104, 107], [140, 149], and [206, 209] – correspond to bad spectral bands and water absorption bands. These components were simply 'passed through' with only a mean shift applied to the bands. The component collections – [3, 57], [58, 103], [108, 139], and [150, 205] – were each transformed by a Karhunen–Loéve transform (KLT) tuned to each spectral region (a mean

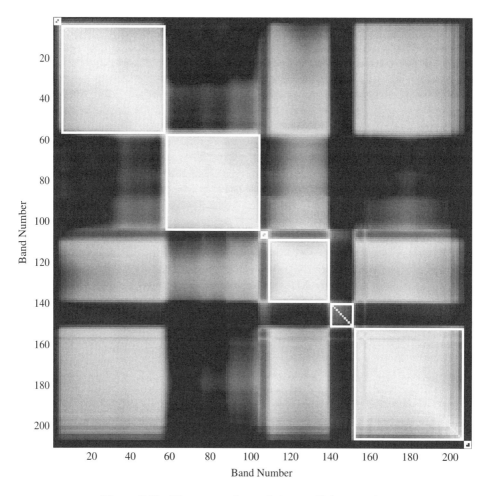

Figure 2.13   Hyperspectral correlation coefficient matrix

shift was also performed). The performance of this compression application is detailed by Shen, Kasner, and Wilkinson (2001) and Hao and Shi (2001).

In this example, multiple-component collections were used for three reasons. First, bad spectral bands and water absorption bands were cut out of the KLT. Second, the KLT was tuned to the different spectral regions (visible and near-infrared). Third, the computational complexity of the KLT was reduced by breaking it into a set of small-dimension transforms.

When the KLT is used for compression, low-energy eigenbands are typically encoded at low fidelity. Code-stream bits are given to those bands containing the most energy. This may be done during application of the transform by simply throwing away the eigenbands, which implies that the KLT matrices are no longer square. The MCT framework accommodates this easily. Alternatively, the eigenbands can be encoded by using explicit quantization or implicit quantization during entropy coding by simply not encoding all bit-planes in the quantization indices (embedded power-of-two quantization).

2.9.2.3   Transform Types

The transform types allowed within the MCT framework fall into two categories: *array-based transforms* and *wavelet-based transforms*. Within each category, the transforms may be implemented in a reversible (integer) or irreversible (floating-point) fashion. Null transforms may be specified within both categories. The multiple-component transform (MCT) marker segment is used to convey the information that the decoder needs to perform inverse array-based multiple-component transforms. Since the Part 1 DC offset and Part 2 VDCO cannot be used in conjunction with the MCT framework, MCT marker segments provide additive offsets (DC level shifts) for both array-based and wavelet-based transforms. The MCC marker segment makes all associations between component collections and inverse transform information necessary to implement a transform stage.

Array-Based Transforms

Array-based transforms may be expressed as linear combinations of their input intermediate components (although this is not strictly true if the transform is reversible). JPEG 2000 Part 2 distinguishes between two subclasses of array-based transforms: *decorrelation* transforms and *dependency* transforms. MCT marker segments convey the necessary transform coefficients. An MCT marker segment carries only one set of array-based transform coefficients (or DC offsets). Thus, in most cases multiple MCT marker segments – typically at least one for DC offsets and one for transform coefficients – are typically used to implement a transform. Given the size of some transform coefficient arrays, more than one MCT marker segment may be needed to convey the transform coefficients alone.

(1) Decorrelation Transforms

The irreversible array-based transform operates on the code-stream components in the same fashion as matrix multiplication operates on an input vector. From the decoder's point of view, the input vector is a given spatial sample (x, y) across all code-stream components. These spatial samples are the values decoded from the two-dimensional spatial processing of the JPEG 2000 decoder. Any DC offsets are applied after the matrix multiplication. An irreversible decorrelation transform may be square or rectangular, and may therefore create fewer, as many, or more output components as input components. In practice, most transforms will create as many or more output components as input components during decoding. However, in some interesting cases fewer components might be created: for example, an array-based decorrelation transform could create an RGB true (or pseudo color) representation of a hyperspectral data set by defining a weighted sum of components.

Reversible decorrelation transforms must have the same number of output components as input components in order to preserve the reversible nature of the transform. Hao and Shi (2001) describe an SERM factorization technique that generates integer-reversible approximations to unitary transforms. The KLT is a multiple-component unitary transform that has been shown to be effective at compressing hyperspectral data sets (Kasner *et al.*, 2000; Shen and Kasner, 2000; Brower *et al.*, 2000; Shen, Kasner, and Wilkinson, 2001, Kasner and Rajan, 2004). Applying the SERM factorization to a KLT transforms it into an integer-reversible transform. Of course, a price is paid in terms of energy compaction and orthonormality of the transform. The numerical stability of the factorization typically degrades with an increased number of components, but it has been shown to be quite effective with up to 80 components. This number of components is more than enough to effect good compression performance. For example, using one 128-component KLT rather than four 32-component KLTs results in similar compression performance. These differences are outweighed by the benefits of a lossless integer transform. For these reasons JPEG 2000 Part 2 includes the SERM factorization technique.

(2) Dependency Transforms

The dependency transform provides a mechanism for DPCM-like prediction transforms. Causality is enforced in the sense that only previously decoded components may serve as the prediction basis for subsequently decoded components. The dependency transform may be implemented in an irreversible or reversible fashion. The reversible approximation of an irreversible dependency transform may be created through appropriate scaling and rounding techniques (Wilkinson *et al.*, 2001). Dependency transforms must have the same number of input and output components. Proper design and implementation of a dependency transform can be a difficult task. An encoder must use decoded components (ones that include the effects of compression) in making its predictions, since it must mimic what the decoder sees when decoding. This can complicate encoder issues such as rate control.

Wavelet-Based Transforms

The wavelet-based multiple-component transform allows use of the Part 1 wavelet transforms, as well as Part 2 arbitrary wavelet transforms. Application of the wavelet transform in the component dimension creates component subbands, which extends the notion of multiresolution decomposition to the component dimension. The wavelet-based transform requires an equal number of input and output components. Only dyadic wavelet decompositions are supported in the component dimension. Therefore the Part 2 arbitrary decomposition techniques do not apply to the component dimension.

The MCC marker segment signals whether a Part 1 wavelet transform is to be used. For other wavelet transforms types, the MCC marker segment points to an ATK marker segment (see Section 2.8) that contains the wavelet kernel for the transform used in the component collection. The MCC marker segment specifies the number of dyadic decomposition levels as well as the phase (reference grid offset in the component dimension) of the wavelet-based transform.

The wavelet-based transform may be applied in either a reversible or irreversible fashion, depending on the type of wavelet transform used. For example, if the 5-3R wavelet transform is used, then the wavelet-based transform is integer reversible. Conversely, if the 9-7I wavelet transform is used, the wavelet-based transform is irreversible. Part 2 arbitrary wavelet filters may also be both reversible and irreversible, so other options exist beyond the 5-3R and 9-7I wavelet kernels.

2.9.2.4 Comments

The MCT transform may be changed on a tile-by-tile basis, in the same manner as Part 1 encoding options. In fact, the MCT transform can be turned off for any given tile.

Multiple-component transforms specified in the main header of the code-stream apply by default to all tiles. The transforms can be overridden on a tile basis by redefining the component collections and transform types at the tile-part header level. When changes are made on a tile basis, the encoder must take care in determining the SIZ and CBD marker segments. The component bit-depths signaled in both marker segments must be the maximum values (neglecting the sign) across all tiles. Typically it is the SIZ marker segment that is of concern, particularly if the MCT is turned off or substantially modified in some tiles. However, issues can arise with the CBD in certain cases.

All components in a given component collection must share the same sampling factors on the reference grid (see JPEG 2000 Part 1 Annex B). This guarantees that a full multicomponent pixel (i.e. missing no components) exists at each spatial (x, y) location.

The combination of the 5-3R wavelet transform in the spatial dimensions and any of the reversible MCT framework transforms makes it possible to losslessly compress multiple-component data. The multi- and hyperspectral user communities hold a widespread misconception that lossless compression of spectral data is not possible. With JPEG 2000, this is clearly not the case. For example, Kasner *et al.* (2000), Shen and Kasner (2000), and Shen, Kasner, and Wilkinson (2001) have examined use of the 5-3R wavelet to create a three-dimensional (3D) lossless wavelet transform. Hao and Shi (2001) investigated the SERM factorization of the KLT, which can be combined with the 5-3R wavelet transform to provide a lossless compression scheme. Kasner and Rajan (2004) analyzed the computational complexity of the 3D wavelet transform versus the KLT/2D wavelet transform relative to increased compression efficiency. Several of these reports also studied the application of trellis-coded quantization in conjunction with the KLT and found superior compression performance.

The MCT framework supports a wide variety of compression techniques. While it is a flexible and powerful tool, it is also very easy to misuse. The designer of an MCT must pay attention to the complexity of the transform and any causality concerns. Increased transform processing typically yields a diminishing return in terms of compression performance. It is also unreasonable to expect all MCT-capable decoders to support all possible transform types or an excessive number of transform stages and component collections. A reasonable starting point for MCT-capable decoders would be to support one transform stage (with multiple-component collections) using either the 3D wavelet transform, with the 5-3R or 9-7I wavelet kernels, and the array-based transforms.

The Digital Imaging and Communications in Medicine (DICOM) standard, jointly developed by the ACR (American College of Radiology) and NEMA (National Electrical Manufacturers Association), has adopted the MCT framework for compression of medical data sets (DICOM, 2006). With the commercial adoption by DICOM and the interest shown by other standards bodies (Geospatial-Intelligence Working Group (GWG) and NTB (NITFS Technical Board)), the number of JPEG 2000 implementations supporting this optional portion of the JPEG 2000 standard should grow.

2.10   Nonlinear Point Transform

Part 2 includes two nonlinear transforms (NLTs) used after the Part 1 decoding processes and after any inverse MCTs (see Figure 2.11). Both are point-wise transforms only. From an encoder's point of view, any nonlinear transform would be applied prior to a multiple component transform. The nonlinear transforms might take a nonlinear sensor's output and make it linear prior to compression; alternatively, a nonlinear transform might map a sensor's response so that a bit allocation matches the human visual system. We consider the nonlinear transform's interaction with the MCT framework below; the transform may also be used with any other Annex in Part 2.

2.10.1 Relationship to Other Annexes

The nonlinear transform, like the MCT, has the ability to alter the bit-depths of the components it operates on. Therefore we must consider its interactions with the MCT framework and the meanings of various marker segments. Using the MCT framework in a code-stream changes the meaning of the SIZ marker segment, which represents the number of code-stream components and their bit-depths.

The CBD marker segment is added to the code-stream to represent the number of decoded image components and their bit-depths. If the nonlinear transform is used with the MCT framework, then the CBD marker segment represents bit-depths of the decode image components prior to application of the nonlinear transform. If a nonlinear transform is applied, but the MCT framework is not used, then the SIZ marker segment once again conveys the number of decoded image components and their bit-depths prior to application of the nonlinear transform. Unlike the MCT, the nonlinear transform cannot change the number of components.

The Rsiz parameter in the SIZ marker segment is modified (see Part 2 Annex A) to indicate use of the nonlinear transform. The nonlinear transform capability is considered 'useful to decode' for Part 2 decoders.

2.10.2 Nonlinear Transform

The nonlinear transform may be applied to all components or to a subset of components. An NLT marker segment is required in the code-stream whenever the nonlinear transform is used. The NLT marker segment indicates the type of nonlinear transform, the non-linear transform parameters, the component(s) to which it applies, and the resulting bit-depths of the transformed components. Two types of transforms are supported: an inverse gamma-style nonlinearity and a look-up table (LUT). For the gamma-style nonlinearity, the NLT marker segment includes a set of equation parameters; with a LUT it includes a set of table parameters.

2.11 Geometric Manipulation via a Code-Block Anchor Point (CBAP)

JPEG 2000 neatly arranges data locally based upon code-blocks, so it would be helpful if image mirroring or $90°$ rotations could be performed without disrupting the blocking. Part 1 code-blocks are anchored at the origin, beginning at even positions and ending at odd ones. To maintain the proper relationship with the reference grid, code-block mirroring would have to begin at an odd position and end at an even one, which, strictly speaking, Part 1 does not allow. This forces geometric manipulation operations to reblock data, with the attendant overhead in memory requirements.

Part 2 solves this problem by specifying a code-block anchor point (CBAP). The CBAP (z_x, z_y) takes on one of four values, (0, 0), (0, 1), (1, 0), and (1, 1), thereby allowing code-blocks to begin at either odd or even positions in both the horizontal and vertical directions. The CBAP (z_x, z_y) is the anchor point for the LL subband code-blocks and precincts, while the other subband code-blocks/precincts have slightly shifted anchor points (Houchin, 2005).

It is important to note that a change in the code-block anchor point does not suffice to cause a geometric manipulation of the image. The CBAP merely facilitates low-memory implementation of the other steps required, including code-block flipping or transposition, recoding, and relocation within the wavelet transform. Manipulations that require transposition cause very small errors in losslessly compressed images. For further details and impacts see Taubman and Marcellin (2002).

In addition to permitting low-memory geometric manipulation, the CBAP enables memory-efficient single-sample overlap (see Section 2.13.1) and other features that benefit from highpass-first code-blocks. Although any CBAP can be encoded in a Part 2 code-stream, the CBAP syntax does not handle simple geometric manipulation of arbitrary decompositions that include extra decomposition of highpass subbands (Section 2.7). Code-streams containing these extra subbands require reblocking.

The CBAP, signaled in the COD marker segment, is backward compatible with Part 1. Rsiz flags the use of Part 2, but does not set any other bits. In fact, this feature is so simple and useful that Part 1 decoders are strongly encouraged to support the CBAP.

2.12 Single-Sample Overlap

While JPEG 2000 Part 1 uses blocks, precincts, and tiles to localize data and reduce memory requirements, tradeoffs remain between totally localized processing and reduced quality at boundaries of localized regions. For example, while tiling completely restricts all computations to one area, artifacts appear at tile boundaries for lower quality compression and small tiles may have excessive tiling overhead. By contrast, code-block and precinct localization reduces these artifacts, but the wavelet transforms cannot be entirely expanded without reference to significant portions of neighboring code-blocks or precincts. Line-based implementations of the wavelet transform, while helpful, still have large memory requirements when the image is extremely wide. The single-sample overlap technique introduced by Canon (Berthelot, 1999a) uses a small amount of overlap between wavelet blocks and a slightly modified wavelet formula to localize computations to fixed size blocks without causing blocking artifacts. The benefits are low-memory implementations, better visual performance, and error containment at the expense of a slight increase in file size.

2.12.1 Theory

2.12.1.1 Blocking

Single-sample overlap works on rectangular regions that overlap by a single row and a single column. When tiles overlap the technique is called tile single-sample overlap (TSSO); when the overlapping occurs inside a tile it is just called SSO.

SSO cells are created by choosing a power-of-2 nominal cell size and forming regions that are anchored at the origin but overlap by one row and column (actual block dimensions are one larger than the 'nominal' dimension). In other words, if 2^d is the nominal cell size in one direction, then the positions at $n2^d$ are both the end point of one SSO cell and the starting point of another SSO cell. The nominal cell size shrinks by a factor of 2 at each successive decomposition level, so during decomposition SSO cells are generated at their largest size on the original sample data and then in shrinking succession on each LL band prior to further decomposition. SSO cells need not be square, but both dimensions must be tied to powers of 2.

Like SSO cells, TSSO tiles have a nominal size and an actual size that is one row and column larger. The SIZ marker segment reports the nominal tile size. The extra actual tile size is used to create a single row or column overlap between tiles, so the last sample of one tile is also the first sample of the adjacent tile. The TSSO tile grid is shifted from the SIZ marker tile offset using a TSSO anchor scheme very similar to the code-block anchor point (see Section 2.11), but this anchor may be chosen independent of the CBAP. TSSO cannot be used unless the nominal tile size and the tile offset meet certain conditions. Roughly speaking, both must be integer multiples of powers of 2, which ensures that the entire tile boundary is composed of lowpass coefficients after the wavelet transform (Taubman and Marcellin, 2002).

2.12.1.2 Wavelet Computation and Encoding

SSO only works in conjunction with WS filters and performs the wavelet transform in a manner very like Part 1 (or a Part 2 WS filter). The key differences are that boundary extension acts as if the SSO cell border were a tile boundary and the transform computation in SSO overlap areas (positions $n2^d$) is different. For reversible transforms such as the 5-3R, the overlap areas pass through the lifting stages entirely unchanged. For irreversible transforms such as the 9-7I, overlap area values are scaled to maintain the correct lowpass normalization during lifting. Since the new transform coefficients in SSO overlap areas are independent of the internal cell content, no extra coefficients are created. The JPEG 2000 encoder does not create any overlapping in code-blocks or precincts when using SSO, but instead proceeds as if SSO had never occurred. These slight changes in the wavelet transform typically cause a slight decrease in coding efficiency.

By contrast, TSSO does not affect the wavelet computation in any way. Any wavelet transform is allowed, but only WS filters are recommended. TSSO tiles are encoded with the tile overlap, so the duplicate sample codings decrease the compression efficiency slightly. However, when WS filters are used, a perceptible visual improvement at tile boundaries accompanies this drop in efficiency (Berthelot, 1999a). On reconstruction only one of the duplicate image samples is retained. TSSO specifies that the left/top tile sample is kept when the horizontal/vertical tile anchor is 0, respectively; otherwise the right/bottom tile sample is kept.

2.12.2 Comments

Beyond ISO/IEC 15444-2:2004, the public literature to date contains very little covering the usage and performance of SSO and TSSO. However, some JPEG committee reports (Berthelot, (1999a, 1999b); Onno *et al.*, 1999) and others explore best practices for memory efficiency and display examples of improved visual quality.

The most memory-efficient usage of SSO occurs when the SSO cells and code-blocks align well. Since SSO does not generate extra overlap coefficients, it is extremely memory efficient at the encode side. The decoder is not quite as efficient, since it needs access to a row and column of coefficients from two neighboring coding blocks before a complete SSO cell is available for reconstruction. It attains maximum efficiency when the required overlap information has already arrived at the decoder, e.g. overlap that occurs above and to the left of the current block for a typical raster scan. Since overlap samples always occupy even or lowpass positions, it is ideal if the first row and column of decoded code-blocks begin on an odd or highpass sample (Berthelot, 1999b; Onno *et al.*, 1999). Standard Part 1 code-blocks always begin with a lowpass sample, but in Part 2 this behavior can be modified by setting the CBAP to (1, 1). SSO memory efficiency also improves when the SIZ tile offset equals the CBAP, because the image boundary code-blocks have full size.

SSO and TSSO can be used either separately or jointly. In either case it is beneficial if the nominal tile size is a multiple of the nominal SSO cell size. Not only does this maximize the SSO coding efficiency, but it also ensures that duplicated TSSO overlap samples have the same reconstruction value (modulo any quantization differences), so the TSSO overlap reconstruction rule may be ignored. At the limit the nominal SSO cell size can be set equal to the nominal TSSO tile size.

2.12.3   Signaling

The fifth least significant bit of Rsiz indicates when either SSO or TSSO is required to decode the code-stream. An extended COD/COC marker segment contains all the SSO/TSSO information. SSO information – whether or not SSO is used and nominal SSO cell dimensions – can change over different tile components. TSSO information – whether or not TSSO is used and tile anchor point – must remain constant across all tiles and components.

2.13   Region of Interest

A region of interest (ROI) is an area of the image that is expected to exhibit higher quality than the rest of the image for some range of decoding bit-rates. Part 1 allows two methods for creating an ROI: maxshift ROI and code-block layering.

Maxshift ROI can handle arbitrary shapes or even isolated high-quality wavelet coefficients, but it forces the entire ROI to appear at maximum quality before much background appears. When used in a lossless code-stream, maxshift ROI can give too much prominence to the ROI. Judicious code-block layering can create ROIs of varying importance that intermingle with the background. However, limitations on code-block size make this a coarse tool that works best on large images; it does not work for code-blocks that mix both background and ROI coefficients.

Part 1 also requires a different RGN marker segment for each component that uses ROI processing, even if the shift is the same in all components. For color imagery, including three RGNs is not a large burden, but for multispectral and especially hyperspectral imagery the overhead can create a problem.

The Part 2 ROI extension has two parts. Primarily, it consists of Ericsson's scaling-based ROI method (Christopoulus, Askelöf, and Larsson, 2000), which is the focus of the remainder of this section. As a side benefit it also allows a single RGN marker segment to signal an ROI covering all components. This all-component signaling ability applies to maxshift ROIs as well, but is not backward compatible with the Part 1 RGN format.

The Part 2 scaling-based ROI has two key features: (1) some background can be included sooner, i.e. before the ROI reaches full accuracy, and (2) different priority levels can be selected for each ROI. However, these benefits come at the cost of ROI shape signaling and extra mask generation at the coder.

As in Part 1, each component is processed independently, so different ROIs can be applied to each component. In fact, maxshift can be used on one component while multiple shape-based ROIs are used in another. However, the maxshift and scaling methods are mutually incompatible, so they cannot be used together on the same tile component.

2.13.1 Theory

For the maxshift method, wavelet coefficients are shifted by one fixed value so that the ROI and background do not overlap. Shifting of ROI coefficients also occurs for the scaling-based method, but the overlap restriction is removed and different ROIs can have different degrees of shift. Each ROI, i, is given a shift, s_i, relative to the background, as shown in Figure 2.14.

Because the decoder does not 'see' gray-shaded boxes like those shown in Figure 2.14, it needs extra information to separate the various ROI and background coefficients. In Part 2, an extended RGN marker segment signals the ROI locations and shapes. For simplicity only two shapes are allowed: the rectangle and ellipse. These shapes are defined in the spatial domain, so both encoder and decoder must construct masks in the wavelet domain indicating which coefficients belong in each ROI (see Figure 2.15). Irregular-shaped regions can be created by combining multiple rectangular and elliptic ROIs and giving all of them the same shift. However, each additional simple ROI adds to the mask generation burden.

Notice that each level of wavelet decomposition causes the ROI to cover a slightly larger portion of the subband area. With multiple ROIs, this expansion can cause different ROI masks to 'overlap' at higher subband levels, so that a single coefficient belongs to several different ROIs. When this occurs, the coefficient is automatically assigned to the ROI with the largest scale factor within the group.

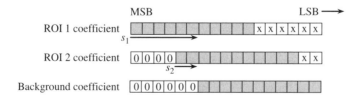

Figure 2.14 Scaling for two ROIs with different scaling values, s_1 and s_2

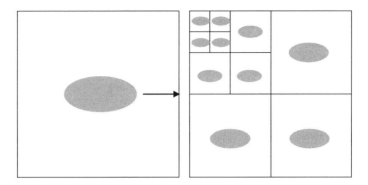

Figure 2.15   ROI spatial domain to subband mask for an elliptical ROI

2.13.2   Implementation Details

2.13.2.1   Mask Generation

To save space the ROI wavelet subband mask is computed from a few shape parameters. This is likely to be the part of ROI implementation most prone to error, as the mask varies according to the wavelet filter length, extension type, decomposition structure, and whether single-sample overlap computations are used. The most generic way to create the mask is to follow the reverse wavelet transform backward, level by level, and determine what wavelet coefficients are used when computing the mask coefficients in the level below. This is *not* the same process as putting the ROI in image space through a forward transform.

The mask for a rectangular ROI can be generated more quickly, since only the upper-left and lower-right corner of the rectangle must be studied. This fast generation will produce the same results as the more involved generic method. Part 2 Annex L describes procedures for both generic mask generation and the fast rectangle speedups.

2.13.2.2   Shifting

In practice the shifts take place in two parts: physical and virtual. If an entire code-block is occupied by one type of data – whether background or all ROIs that share the same scaling value – then all shifting is virtual, i.e. performed entirely via weighting in the layering or rate-control mechanism. If the code-block includes background and one ROI, then the background is physically shifted relative to that ROI only, and the rest of the code-block shift will be virtual. The generic procedure specified in Part 2 gives the correct shifts to apply, based upon the minimum and maximum scaling values of the ROI masks that intersect the code-block.

2.13.2.3   Extra Bits

The bit values at the positions marked 'x' in Figure 2.14 are required for correct processing of the entropy coder, but are promptly destroyed by the decoder during ROI processing. Grosbois, Santa-Cruz, and Ebrahimi (2001) and Ameli, Vaisey, and Jin (2004) point out that these bits can be set in different ways to alter the size of the compressed code-stream, and the natural choice of setting all these bits to zero tends to generate a larger code-stream than necessary. Ameli, Vaisey, and Jin (2004) found it beneficial to set all 'x' bits to zero, except for the most significant 'x' bit following zero-valued coefficients (all gray bits have a value of 0).

2.13.3   Signaling

Part 2 extends the RGN marker to signal shape information when the scaling-based ROI is used, and also simplifies signaling when the ROI is identical across all components. The sixth most significant bit of Rsiz flags the use of this format extension for either reason. When flagged, the extended ROI capability is required to decode the code-stream.

The type of ROI signaled in Srgn is extended to allow the values 1 (rectangle) and 2 (ellipse), in addition to the value 0 indicating the maximum-shift ROI. When a rectangle or ellipse is used, the SPrgn field includes not only the binary shift value s but also location (XArgn, YArgn) and sizing (XBrgn, YBrgn) for the ROI. For a rectangle these parameters are position of the upper-left corner, plus width and height. For an ellipse, the position of the center point plus the half-width and half-height are given. Both the location and sizing information are given in reference grid units.

Component signaling in the extended RGN always uses 16 bits. A component index Crgn of 65535 indicates that the ROI in this marker segment applies to all components.

2.14   Extended File Format: JPX

2.14.1   Encoding versus Interpretation

Readers may believe that this book has discussed 'file format' from the beginning: how an image is converted into sequences of bits that are stored in a file. While the importance of code-stream formation is beyond dispute, equally important is what the application does with the bit sequences it reads back. For example, a user may want to know if the value of 255 in the 'red' channel represents burgundy, crimson, or scarlet. Other questions could address who took the picture or the circumstances under which the image reader has 'permission' to display or distribute the image. In essence, file formats provide all of the information required for an application to 'interpret' the decoded image data under real-world scenarios.

Unfortunately, the interpretation aspect of existing image file standards has often been ignored or minimized, forcing end users and application vendors to restrict their own behaviors to ensure that two different applications (a screen and a printer, for example) would interpret (e.g. render on to a physically viewable medium) the image such that the results were 'close enough' to each other. Even today, over 10 years after digital imaging became commonplace, many applications still require an image file writer to generate a file specific to a target use (e.g. the sRGB colorspace for softcopy display).

While this may create an image file that a specific application can unambiguously and optimally use, the true meaning of the image data has been lost. For example, consider an image of a bride in an ornate white gown standing outside with the sun in the background. While in reality the sun emits many more photons than are reflected off the wedding gown, the display device has no choice but to display both the gown and the sun as pure white.

If an application later desires to 're-expose' the digital image to show the intricate detail in the gown, its performance will be severely limited because the true meaning of those input pixel values has disappeared; they are all just 'white.'

The file formats in JPEG 2000 attempt to break though this complacency and provide mechanisms for the unambiguous specification of both what the image data means and how it should be interpreted. A key point to understand in any discussion of the JPEG 2000 file formats is that if an image file writer can encode the true *meaning* of the image data in the image file, then *all* image readers can extract optimal value from the image. Each individual application can interpret the meaning of the image data in the context of its own specific application domain to provide the best possible results for users in that domain. For example, if that bride/sun image were projected in a dark theater, the gown could be rendered as a series of grays instead of white, allowing the viewer to perceive the greater brightness of the sun properly *and* see the detail in the wedding gown.

2.14.2   File Format Scope

Parts 1 and 2 of the JPEG 2000 standard define two file formats. JP2, in Part 1 Annex I, defines a base compliance level and feature set, targeted at consumer imaging and digital photography. JPX, defined in Part 2 Annexes M and N, extends the capabilities of JP2 to serve the digital photography and commercial printing markets better. However, both formats offer a strong foundation for other application domains and provide the 'hooks' for adding domain-specific information to the file. We will discuss this extensibility further in the discussions on colorspace specifications and metadata.

While the standard appears to define two separate file formats, implementers are strongly encouraged to view JPX as 'the one and only' photography-centric file format in JPEG 2000, and to view JP2 as a compliance level of JPX. In fact, many applications will explicitly seek to write JPX files that are compatible with JP2 readers, as the description of colorspace specification will show. In addition to the JP2 compliance level, Part 2 also defines a JPX baseline compliance level that targets the advanced digital photography and commercial printing market where an image 'file' only contains one image. Other standards bodies and organizations may also define additional compliance levels. To allow readers to interpret JPX files properly in the light of this extensibility, the standard defines a mechanism to help a reader understand what is found within the file (the File Type and Reader Requirements boxes).

2.14.3   Packaging all this Extra Data

The primary purpose of the JPX file format is to provide a flexible container in which an image writer can store and organize all the data needed to provide meaning to the coded pixel values. JPX does this through a data structure called a box; a JPX file is really just a sequence of boxes, as shown in Figure 2.16. JPX also defines a special type of box, called a super box, for which the box contents (D) consists only of a sequence of boxes with no other data fields. Each box contains information about the type of data in the box and the length of the box, allowing a reader to navigate the file even if that particular reader does not understand the format or meaning of the contents of individual boxes.

Obviously, the standard defines several boxes, including those that contain 'header' information such as image height, width, and bit-depth, those that contain other metadata,

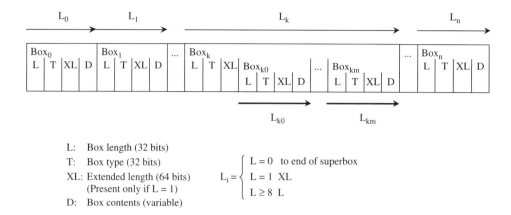

Figure 2.16   Navigating the boxes in a JPX file

and, of course, those that contain the actual coded image data, and in essence closely resemble the respective structures in other image file formats.

2.14.4   Specifying Color in JPX

JPX differs from other file formats in the way it specifies the colorspace of the coded image data. Most other formats take one of two approaches: they either define an enumerated list of specifically defined colorspaces or they allow the writer to use an ICC profile (ICC, 1998). These methods provide workable solutions in many applications, but both can cause problems, especially as images cross application domains; when colorspaces are added to the enumerated list, all applications that seek to support that new colorspace must be revised, and some ICC profiles are too computationally complex for some devices.

In addition, most other formats intermingle information about the way the 'color data' has been transformed to improve coding efficiency with information about interpreting the decoded data. For example, in the original JPEG standard, the use of the multiple component transform to convert RGB values to YCbCr for better encoding is intertwined with the specification of RGB as the colorspace of the image data. However, this caused problems when some digital camera vendors sought to encode values into JPEG files that fell within the YCbCr gamut but outside the sRGB gamut, forcing a revision to the Exif standard.

2.14.4.1   Encoding versus Interpreting Color

In the JPX file format, a clear boundary separates transformations performed to improve coding efficiency from processing performed to interpret (i.e. display or print) the decoded image data when the file is read. Figure 2.17 illustrates this division.

In JPX, the old YCbCr transform, now called the irreversible multiple-component transform, is clearly defined purely as an encoding step. While that transform is optimized for sRGB original data, it can be applied to data in any colorspace (albeit with less optimal

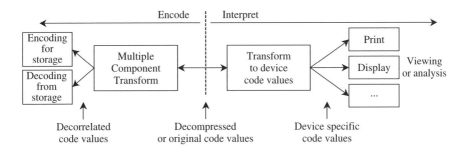

Figure 2.17   Separation of encoding color from interpreting color

results). In addition, the JPX format allows an enhanced component transform where the file specifies a particular gamma curve and custom transformation matrix. In applications that can use the JPX baseline compliance level, this can offer an easy way to improve compression for colorspaces other than sRGB.

2.14.4.2   Color Interpretation

Once the image data are decoded, a reader can begin to interpret the image data. JPX does allow both enumerated colorspaces (including values defined in the standard, values defined by registration, and values selected by individual vendors) and ICC profiles, but extends the color specification data structures in two ways to avoid the disadvantages.

The Restricted ICC Method

The first enhancement to color specification for interpretation is the addition of what JPX calls the restricted ICC method. The chief disadvantage of allowing any ICC profile to be embedded in the file is the computational complexity of some profiles. The restricted ICC method overcomes this disadvantage by defining a subset of the ICC profile standard that only includes the computationally simple profile types: monochrome input profiles and three-component matrix-based profiles. Using these profiles to convert image data to the profile connection space, and then to sRGB or other spaces used in constrained applications, requires at most the capability to apply a 3 × 3 matrix and a series of one-dimensional look-up tables. Restricted ICC profiles are fully compliant ICC profiles and can be processed using either an off-the-shelf ICC processing engine or a less complex, application-specific engine.

Using Multiple Color Specification Methods

The other disadvantage to most color specification systems is that the true meaning of the image data is often lost. Consider a case where a writer may use a nonstandard but still photographic colorspace. To ensure that all applications can properly display the photograph, the writer includes a restricted ICC profile mapping the custom space to the profile connection space. To allow enhanced applications to perform a more accurate display, the writer also includes a more complex ICC profile that includes 3D look-up tables. The writer may then include a third colorspace description, this time using a vendor colorspace code to represent the true meaning of the image data, allowing other applications from that vendor to perform advanced editing operations without data loss. The vendor colorspace method represents a fourth specification option to allow vendors to create new enumerated codes without involving a registration body.

Through these two enhancements, the JPX format provides mechanisms for colorspace specification that balance the flexibility required by a large number of application domains with the desire to maximize the number of applications (including constrained devices) that can make intelligent use of the image.

2.14.5   Metadata

In addition to header and colorspace 'metadata' the JPX format includes structures for embedding 'traditional metadata' in the file and properly associating that metadata with the correct element in the file (specifically needed if the JPX file contains multiple images). To allow for the specification and storage of the metadata fields themselves, the JPX format defines two boxes, one to contain an XML (W3C, 2006) instance document and one to contain metadata encoded using the MPEG-7 binary format (BiM) as defined by ISO/IEC 15938-1 (2002). These two boxes allow any data that can be represented in XML or in BiM to be embedded in the file and then to be extracted and processed by off-the-shelf XML or BiM tools. The location of XML boxes within the file, along with the use of the Association, Label, and Number list boxes, allows a file writer to associate a particular block of metadata with a particular element of the file (e.g. compositing layer) or the file as a whole.

Beyond these generic capabilities to store metadata, the JPX standard (Part 2 Annex N) defines several XML schemas for storing metadata related to digital photography, including information about image creation, content description, processing history, intellectual property, and image identification. ANSI/I3A IT10.2000 (2004) goes beyond the direct definitions in the JPEG 2000 standard and defines additional metadata fields to allow direct conversion of images from the Exif 2.2 format (JEITA CP-3451, 2002) to JPEG 2000 without loss of any metadata.

2.14.6   Other Features

The JPX format allows storage of either a single image or multiple images, as well as storage of instructions for the combination of those images through either composition or animation. The format also permits the encryption or other encoding of image and metadata at a basic level, although applications with serious needs for encryption should look to JPEG 2000 Part 8 for stronger solutions.

2.14.7   Summary

The JPX file format (including the JP2 compliance level) provides a strong foundation for the storage of image data and its associated meaning. Applications that choose to take advantage of the metadata and color specification capabilities can create new, breakthrough applications for sharing images across devices and across application domains.

2.15   Extended Capabilities Signaling

As more features were added in both Part 2 and later sections of the standard, it became clear that the Rsiz field would not be long enough to signal all of the code-stream features that might arise. Moreover, some features in other parts of the standard would extend Part 1 without any reference to Part 2. To anticipate these needs the designers created an extensible capabilities (CAP) marker segment. Amendment 2 (2004) to 15444-2 gives the overall syntax of CAP (ISO/IEC, 2006).

The second most significant bit of Rsiz signals the presence of extra capabilities that require reading a CAP marker segment. This bit can be used in conjunction with more typical Part 1 Rsiz profiles, or with the Part 2 Rsiz. Alternatively, it can be left off entirely if the coder wishes Part 1-only decoders to obtain all the information possible from the code-stream without making use of these extra features. Due to the optional nature of the Rsiz signaling for CAP, decoders that recognize capabilities signaled within CAP may choose to search actively for the presence of CAP. Encoders using CAP should facilitate this parsing by placing the CAP segment as close as possible to the front of the main header.

Use of the CAP segment is not required for the other features described in this chapter.

Acknowledgments

We wouldn't have been able to cover this broad range of topics without access to the standard document itself, as well as relying on memories of many JPEG committee discussions and the publications by the experts in each topic area. Our thanks to our many JPEG committee colleagues who contributed and helped refine and document the ideas in JPEG 2000 Part 2.

References

Ameli, J., Vaisey, J. and Jin, T. (2004) Selecting the don't care bits in JPEG 2000 ROI coding, in *SPIE Proceedings of Visual Communications and Image* Processing, vol. 5308, January 2004, pp. 1383–1390.

ANSI/I3A IT10.2000 (2004) *Photography – Digital Still Cameras – JPEG 2000 DSC Profile*.

Berthelot, B. (1999a) Report on core experiment CodEff1: reduced overlap in SSWT, ISO/IEC JTC1/SC29/WG1 N1189, March 1999.

Berthelot, B. (1999b) Report on core experiment CodEff_Interim2: memory benefits of SSOWT, ISO/IEC JTC1/SC29/WG1 N1466, November 1999.

Bilgin, A., Sementilli, P. J. and Marcellin, M. W. (1999) Progressive image coding using trellis coded quantization, *IEEE Tranactions. on Image Processing*, **8**, 1638–1643.

Brislawn, C. M. and Wohlberg, B. (2004) Lifted linear phase filter bands and the polyphase-with-advance representation, in IEEE 11th Digital Signal Processing Workshop, pp. 29–33.

Brislawn, C. M., Wohlberg, B. E. and Percus, A. G. (2003) Resolution scalability for arbitrary wavelet transforms in the JPEG-2000 standard, in SPIE Proceedings. on Visual Communications and Image Processing, vol. 5150, pp. 774–784.

Brower, B., Lan, A., Kasner, J. and Shen, S. (2000) Multiple component compression within JPEG 2000 as compared to other techniques, in SPIE Proceedings on Applications of Digital Image Processing XXIII, vol. 4115, pp. 544–551.

Christopoulos, C. A., Askelöf, J. and Larsson, M. (2000) Efficient methods for encoding regions of interest in the upcoming JPEG 2000 Still Image Coding Standard, *IEEE Signal Processing Letters*, **7**(9), 247–249.

CJIS (Criminal Justice Information Services) (1997) WSQ Gray-scale Fingerprint Image Compression Specification, Federal Bureau of Investigation Document No. IAFIS-IC-0110(V3), 19 December 1997.

Coifman, R. R. and Wickerhauser, M. V. (1992) Entropy-based algorithms for best basis selection, *IEEE Transactions on Information Theory*, **38**(2), 713–718.

Daly, S. J., Zeng, W., Li, J. and Lei, S. (2000) Visual masking in wavelet compression for JPEG 2000, in *IS&T/SPIE Proceedings of Image and Video Communications and Processing*, vol. 3974, San Jose, CA, January 2000, pp. 66–80.

DICOM (Digital Imaging and Communications in Medicine) (2006) Part 5: *Data Structures and Encoding*, PS 3.5-2006, National Electrical Manufacturers Association.

Fischer, T. R., Marcellin, M. W. and Wang, M. (1991) Trellis-coded vector quantization, *IEEE Transactions on Information Theory*, **37**, 1551–1566.

Fischer, T. R. and Wang, M. (1992) Entropy-constrained trellis-coded quantization, *IEEE Transactions on Information Theory*, **38**, 415–426.

Flohr, T. J., Marcellin, M. W. and Rountree, J. C. (2000) Scan-based processing with JPEG 2000, in *SPIE Proceedings of Applications of Digital Image Processing*, San Diego, CA, July 2000, vol. 4115, pp. 347–355.

Grosbois, R., Santa-Cruz, D. and Ebrahimi, T. (2001) New approach to JPEG 2000 compliant region-of-interest coding, in *SPIE Proceedings of Applications of Digital Image Processing*, vol. 4472, December 2001, pp. 267–275.

Hao, P. and Shi, Q. (2001) Matrix factorizations for reversible integer mapping, *IEEE Transactions on Signal Processing*, **49**(10), 2314–2324.

Houchin, S. (2005) JPEG 2000 Part 2 Draft Corrigendum 4 Study Text v5, ISO/IEC JTC1/SC29/WG1 N3724, August 2005.

ICC (International Color Consortium) (1998) ICC Profile Format Specification 1:1998-09. Available at http://www.color.org/icc_specs2.html.

ISO/IEC (2002) 15938-1 MPEG-7 Systems.

ISO/IEC (2004a) Information Technology – JPEG 2000 Image Coding System – Part 1: Core Coding System, ISO/IEC International Standard 15444-1, ITU-T Recommendation T.800.

ISO/IEC (2004b) Information Technology – JPEG 2000 Image Coding System – Part 2: Extensions, ISO/IEC International Standard 15444-2, ITU-T Recommendation T.801.

ISO/IEC (2006) Extended Capabilities Marker Segment, ISO/IEC International Standard 15444-2:2004/Amd 2.

JEITA CP-3451 (2002) Exchangeable Image File Format for Digital Still Cameras: Exif Version 2.2.

Joshi, R. L., Crump, V. J. and Fischer, T. R. (1995) Image subband coding using arithmetic coded trellis coded quantization, *IEEE Transactions on Circuits System Video Technology*, **5**, 515–523.

Kasner, J. H., Marcellin, M. W. and Hunt, B. R. (1999) Universal trellis coded quantization, *IEEE Transactions on Image Processing*, **8**, 1677–1687.

Kasner, J. H. and Rajan, S. D. (2004) JPEG 2000 enabled client–server architecture for delivery and processing of MSI/HSI data, in *SPIE Proceedings on Airborne Intelligence, Surveillance, Reconnaissance (ISR) Systems and Applications*, vol. 5409, pp. 139–154.

Kasner, J., Bilgin, A., Marcellin, M., Lan, A., Brower, B., Shen, S. and Wilkinson, T. (2000) JPEG-2000 compression using 3D wavelets and KLT with application to HYDICE data, in *SPIE Proceedings on Imaging Spectrometry VI*, vol. 4132, pp. 157–166.

Lepley, M. A. (2001) JPEG 2000 and WSQ image compression interoperability, MITRE Technical Report MTR 00B63. Available at http://www.mitre.org/work/tech_papers/tech_papers_01/lepley_jpeg2000/lepley_jpeg2000.pdf.

Marcellin, M. W. (1994) On entropy-constrained trellis coded quantization, *IEEE Transactions on Communications*, 14–16.

Marcellin, M. W. and Fischer, T. R. (1990) Trellis coded quantization of memoryless and Gauss–Markov sources, *IEEE Transactions on Communications*, **38**, 14–16.

Marcellin, M. W., Lepley, M. A., Bilgin, A., Flohr, T. J., Chinen, T. T. and Kasner, J. H. (2002) An overview of quantization in JPEG-2000, *Signal Processing: Image Communications*, **17**(1), 73–84.

Meyer, F. G., Averbuch, A. Z., and Stromberg, J.-O. (2000) Fast adaptive wavelet packet image compression, *IEEE Transactions on Image Processing*, **9**(5), 792–800.

Onno, P., Mozelle, G., Berthelot, B. and Felix, H. (1999) Testing report: lowpass/highpass filtering convention, ISO/IEC JTC1/SC29/WG1 N1432, September 1999.

Ramchandran, K. and Vetterli, M. (1993) Best wavelet packet bases in a rate-distortion sense, *IEEE Transactions on Image Processing*, **2**(2), 160–175.

Shen, S. and Kasner, J. (2000) Effects of 3D wavelets and KLT-based JPEG-2000 hyperspectral compression on exploitation, in *SPIE Proceedings on Imaging Spectrometry VI*, vol. 4132, pp. 167–176.

Shen, S., Kasner, J. and Wilkinson, T. (2001) New hyperspectral compression options in JPEG-2000 and their effects on exploitation, in *SPIE Proceedings on Imaging Spectrometry VII*, vol. 4480, pp. 154–165.

Taubman, D. S. and Marcellin, M. W. (2002) *JPEG 2000: Image Compression Fundamentals, Standards and Practice*, Kluwer Academic Publishers, Boston, MA.

Ungerboeck, G. (1982) Channel coding with multilevel/phase signals, *IEEE Transactions on Information Theory*, **28**, 55–67.

Wilkinson, T., Kasner, J., Brower, B. and Shen, S. (2001) Multi-component compression in JPEG 2000 Part II, in *SPIE Proceedings on Applications of Digital Imaging Processing XXIV*, vol. 4472, pp. 224–235.

W3C (World Wide Web Consortium) (2006) Extensible Markup Language (XML). Available at http://www.w3.org/XML/.

Zeng, W., Daly, S. and Lei, S. (2000a) Point-wise extended visual masking for JPEG-2000 image compression, in *Proceedings of the IEEE International Conference on Image Processing.*, vol. 1, Vancouver, Canada, September 2000, pp. 657–660.

Zeng, W., Daly, S. and Lei, S. (2000b) Visual optimization tools in JPEG 2000, in *Proceedings of the IEEE International Conference on Image Processing*, vol. 2, Vancouver, Canada, September 2000, pp. 37–40.

Zeng, W., Daly, S. and Lei, S. (2002) An overview of the visual optimization tools in JPEG 2000, *Signal Processing: Image Communication*, **17**(1), 85–104.

Further Reading

Houchin, J. S. and Hauf, C. (2000) JPEG 2000 file format provides flexible architecture for color encoding, in *SPIE Proceedings on Applications of Digital Image Processing XXIII*, vol. 4115, pp. 476–483.

Houchin, J. S. and Singer, D. (2000) JPEG 2000 file format: an imaging architecture for today and tomorrow, in *SPIE Proceedings on Applications of Digital Image Processing XXIII*, vol. 4115, pp. 455–463.

Houchin, J. S. and Singer, D. (2003) File format technology in JPEG 2000 enables flexible use of still and motion sequences, *Signal Processing: Image Communications*, **17** (1), 131–144.

3

Motion JPEG 2000 and ISO Base Media File Format (Parts 3 and 12)

Joerg Mohr

3.1 Introduction

One might wonder why a still picture standard like JPEG 2000 also covers a motion picture representation in two dedicated parts, especially as the Moving Picture Experts Group (MPEG) in the same standardization organization is offering a rich set of video compression tools. However, the application of JPEG 2000-based motion picture representation has already played quite early an important role in the process of standardization. In 1999, the anticipation of a prospering market of JPEG 2000-based digital still cameras (DSCs) also led to the consideration of a wavelet-based video recording format. This format should be equivalent to the simple block-based Motion JPEG formats in still cameras with a video recording function, but without complex motion compensating coding engines such as those present in the MPEG family.

Besides DSCs, the following uses were considered for such a simple wavelet-based video format (Fukuhara, 1999):

- remote surveillance systems;
- digital video recording systems;
- video capture cards.

By creating 'Motion JPEG 2000' as Part 3 of the JPEG 2000 (ISO/IEC, 2007) standardization framework, WG1 intended to avoid the situation encountered with the former Motion JPEG, which was never defined by an international organization and therefore now exists in a plurality of incompatible formats. To reduce complexity, only a simple intrapicture coding scheme was intended.

The JPEG 2000 Suite   Edited by Peter Schelkens, Athanassios Skodras and Touradj Ebrahimi
© 2009 John Wiley & Sons, Ltd

3.2   Motion JPEG 2000 and ISO Base Media File Format

Motion JPEG 2000 does not introduce new coding mechanisms; it is a file format or a *wrapper* for JPEG 2000-based motion picture representation. It belongs to the JPEG 2000 file format family and is derived from Apple's QuickTime file format (Apple, 2007). The MPEG-4 file format, which is standardized within the ISO by WG11 also uses many of the same mechanisms. Instead of maintaining two instantiations of an almost similar file format in different standardization groups, the SC29 Advisory Group on Management (AGM) requested after the 24th WG1 meeting in Stockholm that WG1 and WG11 create a common base text for the MPEG-4 and Motion JPEG 2000 file formats.

The intention was not that those two groups should agree on a common file format standard that would have to cover the different needs and cater for the various applications. Instead, an approach like in object-oriented development (OOD) has been chosen, where subclasses derive from superclasses. In the context of standardization this means that a specific standard derives from a parent standard or a specific standard extends a parent standard. De facto, the parent standard will not be used to define a file format (i.e. in terms of OOD it is 'virtual'), but the derived standards will inherit all its mechanisms and properties and will extend them by appropriate application-specific characteristics.

This approach allowed a common text to be maintained without constraining the working groups to unsuitable compromises. The parent standard is edited by both WG1 and WG11 and is manifested as the 'ISO Base Media File Format' in the respective Parts 12 of ISO/IEC 15444 (ISO/IEC, 2005b) and ISO/IEC 14496 (ISO/IEC, 2005a). Based on this parent standard, both working groups were standardizing 'their' specific file formats, either in Part 3 of ISO/IEC 15444 (ISO/IEC, 2007) or in Parts 14 and 15 of ISO/IEC 14496, respectively. Since its revision from 2003, Motion JPEG 2000 therefore references only the mechanisms and definitions of Part 12.

3.3   ISO Base Media File Format

3.3.1   Boxes

Like the other members of the JPEG 2000 family, the ISO Base Media File Format is formed as a series of objects, called *boxes*. With the boxes it is possible to describe and contain flexibly timed media information for a presentation independent of its encoding in an object-oriented format. Boxes can be nested in other boxes and the standards define exactly which box may or must appear in which parent box. By specification, all data are contained in boxes; no other data are supposed to be within the file.

Boxes start with a header which gives both size and type. The header permits compact or extended size (32 or 64 bits) and compact or extended types (32 bits or full UUIDs with 128 bits). The standard boxes all use compact types (32 bits) and most boxes will use the compact size, which allows enveloping up to 4 GB of data with a box. Another box type, the so-called *Full Box*, has an extend header, which provides additionally an 8-bit version number and a 24-bit wide flag field.

The size gives the entire length of the box, including the header and all contained boxes. This facilitates general parsing of the file. As an example, consider the illustration of a sequence of boxes in Figure 3.1, including one box that contains other boxes. The box types are represented by lower-case four-character codes to identify them. Internally

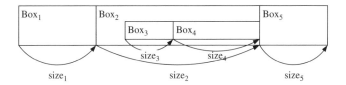

Figure 3.1 Illustration of box lengths. Figures taken from ISO 15444.1:2004–Information technology–JPEG 2000 image coding system are reproduced with the permission of the International Organization for Standardization, ISO. This standard can be obtained from any ISO member and from the Web site of the ISO Central Secretariat at the following address: www.iso.org. Copyright remains with ISO

Figure 3.2 Deriving a specific box from the abstract box class

these are simply 32-bit integers. In the following, a four-character string in single quotes, such as 'abcd', will represent them.

All kinds of boxes in the file structure are derived from the abstract base box class, which consists only of the header with size and type. The base box class and an example for a derivation is shown in the UML diagram for the *Media Data Box in* Figure 3.2, a box that encapsulates the media data.

Therefore, the *Media Data Box* consists of the header structure described above, with *type* set to the value 'mdat' (or 0x6d646174 in the hexadecimal integer reading) and *data* as 'payload.' In the case of very large and long presentations, the *Media Data Box* will probably use the header form with extended size (64 bits), which will be sufficient to contain more than 18 Exabytes.

3.3.2   File Structure

In order to identify clearly a member of the ISO Base Media File Format family and to determine the best use of the file (i.e. how to read and to interpret it), generally the *File Type Box* has to be the first box in the structure. It consists of a *brand field*, which can be considered equivalent to a major version number, a minor version number, and a compatibility list.

As described above, a box can be regarded as a general container for both media-data (the data that forms the presentation) and metadata (data that describes the media-data). With the distinction between those two types of data and a well-defined box order it is relatively easy to subset the media-data or to arrange it in an optimal way for different applications. Consider the following cases of its use:

- Content creation (recording). Different types of media (e.g. audio and video) will be stored either interleaved in one file or in separate files by just appending sample after sample.

At the end of the recording, metadata will be created that references and describes the recorded media-data, but without extensive reordering or copying of the data.

- Local presentation (playback). Often stored on media with slow access time (e.g. CD, DVD), interleaving of audio and video is useful to avoid time-consuming seeking. Other requirements are the ability for random access and support for simultaneous content switching, i.e. choosing different language tracks along the same video-track or otherwise, changing the perspective along the same audio-track, or both.
- Streaming. When a server operates from a file to make a streamed presentation, media-data need to be rearranged to follow the specifications of the protocol(s) used. Often instructions for the protocol need to be interleaved with the media-data so that excessive seeking can be avoided. However, on the other hand it would not be efficient to create a separate file for each possible streaming protocol. Therefore an effective mechanism for handling media-data and streaming information is needed.
- Editing. Media-data of different types and origin will be modified and rearranged to form a new presentation. This can be done by copying the selected media-data to a new file or by creating a small file consisting just of metadata, which references the media in the original files.

One essential mechanism of the file structure is the logical separation of media-data into so-called *tracks*. For media-data, a track corresponds to a sequence of images or sampled audio. Mostly, a presentation (also called a *movie*) consists of two tracks (one of each type), which will be played back simultaneously. Each track has a unique track identifier within the file and an associated handler (see the *Handler Reference Box* example above).

A track does not represent a physical structure for media-data, but is more a logical directory that references to the data. The same data can be referenced by multiple tracks in a file. Otherwise, a track can also link to data that are stored in external files, even without specific order. Special cases are the so-called *hint tracks*, which contain instructions for packaging one or more media tracks into a streaming channel, e.g. RTP (real-time transport protocol) or FLUTE (file delivery over unidirectional transport).

While tracks reference only media-data, these data need to be stored in an organized way. As shown by the cases of its uses above, different applications require different organization of the media, either continuous in one or multiple files or interleaved in patterns dependent on the underlying storage media or protocol needs. So-called *chunks* were introduced, which are just contiguous sets of media-data for one track.

3.4 Motion JPEG 2000

The preferred file extension for Motion JPEG 2000 is 'MJ2,' and in case of internet transmissions the used MIME type is 'video/mjp2.' The brand in the *File Type Box* shall be 'mjp2.' Like in all members of the JPEG 2000 family, the *JPEG 2000 Signature Box* ('jP') has to be the first in the file and therefore precedes the *File Type Box*.

3.4.1 Motion JPEG 2000 Samples

For Motion JPEG 2000, a sample is an individual frame of video or a time-contiguous section of audio (please note that the term *audio sample* in Motion JPEG 2000 may not have the same meaning as in an audio recording context where *audio frame* is used equivalently, as both decompose into a sequence of time-contiguous sample points). Sample data are contained in sample boxes – either in the media part of the same file or optionally in other files.

Additionally, samples are described in the metadata part of the file by *Sample Entry Boxes*. The latter are contained in one or more *Sample Description Boxes* for corresponding video and audio tracks. Motion JPEG 2000 specifies three types of sample entries: 'mjp2,' 'raw' (with space!), or 'twos,' which are described with their relevant parameters in the UML diagram shown in Figure 3.3.

The *Motion JPEG 2000 Visual Sample Entry Box* must consist at least of a Part 1 *JP2 Header Box* (see Section 1.6.3). A Motion JPEG 2000 visual sample itself is contained in a *Contiguous Codestream Box* (type 'jp2c'), which is also defined in Part 1. Compared to still images, the following particularities apply to reflect the special needs for motion picture media:

- If there is no *Field Coding Box* present, or the field count is 1, the sample contains precisely one code-stream box. Otherwise, if the field count is 2, then there have to be two code-stream boxes per sample.
- If two fields are present in the samples, the *JP2 Header Box* applies to the complete image, not to each field individually. Therefore the height as declared in the *JP2 Header Box* and the *Visual Sample Entry Box* applies to the entire de-interlaced image.
- If a *JP2 Prefix Box* exists, the code-streams presented to a decoder are formed by concatenating the prefix with the contents of the individual code-stream boxes. If field coding is used, the same prefix is concatenated with both code-stream fragments. Typically, the prefix will contain a JPEG 2000 main header.

The value *fieldorder* of the *Field Coding Box* describes the order of the two fields and is only relevant if *fieldcount* equals 2. This also applies to *original_fieldorder* in the *MJP2*

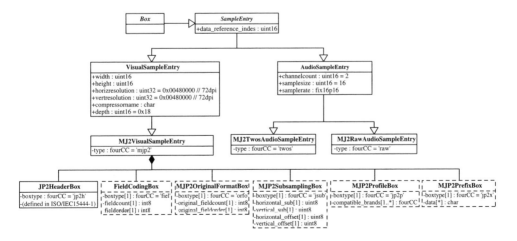

Figure 3.3   Class structure of the Motion JPEG 2000 *Sample Entry Boxes*

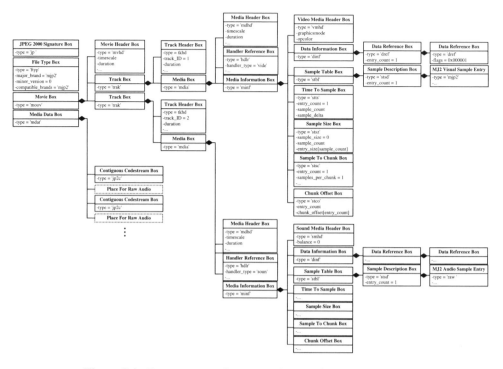

Figure 3.4   Box structure of an exemplary Motion JPEG 2000 file

Original Format Box. The latter is informative only and may assist readers in displaying or transcoding.

For audio tracks, Motion JPEG 2000 specifies only uncompressed samples. If stereo is stored, the data consist of interleaved left/right sample values. The raw format uses unsigned integer sample values with $2^{samplesize-1}$ indicating silence, while 'twos' uses signed integer sample values in two's complement with 0 as the neutral value.

Figure 3.4 shows an example for the box structure of a simple Motion JPEG 2000 file with one video and one audio track. The video and audio samples are organized in interleaved chunks with a granularity of one sample.

3.4.2   Profiles

The ISO Base Media File Format and therefore the derived Motion JPEG 2000 File Format offers such a wide range of flexibility to the user that unlimited playability and interoperability could only be achieved with very high efforts. Generally, ISO standards introduce profiles that restrict the parameters of a coding algorithm to a subset that will be widely understood. Currently, ISO/IEC 15444-3 defines two profiles. One is *unrestricted*, i.e. no limitations superimpose the standard. The *File Type Box* with the brand 'mjp2' signals this to the reader. The second profile is *simple*, which applies the following restrictions to the file:

- Only one video track is allowed. The frame rate of the video must not exceed 30 frames per second. The JPEG 2000 code-streams in the video track must follow Profile-0 (Rsiz = 1).
- At the most a single audio track, using only 8- or 16-bit raw audio, is present. The sample rate of the audio must not exceed 48 kHz.
- The file is self-contained, with all samples being in temporal order.
- If video and audio tracks are present, the media is organized in interleaved chunks with a granularity smaller than a second.
- More than one sample description entries within one track are disallowed.
- Only uniform scaling or right-angled rotation may be applied to the video.
- The profile is indicated in the *File Type Box* by the brand 'mj2s.'

Unfortunately, the simple profile is quite overrestricted for many purposes, as it disallows, for example, multichannel audio, higher frame rates or advanced Part 1 profiles, while the unrestricted profile cannot guarantee at least some level of interoperability. For this reason often other wrapper formats like MXF (Material eXchange Format) (ACR-NEMA, 1994) or DICOM (Digital Imaging and Communications in Medicine) (SMPTE 377 M, 2004) with specific JPEG 2000 restrictions are applied.

At the time of writing two additional profiles began to be standardized within WG1, both addressing the needs of professional film archives (see Section 9.5). While one will address the needs for exact preservation of different kinds of motion picture, the other will be optimized for easy and interoperable access to archived content.

3.4.3 Compliance Points and Testing

Compliance testing is the testing of a candidate product against both the compliance requirements in the relevant standard and the statement of the implementation's capability. For still images, ISO/IEC 15444-4 provides the framework, concepts, and methodology (see Chapter 17). Motion JPEG 2000 makes use of the latter framework of conformance testing and defines a number of so-called *compliance points*, which specify a range of Motion JPEG 2000 properties. To distinguish from profiles, the latter restrict encoders to generate code-streams with a dedicated parameter subset, while compliance points define minimum requirements for decoders (in the MPEG context, the word *level* is used equivalently to *compliance point*).

Currently four compliance points provided for different applications are defined:

- Cpoint-0. With a required image size decoding guarantee of 360×288 pixels, low demands on memory consumption and only 8-bit decoding precision, this Cpoint addresses portable low-cost applications with CIF or QVGA display.
- Cpoint-1 targets standard definition TV (SDTV) applications, demanding an image size decoding guarantee of 720×576 pixels and a decoding precision sufficient for 10-bit imagery.
- Cpoint-2 aims for high definition TV (HDTV) or cinema applications up to 1920×1080 pixels and 12-bit precision. Decoding of a 4th component, e.g. for alpha blending, is required.

- Cpoint-3 requires an image size decoding guarantee of 4096 × 3112 pixels and four components, as well as a decoding precision of up to 16 bits. Digital cinema applications in the acquisition or postproduction workflow are addressed.

It can be seen that Cpoint-2 and Cpoint-3 were intended for cinema applications but differ from DCI parameters (see Section 9.4). However, the latter were specified later in other terms, and – to be precise – define only code-stream profiles; i.e. exact compliance points for testing D cinema decoders were never standardized.

For each Cpoint an executable test set (ETS) is defined. Each ETS consists of motion sequence and single-image code-streams, reference decoded images, and tolerance values for MSE and peak error. An implementation under test (IUT) claiming compliance to a certain Cpoint is obliged to decode the relevant ETS fully and maintain the defined compliance point parameters.

3.4.4   Using Motion JPEG 2000

When using Motion JPEG 2000 some differences to Part 1 have to be considered. One is the potential use of subsampled components in the YcbCr or YUV colorspace. Though Part 1 does not recommend the use of chroma subsampling, this representation is very common in a TV and video context and is therefore reflected in the guidelines section (Annex B) of the Motion JPEG 2000 standard.

Another consideration is the introduction of new temporal artifacts. Of course, motion artifacts that result from exploiting temporal redundancy will not occur, but it has been proven that Motion JPEG 2000 is among other interframe codecs (e.g. H.264 I-frame) sensitive to a kind of temporal noise, the so-called *flickering*: 'Flicker artifacts are distinguishable on moving pictures rather than still pictures: as the phase of the signal changes on each frame, a temporal noise signal is produced which is visually more apparent than in the case of still pictures' (Kuge, 2002). Flickering is especially perceivable at lower bit-rates in addition to regular wavelet compression artifacts in static areas, i.e. parts with low spatial and temporal frequencies where a human observer is very sensitive to content variations. Research has been conducted to reduce the impact of this artifact, mainly by optimizing the postcompression rate control (PCRD) of the EBCOT mechanism. Currently the following schemes have been proposed:

- Frequency weighting, following the contrast sensitivity function (CSF). This scheme has already been introduced in Part 1 of ISO/IEC 15444 (Section J.12) (ISO/IEC, 2004) as a method to improve visual perceived quality. Also the guidelines section of Part 3 recommends the usage of CSF weights with respect to flickering artifact reduction.
- Block quantization based on interframe information (Becker *et al.*, 2004).
- Modified rate control for code-blocks that have been identified to be prone to flickering, i.e. representing mainly static content (Leontaris *et al.*, 2007).

All methods have in common that they affect only the encoder, need no extra signaling to the decoder, and therefore are fully compliant to the standard.

Generally, the interframe nature of Motion JPEG 2000 is reason for less compression efficiency compared to codecs from the MPEG family. However, this characteristic is beneficial where legal issues demand frame-accurate compression or frequent switching of the image source (e.g. between multiple cameras) can reduce the usability of interframe codecs. Beneath interframe coding, Motion JPEG 2000 also offers some features that have to be considered when choosing a compression methodology. If the achievement of extremely low bit-rates is not the primary target of video compression, but scalability, high dynamic range, or even true lossless encoding, Motion JPEG 2000 is often the preferred, if not the only, codec: 'Motion JPEG 2000 (MJ2) is one potential format for longterm video preservation. The format is attractive as an open standard with a truly lossless compression mode' (Pearson and Gill, 2005). Compared with traditional MPEG codecs, Motion JPEG 2000 does not only allow image material to be represented without color subsampling, and with more bits per component, but at higher bit-rates it will also achieve better quality by means of PSNR (Fößel *et al.*, 2003). See also Section 9.5 for archives in the digital cinema context.

Other valuable features of Motion JPEG 2000 are scalability and region of interest (ROI) coding. Figures 3.5 and 3.6 exemplify the principles in surveillance applications.

While scalability is exploited not only in surveillance but also in other motion picture contexts like, for example, digital cinema or archives (see also Chapter 9), ROIs are of special interest when only small portions of the monitored scenery need to be recorded or observed in high quality and the background may receive coarser compression for bit-saving reasons. ROIs can be defined statically before encoding or change dynamically with potential interesting objects. This feature was especially exploited in the WCAM project (see http://www.ist-wcam.org).

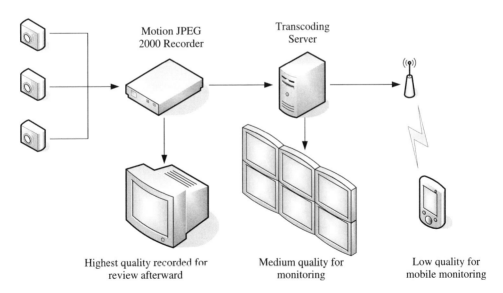

Figure 3.5   Scalability for surveillance applications with Motion JPEG 2000

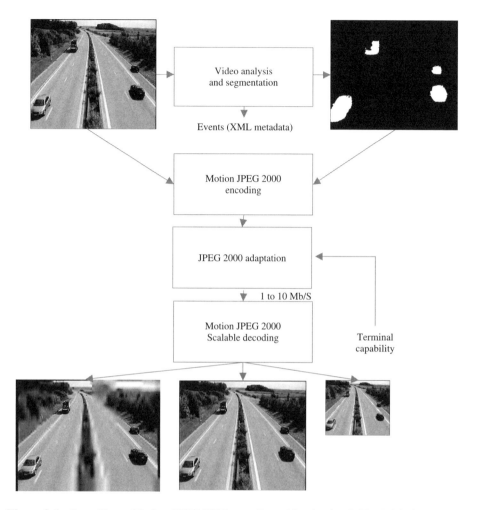

Figure 3.6   Surveillance Motion JPEG 2000 encoding with adaptive ROIs (Nicholson, 2005)

References

ACR-NEMA (1994) ACR-NEMA Digital Imaging and Communications in Medicine (DICOM) Standard V3.0, American College of Radiology, Network Communications Support for Message Exchange.

Apple, Inc. (2007) QuickTime File Format Specification.

Becker, A. *et al.* (2004) Flicker reduction in intraframe codecs, in *Proceedings of the IEEE Data Compression Conference*, pp. 252–261.

Fößel, S. *et al.* (2003) Motion JPEG 2000 for high quality video systems. *IEEE Transactions on Consumer Electronics*, **49**(4), 787–791.

Fukuhara, T. (1999) Proposal of subdivision ISO/IEC 15444-3: Standardization of Motion JPEG 2000.,WG1N1390.

ISO/IEC (2004) Information Technology – JPEG 2000 Image Coding System – Part 1: Core Coding System, ISO/IEC International Standard 15444-1.

ISO/IEC (2005a) Information Technology – Coding of Audio-Visual Objects – Part 12: ISO Base Media File Format, ISO/IEC International Standard 14496-12.

ISO/IEC (2005b) Information Technology – JPEG 2000 Image Coding System, – Part 12: ISO Base Media File Forma, ISO/IEC International Standard 15444-12.

ISO/IEC (2007) Information Technology – JPEG 2000 Image Coding System – Part 3: Motion JPEG 2000, ISO/IEC International Standard 15444-3.

Kuge, T. (2002) Wavelet picture coding and its several problems of the application to the interlace HDTV and the ultra-high definition images., NHK Laboratories Note No. 481.

Leontaris, A. *et al.* (2007) Flicker suppression in JPEG 2000 using segmentation-based adjustment of block truncation lengths, in *Proceedings IEEE International Conference on Acoustics, Speech and Signal Processing*, vol. 1, pp. 1117–1120.

Nicholson, D. (2005) WCAM project overview, WG1 Working Document WG1N3826.

Pearson, G. and Gill, M. (2005) An evaluation of Motion JPEG 2000 for video archiving, in *Proceedings of Archiving 2005*, pp. 237–243.

SMPTE 377M (2004) Television – Material Exchange Format (MXF) – File Format Specification.

4

Compound Image File Format (Part 6)

Frederik Temmermans, Tim Bruylants, Simon McPartlin, and Louis Sharpe

4.1 Introduction

The JPEG 2000 standard defines a family of file formats that enable the encapsulation of one or more JPEG 2000 code-streams. These file formats provide the necessary structures to access the encapsulated code-streams efficiently, along with extra metadata information needed to interpret these streams. Currently the JPEG 2000 standard specifies four file formats: JP2, JPX, MJ2, and JPM.

This chapter gives a description of the multilayer compound image file format (JPM or JPEG 2000 Multi-Layer), which is standardized as JPEG 2000 Part 6 (ITU-T Rec. T.805 | ISO/IEC 15444-6) (ISO/IEC, 2002). It serves as a file format to represent multi-page raster images containing a mixed content, such as photo images, text, or artificially generated figures. However, JPM does not support vector graphics as a content type, clearly distinguishing it from the PDF format. With JPM it is possible to represent a multipage mixed-content document in one or multiple files, compressing different parts with different compression algorithms. Among the supported compression algorithms are JPEG and JPEG 2000 for continuous-tone images and JBIG for bilevel imagery (ITU-T Rec. T.82 | ISO/IEC 11544).

It is well known that JPEG 2000 Part 1 performs very well on pure continuous-tone images. However, for mixed-content typed documents better compression strategies are possible by using JPM. For example, text on a solid background consists of only bilevel image data, which compresses very well with JBIG or FAX G4. Moreover, it is important to compress the text in a lossless mode in order to preserve the sharp edges, while a visually lossless compression mode often suffices for continuous-tone images. Applying specialized compression algorithms or even just different compression settings for each specific type of content in a document often improves the overall compression performance significantly. Many modern document handling applications, including those used for archiving and printing solutions, involve multipage documents with a variety of different image content types. JPEG 2000 Part 6 offers a standardized way for these applications to store, access, and retrieve this type of document efficiently.

Generally, potential applications of JPEG 2000 Part 6 can be found in the area of document management systems that have to deal with raster-based documents of mixed content. The purpose of JPEG 2000 Part 6 is to provide a standardized file format that allows the storage of documents in good quality in combination with an acceptable file size. Efficient compression is important in order to decrease storage costs and to allow an efficient transmission of the files. Such requirements exist in applications of image archiving, accounting, land registers, medical imagery, color and gray scanning, and publishing, to name just a few.

Most contemporary archiving applications use TIFF, and to a certain extent the PDF format, for storing documents of mixed raster content. Applied to color scans, both of these formats frequently produce results that are still too large. This makes them difficult to handle. Therefore, most document archiving is currently done in grayscale or even bilevel black and white (using Fax G4 compression within the TIFF format). However, in many documents, the color reproduction is important, even in cases where one would not expect it: different types of bank deposit slips, for example, often differ only by their color and are as such recognized in that way. Color often contains extremely important information. JPEG 2000 Part 6 typically produces a file size that is much smaller than the corresponding TIFF file and considerably smaller than a PDF file, making archiving of documents in color possible (Boliek and Wu, 2003).

One example of a targeted application of this standard includes the efficient archival of scanned newspapers, as they typically contain mixed content. In order to compress such a scanned document, it is first segmented into different objects such that each object contains either the text or the images. Note that the process of segmentation of the document is not within the scope of JPEG 2000 Part 6. The document can then be saved as a single JPM file. Within this JPM file, all the objects can be encoded with different compression techniques, depending on the content type. In the case where the document consists of multiple pages, all pages can be contained within a single JPM file or linked with each other across multiple JPM files.

Another example of a potential application can be found in the field of online search and retrieval systems (Jung and Seiler, 2003). A search engine could return the results of a user's query in the form of a new, automatically generated, JPM file, rather than returning just a list of links. The table of content of the document is the only local page in the JPM file. All other pages in the JPM file are remote objects, which contain links to certain pages of other JPM documents. Initially, downloading those pages is not necessary. Only when a user chooses to view a specific page is it downloaded and the remote page in the JPM file replaced with a local copy. After the user stops browsing, the local JPM file contains only the relevant pages that were visited.

The applications described here are only examples to illustrate the usefulness of JPEG 2000 Part 6. Other typical applications include document management systems (DMS), digital asset management (DAM), and geographic information systems (GIS). Obviously, one can imagine many other applications in a broad range of application domains where the usage of JPEG 2000 Part 6 could be an optimal choice.

4.2   The JPM File Format

As mentioned in the introduction, the JPM file format is part of a family of JPEG 2000 file formats. As such, a JPM file uses the exact same file format architecture that is specified in JPEG 2000 Part 1 (ITU-T Rec. T.800 | ISO/IEC 15444-1). It consists of a sequence of boxes, where each box comprises three fields: a *Length* field, a *Type* field and a *Data* field. The *Length* field and the *Type* field are both 32 bits, while the *Data* field has a variable size. For each box, the *Type* field determines how to interpret the content inside the *Data* field. A 'superbox' is a box that contains other boxes as its data. The only difference between the JPEG 2000 file formats lies with the allowed box types and their definition.

Figure 4.1 gives a schematic overview of the hierarchical structure of a generic JPM file. Boxes with a solid border are mandatory while boxes with a dashed border are optional. The schema illustrates only the containment relationship between the boxes in the file. A particular order of those boxes in the file is not generally implied. A complete overview of all boxes allowed within this standard is given in Section 4.7.

4.3   Mixed Raster Content Model (MRC)

4.3.1   Introduction

JPM stands for JPEG 2000 Multi-Layer, as previously mentioned. The multilayer aspects of the JPM file format originate from the ITU-T Recommendation T.44 Mixed Raster Content (MRC) specification. It describes a facsimile standard for scanned pages that contain a mixture of text and images. JPEG 2000 Part 6 provides a mechanism, similar to the one in ITU-T Rec. T.44, to compress scanned pages as multiple layers of content. Each layer uses a specialized compression algorithm matching its content characteristics. Layers that contain text data are compressed with a supported facsimile standard such as JBIG, while layers with continuous-tone image data are compressed with a continuous-tone image compression algorithm such as JPEG 2000 Part 1 or even traditional JPEG. By doing so, each layer in a page will be compressed with the algorithm that best matches the specific characteristics and requirements of its content. Consequently, this will result in improved compression performances and visual quality for raster images containing mixed content without having to design new compression methods.

4.3.2   Layout Object Generation

JPM files can consist of one or more pages. A single page can contain multiple layers, called Layout Objects. In turn, these Layout Objects comprise a raster image object, an opacity mask object, or both. The initial state of a page before rendering any Layout Objects is called the *BasePage*. A *PageImage* is an image created by rendering the *BasePage* and all the Layout Objects of that single page. The *BasePage* has the same width and height as the *PageImage* and is either transparent or solid colored. To create the final *PageImage*, the JPM decoder renders all Layout Objects sequentially onto the *BasePage*. The image created by rendering only the first k Layout Objects is denoted as *PageImage$_k$*. The unique Layout Object identifiers of the Layout Objects determine the rendering order. The Layout Object with an identifier value of zero contains a thumbnail of the page and as a consequence should not be rendered to create the *PageImage*.

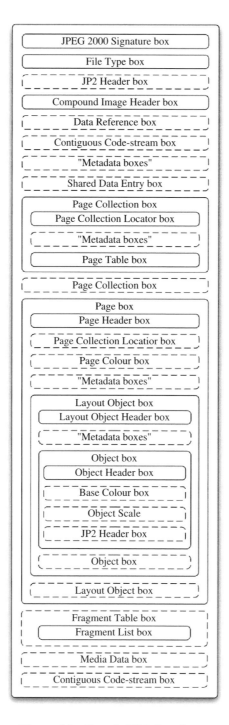

Figure 4.1   Generic JPM file schema

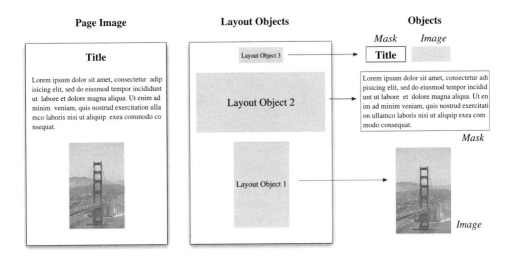

Figure 4.2   Example: a page with three Layout Objects

A JPM decoder renders each Layout Object by masking the raster image with the opacity mask. The opacity mask can be either bilevel or multilevel grayscale. For a bilevel opacity mask, the renderer uses only those pixels that have a value of 'one' at the corresponding coordinates in the opacity mask. For multilevel grayscale opacity masks, the renderer applies a linear blending technique to the corresponding *BasePage*. Figure 4.2 gives a simple example of how different layers make up a *PageImage* (Sharp and Buckley, 2003). Layout Object 1 contains only an image object and renders entirely in the final *PageImage*. Layout Object 2 shows black text through an opacity mask object alone. Finally, an opacity mask object, contained within Layout Object 3, renders the title of the page in the color of its corresponding image object.

In a JPM file, the Layout Objects are contained within a Layout Object box. This is a superbox that contains a Layout Object Header box, an optional Metadata box and one or two Object boxes. The Layout Object Header box contains general information about the Layout Object, such as the unique Layout Object identifier, the width (*LWidth*) and height (*LHeigth*) of the Layout Object, and a vertical (*LVOff*) and horizontal (*LHOff*) offset to determine the position of the Layout Object on the page. At the end of the Layout Object Header box the *Style* field is defined to indicate whether the Layout Object consists of a single image or an image and an opacity mask pair. To specify the image and the opacity mask of a Layout Object two approaches are possible:

1. Two separate objects define the two items.
2. The opacity mask is the last subobject of the image object.

The allowed values of the *Style* field are summarized in Table 4.1.

4.3.3   Layout Generation

The objects of a Layout Object are packed into Object boxes. An Object box has an Object Header box that describes the general properties of its object. The first field in

Table 4.1 Layout Object Header box *Style* field values

Value	Meaning
0	Separate objects for image and mask components
1	Single object for image and mask components
2	Single object for image components only (no mask)
3	Single object for mask components only (no image)
255	User-specified Layout Object

the Object Header box contains the type of object, indicating whether the box contains a raster image or an opacity mask for the Layout Object, or both. The *NoCodestream* field flags the presence of a compressed code-stream. In the case that the Layout Object contains no code-stream, the image is artificially generated by using a single uniform color value, specified in a Base Color box. The *OVoff* and *OHoff* fields respectively specify the vertical and horizontal clipping offsets, as explained in Section 4.3.4. Next, two fields signal the offset and length of the Contiguous Code-stream or Fragment Table box that is associated with the object's image, as explained in more detail in Section 4.4. The last field in the Object Header box is the data reference field, which specifies the data file or resource that contains the Fragment Table box. If the value of this field is zero, then the fragment is contained within this file or the data are contained in a Contiguous Code-stream box within this file. Otherwise, the Fragment Table box is contained within the file specified by this index into the Data Reference box. Apart from the Object Header box and the Base Color box, an Object box can also contain a Metadata box, a Label box, a Scale box, and a JP2 Header box. All of these boxes are optional. Metadata boxes are discussed in more detail in Section 4.6. A Label box can be used to assign a label to an object, which can be used to reference the object, as explained in Section 4.5. The Scale box specifies the scaling of the object before applying it to the page. The JP2 Header box exists when the *NoCodestream* field in the Object Header box indicates that the object contains a code-stream.

4.3.4 Object Clipping and Positioning

As illustrated in Figure 4.3, a JPM decoder applies four steps in order to render and position a Layout Object on a page:

1. First, the JPM decoder scales the object. The scaling factor of the horizontal and vertical directions is independent of each other. Thus, it is possible to change the aspect ratio of the rendered image, encoded in the Layout Object. The parameters that determine the scaling are defined in the optional Scale box of the Object box. The JPEG 2000 Part 6 specification does not specifically describe how any scaling should take place. The most appropriate method may depend on a number of factors. Particular file formats may offer support for generating the required reduced resolution images. The resolution-progressive JPEG 2000 file format often allows the required reduced resolution image to be generated without decoding the complete stream. Even if the precisely required scaled image cannot be generated in this way, decoding a reduced

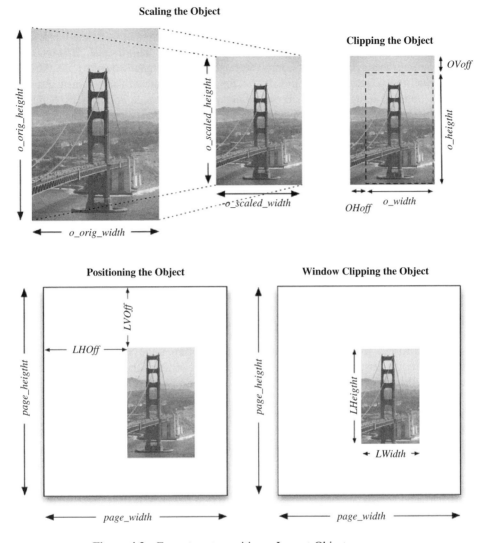

Figure 4.3   Four steps to position a Layout Object on a page

resolution image and then scaling this to the requested size may be significantly more efficient.

2. Second, the JPM renderer clips the Layout Object at the top and left sides.
3. Then, the JPM decoder positions the object on the page. The *LHoff* and *LVoff* fields defined in the Layout Object Header box specify the distance in grid units from the left and top to the upper left corner of the image respectively.
4. Finally, as a fourth step, the JPM decoder clips the object to its layout window (i.e. it clips right and bottom sides), which is bounded by the width and height values in the Layout Object Header box.

4.3.5 Blending

After clipping and positioning the opacity mask and raster image objects, the mask object defines how to blend the image object with the underlying page image. However, just before blending the Layout Objects onto the final page, all images need to use the same colorspace as the page. For example, if the final page image has to be in a grayscale colorspace, then all image objects are first converted to grayscale colorspaces. Similarly, for a color page, all grayscale or bilevel images need conversion to the specific colorspace of that page. Subsequently, after performing all the necessary colorspace conversions, the values in the opacity mask define the relative contribution of the respective pixels in the image object to the final page image. An opacity mask value of zero indicates that only the page image pixel value should be used, while the maximum possible opacity value indicates that only the image object value should be used. Opacity values in between indicate that linear interpolation between the page image and image object pixel values is to be used. JPEG 2000 Part 6 allows for an implementation to reorder or combine one or more of the various clipping, positioning, color conversion, and blending steps freely in order to obtain the same end result more efficiently.

4.3.6 Page Organization and Collections

As previously mentioned, a JPM file typically consists of a sequence of pages. These pages are organized in multiple page collections. JPM allows logical nesting of page collections to form a tree structure. In such a tree structure, pages are leaf nodes and page collections are regular tree nodes. The root of the tree represents the Main Page Collection of the JPM file. Thus, a JPM file has exactly one Main Page Collection that references the other subsidiary page collections. The main purpose of the Main Page Collection is to list all pages comprehensively within the JPM file. Doing so enables efficient and random access to arbitrary pages within the JPM file. The parent page collection of a page or a page collection is also called its respective primary page collection.

Within a JPM file, every page collection is contained within a Page Collection box. A Page Collection box contains a Page Table box and a Primary Page Collection Locator box. The page table is just an array of references to the pages and page collections it contains. The Primary Page Collection Locator box, on the other hand, points to the position of its page collection in the page table of its primary page collection. Pages themselves are packed into Page boxes. Again, a Primary Page Collection Locator box contains a pointer to the relative primary page collection of the page. The Primary Page Collection Locator in pages and page collections enables support for various features such as previous and next page commands.

Figure 4.4 gives an example that illustrates the relationships between page collections and pages (Buckley and Reid, 2003). In the example, PC A is the main page collection that directly or indirectly references all the pages in the file. As shown, PC A directly references pages P0, P8, and P9 and page collections PC B, PC C, and PC E. Consequently, PC A is the primary page collection of these pages and page collections. Thus, the page table of PC A contains the locations of the Page boxes for P0, P8, and P9 and the Page Collection boxes for PC B, PC C, and PC E. The Primary Page Collection Locator boxes of these pages and page collections will point to their corresponding elements in the page table of PC A.

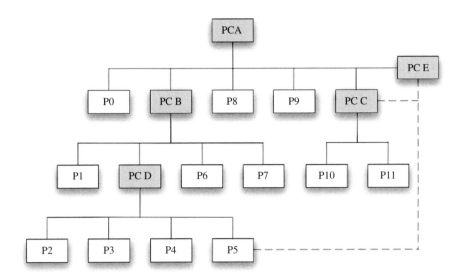

Figure 4.4   A tree of page collections and pages

The order of the pages in a JPM file is determined by the order in which the pages and page collections occur in the page tables of that file. In Figure 4.4, the order of the pages corresponds with the indexes of the pages as used in the example. Page collection PC E is an auxiliary page collection. Auxiliary page collections are page collections that contain pages that meet certain criteria. Such a page collection can, for example, hold all pages that contain a specific keyword. In this example, page collections PC E and PC D both point to page P5. However, PC D is the only primary page collection of page P5, because pages or page collections have only one primary page collection by definition. As mentioned before, pages are represented by Page boxes in a JPM file. Page boxes consist of a Page Header box, containing general information about the page, a Page Collection Locator box that was discussed earlier, an optional Base Color box, describing the base page color, optional Metadata boxes, and one or more Layout Object boxes. Pages occur at the top level of the file format. This means incremental updates to the file are accomplished by simply appending new pages to the end of the file.

4.4   Streaming JPM Files

The code-stream data for the image objects of a page may either be represented by a single contiguous code-stream or be composed of one or more code-stream fragments. For fragmented code-streams, an additional fragment table provides efficient access to the different chunks. A fragment table contains references to sequences of bytes in the current JPM file, in external files, or within Internet resources specified via a URL. This mechanism allows fragments to be placed in an optimal position to support streaming of the data from a server to a client application. It also enables progressive refinement of multiple code-streams simultaneously. A client application may, for example, initially present a rough incomplete version of a served image, eventually filled with thumbnails, and let the user indicate which parts should be downloaded in greater detail first. Fragments start or stop at any arbitrary byte boundary within a code-stream. For each fragment, a length parameter in the fragment table specifies its length. If a fragment does not end at a point in the code-stream where a full code-block or raster-line can be decoded, the decoding application catches the current data until the next fragment is retrieved. The next fragment of the byte-stream is then appended to the residual data of the prior fragment before the decoding process resumes. In this way, the decoder is able to handle code-streams with any type of compression algorithm. Fragment table entries always point to code-streams. Required header information to decode the code-stream, such as the width and the height of the raster image object, the compression type, or the bit depth, is stored in fields in the Object box.

4.5   Referencing JPM Files

The JPEG 2000 Part 6 specification defines a mechanism that allows referencing subelements of JPM files, such as Pages, Images, and Metadata boxes, with URLs. These URLs can be external or internal to the JPM file. As such, the URL specification enables the creation of links within a JPM file. However, it also enables external references, such as web page links, to link to interior elements of a JPM file. A JPM decoder renders just the requested portion of a JPM file when receiving a query from an external source.

The format of the URLs used in JPM is identical to the format used for HTML or XML URLs. A question mark ('?'), appended to the filename, separates the path and filename from the extra parameters. An equal sign ('=') joins attributes and values into pairs. The pairs are all separated by an ampersand sign ('&'). A single Layout Object within a page of a JPM file on path 'path' and with filename 'filename' can be referenced as follows:

```
path/filename.jpm?page=3&obj=32
```

Here, `page=3` references the third page and `obj=32` specifies Layout Object 32 as the target. The metadata of an element in a JPM file can be requested by adding the `type=meta` attribute value pair. In cases where multiple Metadata boxes occur at the same level of a JPM file, an index parameter becomes necessary to indicate which Metadata box to return. For example, when the index is assigned to one, the first Metadata box is returned. In cases where multiple Metadata boxes exist at the referred level but no index parameter is given, the first Metadata box is returned by default. Through the `mtype` parameter it is possible to return the MIME type of the specific Metadata box. The default MIME type is XML. An example metadata reference goes as follows:

```
path/filename.jpm?page=77&obj=19&type=meta&index=1
```

All boxes in a JPM file can have an optional Label box. Labelled boxes can be requested by specifying their label in the URL through the `label` parameter. Page Collections that contain a Label box can be referred to by using the `coll=label` parameter. The main page collection of a JPM file can be referred to by using the reserved label identifier 'main.' However, it is unspecified as to how an application should return the page collection to the client, e.g. as a textual page listing or as graphical page thumbnails. The `type=thumb` or `type=img` attribute value pairs specify whether a page is returned as a thumbnail or a fully rendered image. Finally, JPEG 2000 Part 6 also allows referencing to a specific byte range of a JPM file, via the use of the `off` and `len` parameters. The first parameter specifies an offset in bytes from the beginning of the file and the second parameter specifies the number of bytes to be returned. It is obvious that byte-ranged referencing should be used with care, since the subelements of the targeted JPM file might move due to file growth or modifications. Therefore, generally the methods mentioned earlier in this section are preferable. However, this method can be used in cases where the decoder already has enough information from the file to know where a subelement of interest is located.

4.6 Metadata

An important aspect of the JPM file format is the powerful metadata architecture. Metadata can be attached to any structural element in a JPM file, including the file or document itself, page collections, pages, Layout Objects, and objects. This enables, for example, property rights or vendor specific information to be associated with subregions of the JPM file. The metadata is contained in a Metadata box within the data field of the respective container box. Metadata boxes are optional in a JPM file and can be safely ignored by conforming readers. Whenever there is a reference to metadata in a box definition, it is understood that the box may contain an XML box with metadata or a box that contains vendor-specific information, called a UUID box.

4.7 Boxes

Table 4.2 gives an overview of all boxes that can be used within a JPM file (ISO/IEC, 2002).

4.8 Profiles

A profile defines the legal values or ranges of values for various fields in a JPM file. The profiling mechanism is used to identify subsets of the general architecture for use in specific applications. For example, a profile suited for web browsing of documents uses JPEG 2000 and JBIG 2 compression. Other profiles exist that provide compatibility with the three-layer model found in the MRC and TIFF-FX color fax formats.

4.9 Conclusions

This chapter discussed the multilayer compound image file format called JPM, which is part of a family of JPEG 2000 file formats. As described, the file format can be used for the encapsulation of multipage raster images containing a mixed content, such as photo images, text, or artificially generated figures. One important feature of the file format is that it allows different parts of the image to be compressed with different, already existing, compression algorithms. This enables a much more efficient overall compression for documents with a mixed raster content. This chapter also illustrated multiple applications for JPM, varying from digital archiving systems to online retrieval systems. Technically, it explained how the file format is composed. Starting at the box level, that is the foundation of every member of the JPEG 2000 family of file formats,

Table 4.2 Overview of boxes used within a JPM file

Box name	Type	Superbox	Comments
Base Color	'bclr' (0x6263 6C72)	Yes	This box contains all the information specifying the color of a page or of an object for which no image data are available.
Base Color Value	'bcvl' (0x6263 766C)	No	This box specifies the component values for the color of a page or of an object for which no image data are available.
Color Specification	'colr' (0x636F 6C72)	No	This box specifies the colorspace of an image.
Compound Image Header	'mhdr' (0x6D68 6472)	No	This box contains the compound version.
Contiguous Codestream	'jp2c' (0x6A70 3263)	No	This box contains a valid and complete JPEG 2000 code-stream.
Cross-Reference	'cref' (0x6372 6566)	No	This box specifies that a box found in another location (either within the JPM file or within another file) should be considered as if it was directly contained at this location in the JPM file.
Data Reference	'dtbl' (0x6474 626C)	No	This box contains a set of pointers to other files or data streams not contained within the JPM file itself.
File Type	'ftyp' (0x6674 7970)	No	This box specifies file type, version, and compatibility information, including specifying if this file is a conforming JPM file.
Fragment List	'flst' (0x666C 7374)	No	This box specifies a list of fragments that make up one particular code-stream within the file.
Fragment Table	'ftbl' (0x6674 626C)	Yes	This box describes how a code-stream has been split up or fragmented and then stored within the file.
Free	'free' (0x6672 6565)	No	This box contains data that are no longer used and may be overwritten when the file is updated.
Image Header	'ihdr' (0x6968 6472)	No	This box specifies the size of the image and other related fields.
JP2 Header	'jp2h' (0x6A70 3268)	Yes	This box contains a series of boxes that contain header-type information about a file containing a single image.
JPEG 2000 Signature	'jP\040\040' (0x6A50 2020)	No	This box uniquely defines the file as being part of the JPEG 2000 family of files.
Label	'lbl\040' (0x6C62 6C20)	No	This box specifies a textual label for a Page box or a Page Collection box.

Table 4.2 *continued*

Box name	Type	Superbox	Comments
Layout Object	'lobj' (0x6C6F 626A)	Yes	This box contains the information and image data needed to composite a mask-image object pair.
Layout Object Header	'lhdr' (0x6C68 6472)	No	This box describes the properties of the Layout Object and assigns each a unique identifier within the page.
Media Data	'mdat' (0x6D64 6174)	No	This box contains generic media data, which is referenced through the Fragment List box.
Object	'objc' (0x6F62 6A63)	Yes	This box contains all the image data and information for one object in the Layout Object.
Object Header	'ohdr' (0x6F68 6472)	No	This box describes the properties of an object, identifying it as a mask, an image, or a combined mask/image object, and assigns each object a unique identifier within the file.
Object Scale	'scal' (0x7363 616C)	No	This box describes the scaling for an object before applying it to the page.
Page	'page' (0x7061 6765)	Yes	This box contains all the information needed to image a page including references to code-stream data.
Page Collection	'pcol' (0x70636F6C)	Yes	This box groups together the locations of a set of pages so that they are treated as related and associated with each other.
Page Header	'phdr' (0x7068 6472)	No	This box describes the properties of a page and gives the number of Layout Objects in the page.
Page Table	'pagt' (0x7061 6774)	No	This box gives the locations of the pages in a page collection and enables random access of pages in a file.
Primary Page Collection Locator	'ppcl' (0x7070 636C)	No	This box gives the location of the primary page collection for the page collection.
Shared Data Entry	'sdat' (0x7264 6174)	Yes	This box contains a box that can be referenced by an identifier from multiple places within a file.
Shared Reference	'sref' (0x7372 6566)	No	This box can be used to insert a box in the file by reference to a previous occurrence of the box in the same file.
UUID	'uuid' (0x7575 6964)	No	This box contains vendor-specific information.
UUID Info	'uinf' (0x7569 6E66)	Yes	This box contains additional information associated with a UUID.
XML	'xml\040' (0x786D 6C20)	No	This box contains vendor-specific information in XML format.

over to the organization of objects, pages, and page collections, and finishing with the powerful metadata architecture. Therefore, it is true that the compound file format of JPEG 2000 offers a rich set of features and functionalities that are usable by a wide range of applications.

References

Boliek, M. and Wu, G.K. (2003) JPEG 2000-like access using the JPM compound document file format, in *International Conference on Multimedia and Expo (ICME '03)*, vol. 1, pp. 357–360.

Buckley, R.R. and Reid, J.W. (2003) A JPEG 2000 compound image file reader/writer and interactive viewer, in *Proceedings of SPIE on Applications of Digital Image Processing XXVI, 2003*, vol. 5203.

ISO/IEC (2002) JPEG 2000 Information Technology – Image Coding System – Part 6: Compound Image File Format, FDIS ISO/IEC 15444-6.

Jung, K. and Seiler, R. (2003) Segmentation and compression of documents with JPEG 2000, *IEEE Transactions on Consumer Electronics*, **49**(4).

Sharpe, L.H. and Buckley, R. (2000) JPEG 2000.jpm file format: a layered imaging architecture for document imaging and basic animation on the web, in *Proceedings of SPIE on Applications of Digital Image Processing XXIII*, vol. 4115.

5

JPSEC: Securing JPEG 2000 Files (Part 8)

Susie Wee and Zhishou Zhang

5.1 Introduction

5.1.1 Overview

JPSEC, Part 8 of the JPEG 2000 family of standards (ISO/IEC, 2007), specifies ways of applying security to JPEG 2000 coded images. JPSEC provides three types of security tools in the normative part of the standard: confidentiality, authentication, and integrity. While these security tools are common to many security applications, JPSEC's unique contribution lies in defining how these tools are applied to media. Specifically, JPSEC applies security tools to JPEG 2000 images in a manner that respects the structure of the coded media; in other words, JPSEC specifies how to apply security tools to JPEG 2000 images in a media-aware way.

5.1.2 Media-Aware Security

The traditional way to apply security to media is to treat the media file like any other data file and secure the entire file in a media-unaware way. If a security tool such as encryption is applied in a media-unaware way, then any structure in the media-data would be lost.

However, some structure in the media-data can be quite useful. For example, scalable coding methods code images into a bit-stream that has a structure that makes it easy to access a low-resolution version of the image, without requiring one to decode the entire high-resolution bit-stream. If these image data are encrypted for confidentiality in a media-unaware way, then the ability to extract the low-resolution version would be lost,

The JPEG 2000 Suite   Edited by Peter Schelkens, Athanassios Skodras and Touradj Ebrahimi
© 2009 John Wiley & Sons, Ltd

or it would require you to decrypt the entire bit-stream and then extract the low-resolution data. However, once the image is decrypted, it is no longer secure.

On the other hand, if media-aware security tools are used, then security can be applied in a way that preserves the useful structure in the media. JPSEC recognizes the fact that media-data actually have some useful structure to them and specifies how to apply security tools to JPEG 2000 images in a manner that considers the structure of the media.

5.1.3 Scalable Coding of Media: The Structure of JPEG 2000 Image Data

While JPEG 2000 provided compression improvements over JPEG and other image coding standards, one of the biggest advantages of JPEG 2000 over other image coding standards is its built-in scalability. The JPEG 2000 image coding standard was designed to be scalable, both in the compression methods that it employs and the structure in which the scalable data are organized in the bit-stream (ISO/IEC, 2000).

Scalable coding methods code media (images, video, or audio) in a manner that makes it easy to extract and decode different versions of it. For example, a scalably coded image can easily be decoded in high or low resolution. Decoding the image in high resolution involves decoding the entire bit-stream. Decoding the image in low resolution simply involves extracting the segments of the coded media data that contain the low-resolution data and decoding only those segments. This greatly decreases the computing power and memory needed to decode a low-resolution image.

JPEG 2000 was designed in a way that makes it very easy to extract and decode a resolution, a tile, a color component, or a quality layer of the coded image. This can be done by simply scanning the bit-stream, identifying and extracting the desired segments of the bit-stream, and decoding those segments. This ability of transcoding to a lower fidelity version of the image by simply extracting the appropriate portions of the image is very useful in many applications.

5.1.4 Example Application for Scalable Images

Suppose that a server stores a very large, high-resolution image and a client with a smaller display would like to look at and virtually navigate around this image. Because the client's display resolution is much smaller than the original image resolution and because the bandwidth between the client and server may be limited, the options are to serve the client a small portion of the image in full resolution or the entire image in low resolution. In order to do this, the server would have to extract a portion of the image or extract a low-resolution version of the image.

This can be achieved in different ways. If regular image coding is used, then the server would have to decode, process (select an area or downsample the image), and encode the image or transcode it accordingly. On the other hand, if scalable image coding is used, then it is very easy to extract portions of the image in different resolutions. Transcoding to lower resolutions or smaller tiles simply involves extracting the appropriate set of coded data. This requires very little computation, so it allows the server to support simultaneous image streaming sessions for many clients.

5.1.5   Applying Security to Media

A question that arises is what happens if end-to-end security is required for the application? For example, what if the application requires the image to be encrypted at the source and decrypted only by people who are allowed access? If this is required, then when the media-data are transported between the sender and receiver, it must remain encrypted at all times, including when it is stored on the server.

When the media-data are encrypted, what happens to the nice property of being able to stream adaptively portions of the high-resolution original image to lower-resolution clients? If the media is encrypted in a media-unaware way, then this property is lost, or the only way to stream adaptively is by decrypting the image, but this breaks the end-to-end security of the system.

On the other hand, if media-aware security is used, then the security tools can be applied to the media-data in a media-aware way in order to preserve the structure of the protected media and allow the server to stream the appropriate portions of the protected media data adaptively (Wee and Apostolopoulos, 2001a).

5.1.6   JPSEC Media-Aware Security Tools

As mentioned earlier, JPSEC was designed to provide media-aware security tools for JPEG 2000 images. It recognizes the structure of the JPEG 2000 image data and secures the media-data within that structure. Specifically, it recognizes where the media-data are located and which parts of the data correspond to which image components (tile, resolution, quality layer, color component, or image subband). It then allows security tools to be applied to subsets of the image data and specifies the signaling data that must be included in the protected bit-stream to allow the protected subsets of data to be extracted. By using this approach, JPSEC simultaneously allows mid-network transcoding and end-to-end security (Wee and Apostolopoulos, 2001b, 2003).

5.2   JPSEC Security Services

5.2.1   Overview

The JPEG 2000 Part 8 Security standard (JPSEC) incorporates the following six security service requirements into the standard:

1. Confidentiality via encryption and selective encryption.
2. Integrity verification (including data and image content integrity, using fragile and semi-fragile verification).
3. Source authentication.
4. Conditional access and general access control.
5. Registered content identification.
6. Secure scalable streaming and secure transcoding.

The 6th service is a new security service requirement (i.e. a nontraditional security service) adopted by the JPSEC committee. The JPSEC standard was defined in a way that allows significant flexibility and a high level of security through the use of protection tools and associated signaling that are applied to JPEG 2000 coded images. These protection tools include decryption, authentication, and integrity:

- *Confidentiality* ensures that all or part of a JPEG 2000 stream is only accessible by those authorized to have access.
- *Integrity* allows one to verify the integrity or correctness of all or part of a JPEG 2000 stream.
- *Authentication* allows one to verify the authenticity of all or part of a JPEG 2000 stream.

JPSEC does not invent its own security methods, but it leverages those that are established and proven as secure by the broader security community. Building upon this, JPSEC's key innovation lies in specifying how security methods are applied to scalable media, in particular to JPEG 2000 imagery. JPSEC allows JPEG 2000 images to be protected in a manner that respects the structure of the JPEG 2000 data. We now briefly discuss each security service.

5.2.2   Confidentiality Service

The JPSEC confidentiality service allows one to ensure that all or part of a JPSEC stream is only accessible by those authorized to have access. The JPSEC confidentiality service is based on cryptography – the JPEG 2000 image data are encrypted with an encryption key; and only those who have the appropriate decryption key can decrypt the stream. Ensuring confidentiality requires using a secure encryption/decryption method and a secure key exchange method. JPSEC does not define its own encryption methods; rather, it provides a way to specify the decryption method that must be used. Furthermore, JPSEC does not define or specify key exchange methods; rather, it provides a way to specify information about the decryption key needed to decrypt the JPSEC-protected stream.

5.2.3   Integrity Service

The JPSEC integrity service allows one to verify the integrity or correctness of all or part of a JPSEC stream. The JPSEC integrity service uses hash-based authentication without a key to generate a MAC value that can be used to verify the integrity of the JPSEC data. Note that this service does not verify authenticity, as it does not protect the stream from malicious attacks. Rather, the integrity service verifies correctness and thus helps protect the JPSEC stream from inadvertent bit errors that may occur.

5.2.4   Authentication Service

The JPSEC authentication service allows one to verify the authenticity of all or part of a JPSEC stream. The JPSEC authentication service uses standard key-based authentication methods. This service provides a way to specify the authentication/verification method that must be used to authenticate the stream. This service requires using a key to protect the JPSEC stream from malicious attacks; thus it also provides a way to specify information about the key that must be used to authenticate the stream.

5.3   JPSEC Architecture

JPSEC provides security services for JPEG 2000 by defining a number of protection tools that can be applied to JPEG 2000 bit-streams (Dufaux *et al.*, 2004; Wee and Apostolopoulos, 2004). A JPSEC system consists of a JPSEC creator, JPSEC bit-stream, and JPSEC consumer. The JPSEC creator can apply one or more protection tools to an image. The resulting JPSEC bit-stream contains signaling information for the protection tools and the modified data that may have resulted from their application. The signaling information is placed in an SEC (security) marker segment that is added to the JPEG 2000 header, and includes the parameters that a JPSEC consumer needs to interpret and process the protected stream. The JPSEC bit-stream contains three general types of information to describe to the JPSEC consumer: the what, where, and how of the applied security services.

5.3.1   What Security Service Is Provided?

The JPSEC syntax has three types of security tools: template tools, registration authority tools, and user-defined tools. The *template tools* are defined by the normative part of the JPSEC standard and are discussed in this chapter. They have an identifier that specifies which protection method template is used. JPSEC provides templates for decryption, integrity, and authentication. The *registration authority tools* are registered with and defined by a JPSEC registration authority and have an ID number that is specified in the syntax. The *user-defined tools* are defined by a user or application. JPSEC reserves a set of ID numbers that can be used by private applications. However, ID collisions may occur if the same ID number is used by different JPSEC applications, so the user must be careful in defining the use and scope of these streams. Both the registration authority and user-defined tools enable the application of proprietary protection methods, e.g. new techniques or classified government security techniques. The remainder of this discussion focuses on JPSEC template tools as defined by the normative part of the standard.

5.3.2   Where Is the Security Tool Applied?

JPSEC uses a zone of influence (ZOI) to describe where the security tool is applied. The ZOI functionally describes how to apply tools to the stream, and also corresponds to valuable metadata about the coded and protected image. The ZOI describes the coverage area of each JPSEC tool. This coverage area can be described by image-related or non-image-related parameters. Image-related parameters can specify parameters such as resolution, image area, tile index, quality layer, or color component. Non-image-related parameters can specify areas such as bit-stream segments or packet indices. In cases where image-related and non-image-related parameters are used together, the ZOI describes the correspondence between these areas. For example, the ZOI can be used to indicate that the resolutions and image area specified by the image-related parameters correspond to the bit-stream segments specified by the non-image-related parameters. This allows the ZOI to be used as metadata that signals where certain parts of the image are located in the JPSEC bit-stream. This is especially useful when encryption is used because it conveys some information about the protected data even though the image data are no longer accessible in the protected JPSEC stream. This makes the ZOI one of the most powerful features of JPSEC.

5.3.3   How Is the Security Tool Applied?

While the identifier describes what security services are used and the zone of influence describes where the security tool is applied, further details are needed to instruct a JPSEC consumer how to consume the protected stream. JPSEC uses template and processing parameters for this task. The template parameters describe the detailed parameters of the template tool. For example, while the IDs indicate that decryption and authentication are to be applied, the template parameters would indicate that the JPSEC consumer should use AES decryption in counter mode for the decryption and HMAC with SHA-1 for the authentication.

JPSEC also specifies the processing parameters, including the processing domain and granularity, with which the tools are applied. For example, the processing parameters can instruct a JPSEC consumer to apply the specified decryption method with a granularity of a resolution layer and to the domain of packet bodies only. With this information and with the access keys, a JPSEC consumer can correctly decrypt and decode the portions of the data that it is allowed to access.

In the next section, we will describe the framework and architecture of the JPSEC standard in detail.

5.4   JPSEC Framework

5.4.1   A JPSEC System

A JPSEC system consists of a JPSEC creator and a JPSEC consumer. The JPSEC creator creates a JPSEC stream by applying one or more security tools to raw images, JPEG 2000 image data, or another JPSEC stream (Figure 5.1). When the input is a raw image, the JPSEC creator integrates a JPEG 2000 encoder with a set of security tools. When the input is JPEG 2000 image data, the JPSEC creator applies a set of security tools to the JPEG 2000 image data. When the input is a JPSEC stream, the JPSEC creator applies one or more additional security tools to the JPSEC stream.

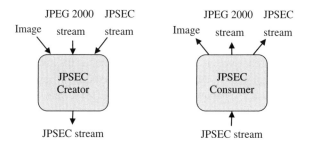

Figure 5.1   A JPSEC creator produces a JPSEC stream by protecting an input image, JPEG 2000 stream, or JPSEC stream. Authorized JPSEC consumers can apply the appropriate protection methods to the JPSEC stream and produce the image, JPEG 2000 stream, or JPSEC stream

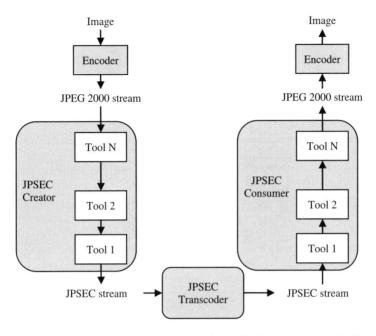

Figure 5.2 JPSEC creators and consumers can apply multiple protection tools. These tools can be applied in a manner that allows a JPSEC transcoder to transcode the JPSEC stream securely without unprotecting it

Likewise, the JPSEC consumer performs security services by applying the specified set of protection tools to the JPSEC stream (Figure 5.2). The output of a JPSEC consumer can be another JPSEC stream, JPEG 2000 image data, or a raw image. In the case of confidentiality services, the JPSEC consumer allows authorized users to decrypt a specified part of the JPSEC stream. In the case of integrity or authentication services, the JPSEC consumer can verify the integrity or authenticity of the appropriate part of the JPSEC stream.

5.4.2 JPSEC Stream

A JPSEC stream is similar in structure to JPEG 2000 coded data (see Figure 5.3). JPEG 2000 coded data contains a header that contains a number of marker segments that contain parameters of the coded stream and data that contain the actual coded data itself organized by tiles. Markers are used to delineate various subparts of the stream. The JPEG 2000 coded data are structured as follows:

- Start of code-stream marker (SOC)
- Header
  - Marker segments
    - SIZ
    - COD
    - QCD

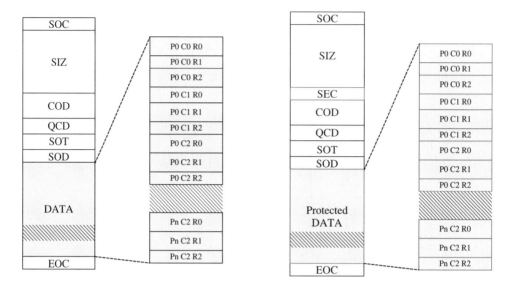

Figure 5.3 The structure of a JPEG 2000 bit-stream (left) and a JPSEC bit-stream (right) are shown. Each PCR block is further split into layers

- Data (repeated if multiple tiles are used)
  - Start of tile (SOT)
  - Start of data (SOD)
  - Coded image data
- End of code-stream marker (EOC)

A JPSEC stream is similar, but adds a security marker segment (SEC) in the header of the JPEG 2000 data and appropriately modified data for the applied security tools. The SEC marker segment is placed between the SIZ and COD marker segments. Thus, a JPSEC stream is structured as follows:

- Start of code-stream marker (SOC)
- Header
  - Marker segments
    - SIZ
    - SEC: contains the JPSEC tools that are applied to this JPSEC stream
    - COD: possibly protected
    - QCD: possibly protected
- Data (repeated if multiple tiles are used): possibly protected
  - Start of tile (SOT)
  - Start of data (SOD)
  - Coded image data: possibly protected
- End of code-stream marker (EOC)

5.5   What: JPSEC Security Services

5.5.1   Overview

As mentioned earlier, JPSEC provides three types of security services for JPEG 2000 images: confidentiality, integrity, and authentication. JPSEC leverages security methods that are established and proven as secure by the broader security community, but then specifies how these established security methods are applied to JPEG 2000 images in a media-aware way that respects the structure of the JPEG 2000 data.

5.5.2   Confidentiality Methods

Confidentiality is performed by encryption/decryption methods with public and private keys. The encryption/decryption methods supported by JPSEC include:

- block cipher (ECB, CBC, CFB, OFB, CTR), with or without an Initialization Vector (IV) and with or without padding (ciphertext stealing, PKCS#7-padding);
- stream cipher (ECB, CBC, CFB, OFB, CTR), with or without an IV and with or without padding (ciphertext stealing, PKCS#7-padding);
- asymmetric cipher.

5.5.3   Integrity Methods

The JPSEC integrity service uses hash-based authentication without a key to generate a MAC value that can be used to verify the integrity of the JPSEC data. The integrity verification methods supported by JPSEC include:

- hash-based integrity verification (SHA-1, RIPEMD-128, RIPEMD-160, MASH-1, MASH-2, SHA-224, SHA-256, SHA-384, SHA-512, WHIRLPOOL).

5.5.4   Authentication Methods

The JPSEC authentication service uses standard key-based authentication methods. The authentication methods supported by JPSEC include:

- hash-based authentication (SHA-1, RIPEMD-128, RIPEMD-160, MASH-1, MASH-2, SHA-224, SHA-256, SHA-384, SHA-512, WHIRLPOOL);
- note that these are the same methods as supported by the JPSEC integrity service, but the JPSEC authentication service requires using a key with these methods;
- cipher-based authentication (CBC-MAC Mac Algorithm 1, CBC-MAC Mac Algorithm 2, CBC-MAC Mac Algorithm 3, CBC-MAC Mac Algorithm 4);
- digital signature (RSA, Rabin, DSA, ECDSA).

5.5.5   Key Template

The JPSEC confidentiality and authentication tools require the use of a key. For this reason, the JPSEC stream must contain information about keys that must be used by a valid

Table 5.1 Decryption and authentication templates

Decryption template		Authentication template	
Block cipher		**Hash-based authentication**	
Cipher	DES, 3DES, AES	Method	HMAC
Block cipher mode	ECB, CBC, CFB, OFB, CTR	Hash function	SHA-1, RIPEMD-160, SHA256
Padding mode	Ciphertext stealing, PKCS#7	Key template	Application dependent
Block size	Cipher dependent	Size of MAC	Variable
Key template	Application dependent	MAC value	Signal dependent
Initialization vector	Variable	**Cipher-based authentication**	
		Method	CBC-MAC
Stream cipher		Block cipher	Cipher ID
Cipher	RC4	Key template	Application dependent
Key template	Application dependent	Size of MAC	Variable
Initialization vector	Variable	MAC value	Signal dependent
		Digital signature	
Asymmetric cipher		Method	RSA, Rabin, DSA, ECDSA
Cipher	RSA	Hash function	Hash ID
Key template	Application dependent	Key template	Application dependent
		Digital signature	Signal dependent

JPSEC consumer. JPSEC includes a key template that can be used to specify key information for each JPSEC tool that requires a key (see Table 5.1). Keys can be represented with:

- X.509 certificate (ISO/IEC 9594−8)
  - Encoding rules DER (RFC3217) or BER (RFC3394)
- URI for a certificate or secret key

5.6 Where: Zone of Influence (ZOI)

JPSEC's main contribution lies in applying standard security methods to JPEG 2000 images in a media-aware way. As mentioned in the previous section, the JPSEC confidentiality service, integrity service, and authentication service describe how standard security methods are applied to all or part of a JPSEC stream or, in other words, they specify how to protect all or part of a JPEG 2000 image. This requires finding a way to represent the portion of the JPEG 2000 image data that is protected. JPSEC defines a structure called the zone of influence (ZOI) to represent this.

5.6.1 Description Classes

The JPSEC ZOI can refer to JPEG 2000 image data using two types of descriptions – image-related descriptions and non-image-related descriptions. These descriptions can be used to describe which part of the JPEG 2000 image data is protected.

5.6.1.1 Image-Related Descriptions

These allow one to specify parts of a JPEG 2000 image through its basic JPEG 2000 image coding structures, specifically by:

- tile (T);
- resolution (R);
- quality layer (L);
- color component (C);
- precinct (P);
- TRLCP tag (a structure defined by JPSEC);
- subband;
- code-block;
- region of interest (ROI);
- bit-rate.

5.6.1.2 Non-Image-Related Descriptions

These allow one to specify part of a JPEG 2000 data stream through its bit-stream structure, including:

- packets (JPEG 2000 packets);
- byte ranges (padded and unpadded);
- TRLCP tags.

5.6.1.3 Additional Non-Image-Related Descriptions

These include:

- distortion value;
- relative importance.

5.6.1.4 Combined Descriptions

These descriptions can be used alone or together. For example, when using one type of description alone, one can specify that 'the lowest-resolution component of the image is encrypted' or that 'byte range 1254 through 1386 of the JPSEC stream is encrypted.'

In addition to describing which part of the JPEG 2000 data is protected, the ZOI also allows one to provide information about the protected data by using the image-related and non-image-related descriptions together. For example, if the ZOI specifies 'tile 4' and 'byte range 1634 through 1784,' then it signals that 'tile 4 is represented in byte range 1634 through 1784.' The concept of providing information about parts of a protected stream is a fundamental property of JPSEC and represents the significance of the ZOI beyond its referral property.

These descriptions can be used in conjunction with the other non-image-related descriptions or image-related descriptions to provide metadata about the protected data. This metadata can be used as hints by other components in a JPSEC system. For example, a streaming server that adaptively sends JPSEC streams over a bandwidth-limited network may choose a subset of the JPSEC data to send based on relative importance. If the data are encrypted, it would be impossible for the streaming server to determine the distortion value or relative importance of part of the data, so the ZOI allows a way to represent it explicitly. This is also useful for data that are not encrypted because it simplifies the computing processing that would be necessary to infer this kind of information, which can be prohibitive for a highly loaded streaming server.

5.7 How: Processing Domain and Granularity

Once the security method and zone of influence are specified, the next piece of critical information is how the security method is applied to the data in the zone of influence. JPSEC defines the processing domain and granularity in which to apply the security method.

The processing domain specifies where the security method is applied. Values can include:

- pixel domain;
- wavelet domain;
- quantized wavelet domain;
- code-stream domain.

If the processing domain is the wavelet domain or quantized wavelet domain, then the processing domain further specifies:

- Protection is applied on the sign bit.
- Protection is applied on the most significant bit.

If the processing domain is the code-stream domain, then the processing domain further specifies:

- Protection is applied on both the packet header and packet body.
- The protection method is applied on the packet body only.

Furthermore, the granularity specifies the unit of protection to which the protection method is applied. The granularity specifies the following:

- Processing order
  - ○ Order specified by the zone of influence image-related parameters
  - ○ Order specified by the zone of influence non-image-related bit-stream parameters
  - ○ Order specified by the zone of influence non-image-related packet parameters
  - ○ Tile−resolution−layer−component−precinct
  - ○ Tile−component−precinct−resolution−layer
  - ○ Tile−layer−resolution−component−precinct
  - ○ Tile−precinct−component−resolution−layer
  - ○ Tile−resolution−precinct−component−layer
- Granularity level
  - ○ Tile
  - ○ Tile part

- Component
- Resolution level
- Layer
- Precinct
- Packet
- Subband
- Code-block
- Total area identified in the ZOI
- Item identified in the non-image-related ZOI
- Zone identified in the non-image-related ZOI

Figure 5.4 depicts the protection of the middle- and high-resolution components of the stream and Figure 5.5 depicts the protection of the low- and high-quality layers of the stream.

A JPEG 2000 code-stream can be encrypted in a number of ways, where each encryption method has different implications on the privacy and transcoding flexibility of the protected code-stream and on the complexity requirements of JPSEC creators, streamers, transcoders, and consumers. These requirements are especially critical for servers that adaptively stream and transcode large numbers of streams and thin clients that have limited device capabilities.

Three example encryption methods that have dramatically different implications are (1) file-based encryption, (2) JPEG 2000 packet-body encryption (different from network packets), and (3) segment-based encryption with metadata. Table 5.2 summarizes the trade-offs of these methods in terms of privacy, transcodability, create/consume complexity, and transcoding complexity. File-based encryption provides the highest level of protection and has the lowest encryption/decryption complexity, but it does not allow secure transcoding,

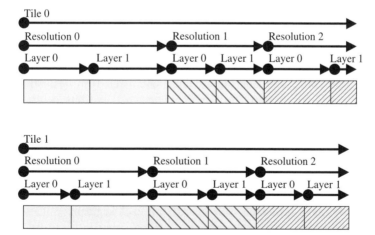

Figure 5.4 The middle- and high-resolution components of the stream are protected separately. In this example, the zone of influence specifies protecting the Resolution 1 and Resolution 2 components. The processing domain is the code-stream domain, the processing order is Tile–Resolution–Layer–Color–Precinct, and the granularity is Resolution

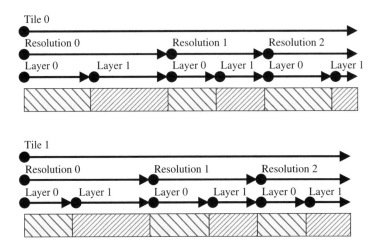

Figure 5.5 The low- and high-quality layers of the stream are protected separately. In this example, the zone of influence specifies protecting the Layer 0 and Layer 1 components. The processing domain is the code-stream domain, the processing order is Tile–Resolution–Layer–Color–Precinct, and the granularity is Layer

Table 5.2 JPSEC encryption modes have different implications on the privacy and transcodability of protected code-streams and the complexity requirements of JPSEC creators, consumers, and transcoders

Method	Privacy	Secure transcodability	Create/consume complexity	Transcode complexity
File	Most	No	Low	N/A
Segment	Middle	Partial	Middle	Very low
J2K packet body	Least	Full	High	High

i.e. transcoding without decryption. At the other extreme, packet-body encryption provides full transcoding flexibility similar to transcoding unprotected JPEG 2000 code-streams, since only the coded image data itself is encrypted, and the bits describing the coding parameters are left unencrypted. However, the creation, consumption, and transcoding complexity are relatively high. Segment-based encryption with metadata hits the middle ground in terms of privacy, transcodability, and create/consume complexity. Meanwhile, it enables the lowest complexity transcoding, and thus is most suitable for transcoding large numbers of streams. Note that file-based encryption does not use metadata; packet-based encryption uses only the metadata in the JPEG 2000 header, while the segment-based approach uses additional metadata, such as ZOIs for each segment.

5.8 JPSEC Examples

The following examples describe how the ZOI and key templates can be used to perform basic security services such as encryption and authentication on a JPEG 2000 coded image. This section provides examples of JPSEC streams using different normative tools from the standard.

5.8.1 Example: Encryption by Resolution

An image is coded with JPEG 2000 and has three resolutions. In this example the first resolution is not encrypted in order to provide preview capability, and the second and third resolutions are encrypted with keys k1 and k2, respectively. The input image in this case is coded in RLCP progression order, and has one tile, three resolutions, three layers, N_c components, and N_p precincts (the number of components and precincts is not significant in this specific example). Encryption is performed using AES in CBC mode without padding (using cipher-text stealing), using key k0 to encrypt resolution 1 and key k2 to encrypt resolution 2, and resolution 0 is left unencrypted.

JPSEC signals how a JPSEC consumer should decrypt the JPSEC code-stream. First, the tool template ID for the decryption template is signaled. Two ZOIs are specified for resolution 1 and its corresponding byte range B0–B1, and for resolution 2 and its corresponding byte range B2–B3. The decryption template parameters identify that AES encryption is applied without padding (using cipher-text stealing). The keying information and the fact that different keys are applied to different resolutions are signaled with the key information parameters. The key granularity is particularly specified as resolution so each resolution has a different key, where the processing order is signaled as TRLCP. The key information for each resolution is contained in the value list of the keys. The encryption is performed on the code-stream, encrypting both the packet headers and packet bodies. The encryption granularity is resolution, where the processing is performed in TRLCP ordering, which is the same ordering as the original code-stream. Since the two resolutions are encrypted separately, two initialization vectors (IVs) are required, which are contained in the value list.

Note that a packet's cipher text results are specified by the processing order and therefore are independent of the input code-stream progression order. However, the placement of the encrypted packets in the output code-stream follow the ordering of the input code-stream packets. Various parameters are given in Tables 5.3 to 5.7.

Table 5.3 SEC marker segment for the encryption example

Parameter		Meaning
SEC		SEC marker
L_{SEC}		Length of SEC marker segment is 130 bytes
Z_{SEC}		Index of this SEC marker segment
P_{SEC}	F_{PSEC}	SEC flags
	N_{tools}	Number of security tool is one
	I_{max}	Maximum tool instance index is zero
T		JPSEC normative tool
I		Tool instance index
ID_T		Decryption template
L_{ZOI}		Length of ZOI is 23 bytes
ZOI		Zone of Influence for this tool (see Table 5.4)
L_{PID}		Length of P_{ID} is 94 bytes
P_{ID}		Parameters for this technology (see Table 5.5)

Table 5.4 ZOI example

Parameter			Derived meaning
NZ_{ZOI}			Number of zones is one
$Zone^0$	DC_{ZOI^1}		The byte aligned segment follows
			Image-related description class
			Resolution is specified
	DC_{ZOI^2}		The byte aligned segment does not follow
			Non-image-related description class
			Byte ranges after the SOD marker are specified
	$P_{ZOI^{0,1}}$	M_{ZOI^1}	The byte aligned segment does not follow
			The specified zones are influenced by the JPSEC tool
			Single item is specified
			Index mode
			Izoi uses 8-bit integer
			Izoi is described in one dimension
		I_{ZOI}	Resolution 1 is specified
	$P_{ZOI^{0,2}}$	M_{ZOI^2}	The byte aligned segment does not follow
			The specified zones are influenced by the JPSEC tool
			Single item is specified
			Range mode
			Izoi uses 16-bit integers
			Izoi is described in one dimension
		$I_{ZOI^{21}}$	Starting byte location is 12 748 (bytes) (B0)
			Ending byte location is 41 960 (bytes) (B1)
$Zone^1$	DC_{ZOI^1}		The byte aligned segment follows
			Image-related description class
			Resolution is specified
	DC_{ZOI^2}		The byte aligned segment does not follow
			Non-image-related description class
			Byte ranges after the SOD marker are specified
	$P_{ZOI^{0,1}}$	M_{ZOI^1}	The byte aligned segment does not follow
			The specified zones are influenced by the JPSEC tool
			Single item is specified
			Index mode
			Izoi uses 8-bit integer
			Izoi is described in one dimension
		I_{ZOI^1}	Resolution 2 is specified
	$P_{ZOI^{0,2}}$	M_{ZOI^2}	The specified zones are influenced by the JPSEC tool
			Single item is specified
			Range mode
			Izoi uses 32-bit integer
			Izoi is described in one dimension
		I_{ZOI^2}	Starting byte location is 41 966 (bytes) (B2)
			Ending byte location is 135 425 (bytes) (B3)

Table 5.5　P_{ID} example

Parameter		Meaning
T_{ID}		Decryption templates (see Table 5.6)
PD		Pixel domain is not used
		Wavelet coefficient domain is not used
		Quantized wavelet coefficient domain is not used
		Code-stream domain is used
F_{PD}		Only packet body is encrypted
G	PO	Processing order is tile–resolution–layer–component–precinct
	GL	Unit of protection is resolution level
V	N_V	Number of values in the value list V is 2
	S_V	Length of each V_n is 16 bytes
	V1	Initialization vector value for R1
	V2	Initialization vector value for R2

Table 5.6　Decryption template example

Parameter		Derived meaning
ME_{decry}		Marker emulation has occurred
CT_{decry}		Block cipher (AES)
CP_{decry}	M_{bc}	CBC mode; bits are not padded
	P_{bc}	Ciphertext stealing
	SIZ_{bc}	Block size (16 bytes, 128 bits)
	KT_{bc}	Key template (see Table 5.7)

Table 5.7　Key template example

Parameter		Derived meaning
LK_{KT}		Length of key is 128 bits
KID_{KT}		URI for secret key
G_{KT}	PO	Processing order is tile–resolution–layer–component–precinct
	GL	Unit of protection is resolution level
V_{KT}	N_V	Number of values in the value list V is 2
	S_V	Length of each V_n is 19 bytes
	V1	Secret key for resolution level 1 can be retrieved from https://server/key1
	V2	Secret key for resolution level 2 can be retrieved from https://server/key2

5.8.2　Example: Authentication by Resolution

In this case authentication is applied to the same JPEG 2000 coded image as above. In this example all three resolutions and three layers per resolution are authenticated, where the authentication of each resolution uses a different key. Since there are three resolutions there are three keys, and since there are three layers per resolution there will be three

MAC values per resolution. Thus, there will be a total of nine MAC values for the entire JPSEC image. Specifically:

- Resolution 0 has MAC values M0, M1, M2 (one for each layer) using *key0*.
- Resolution 1 has MAC values M3, M4, M5 (one for each layer) using *key1*.
- Resolution 2 has MAC values M6, M7, M8 (one for each layer) using *key2*.

This example illustrates how authentication can be signaled as well as the flexibility provided by the ZOI and granularity tools. As in the previous example, the input image is coded in the RLCP progression order, and has one tile, three resolutions, three layers, N_c components, and N_p precincts (the number of components and precincts is not important in this specific example). Authentication is performed using HMAC with SHA-1.

JPSEC signals how a JPSEC consumer can verify or authenticate the JPSEC protected content. First, the tool template ID for the authentication template is signaled. Then the ZOI is used to signal that there are three resolutions and the associated byte ranges for each resolution. The authentication template parameters signal that HMAC is applied using SHA-1. The key information template provides information about the keys including that the key granularity is resolution and suppling the information for each of the three keys in the value list for the keys. The processing domain for authentication is specified as the code-stream including packet headers. The tool granularity for authentication is specified as the layer; therefore there are three MACs for each resolution, for a total of nine MAC values. The value list contains the nine MAC values. The processing order for the above was identified as TRLCP, which is the same as the original code-stream order.

Note that the use of processing order in the granularity field ensures that the same MAC values would result independent of the code-stream progression order. While this example demonstrated the use of MACs, the same approach can be used to signal the use of multiple digital signatures. Various parameters are given in Tables 5.8 to 5.10.

Table 5.8 The SEC marker segment

Parameter		Derived meaning
SEC		SEC marker
L_{SEC}		Length of the SEC marker segment
Z_{SEC}		Index of this SEC marker segment
P_{SEC}	F_{PSEC}	SEC flags
	N_{tools}	Only one tool is used in this code-stream
	I_{max}	The maximum tool instance index is 0
$Tool^0$	T	JPSEC normative tool
	I	Tool instance index
	ID_T	This normative tool uses an authentication template
	L_{ZOI}	Length of ZOI is 32 bytes
	ZOI	The covered zone of the image (see Table 5.9)
	L_{PID}	Length of P_{ID} is 108 bytes
	P_{ID}	Parameters for JPSEC tool (see Table 5.10)

Table 5.9 ZOI signaling

Parameter			Derived meaning
NZ_{ZOI}			Number of zones is 1
$Zone^0$	$DC_{ZOI}{}^1$		The byte aligned segment follows
			Image-related description class
			Resolution levels are specified in order
	$DC_{ZOI}{}^2$		The byte aligned segment does not follow
			Non-image-related description class
			Byte ranges are specified
	$P_{ZOI}{}^{0,1}$	$M_{ZOI}{}^1$	The byte aligned segment does not follow
			The specified zones are influenced by the JPSEC tool
			Single item is specified
			Range mode
			Izoi uses 8-bit integer
			Izoi is described in one dimension
		$I_{ZOI}{}^1$	The beginning of the range is 0
			The end of the range is 2
	$P_{ZOI}{}^{0,2}$	$M_{ZOI}{}^2$	The byte aligned segment does not follow
			The specified zones are influenced by the JPSEC tool
			Multiple items specified
			Range mode
			Izoi uses 32-bit integers
			Izoi is described in one dimension
		N_{ZOI}	Number of I_{ZOI} is 3
		$I_{ZOI}{}^1$	Starting byte location is 104 (bytes)
			Ending byte location is 12 762 (bytes)
		$I_{ZOI}{}^2$	Starting byte location is 12 768 (bytes)
			Ending byte location is 41 980 (bytes)
		$I_{ZOI}{}^3$	Starting byte location is 41 986 (bytes)
			Ending byte location is 135 445 (bytes)

5.8.3 Example: Combining Encryption and Authentication

This section describes a JPSEC system that uses AES encryption applied in CTR and CBC modes, and authentication with HMAC using SHA-1. Using a secure transcoder, transcoding is performed in the encrypted domain where the input encrypted JPSEC is transcoded to produce another encrypted JPSEC at the desired bit-rate, which can be decrypted and decoded by the client to produce an image with the desired resolution, spatial location, or fidelity.

Figure 5.6 shows an example of how JPSEC can be used for decryption and authentication. First, the received data are decrypted with the key. Then, the decrypted data can be authenticated with the MAC values placed in the JPSEC header by the JPSEC creator, and with the associated authentication keys. This allows a user to verify that the data it received was not intentionally or unintentionally modified.

This stream is decrypted and then verified using the parameters specified in the JPSEC header. Figure 5.7 shows the type of information that is placed in the JPSEC header. The

Table 5.10 P_{ID} signaling parameters

Parameter						Derived meaning
T_{auth}	M_{auth}					Authentication methods: hash-based authentication
		P_{auth}	M_{HMAC}			HMAC is used for authentication
			H_{HMAC}			SHA-1 is used for hashing
			KT_{HMAC}	LK_{KT}		Length of the key in bits
				KID_{KT}		KI_{KT} contains the URI for the private key
				G_{KT}	PO	The order is tile–resolution–layer–component–precinct
					GL	Key granularity is resolution
				V_{KT}	N_V	There are 3 keys in the list
					S_V	Size of each key is 8 bytes
					VL	The first key is *key0*, for resolution 0
						The second key is *key1*, for resolution 1
						The third key is *key2*, for resolution 2
			SIZ_{HMAC}			Size of MAC is 20
PD						The byte aligned segment does not follow
						Pixel domain is not used
						Wavelet coefficient domain is not used
						Quantized wavelet coefficient domain is not used
						Code-stream domain is used
F_{PD}						Both packet header and body are encrypted
	G	PO				The order is tile–resolution–layer–component–precinct
		GL				Tool granularity is layer
	V	N_V				There are 9 MACs (3 MACs per resolution)
		S_V				Size of each MAC is 20 bytes
		VL				The first MAC is $M0$
						The second MAC is $M1$
						The third MAC is $M2$
						The fourth MAC is $M3$
						The fifth MAC is $M4$
						The sixth MAC is $M5$
						The seventh MAC is $M6$
						The eighth MAC is $M7$
						The ninth MAC is $M8$

decryption parameters are shown on the left and the authentication parameters on the right. Note that in this example the decryption parameters specify a single key that is used for the entire image and a set of initialization vectors that can be used to refresh the encryption with the granularity of a JPEG 2000 quality layer. Also note that the zone of influence specifies the image-related parameters of a resolution, but it also includes the byte ranges of the encrypted bit-stream that correspond to each of the resolutions. This metadata provided by the zone of influence allows an untrusted transcoder to extract and send lower resolution components even though the data itself is encrypted. The authentication parameters specify that the packet bodies of each resolution can be authenticated and that they contain the computed MAC values that a JPSEC consumer can use to perform the verification. Thus, a JPSEC consumer can use this information to verify that it received

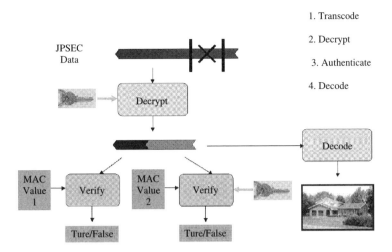

Figure 5.6 Example showing how a received JPSEC stream is decrypted, authenticated, and finally decoded for display. In this example two MACs are associated with two different parts of the JPSEC stream

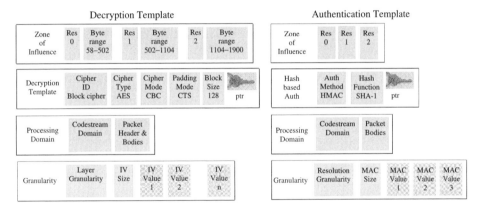

Figure 5.7 Example of information contained in encryption (left) and authentication (right) templates for a JPSEC image providing access to three resolution levels and authentication of these levels

a valid version of the image, even in the event that it received a transcoded version of the original protected bit-stream.

JPSEC was designed to provide confidentiality and authentication in a manner that enables the two seemingly conflicting properties of end-to-end security with secure transcoding at mid-network nodes. This is achieved by designing the JPSEC protection methods and signaling syntax to enable an entity to adapt or transcode the resulting JPSEC-protected stream securely without requiring unprotecting or decrypting the content. Furthermore, this method also provides authentication that the transcoding was performed only in a valid and permissible manner. This enables a potentially untrusted server or mid-network caching server or proxy to perform secure transcoding and enables a JPSEC consumer to authenticate that the received content was transcoded in a valid and permissible manner.

5.9 Summary

The JPSEC standard was created to provide security for JPEG 2000 images in a manner that respects the media attributes of the JPEG 2000 stream. The JPSEC architecture was designed with a system perspective in mind and thus considers the interlayer interaction between compression, security, file format, and packetization. The resulting standard described in this chapter provides the features listed below:

- *Secure transcoding.* Secure transcoding is defined as the capability of adapting a protected media object without requiring one to unprotect it. For example, secure transcoding enables an encrypted image to be transcoded without requiring decryption. JPSEC also allows secure transcoding to be performed in a rate-distortion optimized manner by allowing metadata to be included about the encrypted data. It also allows the end user to verify that the adaption was performed in a valid and permissible manner.
- *Digital signatures.* JPSEC allows semantically meaningful portions of the media to be signed with digital signatures using the private key in a public/private key pair, while allowing the end user to use the public key to verify who created the image and verify that it was not altered. In addition, if secure transcoding was used, JPSEC allows the user to verify that the adaptation (secure transcoding) was performed in a valid and permissible manner.
- *Authentication.* Message Authentication Codes (MACs) can be computed for semantically meaningful portions of the media (using a secret key) and the end user (who also has the same secret key) can verify who created the image and that it was not altered. In addition, the JPSEC image can be adapted and the end user can verify that the adaptation was performed in a valid and permissible manner. Note that for authentication (digital signatures or MACs) the JPSEC creator determines the modifications that are allowed; the end user can then authenticate that only valid and permissible modifications have occurred.
- *Optimized streaming of encrypted data.* JPSEC allows rate-distortion information, i.e. rate-distortion hints, to be embedded into an encrypted JPSEC file. This enables intelligent, content-secure processing of an encrypted JPSEC file. These hints include the distortion that would be incurred if a JPSEC segment is not available at the decoder, as well as the tile, resolution, quality level, color component, and precinct attributes associated with different parts of the JPSEC bit-stream. These hints enable rate-distortion optimized secure transcoding to be performed, i.e. rate-distortion optimized transcoding without knowledge of the content.
- *JPEG 2000 'compliant' encryption.* The encryption can be performed in a manner that provides zero overheads (no data expansion) and may be performed in-place and using well-studied encryption primitives such as the Advanced Encryption Standard (AES), while preventing illegal bit-stream combinations in the stream. An interesting and useful consequence of the above, plus other design tricks, is that the encryption can be performed in a manner that creates encrypted bit-streams that can be informally referred to as being JPEG 2000 compliant and can be decoded by a JPEG 2000 decoder

(as opposed to a JPSEC decoder) to produce a noisy-looking image, but without causing the JPEG 2000 decoder to crash. This enables the design of encrypted bit-streams to be decoded/displayed by a user who does not have a JPSEC decoder or the key.

References

Dufaux, F., Wee, S., Apostolopoulos, J. and Ebrahimi T. (2004) JPSEC for secure imaging in JPEG 2000, in *SPIE Proceedings on Applications of Digital Image Processing XXVII*, Denver, CO.

[ISO/IEC (2000) Information Technology – JPEG 2000 Image Coding System – Part 1, ISO/IEC International Standard 15444-1, ITU Recommendation T.800.

ISO/IEC (2007) Information Technology - JPEG 2000 Image Coding System – Part 8: Secure JPEG 2000, ISO/IEC International Standard 15444-8, ITU Recommendation T.807.

Wee, S. and Apostolopoulos, J. (2001a) Secure scalable video streaming for wireless networks, in *IEEE International Conference on Acoustics, Speech, and Signal Processing (ICASSP)*.

Wee, S. and Apostolopoulos, J. (2001b) Secure scalable streaming enabling transcoding without decryption, in *IEEE International Conference on Image Processing (ICIP)*.

Wee, S. and Apostolopoulos, J. (2003), Secure scalable streaming and secure transcoding with JPEG 2000, in *IEEE International Conference on Image Processing (ICIP)*.

Wee, S. and Apostolopoulos, J. (2004) Secure transcoding with JPSEC confidentiality and authentication, in *IEEE International Conference on Image Processing (ICIP)*.

Further Reading

ISO/IEC (1989) Information Processing Systems – Open Systems Interconnection Reference Model – Basic Reference Model – Part 2: Security Architecture, ISO/IEC International Standard 7499-2.

ISO/IEC (1994) Information Technology – Security Techniques – Hash-Functions, ISO/IEC International Standard 10118-1,2,3,4.

ISO/IEC (1996) Information Technology – Security Techniques – Key Management, ISO/IEC International Standard 11770-1,2,3:1996.

ISO/IEC (1997) Information Technology – Security Techniques – Entity Authentication – Part 1: General, ISO/IEC International Standard 9798-1:1997.

ISO/IEC (1998) Information Technology – Security Techniques – Digital Signatures with Appendix – Part 1: General, ISO/IEC International Standard 14888-1:1998.

ISO/IEC (1998) Information Technology – Security Techniques – Digital Signatures with Appendix – Part 3: Discrete Logarithm Based Mechanisms, ISO/IEC International Standard 14888-3:1998.

ISO/IEC (1999) Information Technology – Security Techniques – Message Authentication Codes (MACs) – Part 1: Mechanisms Using a Block Cipher, ISO/IEC International Standard 9797-1:1999.

ISO/IEC (2001) Information Technology – Guidelines for the Management of IT Security – Part 4: Selection of Safeguards, ISO/IEC International Standard TR 13335-4:2001.

ISO/IEC (2002) Information Technology – Security Techniques – Cryptographic Techniques Based on Elliptic Curves – Part 2: Digital Signatures and Part 3: Key Establishment, ISO/IEC International Standard 15946-2,3:2002.

ISO/IEC (2002) Information Technology – Security Techniques – Digital Signature Schemes Giving Message Recovery – Part 2: Mechanisms Using a Hash-Function, ISO/IEC International Standard 9796-2.

ISO/IEC (2004) Information Technology – Security Techniques – Cryptographic Techniques Based on Elliptic Curves – Part 4: Digital Signatures Giving Message Recovery, ISO/IEC International Standard 15946-4:2004.

ISO/IEC (2004), Information Technology – Security Techniques – Management of Information and Communications Technology Security – Part 1: Concepts and Models for Information and Communications Technology Security Management, ISO/IEC International Standard 13335-1:2004.

ISO/IEC (2005) Information Technology – Security Techniques – Encryption Algorithms – Part 2: Asymmetric Ciphers, Part 3: Block Ciphers, and Part 4: Stream Ciphers, ISO/IEC International Standard 18033-2,3,4.

6

JPIP – Interactivity Tools, APIs, and Protocols (Part 9)

Robert Prandolini

6.1 Introduction

Part 9 of the JPEG 2000 family of standards is an interactive imagery protocol called JPIP (ISO/IEC 15444-9) (ISO/IEC, 2005). It was developed to provide efficient dissemination of imagery for those classes of applications with dynamic interactions, which is of partic-ular importance when the imagery is large and on a remote server. A common example of interactive imagery is a user browsing a large geospatially registered image mosaic over the Internet. This user is browsing the image because of interest in obtaining some information from an interactive session, e.g. to gain an appreciation of the route from a hotel to the conference venue for an upcoming business trip. In this browsing session, the user will determine a suitable route and note navigation features (buildings, overpasses, roads). The user typically changes scale (zooms in or out) and pans the view of the image while browsing. When the task is completed, the amount of imagery actually seen by the user is only a fraction of the image base on the host server. To download this image base to the user would be wasteful, and often impractical. It is much better to download only that imagery of interest to the user. This is the prime motivation behind JPIP, which is a protocol for signaling the user's interest to an image server and for the server to respond with only that portion of interest from the imagery.

In imagery compression, a region of interest (ROI) is by necessity defined *a priori* to encoding – it is a region that has a higher image quality on decoding than that of the background region. The entire point of JPIP was to provide interactive ROI (IROI) such that a remote image server can efficiently disseminate the imagery data of interest, steered by the user's dynamic requirements. Clearly, IROI are defined *a posteriori* to encoding. JPIP is not a form of compression in the normal sense, but it will be shown later how JPIP

The JPEG 2000 Suite   Edited by Peter Schelkens, Athanassios Skodras and Touradj Ebrahimi
© 2009 Commonwealth of Australia

can be used to create highly compressed JPEG 2000 imagery optimized for the user's interests.

JPIP is an enabling protocol for optimally efficient interactive dissemination. The ideal is the transmission of only that imagery data which meets the user's needs at that time, and where that data has not been sent in prior image data transactions and cached by the user's image browser system. JPIP was also developed to enable a responsive experience for the user's interactive image browsing session. Note that JPIP need not be implemented to achieve optimally efficient interactive dissemination. JPIP implementations can trade off dissemination efficiency for user responsiveness, but it is important to note that JPIP can be implemented for very efficient interactive dissemination.

To demonstrate the fundamental objectives behind JPIP and introduce some terminology, the following discussion returns to the large image browsing use case in more detail. A 'server' hosts a large image and the user connects to it for a 'session.' The user's application that interfaces to the server is the 'client.' Assume that this is the first ever session by the client browsing this 'logical target' image. It is called a 'logical target' because in theory, and often in practice, the server can perform transcoding. For example, the server can offer multiple representations of the same image for different return media types. This possibly includes re-indexing of the code-blocks due to it changing the size of the precincts to enable a finer-grain random access into the image. The server also, by its own devices, must index metadata boxes. The server is obliged to offer to serve a 'logical target' with consistent indexing.

Typically, the first 'request' from the client is for an overview of the entire image. Now this client has a limited display size on which to render the imagery, and so the server need only send a 'response' consisting of that imagery data required for this display size. Hence, this overview image for this first request need only be a low-resolution version of the original image. Ideally, it would be the same resolution size as the client's display. Since available resolutions are dyadic in JPEG 2000, the image data for the closest available resolution level to the size of the display would be an appropriate response for the client.

For any device rendering a JPEG 2000 image larger than its display resolution, there is available a full-resolution level of the image that shows the details in their highest fidelity, but of course with a limited spatial extent. When the scale difference between the image and display is significant, this is often called the 'soda-straw' view. In this particular example use case, the user needs to zoom into the image to locate some feature. However, the exact location of this feature is not discernable at this first overview scale. Some navigation and searching will be required, and so the user will zoom and pan interactively about the image in this feature hunt. The view at each step is an IROI. When the user zooms into a region, the spatial extent of the IROI is reduced, while the resolution of the details that can be displayed in this IROI increases. When the user pans, the resolution scale is constant, while the location of the spatial extent changes. Finally, the user ends the feature hunt with some soda-straw view of the feature of interest. The spatial extent of each IROI along with the required resolution scale (e.g. the client's display size) is termed the JPIP 'view-window.'

It is important to note that in this example use case, the timely dissemination of the entire large image over a capacity constrained network is not possible – thus the motivation for another approach. One early technique was screen-scraping. Here the user sends various view-window requests – specified by the IROI spatial extent and desired resolution – and receives response data such as a JPEG image. (Note that for backward compatibility with legacy viewers, the JPIP standard also supports a screen-scraping mode using JPEG as the media return type.) Screen-scraping is a more efficient interactive dissemination technique than simply sending the whole large image file, but it is suboptimal in two ways.

First, the system is not actually disseminating the original source image data. It may be a multi-resolution image pyramid source (such as multi-resolution TIFF or FlashPix), but this has a 33% data overhead for the image pyramid. To reduce the 33% overhead, a Laplacian pyramid (Burt and Adelson, 1983) could be used, but this leads to wavelet compression (Mallat, 1989), then to JPEG 2000 compression (Taubman and Marcellin, 2002), and finally to the JPIP technology. Many screen-scraping implementations are simple and designed for web-browsers that display and discard the imagery. Thus when panning across the image, even by just a single column of pixels, an entire new JPEG is screen-scraped by the server. Likewise, when zooming into a view-window, there is no reuse of the low-resolution data from previous view-windows. This lack of data reuse is a consequence of the transacted data, the JPEGs, being transformed and abstract views of the original source data. Responses are incompatible – previous responses are not readily reused.

The second shortfall with screen-scraping is that it is based on tight synchronous request–response transactions. Thus the user waits for the response JPEG before further interacting with the image. This can result in poor responsiveness and hinder the interactive experience when the network capacity is low or the delay is high. Early Internet map services using screen-scraping on dial-up modems are an example – they had very poor responsiveness.

Both of the above shortfalls can be addressed by having the imagery transaction data being (a) compatible subsets of the entire compressed image data rather than a transcoded representation, (b) cached by the client, and (c) having an asynchronous or loosely coupled request–response signaling. JPEG 2000 Part 1 (ISO/IEC 15444-1) (ISO/IEC, 2004) is an image compression system that can allow the transaction data to be compatible subsets of the original compressed image data. This is because the image is compressed into independent embedded data sets from spatial code-blocks from a hierarchy of resolution subbands. It is the scalability of JPEG 2000 Part 1 that allows the possibility for the compressed data to be the compatible transaction data in JPIP. Subsets of the compressed image data can be random-accessed independently on the basis of resolution, spatial extent, quality layers, and components. (It is also possible to access metadata in a similarly independent manner, although this is dependent on its structure; see below.)

Returning to the use case, assume that the transaction data are subsets of the JPEG 2000 compression data of the served image. Before a zoom operation, consider that the user may have already received all of the relevant low-resolution data from the current or a prior view-window request. When they zoom into the image, they will only need that higher resolution data from the new zoomed view-window. On each pan, the user only needs the imagery data from the new spatial extent and at the same resolution level. The difference between the imagery data required to render a new view-window and that already held by the client can be small. If some or all of the higher resolution or new spatial extent data had been sent previously, then that data should not be sent again. The simplest meaning of optimal dissemination is that imagery data immediately relevant to the user should be sent, but only sent once.

Hence it is clear that the client will require some form of cache to hold data from previous requests if it is going to transact with the server efficiently. The benefit of having a cache is not only the enabling of optimal dissemination, but also the increased responsiveness from the interactive experience. When zooming into a view-window, a user's application can immediately render that imagery data already held in its cache. Over time, depending on the network capacity and latency, more imagery data may be received, added to the cache, and then incrementally decoded and rendered. This provides a 'continual quality update' experience. The boon here is that the user need not wait until the current response is received (either in full, in part, or at all) before applying judgment and deciding on the next pan-zoom operation. For example, if the user's task involved locating a feature in the centre of the display and at full resolution, then once the feature is found from the initial hunt, a series of zoom and minor pan operations can be done using only that data in the cache. The user still waits for the response to the final view-window, but the intermediate steps getting to this point requires no waiting at all. The user's experience will be a much higher level of responsiveness than that from a screen-scraping implementation. Changing the user's view-window before receiving all (or any) of the previous response is to 'preempt' the previous request.

The system described so far is the basic design for session-based JPIP and is shown in Figure 6.1. As in all engineering, there is a compensatory price for such benefits, which in this case is higher complexity. Consider the following possible permutations:

1. The cache may preexist from an earlier session. To optimize dissemination, the client should inform the server of its preexisting cache contents.
2. On each request, there will be a subset of imagery data applicable for the task of rendering the view-window on the display at the specified resolution and quality. For the server to decide what portion of this subset needs to be sent (i.e. data not in the client's cache), either it must be signaled explicitly or it must keep a log of the assumed client 'cache contents.' In JPIP, this server-side log of assumed cache contents is called the 'cache model.' JPIP does have explicit cache model signaling, but its usage is occasional (e.g. the preexisting cache example above in 1). The use of a cache model is preferred as it lessens the payload (size) of view-window requests. A server implements a cache model within a session.
3. The cache model could be based on the simple assumption that the client has cached all transacted data sent, unless told otherwise. Since all clients have finite memory resources, it is not possible to guarantee that all received data will be cached or stay cached. Thus the server should be informed at times when the cache is purged or the data are not cached.
4. A further complexity arises when one considers the possibility of response data being undelivered by the network. (The client may be able to analyze its cache to discover

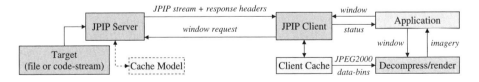

Figure 6.1   Client–server interaction in JPIP

the fact that it is missing data. Since JPEG 2000 embedded data are contiguous, any 'holes' will indicate missing data.) In this case, the client may need to signal its cache contents more frequently to the server.

5. Lastly, consider the case when the client does not implement, or has but will cease to maintain, a cache. Should such an implementation be permitted? While not optimal for interactive sessions, such a single-shot mode of operation could be suitable for some applications. (An example would be a printer tasked to print a view-window from an image.) These are termed 'stateless requests,' because the server will not maintain a cache model after processing the request. Note that stateless requests and requests within a session are mutually exclusive.

The last issue to be addressed here is that of asynchronous or loosely coupled request–response pairs. Already discussed is the possibility of the user preempting a request prior to receiving its response, and how this improves the responsiveness for the user. Consider the possibility of loosening the optimal dissemination objective and allow the server to send first that data required for the current view-window request, and then preemptively to use otherwise idle network capacity to send other data it considers useful. An example is to send data for a larger spatial extent centered by the current view-window, plus data from the next higher resolution level from within the current view-window, on the basis that the next request is likely to be a pan or zoom operation on the current IROI. Another example is to send metadata (say the title and abstract) that is spatially related to the view-window request. Except for those capacity challenged networks, this preemptive data will cost little in terms of the timely dissemination of the IROI response data, because it will mostly use idle capacity, but it can significantly improve the responsiveness of the interactive browsing experience for the user.

There is more to be considered for this asynchronous or loosely coupled request–response pairs than just preemptive requests and preemptive response data providing improved responsiveness. In a typical HTTP GET request, the response is often a static web-page; the response is identical for identical requests. This is why a local or intermediate HTTP proxy can improve web-performance. In JPIP, however, the client cache and the server cache model enable efficient data dissemination. If any portion of the imagery data (subset C) relevant to a view-window request (subset V) was in the client cache, then only its complement should be sent (subset V-C). As the relevant response data are disseminated, possibly in incremental portions, any subsequent identical view-window requests would produce a different server response. When all of the relevant data are finally in the cache, then all subsequent identical view-window requests should returned no data. (The complement is empty.) Clearly, request–response pairs are based on the context of the cache and the cache model, and so cannot leverage any intermediate HTTP proxy. (Note that a form of JPIP can be implemented to use HTTP file-servers and HTTP proxy infrastructure. See JPIP indexing below.)

When the JPIP standard project commenced in the early 2000s, such asynchronous coupling between requests and responses was considered an unconventional approach. HTTP experts held a fixed response–request behavior paradigm. The example use case developed above broke this model, which became a part of the revolutionary JPIP. Since then, asynchronous behavior has become a matured concept (e.g. AJAX[1] programming).

---

[1] Asynchronous JavaScript and XML (AJAX) is a web technology for interactive web applications.

JPIP was also completely different and revolutionary in another way. All compression systems to date use parameters *a priori* to compression, typically compressed for PSNR^2 optimized encoding or perhaps for a superior PSNR ROI over the background. By allowing the user to interact with the imagery in a dissemination efficient manner, the resultant *a posteriori* data subset – i.e. the contents of the client-side cache – is a highly compressed version of the original image. If the purpose of the user was to capture in high resolution a particular route and navigational features from an image-map, then a JPIP session provides an effective means to truncate the original compressed image data, postcompression. The resultant client-side cache has been optimized by the interactions of the user through the user's selection of view-windows. In our example use case, the typical ratio of the source image file-size to the size of the resultant cache could be more than 2000:1. No algorithmic or otherwise *a priori* compression system can achieve such high compression ratios and provide high-quality imagery so completely customized for the user's requirements.

6.2   Data-Bins

6.2.1   Streaming Data

From the introduction it should be clear that JPIP is based on the concept of disseminating a subset of the original data, such that this will meet the dynamic requirements of the user. This is the basis for efficient interactive dissemination, which is possible only when the original data are decomposed into suitably compatible subsets. Compatibility is the ability for the data subsets to work together. When the client specifies a view-window to the server, the response data are the difference between the subset of data that is required to satisfy the view-window request and the subset of data already held in the client-side cache. If the subsets of data were incompatible, then it would not be possible to compute such a difference and therefore have efficient interactive dissemination. Tiling in Flashpix is an example of compatible decomposition in the spatial domain as adjacent tiles can work together to provide a larger picture. On the other hand, Flashpix is also an example of incompatible decomposition in the resolution domain because each resolution level does not operate (addition or subtraction) with the other levels. Screen-scraping is another example of interactive imagery dissemination using noncompatible image compression, such as the JPEG baseline, whereas JPIP, using JPEG 2000 compressed imagery, disseminates truncated portions of independent embedded data-streams.

The different resolution levels in the JPEG 2000 decomposition do interoperate, as do the tiles and precincts that decompose the image data spatially. The possible decomposition modes for still imagery are spatial, resolution, pixel quality, and components. JPEG 2000 compression is a decomposition of the imagery in each of these domains. Each component of each tile of the image is independently decomposed into resolution levels by the discrete wavelet transform generating multiple subbands. Each subband is independently decomposed into code-blocks (grouped into precincts). Each code-block is independently encoded into an embedded bit-stream, which can be incrementally accessed in bytes (grouped by precincts and segmented into quality layers). The degree of scalability in JPEG 2000 is high because it supports a large dynamic range with a fine-grain level of access into each of the possible decomposition domains.

---

^{2} The peak signal-to-noise ratio (PSNR) is a common objective measure of image quality.

The encoding of data, when encoded at two different rates of compression, having resultant encoded bit-streams of length L and M, where $L > M$, and where the first M bits of the first bit-stream are identical to the second bit-stream, is called embedded encoding. Simple truncation of a JPEG 2000 code-stream[3] to obtain different compression rates is a feature of JPEG 2000. In JPEG 2000, a possible basis for defining subsets for use in efficient interactive dissemination are tiles (concatenated tile parts) and, for finer granularity, the precinct-based code-block embedded bit-stream (concatenated packets).

Tile-based interactive dissemination in JPIP uses the compressed image tile data, called JPT-stream media. Finer grain spatial access is available via precincts and code-blocks. Precincts can be as large as the tile or a small as a code-block. Since a server can readily transcode a code-stream for different precinct size simply by reindexing the code-blocks and precincts, precinct-based interactive dissemination was chosen for a spatially finer scale response data. Based on JPEG 2000 packets, it is called JPP-stream media. The JPT- and JPP-stream data are abstractly termed 'data-bins.' Each set of concatenated code-stream tile parts from a tile form a JPT-stream data-bin. Each set of concatenated code-stream packets from a precinct form a JPP-stream data-bin.

In JPIP the server transmits in its response those data-bins relevant to the view-window request. However, this is not all that is required for the client to be able to render the image. To be able to understand the context of where these JPT- and JPP-streams fit in the imagery, the client will also need header information. Convenient packagings of this type of information are the JPEG 2000 code-stream main header, tile-part headers, JP2 header, and others. Metadata boxes can encode additional information within the imagery, which may, or may not, be associated with any specific spatial position. Each image data-bin, header data-bin, and metadata-bin is an independent package of data, and their construction is discussed next.

6.2.2   Defining Data-Bins

Header data-bins and metadata-bins consist of that data inside the JPEG 2000 headers or metadata boxes. There is only one main header data-bin per code-stream (denoted 'Hm') and it always has an in-class identifier of 0. (This redundant 0 index was implemented for the sake of consistency with other data-bin in-class identifiers.) The main header data-bin is the concatenation of all markers and marker segments in the main header, starting from the start of the code-stream (SOC) marker. Note that it does not include the start of tile part (SOT), start of data (SOD), or end of code-stream (EOC) markers and segments. Metadata-bins (denoted 'M') are the collection of boxes from a JPEG 2000 family file. It will be shown later that servers can manage complex superbox structures by reorganization using placeholders, but a consistent principle is that ultimately the metadata are packaged into some congruous set of metadata-bins. Unlike other data-bin indexing, the in-class identifier for metadata-bins is at the discretion of the server. The server is required to maintain consistency in its indexing for a particular logical target, and metadata-bin 0 must be the root of the logical target.

Packaging the compressed imagery data into data-bins is a central concept in JPIP. The two JPIP media types are not compatible with each other. Either the tile-based JPT-stream

---

[3] The code-stream is the interleaved embedded code-block bit-streams grouped by precincts into packets of quality layers and forming tile parts.

or the precinct-based JPP-stream media is used per logical target. The data disseminated using the main header data-bin, plus either tile data-bins (JPT-stream media), or tile header data-bins and precinct data-bins (JPP-stream media), is called the 'incremental code-stream' because it allows the incremental dissemination of the imagery data. (Note that other response data types are possible including the 'raw' media type, which is the code-stream accessed via byte-range requests.)

Tile data-bins (denoted 'T') are used in the JPT-stream media type and consist of the concatenation of all the code-stream tile parts belonging to the tile, including all marker segments. Tile data-bins are indexed by the in-class identifier (t) numbered in raster order starting from 0 to $num\_tiles - 1$, as per the numbering of tiles in JPEG 2000 Part 1. Since the JPT-stream disseminates the tile parts, which includes the tile headers, the tile header data-bins are not required and are not available from a JPT-stream logical target.

Each precinct data-bin (denoted 'P') corresponds to a single precinct in a code-stream and is indexed by the in-class identifier:

$$I = t + (c + s \times num\_comp) \times num\_tiles,$$

where

t is the index (starting from 0) of the tile to which the precinct belongs;

c is the index (starting from 0) of the image component;

s is a sequence number which identifies the precinct within its tile component; and *num_comp* and *num_tiles* are the number of components and tiles respectively.

In its tile component, each precinct has a sequence number (s) starting from 0 for the top-left precinct in the LL subband. It increments for each precinct in raster-scan order in the LL subband, and then increments for each precinct in raster-scan order in the next higher resolution level, and so forth through each successive resolution level.

Each precinct data-bin is the concatenation of all code-stream packets belonging to the precinct, complete with all relevant packet headers. Note that when packet headers are packed into either the main header or the tile-part headers (i.e. PPM or PPT markers were used), then the precinct data-bin consists of only the packet bodies. A client receiving JPP-stream media will also require the tile header information to be able to decode the precinct data-bins. Tile header data-bins (denoted 'H') are only used for the JPP-stream media type. Again the in-class identifier for tile header data-bins is the tile number (t), and its content is the concatenation of all markers and marker segments belonging to the tile. It does not include any start of tile (SOT) marker and segment because this is redundant information on the tile index and data length. Also, progression order change (POC) markers and segments are not included since there is no progression ordering with JPP-stream media – knowledge of the interleaving of packets in the target image is not required to decode the precinct data-bins. The tile header data-bin does contain all other markers and segments from the tile-part headers. Of particular importance are the packed packet headers (PPT), if these were used, since they are the precinct packet headers.

6.2.3 Defining Metadata-Bins

Returning to metadata-bins, recall that the server enumerates metadata box indices by its own methods. All metadata could conceivably be included in the root metadata-bin 0, and thus metadata-bin 0 would be the entire logical target, since all JPEG 2000 boxes are held in the root box. Clearly, it is sensible for the server to represent the JPEG 2000 file in a more appropriate structure. The server is allowed to decompose in a hierarchical manner the structure of the file. However, there is to be no more than one consistent structure for serving a logical target. The standard provides the following example.

Figure 6.2 is an example of the box structure for a simple JP2 file. The server can restructure the file as shown in Figure 6.3. Metadata-bin 0 represents the top-level of the original file. Metadata-bin 1 is the contents of the jp2h box from the original file, which now does not reside within the root metadata-bin. Instead, the server has reorganized metadata-bin 0 using the placeholder box to point to metadata-bin 1 for the contents of the jp2h box. The placeholder box is a new JPIP box type and identifies the size and

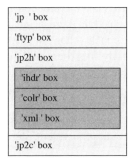

Figure 6.2   A sample JP2 file (after ISO/IEC 15444-9). Figures taken from ISO 15444.9: 2005 – Information technology – JPEG 2000 image coding system: interactivity tools, APIs and protocols are reproduced with the permission of the International Organization for Standardization, ISO. This standard can be obtained from any ISO member and from the Web site of the ISO Central Secretariat at the following address: www.iso.org. Copyright remains with ISO

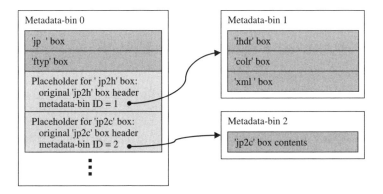

Figure 6.3   A sample JP2 file divided into three metadata-bins (after ISO/IEC 15444-9). Figures taken from ISO 15444.9:2005 – Information technology – JPEG 2000 image coding system: interactivity tools, APIs and protocols are reproduced with the permission of the International Organization for Standardization, ISO. This standard can be obtained from any ISO member and from the Web site of the ISO Central Secretariat at the following address: www.iso.org. Copyright remains with ISO

type of box to which it is pointing. The placeholder includes the original box header (in this case the jp2h box header) so that a client with the root metadata-bin knows that metadata-bin 1 contains the jp2h box contents. Note that since the jp2h box is a superbox, within metadata-bin 1 are the complete headers and the contents for the ihdr, colr, and xml boxes.

Placeholders can also point to code-streams, which can then be accessed by 'raw' byte-range requests. In this example, metadata-bin 2 has the raw contents of the jp2c code-stream but no headers, since a code-stream box is not a superbox. This is also called a 'raw code-stream' representation and provides an alternative method for the client to access the imagery data. Raw code-streams are served from metadata-bins. Contrast this with the incremental code-stream, i.e. the JPT-stream or JPP-stream media type. The placeholder provides a code-stream identifier (index) for the incremental code-stream. A server would offer the imagery as an incremental code-stream (typical) or as a raw code-stream, or both. However, note that there is no guarantee that these separate representations will be consistent with each other. Whenever a server offers an incremental code-stream from a JPEG 2000 family file, there will be a placeholder box containing the code-stream identifier with the value corresponding to its code-stream index in the JPEG 2000 family file. Placeholders can also point to 'stream equivalent' metadata-bins, which are boxes that act as an equivalent representation of some contents. An example is a server providing a more convenient stream equivalent when some content is cross-referenced to an external source (e.g. a URL). The equivalent stream is offered when it is a more appropriate representation.

6.3   JPIP Basics

6.3.1   Sessions, Channels and Cache Models

There are two types of JPIP requests that a client may use. The first are 'stateless' requests. State in JPIP refers to the existence of a client cache model that is persistent on the server over multiple requests within a session. The client may have a cache, indeed this is recommended, but in stateless requests the server does not maintain a persistent cache model.

Recall the earlier example of a printer's single-shot request; this is an example use case for a stateless request. To make this example more realistic, the client is on a memory-constrained mobile device, the task at hand is to print some segment of a large image as selected by the client, and the data of this printed image are beyond the available memory of the device. After the user selects the region of interest to be printed, the client sends to the printer (a) the required view-window (region and resolution parameters) defining the imagery for printing, plus (b) the identity of the server and logical target, and (c) the relevant contents of its JPIP cache. The printer may already have imagery data from this target in its own cache. The printer adds the client data to its own cache. The printer then uses the provided view-window to request imagery from the server. Since the printer is going to make one single request, the server need not be troubled about maintaining a cache model after this transaction, and so the printer would use a stateless request. To be dissemination efficient, this single stateless request identifies the relevant cache contents the printer already holds, so that the server need not send redundant data.

Only those complementary data to the printer's cache need to be sent by the server for the task of printing the image region at the appropriate resolution.

For efficient dissemination with multiple stateless requests, the server does not have the overhead of maintaining a persistent cache model; however, the client would have to signal its relevant cache contents for each stateless request. This potentially leads to server implementations with lower memory requirements, but it also obliges (for efficient dissemination) clients with higher overheads for computing relevant cache contents and larger request message payloads. It is important to note that servers still implement temporary cache models for stateless requests and thus there is no reduction in server computation. The server-side computation – identifying view-window data from the image and subtracting the relevant cache model to obtain the minimum response data – is the same for both stateless requests and those requests in a session. This is an important point to understand for implementers who are concerned about server performance; stateless JPIP transactions only mean that the server's cache models are not persistent.

The best approach for multiple requests – indeed nearly all interactive use cases – is to use requests within a session where the server maintains a persistent model of the client's cache. The benefit of a session is to remove from the client the requirement to inform the server of the relevant cache contents for each and every request. The assumption is that unless informed otherwise, the client caches all data sent to it within the session. The server can respond to requests in an efficient manner by sending only that subset of the image data (a) relevant to the view-window request and (b) not already sent to, and cached by, the client.

A session between a client and a server includes one or more logical targets, with a single data return type for each logical target, a server-side cache model for each logical target, and one or more 'JPIP channels.' A JPIP channel is a queue for JPIP client requests, plus their one-to-one mapped server responses, via an associated data transport. Only one JPIP request is active at any time for each JPIP channel. Multiple channels for the same target, and within a session, permit multiple concurrent requests to be put to the server, with the intention of them being serviced simultaneously using a single server-side cache model. As an example, consider a user who wishes to view an IROI with good fidelity, and the remaining background with lower fidelity, because there is a *very* capacity constrained network connection to the server. The IROI might well be presented as a highlight brush icon – wherever it sweeps the image quality improves. Alternatively, it may be presented as a magnifying glass icon – wherever it hovers the image at its center is presented at a higher scale. In both cases, the background remains at the base resolution and its quality improves gradually over time. Serving the imagery data for this example can be achieved by using a single session and two JPIP channels. The lower priority channel is for the whole image and the higher priority channel can be for the IROI. The IROI channel can be requested with a higher bandwidth capacity or bandwidth slice preference parameter than that for the background request.

It is through the tracking of transacted data by the server that efficient interactive dissemination is possible. It is now worth noting some of the finer aspects of cache modeling. First, the server may assume that the client has cached all of the disseminated data, and it may remove from the response data that which it believes has already been sent, but it need not do so. In JPIP, the server may not implement optimally efficient dissemination. The degree of dissemination efficiency by a server is but one of its implementation performance characteristics, along with response speed, memory usage, and its preemptive disposition. To save on processing and improve response speed, a server may not implement, or it may only partially implement, the cache model; i.e. the implemented cache model may be a subset of the total transacted data. To improve the interactive experience, the server may disseminate more data than that which meets the view-window request. However, while the implemented cache model is allowed to be less than the true record of the transacted data, it is not allowed to erroneously represent more. Typically, any competitive implementation is going to strive for dissemination efficiency by using the cache model to eliminate the dissemination of redundant data, and perhaps via a user selectable system option parameter, trade off optimal efficient dissemination for improved responsiveness using supplementary preemptive response data.

Use case examples where the client needs to inform the server on the status of its cache are clients that utilize a cache file obtained from another user or a previous session (e.g. the printer example above) and clients that need to purge their cache due to memory limits. Thus a server needs to be able to 'add' to and 'subtract' from its cache model when so informed by the client. Clearly, a net subtractive cache model manipulation can only be applicable within sessions. While additive data signaled in the client's request and response data sent by the server is to be added to the cache model, the former must be added prior to computing the response data. For a cache model manipulation request, the portions of the data-bins indicated by the request are added to, or subtracted from, the cache model prior to mustering the response data. The portions from data-bins disseminated are added to the model after the response dissemination is actioned.

The cache model is not, of course, the actual data disseminated. Rather, it is the identity of the data-bins and its byte-range assumed to be with the client. Since all data-bins and metadata-bins are delivered incrementally, the byte-range can simply be those bytes in the bin from byte 0 to some B bytes. For example, Hm:30 is the first 30 bytes from the main header data-bin and M1:20 is the first 20 bytes from metadata-bin 1. For JPT-stream media-type sessions, T1:10 is the first 10 bytes from the tile 1 data-bin. For JPP-stream media-type sessions, H4:20 is the first 20 bytes from the tile 4 header data-bin and P5:0 is zero bytes from the precinct 5 data-bin. An alternative pointer is the quality layer boundary; e.g. P5:L2 are the first two layers (i.e. packets) from the precinct 5 data-bin (precinct index $I = 5$).

It is to be emphasized that the server *may* add to the cache model that indicated in an additive cache model manipulation request, but it *shall always* subtract at least, if not exactly, that indicated in a subtractive cache model manipulation request. A server *may* add to the cache model the number of bytes disseminated in a response. The server *may* subtract from the relevant data mustered for a view-window request that data it believes to be redundant because it is identified in its cache model. If the server-side cache model is not exactly accurate in matching the data in the client-side cache, at the worst the cache model will be *less* than the client cache and therefore some redundant data may be re-sent. Except for the waste of network resources, this is generally of little consequence. On the other hand, if the cache model were allowed to indicate *more* (e.g. B_s bytes in a data-bin) than that actually in the client cache (say B_c bytes, $B_c \langle B_s \rangle$), the result would be that the server may not send some relevant bytes, thus leaving holes (i.e. the byte range $B_c + 1$ to B_s in the data-bin). Since data-bins are embedded bit-streams, the client would not be able to decode any data beyond the hole.

Servers of logical targets containing a large number of code-streams, such as a motion JPEG 2000 file, are likely to implement a partial cache model because of practicalities. For example, once a frame from a video has been disseminated, the cache model will possibly not be used again unless the client revisits that segment of the video. Thus, after some time, the server may decide to discard the cache model of old code-streams. (For some applications, the server implementation may not even use cache models.) In addition, clients may not even cache all the data received. To avoid significant subtractive cache manipulation communications, the concept of 'model set' (`mset`) was created for the JPIP standard. Using the model set, a client informs the server to limit the number of code-streams with cache models. The server can respond to a model set request by modifying the model set to inform the client what code-streams it is actually modeling. Clients will then know if cache model manipulation requests will be observed by the server; i.e. servers will only be cache-modeling those code-streams in its model set. In the absence of any model set, clients may assume that the server is maintaining a cache model for all code-streams involved in the session.

The basic caching assumptions and rules above will work properly with many server–client transactions, but be aware that requests and responses may be delivered out of order and possibly not at all. For example, while TCP-based transport does guarantee reliable and in-order delivery, the UDP transport does not have such guarantees. Implementers need to consider the behavior of servers and clients to make them robust. Transaction resilience will be discussed in the next section on the behaviors of JPIP clients and servers.

6.3.2 Behaviors

Already discussed are the principles for efficient dissemination using cache modeling, asynchronous or loosely coupled request–response signaling, preemptive data, preemptive requests, and the relaxation of requirements on servers to model client caches exactly. There are two further behavioral aspects of JPIP to be addressed. The first is the permission for servers to modify view-window requests and the second is JPIP transaction resilience.

JPIP data-bins are an abstraction of the data contained in the JPEG 2000 file. Usually some 'handle' implements this abstraction and in practice this handle is a data structure that occupies memory. Active data-bins are those bins that map to the client view-window request. The cache model consists of those data-bins that are active and those that were previously active but have become inactive. The status of inactive data-bins can simply be saved to disk, but typically active data-bins are processed from the main memory. Poor management of the cache model can lead to disk memory swapping when active data-bins are on disk rather than in the main memory. There can also be significant disk thrashing between the active data-bin pointers and the actual imagery data if these are not in the main memory. To avoid this, implementations often use other handles to the imagery within the main memory. Thus there can be considerable resources required by the server. As an example, one implementation of JPIP serving large imagery has been found to require 10 to 20 Mbytes of memory per client per code-stream cache model.

Of course, servers are resource limited. Too many active data-bins can exhaust the main memory resources. The approach taken in JPIP is to allow a server to manage its resources and reduce the set of active data-bins. It does this by reducing the view-window request parameters. How it determines this is implementation specific – it is not addressed in the JPIP standard – but a typical approach is to reduce the spatial extent of the view-window region of interest. Whenever the server does modify the request parameters, it must inform the client of the modifications using headers in its response. Sometimes, particularly for the first request, this is done with no or limited response imagery data. The purpose of signaling these modifications is to cue or hint to the client more appropriate values for its future request parameters, and the client should use these in subsequent requests on that target image.

Implementers should appreciate this behavior and the rationale behind it. The principle is that a client cannot issue a binding request that is unreasonable for a server to comply. If this principle was not fundamental to JPIP then servers could be open to inadvertent denial of service attacks. Modifying view-window parameters is a more graceful escape for a server than a 'service unavailable' response message. This same principle is also behind the many subclauses in the standard disallowing usage of multiple code-stream qualifiers, data-bin ranges, and wildcards. For example, it is unreasonable for a client opening a large motion JPEG 2000 video to expect the server to have cache models for each code-stream.

An early user of JPIP did not understand this principle and incorrectly believed that the server limit on view-window size was some bound within the standard. i.e. that the standard defined the limit behavior. They were using a Software Development Kit (SDK) implementation without appropriate consideration to configuring it for their application. One possible solution would have been to increase the server resources from those set by default in the SDK. Another would have been to use the JPIP client preferences to request processing of the full-window. This preference can be either 'progressive' or 'full-window.' 'Progressive,' the default mode, allows the server to modify the view-window request parameters, but on the implied condition that it serves the data in a progressive manner; i.e. data from each relevant data-bin are served in (approximately) quality layer priority. 'Full-window' forces the server to respect the view-window request, but it may disseminate data in a nonprogressive manner. For example, the server may process a request by performing the equivalent of dividing the view-window region into multiple view-window subregions, and then mustering and serving the response data sequentially from these multiple view-window subregions. The impact of this on the client is that data are not delivered quality layer progressively across the full view-window, but the client will receive a complete response for the original view-window region.

The second aspect to JPIP behavior is that JPIP does not depend upon assumptions of guaranteed delivery or order for requests and responses. The standard provides instructions on HTTP, TCP, and UDP-based message transports (discussed later), but the philosophy behind JPIP is indifferent to the message transport. (One could implement JPIP using e-mail!) Since the server's response data are dependent on the cache model, undelivered subtractive cache model manipulation requests can leave holes in the client cache with respect to the server-side cache model, as can undelivered data responses to the client. Out-of-order cache model manipulation requests can also potentially create holes. Recall that a hole is where the server-side cache model points further into a data-bin than actually cached on the client, and a client cannot decode any data beyond a hole.

There are two possibilities when a discrepancy between the cache model and cache exists. The first is when a subsequent view-window request relevant to the mismatched data-bin generates response data; these data will be subsequent to the hole. When this response is received, there will be an obvious exposed hole in the cache since the client will be expecting contiguous data for the data-bin. In this case, the client can make its next view-window request relevant to this data-bin include a subtractive cache model manipulation that identifies the last contiguous byte of the data-bin in its cache (e.g. -P54:234 means that the client has at most the first 234 bytes of precinct data-bin 54).

The second possibility is where the server responds with no response data and a 'window done' (or 'image done') response code. This will result in a hidden hole, since the server thinks the data-bin is complete in the client's cache and will not send any more data, whereas the reality is that the client cache is not complete. For the client, there may be no obvious method of detecting this state by inspection of its cache contents. The main, tile, and packet headers could be used to determine expected lengths of the data-bins, but if the headers are part of the missing data then this is problematic. The following handshake between the client and the server is a simple solution. After receiving a 'window done' response code, the client repeats the view-window request (assuming that it takes account of any server view-window parameter modifications – this view-window request must be relevant to the data-bin with the hidden hole) and includes a subtractive cache model manipulation that identifies the length of the relevant cached data-bins. The response to this will either be another 'window done' with no response data (i.e. the data-bins do not have any holes) or it will include response data for those data-bins with hidden holes.

As a final note, recall that the only adverse consequence when the cache has more data than that indicated by the cache model is the possible retransmission of redundant data. Along similar lines to other aspects of JPIP, there is no strong requirement for servers strictly to follow additive cache operations with their cache model; cache models need not be implemented with high accuracy. It is implementation specific how accurate or approximate the server instantiates the cache model with respect to additive operation. Of course, the more accurate the model, the more potential there is for dissemination efficiency (i.e. less redundant data), but there are no specified requirements for implementations except that a server must accurately implement subtractive cache model operations, and do so before formulating its response.

6.3.3 Compliance

Whenever a client makes a request, the server has the ability to modify the request as it sees fit, but it is obliged to inform the client of the modifications via its response. Initially, this is somewhat difficult to grasp by implementers. It also makes the meaning of compliance subtle. The reason for this approach is that the server must be able to control its resources and performance, as described in the previous section. A JPIP standard where the client could demand certain functionality with guarantees would have been unwieldy for both the standard's documentation and any implementation. The JPIP approach is not to be prescriptive, but to allow implementers freedom. This moves the performance implications of specific 'use case scenarios' from the purview of the JPIP standard to the domain of the implementers. It becomes a point of market differentiation how well an implementation performs for a given scenario.

There are many details regarding compliance, but the following are the main issues:

- The client makes properly formed JPIP requests and correctly decodes JPIP responses.
- The server processes JPIP requests appropriately and produces properly formed JPP- and/or JPT-stream responses.
- Servers inform clients of modifications to their requests via the response headers.
- Servers implement subtractive cache manipulations properly.
- A client uploading to a server produces properly formed JPP- and/or JPT-streams and the server processes them accordingly.
- Client and server implementations obey all the 'shall' statements in the standard as required by their profile. (JPIP profiles are presently being developed. Clearly, a JPT-stream-only server or client need not observe 'shall' statements for JPP-stream implementations.)

Implementations are customizable for different use cases, and their performance is the basis of market differentiation. Implementers of JPIP must appreciate this and understand the engineering tradeoffs possible. For example, a TCP-based JPIP system on a quality-of-service challenged network is a poor technical design. A UDP-based JPIP system implemented for robust dissemination and tuned for network characteristics will do the job better. However, on a typical IP network the TCP system may be more efficient and perform better.

The two challenging aspects of JPIP compliance is that (a) there are many 'may' statements in the standard and (b) there are many different responses that are 'appropriate' for most JPIP requests. The following is probably the best and most succinct expression of the JPIP philosophy on compliance. A server that responds to a nonempty region-size view-window request, with a correctly formed legal response but containing no image data, may be compliant, but one would not expect it to compete well in the market. A final note: JPIP implementations are to JPIP compliance what a bowl of fruit is to the question 'what's to eat?' Therefore, take care not to compare apples with oranges.

6.4   Client Request–Server Response

This chapter cannot exhaustively discuss JPIP client requests and server responses, but it will cover some of the most common usage. Implementers must refer to the JPIP standard for the full details. JPIP requests are formed by concatenation of request fields of the following types:

- target identification fields;
- session and channel management fields;
- view-window request fields;
- server control request fields;
- cache management request fields;
- metadata request field;
- client capabilities and preference fields.

JPIP responses consist of:

- a status code and reason phrase that is typically transport specific;
- JPIP response headers;
- JPIP response data body;
- a transport specific end-of-response (EOR) header and message containing a JPIP defined reason code.

The response headers provide JPIP messages from the server to the client. Recall that the JPIP philosophy is that the server is free to respond to most JPIP requests in any way it desires. The server is obliged to inform the client of the modifications via the response headers. Thus there is a correspondence between most request fields and the server's response headers. The response data are JPT- or JPP-stream media for the incremental code-stream. Other data types are 'raw' and media types such as JPEG and TIFF.

6.4.1 Target, Session and Channel Identification

The purpose of the target field is to identify the original named resource.' This is in the first request by the client and specifies the name of the target image, usually by a *path*. The server assigns a target ID to this named target and informs the client by the target ID response header. This target ID identifies the 'logical target' and the server guarantees to use it to represent the original named target consistently. This is required since the server may transcode the JPEG 2000 decomposition of the original named target image and structure the metadata according to its own devices. If the server does change its representation of the named target, then it will have to assign a different target ID for each different representation. A server that was not constant in its representations of target images should log the representation of the source for each target ID. Even servers with constant representation methods – e.g. one that does not transcode the served image – may have version updates for a named image resource. Thus it is good practice for the server to generate a unique target ID for each combination of original named resource, source version, and representation.

The purpose of the target ID is so that client cache contents can be reused consistently within and between sessions where the target ID is the same. Consider a client that obtains from a local repository the logical target's name (path), cache contents (from some other user's previous session), and the target ID associated with that cache. The client would make an initial request using the target (path) and target ID request fields. If the server can respond with consistency, it would return no target ID response header – implying no change in the target ID value. This will be the case if the original logical target source is unchanged and the server remembers how it represented the file for serving that target ID. The client can use this cache in its session, which could be a considerable saving in dissemination data. On the other hand, imagine that the original image file was replaced by a new updated version. The server will not be able to respond consistent with the first target ID, and it will instead respond with a new target ID value. In this case, the client knows to discard the old cache contents.

A request is stateless unless it includes a valid `channel-id` or a request for a new channel. The client requests a new JPIP channel using the `cnew` request field and specifying one or more transports, such as 'http,' 'http-tcp,' or 'http-udp.' (For example, the 'http-tcp' mode consists of the JPIP response headers in the primary HTTP reply paragraph and the JPIP response data body transported by the TCP connection.) If the server can comply with this request, using one of the indicated transports, it will return a new channel identifier in the response header, and the response data body for this request will be the first data body in the new channel. The response can also include optional transport parameters. For example, the server may direct the client to send all future requests using the identified path, such as /jpip.cgi for the server's JPIP common gateway interface. The server would indicate which of the requested transports it is using for this channel (e.g. http-tcp) and, if required, the relevant port.

Once a channel is established, the channel ID is the link to the session, since JPIP does not have any session identification signaling. The channel ID also associates the logical target ID with the session's cache model(s). New channels may be added to the session, on either the same logical target or new targets, but all channels in a request must belong to the same session. Multiple channels on the same target ID can be used to implement flow control for different tasks, e.g. to prioritize foreground over background requests. The client will use one cache for both channels and the server implements a single cache model.

The following is an example establishing two channels on a logical target in a session. The first request starts the session for image1.jp2. The channel is the 'http' type and the media type is the JPP-stream. The server responds with a unique target ID. (For demonstration purposes, the example uses a simple scheme to code the image file, its date, a default representation flag – i.e. no transcoding – and a pseudo-universal unique ID. An implementation would take these and other parameters and hash them to form a true UUID code). It also responds with a channel ID. (Again, for demonstration purposes, it consists of an explicit session code, client code, and channel code.) The client then requests a new channel on the same target using the http-tcp transport system. (The request explicitly uses the request ID field, qid, set to 1. Implementers can use the request ID to ensure that request–response pairs are processed in order.)

First request–response pair:

```
Client> target=image1.jp2&cnew=http&tid=0&type=jpp-stream
Server> JPIP-tid: image1_2007_12_25_rep_default_uuid_code_123
Server> JPIP-cnew: cid=Session01_Client01_Channel01,
     path=jpip.cgi,transport=http,
Server> JPIP-type: jpp-stream
```

Second request–response pair:

```
Client> cid=Session01_Client01_Channel01&cnew=http-tcp&qid=1
Server> JPIP-cnew: cid=Session01_Client01_Channel02,
     path=jpip.cgi,transport=http-tcp,auxport=80
Server> JPIP-type: jpp-stream
Server> JPIP-qid: 1
```

Each channel on the server has its own request queue that may be processed on a 'first-in' basis, unless requests have been given a request ID (qid) by the client. When the client does use a request ID, the server shall not process this request until it has processed all those requests with a smaller request ID. Note that when processing requests the server is not obliged to complete each queued response in full. Certain parts of a request, such as subtractive cache manipulations, must be completed, but the server need not complete the mustering nor the dissemination of the requested imagery data. (The exception is a request preceding a `wait=yes` request, which will be discussed later.) The request ID value is a unique request field because it is the only one that the servers cannot modify, yet it must return the same request ID value.

6.4.2   View-Window Region

The view-window is the interactive region of interest within the target image. The view-window is the inner region shown in Figure 6.4, defined by the region size parameters `sx`, `sy` with offset `ox`, `oy` relative to the frame size for the whole image code-stream given by `fx` and `fy`. The frame size is the size of the whole image relative to the resolution of the region size. This is because a client making its first request will not know the image size, and consequently the sizes of available resolution levels. For example, a client with a 640 × 480 display and wanting the whole image would make a request:

```
fsiz=640,480&rsiz=640,480&roff=0,0.
```

If the client wished to view the top-right quarter of the image, then the request would be

```
fsiz=1280,960&rsiz=640,480&roff=640,0.
```

These requests can be made without knowledge of the size of the original image.

It is expected that the requested resolution will often not be available; the server can respond with either of the closest higher or lower resolution levels available from the code-stream. As shown in Figure 6.5, the suitable region parameters `sx'`, `sy'` and `ox'`, `oy'` are relative to the suitable code-stream image resolution frame size parameters `fx'` and `fy'`. The client may request the direction of the rounding to the next available resolution level, but it must be prepared for return data that does not match the request

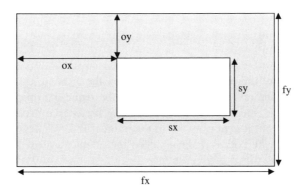

Figure 6.4   A desired view-window within an image (after ISO/IEC 15444-9). Figures taken from ISO 15444.9:2005 – Information technology – JPEG 2000 image coding system: interactivity tools, APIs and protocols are reproduced with the permission of the International Organization for Standardization, ISO. This standard can be obtained from any ISO member and from the Web site of the ISO Central Secretariat at the following address: www.iso.org. Copyright remains with ISO

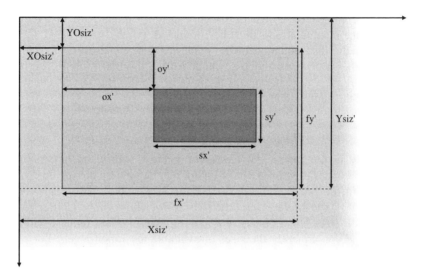

Figure 6.5 A desired view-window with respect to the subsampled reference grid (after ISO/IEC 15444-9). Figures taken from ISO 15444.9:2005 – Information technology – JPEG 2000 image coding system: interactivity tools, APIs and protocols are reproduced with the permission of the International Organization for Standardization, ISO. This standard can be obtained from any ISO member and from the Web site of the ISO Central Secretariat at the following address: www.iso.org. Copyright remains with ISO

parameters exactly. (Options are 'round-up,' 'round-down,' and 'closest.') Whenever the server modifies request parameters, it notifies the client via the response headers.

The size of the suitable code-stream image resolution is given by

`fx'=Xsiz'-XOsiz'` and `fy'=Ysiz'-YOsiz'`,

where

$$XOsiz' = \left\lceil \frac{XOsiz}{2^r} \right\rceil; \quad YOsiz' = \left\lceil \frac{YOsiz}{2^r} \right\rceil; \quad Xsiz' = \left\lceil \frac{Xsiz}{2^r} \right\rceil; \quad \text{and } Ysiz' = \left\lceil \frac{Ysiz}{2^r} \right\rceil.$$

`XOsiz` and `XOsiz` are the offsets of the image on the JPEG 2000 canvas. The server determines the resolution level r in order to match the requested image size (`fx` and `fy`) as closely as possible, subject to any client rounding preference. Note that r must be an integer no less than 0, but it is not limited to the number of DWT levels used to compress any tile component in the code-stream – indeed the client may not know the limit. The suitable region size and offset is given by

$$ox' = \left\lfloor ox \cdot \frac{fx'}{fx} \right\rfloor; \quad oy' = \left\lfloor oy \cdot \frac{fy'}{fy} \right\rfloor; \quad sx' = \left\lceil (sx + ox) \cdot \frac{fx'}{fx} \right\rceil - ox';$$

$$sy' = \left\lceil (sy + oy) \cdot \frac{fy'}{fy} \right\rceil - oy'.$$

6.4.3 View-Window Requests

Table 6.1 shows the most common view-window request parameters. If the frame size request field is omitted from a view-window request, then the response will have only the main header. Clients may use such an initial request to obtain the main image header, which is useful for formulating subsequent requests. If the region offset is not present, the default is the top-left corner (0,0). If the region size is not present, then the default is the region that extends to the lower-right corner of the image. If the region size is empty (0,0) the response should not include any compressed imagery data. This is approximately equivalent to omitting the frame size and the server may respond with the main header.

For the client to request imagery data, it should use the frame size, region offset, region size, and optionally the component request fields. Table 6.2 shows some of the frequent response headers. The server should reply with corresponding response headers only when it has modified a value. In the example below, the client has a 640 × 480 display, and requests the top-right quarter of the image. It does not know that the image is actually 4096 × 4096 with three resolution levels. The server's response selects the lowest resolution level at the top-right corner. Note that the region offset is modified, but not the region size. In the second request for the top-left corner of the image, the client now knows the appropriate frame size and the server does not need to modify any parameters. The only response header is the request ID, which, unlike other headers, must not be modified and must be retuned.

Table 6.1 View-window request fields

Frame size	"fsize" "=" fx "," fy ["," round-direction]
Region offset	"roff" "=" ox "," oy
Region size	"rsiz" "=" sx "," sy
Components	"comps" "=" 1#uint-range
Region of interest	"roi" "=" region-name
Layers	"layers" "=" uint

Table 6.2 View-window response headers

Frame size	"JPIP-fsize: " fx "," fy
Region offset	"JPIP-roff: " ox "," oy
Region size	"JPIP-rsiz: " sx "," sy
Components	"JPIP-comps: " 1#uint-range
Region of interest	"JPIP-roi: "
	("roi" "=" region-name ";"
	"fsiz" "=" uint "," uint ";"
	"rsiz" "=" uint "," uint ";"
	"roff" "=" uint "," uint ")
	/ ("roi=no-roi")
Layers	"JPIP-layers: " uint

First request–response pair:

```
Client> cid=Session01_Client01_Channel01&qid=1&
        fsiz=1280,960&rsiz=640,480&roff=640,0
Server> JPIP-fsiz:1024,1024
Server> JPIP-roff:512,0
Server> JPIP-qid:1
```

Second request–response pair:

```
Client> cid=Session01_Client01_Channel01&qid=2&
        fsiz=1024,1024&rsiz=640,480&roff=0,0
Server> JPIP-qid:2
```

The ROI request field specifies one or more named regions of interest in the target to define the view-window(s). The available ROI may be defined in the ROI description box within the logical target or by implementation-specific means by the server. The special region name 'dynamic' is a nonconstant spatial view-window implemented by the server. As an example of a dynamic region, consider a surveillance camera system to capture a still image of the faces of people moving through an airport queue. The dynamic ROI here would be the bounding box around the face as detected by the server. Since the people are moving through the queue, this region will be different from one face to the next. For ROI requests, the server's response header can identify the ROI name, frame size, region size, and offset. If the server is unable to fulfill the ROI request, its response is

```
Server> JPIP-roi:roi=no-roi.
```

The `layers` request is grouped in the standard with the view-window request fields, but it works to limit the response data as per other server response controls. The reasoning for its inclusion as a view-window request is that the client uses it to specify the quality layers of interest. These are the JPEG 2000 quality layers given by the precinct packets. By default, all quality layers are of interest; when the layers request field is used, the server should limit the response data relevant to the view-window to no more than the specified layers.

6.4.4 Server Response Controls

The data limiting and server control parts of a request ask for certain behavior by the server. The purpose of data limiting a response is to improve response data flow-control performance. By providing these controls, the client and server can tune the size of the response data for their network conditions. On congested or quality-of-service challenged networks it is good practice to use a sensible value for the maximum response length request field. Without any data limiting request field, the server will transmit to the client until either all the relevant data are disseminated (window done or image done) or the response is preempted by the arrival of a new request (that does not use `wait=yes`; see below). Note that though the `layers`, `quality`, and alignment requests can limit the response data, they can be ineffective. For example, when the image has only one quality layer or is very large, the response may still be significant. The safest practice is to use the maximum response length request field and set it to a sensible value.

The client's suggested length for the response is given in bytes. If this limit is too small, the server may not be able to respond at all without exceeding the limit. In this case, the server should respond with no response data and use the `JPIP-len:` response header to indicate to the client the smallest value appropriate to allow it to respond. When the client uses this or a larger value, the server responds with approximately no more than this data limit. It is implementation specific how approximate. If `len=0`, the server responds with response headers but no data. The `quality` request field is similar to the layer limiting request field, except the server uses an abstract 0 (lowest) to 100 (highest) scale to define this quality. (These are not the JPEG 2000 quality layers.) The means for this abstraction is server implementation specific, and the server may not offer this feature at all. Byte-range chunking of data-bins is a relatively simple mode for implementing the 100-point quality score. More complex would be for the server to analyze the imagery for PSNR quality points and map these to the 100-point quality score. This abstract quality control was envisaged to provide finer granularity access for cases where an image had few quality layers. The server can include a quality response header to inform the client what quality the response data will provide, assuming that the response is completed without interruption (preempted). It should do this if the quality achieved is not as requested. This can occur, for example, if the response data limit was reached at a point of less quality than that requested. If the server does not support this quality abstraction, then it shall return a `-1` value when the client uses the `quality` request field. Note that the quality value only refers to the view-window used in the request.

Alignment is another data limiting option. The client can request the server to terminate its response at a 'natural boundary.' For precinct data-bins, this is the end of a packet (i.e. a quality layer). For tile data-bins, this is the end of a tile part. For all other header data-bins and metadata-bins, this is the end of the header or box. The server responds when requested with `align=yes` by first determining the potential response data, including any maximum response length limits. For each data-bin in the potential response that crosses one or more natural boundaries, only the data up to the last natural boundary are disseminated. Alternatively, if this amount of data does not cross a natural boundary, then all these data are disseminated. This means that the response can still be partial packets or tile parts, even though the alignment request field is set.

By default JPIP requests are preemptive, which is the most common means for flow control by clients. Preemptive requests enable a user on a JPIP application to browse interactively with imagery without awaiting a server to complete its response for each and every request. Indeed, as a user dynamically adjusts their IROI, earlier queued requests are simply expired regions of interest. The server should quickly process them with a nil data response. With the strict exception that the server must process cache model request fields within every request it receives, it can otherwise terminate its data mustering and dissemination processing for a previous request according to its own policies and devices. This is implementation specific; however, it is not to treat a preemptive request as an interrupt to its processing. The only requirement by the standard is that the server must properly and legally complete any messages and response data packing that are already in transit, but it can curtail dissemination of the mustered response data at any point. For example, a good implementation would have the server terminate the response data at the next earliest natural boundary of the data-bins if the alignment request field was set. This indeterminate behavior policy was used because servers will have significant computation and data mustering processing when formulating a response. Without a flexible policy for the server's management of preemptive requests, there would be considerable wasted overheads on server resources when a client transmitted frequent preemptive requests.

If a request is not preempted and processed long enough by the server, then after sending all the requested data the server may elect to disseminate preemptive response data. Consider the possibility of loosening any optimal dissemination objective and allow the server to send first that data required for the current view-window request, and then preemptively to use otherwise idle network capacity to send other data it considers useful. An example is to preemptively send data for a larger spatial extent centered by the current view-window, plus data from the next higher resolution level from within the current view-window, on the basis that the next user request is likely to be driven by a pan or zoom operation on the current IROI. Another example is to preemptively send metadata (say the title and abstract) that is spatially related to the view-window request. Except for those capacity challenged networks, this preemptive data will cost little in terms of the timely dissemination of the IROI response data, because it will mostly use idle capacity. However, it can significantly improve the responsiveness of the interactive browsing experience for the user.

In JPIP, a server can always transmit data it believes appropriate, and disseminating preemptive data is usually more advantageous than an idle server-side response data queue and idle network capacity. This is part of the reasoning behind JPIP *not* having tightly coupled and synchronized client–server interactions. Although note for servers deployed on very capacity-challenged networks, it is important to implement a tighter policy (i.e. minimal preemptive server behavior) to optimize dissemination efficiency. The JPIP standard allows for an implementation tradeoff between highly efficient dissemination and more responsive, but slightly less dissemination efficient, server interaction.

A client may signal the server not to preempt its last previous request by using the wait=yes request field. When a request with wait=yes arrives, the server shall complete the current response to the previous request entirely while queuing all newly arrived requests with wait=yes. Unless clients use the request ID, the server may simply assume the request order is the received order. If received requests were out of order, there may be adverse consequences, particularly when subtractive cache manipulations are involved. Therefore, it is good practice to use the request ID if the wait field is going to be used within the session. While a request with wait=yes is queued by the server, any new preemptive request (wait=no) will remove it from the queue and also curtail the current request. This is very useful when a request that is followed by a request with wait=yes is generating a data flood that the client wishes to terminate. (This is also an example where the client should have been using the response data limit field.) A client using the request ID with wait=yes can queue multiple requests on the server until the server reaches its queuing limit. In such a condition the server would respond with a '503 service unavailable' status code and reason phrase. It is good client practice not to issue too many queued requests.

In the first view-window request on an image in a session, the client can specify the media types it wishes to receive using the image return type request field. The image-return-types are media-types specified in RFC 2046 (e.g. image/jp2, image/jpeg) and JPIP reserved values of jpp-stream, jpt-stream, and raw. There is an option to specify the use of extended precinct and tile data-bin headers – this makes the response data's headers independently self-describing for more error resilient signaling. If the server can respond using one of the requested media types, it will reply using either the `JPIP-type:` response header in the JPIP response header, or via other means, such as an HTTP content type header. In a session, once the return type has been determined, there is no further need to use this request field, and indeed it is illegal to use the request field to specify a different return type. Use a media type like JPEG for simple screen-scraping applications via JPIP requests. The `raw` media type is for direct access to byte-ranges in the raw code-stream.

6.4.5 Cache Management

The accuracy of the cache model is important for efficient dissemination. Recall that a stateless request is one where the server does not maintain a cache model after processing the request. However, the server still implements a temporary cache model, but after the request is processed, the cache model is scrapped. Thus both stateless and session-based requests can include cache model instructions. Clients with JPIP sessions must inform the server of purges to their cache. Clients that add data to their cache from sources other than a JPIP channel, and wanting efficient dissemination, should inform the server of these data when it is relevant to a request. This is achieved using the `model` request field. It is a list of data-bins and metadata-bins, qualified by a code-stream or code-stream range, and each identifying that data added to, or subtracted from, the client's cache. (Note for brevity that metadata-bins are a type of data-bin and are implied in the following discussion.)

Table 6.3 shows the ABNF syntax for the cache management request fields. In a session, the model request field modifies the cache model on the server. In a stateless request, the server's cache model starts empty and the model information can only be net additive. The server processes the cache model instructions in the order listed in the request and always prior to formulating its response. Subtractive cache manipulations are prefixed by a minus sign before the *bin-descriptor*. The model request may reference any

Table 6.3 Cache management request fields

Cache model	`"model" "=" 1#( [codestream-`
	`qualifier "," ]`
	`["-"] 1$bin-descriptor )`
	`codestream-qualifier="["1$(codestream-range)"]"`
	`codestream-range = first-codestream`
	`[ "-" [last-codestream] ]`
	`bin-descriptor = explicit-bin-descriptor /`
	`implicit-bin-descriptor`

Tile-part model	`"tpmodel" "=" 1#( [codestream-`
	`qualifier ","] "-" tp-descriptor)`
	`tp-descriptor = tp-number / tp-range`
	`tp-range = tp-number "-" tp-number`
	`tp-number = tile-number "." part-number`

Stateless request needs	`"need" "-" 1#( [codestream-` `qualifier "," ]bin-descriptor )`
Stateless tile-part needs	`"tpneed" "=" 1#( [codestream-` `qualifier "," ] tp-descriptor )`

Table 6.4 Explicit data-bin descriptor

```
explicit-bin-descriptor = 1*explicit-bin [":"
    (number-of-bytes / number-of-layers )]
explicit-bin = codestream-main-header-bin
    / meta-bin
    / tile-bin
    / tile-header-bin
    / precinct-bin
number-of-bytes = uint
number-of-layers = %x4c uint                                    ; "L"
codestream-main-header-bin = %x48 %x6d                          ; "Hm"
meta-bin = %x4d bin-uid                                         ; "M"
tile-bin = %x54 bin-uid                                         ; "T"
tile-header-bin = %x48 bin-uid                                  ; "H"
precinct-bin = %x50 bin-uid                                     ; "P"
bin-uid = uint / "*"
```

Table 6.5 Implicit data-bin descriptor

```
implicit-bin-descriptor = 1*implicit-bin
    [":"number-of-layers ]
implicit-bin = implicit-bin-prefix
    (data-uid / index-range-spec)
implicit-bin-prefix =    %x74     ; t - tile
                     / %x63     ; c - component
                     / %x72     ; r - resolution
                     / %x70     ; p - position
index-range-spec = first-index-pos
    "-" last-index-pos
first-index-pos = uint
last-index-pos = uint
bin-uid = uint / "*"
```

data-bins, both inside and outside the view-window. A model instruction on data-bins outside the view-window may not affect this response, but it may affect subsequent responses in the session.

The data-bins are identified either explicitly (Table 6.4) using their in-class bin index, or implicitly (Table 6.5) by their membership of a specified JPEG 2000 tile, component, resolution level, and position description. The explicit version identifies the data-bins by their class type descriptor (tile T for the JPT-stream; tile header H and precinct P for the JPP-stream) and their in-class bin index. A special case is the main header labeled as "Hm" with no index. The implicit data-bin descriptor refers to the data-bins by their association with the JPEG 2000 tile, component, resolution level, and position.

Use of the wildcard (*) means all those data-bins of the specified class are relevant to the view-window. It is restricted to stateless requests and limited by the view-window for two reasons. First, a client could inadvertently make more data-bins active on the server than it can manage. Since the server can modify the size of the view-window, limiting the wildcard to the view-window keeps the control of server resource requirements with the server. Second, the standard does not define the method for limiting active data-bins in the view-windows. Therefore, using the wildcard within a session could lead to cache model errors, particularly for subtractive manipulations. Such cache model errors may affect the current and future responses. Therefore the use of wildcards in cache model requests within a session is not allowed. On the other hand, the processing of stateless requests with wildcards in cache model fields starts with an empty cache model, can only have a net additive model manipulation (net positive can be after first addition and then subtraction), and is limited to the data-bins relevant to the view-window in the request. For example, the client could tell the server that it has all metadata-bins except metadata-bin 5 using `model=M*,-M5` in a stateless request. Now the fact that this may not be true (the client may not have all other metadata-bins) does not matter, since the stateless request is completely concluded once the server finalizes its response. Hence this is an example of how a client could specifically request one data-bin of interest.

Each data-bin can be qualified by a number of bytes. An additive bin-descriptor qualified by B bytes means that the client has, at the least, the first B bytes of the data-bin. The server should add this to its cache model. A subtractive bin-descriptor qualified by B bytes means that the client does not have data after B bytes of that data-bin. Typically, the client is using the subtractive model request because it has removed from its cache the data from $B + 1$ byte in the data-bin. However, the server shall not imply from this that the client does have the first B bytes; it shall only remove from its cache model the data from the $B + 1$ byte, but it shall not add anything to the cache model. Precinct data-bins can also be qualified by the number of JPEG 2000 quality layers (packets). This indicates the number of layers the client has at the least (additive) or at the most (subtractive). The server will add to, or subtract from, the data-bin according to the last byte in the specified layer. For example, `model=-H*,-P*:L3` tells the server that the client does not have any view-window relevant tile headers nor precinct data after the third layer. Where no byte or layer qualification is used, the default is the entire data-bin.

A code-stream identifier may prefix a set of data-bins, where the code-streams each contain the identified data-bins. Refer to the standard for details on using the model request for multiple code-streams and the default conditions, but basically the code-stream identifier takes precedence. If absent, the default is the value in the code-stream request field (see below); if that is absent, it is code-stream 0. Finally, if the *last-codestream* is not present in the *codestream-range* of the qualifier, and the optional hyphen ("-") is used, then the code-stream range is from the *first-codestream* and all subsequent code-streams. For example, the following stateless model request:

```
stream=0-9&model=Hm,H*,[0-8],P*,[5-],-P*:L3,[9],P0:800
```

identifies that the view-window request is limited to the first 10 code-streams, and the client has the main header and all the tile header data-bins relevant to the view-window. The server should add these to the cache model. Next identified are all precinct data-bins, relevant to the view-window, for the first 9 code-streams. The server should add these to the cache model. Next, the client identifies a subtractive model manipulation for code-streams 5 to 9. The client does not have any of the layers after the third, for the precincts relevant to the view-window, in these code-streams. The server shall subtract these if they are in its cache model (i.e. only for code-streams 5 to 8). Note that this example demonstrates why the server shall not interpret any additive implications from a subtractive instruction – the subtractive instruction for code-stream 9 does not imply that the client actually has the first three layers in those precincts. Lastly, the client says it does have the first 800 bytes of the first precinct in code-stream 9 and the server should add them to its cache model.

The implicit data-bin descriptor refers to data-bins belonging to tile components, resolution levels, and precinct position. The indexes for tile, tile part, component, and resolution start from 0 and are the same as those used in JPEG 2000 Part 1. The indexes for position are the precincts counted from 0 in raster-scan order in the tile–component–resolution. Wildcards and index ranges can be used in stateless requests and are limited to those items relevant to the view-window. The index range for tiles and precinct position makes a rectangular region with the top-left corner specified by the first indexed item and the bottom-right corner specified by the last indexed item. If a tile, component, resolution level, or precinct position index value is missing, then it means all items relevant to the view-window. For example, `model=t0c*r0-2p*:L2` identifies the first two layers of all data-bins (relevant to the view-window), from all components, in the first three resolution levels of the first tile. This is equivalent to `model=t0r0-2:L2`. Note that here only the number of layers can specify the depth into a data-bin; byte depth is illegal usage for the implicit form. The constraints of using ranges and wildcards to only stateless requests and byte depth to the explicit form leads to the summary in Table 6.6. For all model requests that have a channel ID (i.e. are in a session), the wildcard and index range are not allowed.

The JPT-stream deals with tile data-bins consisting of the tile parts. In addition to the model request field, a client using JPT-stream media can also specify cache model manipulation using the `tpmodel` request to add or subtract tile parts or a range of tile parts. Tile parts are specified by *tile-number.part-number*. The range of tile parts works on a block arrangement, such that

$$tn_A.tp_M - tn_B.tp_N = \bigcup_{i=A}^{B} \bigcup_{j=M}^{N} tn_i.tp_j.$$

For example, 0.0–1.2 are tile parts {0.0, 0.1, 0.2, 1.0, 1.1, 1.2}. There are no other parts from tile 0 or tile 1, nor parts from any other tiles in this set.

While model and tile-part model requests add or subtract from the cache model on the server, they do not directly request specific data, as the data disseminated is a function

Table 6.6 Cache descriptor option summary (after ISO/IEC 15444-9). Tables taken from ISO 15444.9:2005 – Information technology – JPEG 2000 image coding system: interactivity tools, APIs and protocols are reproduced with the permission of the International Organization for Standardization, ISO. This standard can be obtained from any ISO member and from the Web site of the ISO Central Secretariat at the following address: www.iso.org. Copyright remains with ISO

| Form type | Wildcard | | Index range | Number of layers (e.g. ':L3') | Number of bytes (e.g. ':256') |
	Stateless	Session			
Explicit form	Allowed	Not allowed	Not allowed	Allowed	Allowed
Implicit form	Allowed	Not allowed	Allowed only for stateless	Allowed	Not allowed

of the cache model and the view-window on the image data. The need and tpneed request fields is a direct way to specify the client requirements for those portions of the data-bins listed and relevant to the view-window. They are only allowed in stateless view-window requests, are not allowed with a model or tile-part model request, but their syntax is otherwise similar to the model fields, except that the subtractive hyphen "–" is not allowed. The request is moderated by the view-window and the data-bins requested must be relevant to the view-window.

Recall that in the stateless request the server initializes an empty temporary cache model. Both model and need requests manipulate this temporary cache model. Recall that to specifically request a data element using the model field would consist of the entire list of all those data-bins relevant to the view-window, less the request element. Need requests are actually a shorthand means for implementing this type of model plus the view-window request. For a need request, the server adds all relevant data-bins relevant to the view-window request to the cache model, and then subtracts from this set those portions listed as 'needed.' The response data will be that data needed, subject to any data limits.

As an example, consider a target code-stream with several layers, one tile, three components, and an available LL resolution level of 1024×960. The client has a cache containing the first three layers of the LL subband and only wants the next layer in the first precinct. The following will request this next layer:

```
Client> target=image1.jp2&
        tid=image1_2007_12_25_rep_default_uuid_code_123&
        type=jpp-stream&fsiz=1024,960&layers=4
        model=Hm,H*,r0:L4,-t0c*r0p0:L3
```

To obtain the 0 precinct's next layer, rather than using the model request field above to specify a hole in the cache model at t0c*r0:L4 plus the layers=4 request field, the request need=t0c*r0:L4 would have achieved the same result. Thus the need and tile-part tpneed requests are sometimes a more direct means of specifying the data to be requested.

6.4.6 Metadata Requests

The philosophy of JPIP is that in general the server is free to respond to most JPIP requests in any way it desires; however, this response should be enough to enable the client to get the most out of the interaction. (How implementers interpreted this – deciding on their engineering tradeoffs – will be a key source of any market differentiation.) Metadata can have a complex structured in the JPEG 2000 family, since it is so very extensible. An effective JPIP server would analyze the metadata structure and reconstitute it to improve the interactive experience for users. Recall that when the server does this, it must remember this representation and keep it fixed for that target ID.

To demonstrate an extreme case, consider a server that presents all the metadata structure in one single superbox metadata-bin. In this case, the only means for a client to discover content is to byte-range seek through the metadata-bin. The client could byte-range seek to each metadata box header to build the skeleton of the metadata structure, but this could be at the cost of many request–response cycles. Therefore it is very desirable that instead the server intelligently restructures the metadata boxes into

Table 6.7 Metadata request fields

Metadata request	"metareq" "="
	1#("[" 1$(*req-box-prop*) "]"
	[*root-bin*] [*max-depth*])
	[*metadata-only*]

req-box-prop = box-type [limit]
            [metareq-qualifier] [priority]
limit = ":" (uint / "r")
metareq-qualifier = "/" 1("w"/"s"/"g"/"a")*
priority = "!"
metadata-only = "!!"
root-bin = "R" uint
max-depth = "D" uint

an appropriate metadata tree. This may be equivalent to the structure in the source file, but is it typically appropriate to make a different hierarchical set of groupings suitable for interactive dissemination. The result should be a skeleton consisting of a top-level table of contents, with some entries having a few levels of subheadings. Using the metareq request field (Table 6.7), the client can quickly discover the existence of important metadata boxes plus bracketed subgroups, which can subsequently be opened should the client so desire.

When the client does not use the metareq request field, the client implicitly is requesting whatever metadata boxes may be required from the file in order to utilize the imagery response from the code-stream. The standard defines a list of the minimum box contents implied, such as the entire metadata-bin 0 box (restructured with placeholders and stream equivalents), all immediate subbox headers from the JP2 header ('jp2h') and code-stream header ('jpch') superboxes, and the entire contents of other boxes such as the image header ('ihdr') found within these superboxes. When using a view-window request, the client identifies one or more code-streams, with the default being the first code-stream (code-stream 0). Even when the frame size request field is omitted, the response includes the relevant code-streams main headers.

If the client explicitly does not want to access the imagery in a request, but only wants to access the metadata relevant to a view-window, it specifies *metadata-only=!!* in the metareq request field. A server does not include any of the above implicit imagery metadata, such as image headers, for a *metadata-only* request. For JPT-stream and JPP-stream media types, this means that the response data will consist of only metadata-bins. A use case example for metadata-only requests might be a number of synchronized geospatial-based image servers on a network, of which one contains the revenue generating advertising. The client establishes a session with the closest image server on the network. This will be the server for the imagery, but it will proxy for the advertising server, which is much further away. This structuring is appropriate, since imagery can be a rather large network load, while advertising metadata are relatively small. This closest interposing server will make metadata only requests to the far-server. To the user, this dual-server architecture is more responsive than simply using the single far-server.

Each metadata request is relative to the specified *root-bin* value, which defaults to metadata-bin 0 if not specified. All the metadata boxes within this *root-bin* are requested unless the *max-depth* value is specified, which limits the request to the number of levels traversed from the *root-bin*. The *req-box-prop* field specifies the *box-types* of interest; * is the wildcard for all boxes. (The box types are defined in other JPEG 2000 family standards.) The *limit* field allows the client to restrict the type and amount of information. It is either :*uint*, which limits the response to the entire box headers plus the first *uint* bytes of their contents for boxes of the specified type, or :*r*, which recursively includes only the box headers of the indicated box types, plus the box headers for all descendant boxes, regardless of their box type. Thus :*r* is a means for obtaining the skeleton of the box structure. Note that while *root-bin* and *max-depth* limits the depth in the metadata box hierarchy, when a nonzero byte limit is specified, the response may exceed this depth. In other words, the byte limit overrides *max-depth* if the byte-depth into the relevant boxes encroaches further than the *max-depth* limit.

After using the recursive option in the first few metadata requests, the client has a fair idea of what is available from the server. Further metadata requests can then access specific boxes of interest. The *metareq-qualifier* is a tool for limiting the boxes to specified box type(s). The 'window' qualifier (w) limits the context to all boxes associated with the specified view-window region. Of note is that boxes can be associated with the resolution level and specified components, as well as the spatial extent. The 'stream' qualifier (s) limits the context to all boxes associated with the code-stream(s) of interest and not associated with the view-window region. The 'get the rest' qualifier (g) includes all boxes relevant to the requested view-window and not included by the 'window' and 'stream' context – the sets of metadata boxes from w, s, and g contexts are mutually exclusive. The 'all' qualifier (a) requests all boxes in the logical target. Note that the set of all boxes (a) may be greater than the union of the sets from context w, s, and g. The *priority* flag (!) makes the metadata response have priority over the imagery response data.

As per other JPIP server behavior, the server shall send a metadata request response header if it modifies the *max-depth, limit, metareq-qualifier*, or *priority* values from that in the client's metadata request.

6.5 Advanced Topics

6.5.1 JPIP Proxy Server

JPIP can optimize the data transactions between a client and a server for interactive image browsing. It is possible to improve the dissemination efficiency further for a community of users by interposing a local JPIP proxy server as shown in Figure 6.6. To the remote server the JPIP proxy server functions as a client; to a local client it is the server. It maintains its

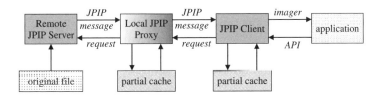

Figure 6.6 Interposing a JPIP proxy server

own cache of imagery data. It receives JPIP requests from a local client and forms its own JPIP response using data from its cache and its model of the client's cache. Either it is already in a session with the remote server for this logical target or it establishes one. The remote server models the JPIP proxy server's cache; it does not model the cache of any client of the JPIP proxy server. The JPIP proxy server passes the client's view-window requests through to the remote server, minus any cache management fields, because they are only relevant to the client's cache model on the JPIP proxy server. If the client's JPIP request can be fully satisfied from the local JPIP proxy server's cache, then no data needs to be sent from the remote server to the local JPIP proxy server. This would happen if the portion of imagery from the client's view-window were already cached on the JPIP proxy server, e.g. cached from other prior local client session(s). Data will only be downloaded from the remote server to the local JPIP proxy server if this is additional data (to that in the JPIP proxy server's cache) and is required to fulfill the client's view-window request. The use of a local JPIP proxy server in this configuration allows the data transactions to be optimized for dissemination efficiency for the local community of users. This will be especially beneficial to communities using remote server connections with limited network capacity.

6.5.2 Upload and Version Control

Clients can upload an image to a server using JPIP. This may be an entire image, but of more interest is the possibility to merge uploaded imagery with an existing image. An example use case is users updating collaborative imagery. This could be limited to adding or editing metadata only, such as an annotation layer, or adding enhanced imagery over the base imagery, or possibly replacing a portion of the original base imagery. In any case, the purpose is to allow other JPIP users to see these updates. To do this, the named source is kept constant; however, there must be a new target ID for each new version.

In use cases where there are frequent updates or where the desire to optimize dissemination is a high priority, the issuing of new target IDs when only a portion of the imagery has changed would be wasteful on network capacity. This issue was not addressed in the JPIP standard, but what follows are possible innovative approaches. To first address frequently updating metadata (say vector or text overlays), it is sensible to have the base imagery and the metadata overlays separated via different target IDs. Clients would cache the unchanging base imagery, but depending on the frequency of changes, they may or may not cache the overlay metadata. For example, one might not want to cache for 'white board' collaborative sessions.

For a use case where the base imagery data were also updated, there is a number of possible approaches. The JPEG working group, or another standards body, will possibly address this via some standard in the future, but at the moment the JPIP standard requires the target ID to change. The spatial granularity of updates needs to be considered. Since a JPIP upload is the update of some view-window spatial region, but is across all resolution levels and components within the region, tiles could be one logical basis for spatial access. Each version of the image will have a unique target ID. However, it is proposed that within a metadata table there is a list for each tile of all the previous target ID versions that are unchanged from the current version. A client with a JPIP cache containing these previous versions, and wanting to save on dissemination costs and thus improve responsiveness, would find this table, determine what tiles in its cache are unchanged, associate them with the current target ID, and use a cache model request to inform the server of the extent of this tile data in its cache. Now efficient dissemination is possible. Only those changed tiles, plus any additional data for unchanged tiles (if they were not complete in the cache) would form the data response for the client. Note that doing a similar equivalence table for precincts would also work, but the proposed version for tile equivalence tables would be much simpler and compact, and thus possibly more dissemination efficient.

6.5.3   JPIP on Multiple Code-Streams

Multiple code-streams exist in JPX, MJ2, and JPM JPEG 2000 family files. The code-stream request field (`stream`) identifies those code-streams of interest from such JPEG 2000 multi-code-stream files. When using a Motion JPEG 2000 (MJ2) file, the client uses this request field to identify groups of code-streams and their sampling rate. An example use case is to request imagery for a fast-forward viewing of the first 3 minutes from an MJ2 video (assume it is 50 frames per second progressive) by requesting every 10th code-stream from 0 to 9000:

```
stream=0-9000:10.
```

A code-stream context was developed for more complex arrangements of multi-code-streams, such as JPX and MJ2 video tracks. If the code-stream context field is supplied, the requested view-window includes each of the code-streams that are associated with the context, in addition to any code-streams requested in the code-stream request field. The context can be identified using the coordinate remapping transformation given by a specified compositing instruction within a JPX composition box in the JPX logical target. The server indicates that it is processing the request using the JPX composition context, with possible modifications, via its code-stream context response header.

When the optional '+ now' suffix is used in an MJ2 context request, the context includes all code-streams in a live video-stream starting from those code-streams captured at the time the request was received by the server. Implementers should interpret this as real-time delivery for applications such as live broadcasts. If '+ now' is not specified, then the context is all code-streams belonging to the specified MJ2 video track. View-windows are by default mapped on to code-streams, but JPIP provides means to use the geometric transformation in MJ2 to map view-windows on to the track presentation or the movie composition. When 'track' is specified, the view-window request parameters (frame size, offset, and region size) are not relative to the code-streams, but are relative to the smallest bounding rectangle that contains the track's presentation and its desired presentation size. The geometric transformation described by the MJ2 track header box is applied to obtain a code-stream-based view-window for each code-stream.

When 'movie' is specified, then the view-window is on the smallest bounding rectangle that contains the movie. The composition of the movie is given by the movie header box, and the combined movie and track transformations map a constant movie view-window on to tracks and then on to code-streams. The server indicates that it is processing the request using the movie or track composition context, with possible modifications, via its code-stream context response header. The client can request the server to sample those code-streams given by the code-stream and the code-stream context, at the rate indicated in the sampling rate request field. The server should then deliver, on average at the desired rate, a regular subsample from all groups of code-streams.

The delivery rate request field is a server control request applicable to multiple code-stream sources. The client can use the delivery rate request field to request delivery of the view-window on multiple code-streams at a `rate-factor` times the normal playback rate of the source. This request is intended for the likes of Motion JPEG 2000 files that have some 'source timing' associated with the multiple code-streams. For example, to request review of a view-window at 1.5 times the video source time of 30 frames per second, the request would include `drate = 1.5`. The server should deliver the multiple code-streams referenced in the request and serve them at a rate of 45 frames per second from the original source.

The server may and should model the cache for code-streams specified in the code-stream or code-stream context request fields. The model set (`mset`) request is a type of cache management request, and is only applicable to within a session on multiple code-stream sources. It informs the server of the set of code-streams that the client intends caching. If the server is prepared to model these code-streams, then it need not issue an `mset` response header. It can model less code-streams, but it must inform the client which code-streams it will model using the `mset` response header. The request containing the model set request field, and subsequent client requests, may specify additional code-streams. If the server is prepared to serve these, then the model set increases. If the server will not serve any of these additional code-streams, it must signal this using the appropriate response header (e.g. code-stream, code-stream context). The server may also issue a model set response header to confirm explicitly those code-streams for which it has a cache model. The server may drop cache models at any time, but must inform the client of the updated list of code-streams modeled.

6.5.4 Advanced Behaviors

There are four communications paths in JPIP:

- the forward (client to server) client JPIP request connection;
- the return (server to client) server JPIP response header connection, which is a server acknowledgement receipt of the client request);
- the forward (server to client) server JPIP response data connection;
- the return (client to server) client JPIP response data receipt acknowledgment connection.

A server response is in two parts. The first (response headers) is used to signal the server's modification to the request and by implication acknowledges receipt of the request. The second part (response data) is the actual imagery data (JPT- and JPP-stream data types). These may both be together on the same transport, as in the 'http' transport. Here the server's JPIP response header forms an HTTP header entry in its HTTP response, while the JPIP response data forms the HTTP body. Alternatively, separate transports can be used, as in the 'http-tcp' transport mode, where the server's response header is in its HTTP response (as per the 'http' mode) and the response data are transported on an auxiliary TCP connection.

TCP is known to be inappropriate for many applications where packet loss is not caused by network congestion (which is the assumed cause of time-outs in TCP behavior); many example use cases are on error-prone wireless networks (Prandolini *et al.*, 2000). In these instances, it is more appropriate for the streaming imagery data to be transported on UDP. Note that UDP is a best-effort transport and it does not guarantee delivery. Regrettably, the JPIP standard is only informative on UDP implementations.

Since the client request and server response headers are small relative to the streaming image data, one suitable approach is to implement what is termed the 'http-udp' method, where the request and response headers are carried on HTTP (as before in the 'http-tcp' mode) and the data are transported on an auxiliary UDP connection. In this case, the request and response header connection paths are reliable and guarantee delivery of requests to the server and response headers to the client. Since the response data on the unreliable UDP transport may be lost, the following signaling and handshaking should be implemented:

- Each request shall include a request ID.
- Normally JPIP assumes that, unless otherwise stated, the message infers identity based on the previous message. If the transport is unreliable, this should not be practiced. Every response data connection packet (not to be confused with a JPEG 2000 packet) shall consist of a whole number of JPIP-stream messages (i.e. no fragmentation across packets) with at least the first message including a complete header with an explicit code-stream identifier and JPIP-stream extended in-class code identifiers.
- Each data response will include a request ID header and end with an EOR message, even if there is nil response data for that request ID. This ensures that the client can properly match response data to requests. (Preempted requests will have empty responses, but their EOR must still be sent.) With the exception of the EOR message, the packet will belong to one request only. The exception is that multiple EOR messages can be in the start of a packet. This allows multiple preempted requests preceding the next nonempty response to be counted. The request ID used in this packet is that for the nonempty response.
- Multiple JPIP response data messages within a connection packet must belong to the same request. The one request ID is to be in the packet's header and the message ends with an EOR. Pay attention to the network's maximum transmission unit (MTU) to avoid possible connection packet fragmentation by the network. This may affect performance depending on the statistics for fragmented packet losses.
- Since response data are likely to be carried on multiple connection packets, each connection packet shall have both the request ID and a packet sequence number (starting from 0). The last packet has the EOR message. In this way the client will know it has all data for a response when it has the set of connection packets numbered 0 to $N-1$ with the same request ID and where the $N-1$ packet includes the EOR message.
- Clients shall acknowledge receipt of each connection packet by sending an acknowledgment consisting of at least a copy of the connection packet header (this includes the request ID and packet sequence number) via the return connection path.

Using the above signaling and handshaking, the client and the server will be able to detect missing response data. The server is not required to retransmit packets it considers possibly lost, but may do so depending on policy. For example, retransmitting image main and tile headers would be a good policy since the client cannot decode any imagery data without them. The server may possibly compute the consequences for its cache model from missing packets. It may also be a good policy for the client on detecting packet losses to use subtractive cache manipulation requests to inform the server explicitly of its cache contents.

There may be quality of service challenged networks where packet losses are significant to the point where HTTP cannot be used for the client's request nor the server's response header messages. The approach then is to implement JPIP purely on UDP. In this case, the request connection path and the response data path are not reliable. It is important to note that with UDP the connection packets may be lost, or possibly delivered with errors and out of order. (Note that UDP has a low-level error detection that normally causes packet loss at the routers when errors are detected.) The previous signaling and handshaking for the unreliable data connection is used, along with the following additional signaling, to address the possibility of failure for the delivery of client requests and server response headers:

- Each request is transported using one or more connection packets. Each connection packet shall include its request ID and a connection packet sequence number (starting from 0). Each request shall be terminated using an end of request signal. This will allow the server to know when it has all connection packets for a complete request message.
- Some client requests are important for cache modeling, such as those with subtractive cache model manipulation. Due to the unreliability of the request transport, the client shall signal the importance of a request using a 'precedence' flag in all the connection packets for an important request. This allows the server to detect the existence of such important requests from the receipt of any one connection packet associated with this request. When the existence of a request is detected with this precedence flag set, the server is not allowed to process any subsequent requests until this important request has been processed.
- All other nonimportant requests shall have the precedence flag cleared in each request connection packet, and will also have a 'last important request ID' field in at least one of its connection packets. The first request (i.e. 0) is always an important request and shall have its precedence flag set. All subsequent nonimportant requests shall have the last important request ID set to 0, until another important request is issued.
- As in the 'http-udp' case, each response packet shall have the request ID and the last response packet shall contain an EOR.
- The server shall always send a response for each request ID, even if it is missing any of the connection packets for a request. If a server is missing part of a nonimportant request and is preempted by a subsequent request, it shall respond with nil response data and include an EOR with the reason code 'window changed.' (This is normal behavior for any JPIP server – not just those with unreliable connections.) Receipt by the client of any response connection packets containing the request ID will at least acknowledge to the client receipt by the server of at least some part of the request.
- For each complete request that arrives at the server, the server shall send one or more response connection packets, each containing the request ID, response packet sequence

number, and an EOR in the last packet. This shall be done even if the request arrives after a subsequent request (higher request ID) has been processed.

- If the server has received part of an important request, but is missing some packets, after a time-out period it shall respond with the EOR reason code 'response limit reached.' Clients receiving such an EOR reason code shall retransmit the contents of the important request, but using a new request ID.
- If the server receives nonimportant requests, which indicates the last important request ID, and where the server has not received any connection packets for this last important request, then the standard says that it shall wait until it receives at least one of these last important request connection packets. The implication is that after a time-out period, the clients should retransmit important requests until acknowledgment is received from the server; otherwise the server will idle. The client need not stop sending other nonimportant requests. Also implied is that after sending a response to an important request, the server should retransmit this response if subsequently one of these important request connection packets are re-received; otherwise the client may be idle waiting for an acknowledgment receipt of this important request. Not suggested in the standard, but made here, is that after a time-out period the server should respond with the EOR reason code 'response limit reached' using this missing important request ID. Clients receiving such an EOR reason code shall retransmit the contents of the important request, but using a new request ID.
- Note that there is always a one-to-one mapping of requests and responses. Once the client forms a request, and once the server forms a response, there can be retransmissions of these messages, but they cannot be changed.

6.5.5   JPIP Indexing

The standard describes details for indexing the structure of a JPEG 2000 Part 1 file. This may be of use to implementers, particularly those applications using HTTP file servers (i.e. non-JPIP image servers) and clients that are more intelligent. Intelligent clients could seek via the byte-range requests on the 'raw' code-stream and thus exploit benefits from existing HTTP infrastructure. Through a series of interactions, the client will gain an understanding of the image structure and then it will be able to determine and seek the data required to fulfill its view-window requirements. Another related and interesting approach was proposed by Ortiz, Ruiz, and Garcia (2004) that would make the response to view-window requests constant such that HTTP proxy infrastructure could be exploited. These topics were embryonic at the time of writing the standard and recent work suggests innovative means for lower complexity realizations of the JPIP interactive imagery paradigm using legacy infrastructure.

6.6   Conclusions

This text has presented the essential elements and concepts of JPIP, but there are many more details regarding the standard that space does not allow here. The purpose of JPIP was to standardize a method that allowed efficient (possibly optimal) interactive dissemination and a high degree of responsiveness. JPIP exploits the scalability of JPEG 2000 to provide an enabling interactive technology. The fundamental aim of JPIP was to allow an imagery server to provide data to a client, customized for and by the user's interactivity. In a sense, the *a posteriori* JPIP cache on the client is a highly truncated, and therefore a highly compressed, version of the *a priori* compressed source data. Key JPIP principles are asynchronous messaging, client and server freedoms, preemptive behaviors, and transport independence. The intention of the designers of JPIP was to allow for a wide range of implementation possibilities tuned for various use case scenarios.

There is still work to be done on JPIP for multiple code-streams (e.g. Motion JPEG 2000), JPIP proxy systems, the upload feature, error resilience (e.g. nonequal error protection policies for UDP), and exploiting the potential from JPIP indexing. These topics may find themselves subjects for further JPEG standard activities, where experts will come together to make interoperable their advanced techniques. Through standardization, multiple vendors can compete in the market place with their various implementations. Though interoperable, the performance of various implementations, in the context of their specific application use case, is what differentiates and what ultimately counts.

Acknowledgments

The JPIP standard derives from imagery and dissemination research by Dr Mark Grigg of the Defence Science and Technology Organisation (DSTO) in 1998. This original client–server architecture was extended to JPEG 2000 by DSTO, and then by Dr David Taubman of the University of New South Wales for the Kakadu Software SDK. Many thanks are owed to Dr Taubman for his contribution to the JPIP standard, as well as co-editors Scott Houchin and Greg Colyer. Michael Gormish and the other numerous members of the JPEG Working Group are appreciatively acknowledged for their support and contributions. Financial support for the Australian contribution to JPIP was provided by DSTO and the Defence Imagery and Geospatial Organisation, both organizations of the Australian Department of Defence.

References

Burt, P. J. and Adelson, E. H. (1983) The Laplacian pyramid as a compact image code, *IEEE Transactions on Communications*, **COM-31**, 532–540.

ISO/IEC (2004) Information Technology – JPEG 2000 Image Coding System – Part 1: Core Coding System, ISO/IEC International Standard 15444-1:2004.

ISO/IEC (2005) Information Technology – JPEG 2000 Image Coding System – Part 9: Interactivity Tools, APIs and Protocols, ISO/IEC International Standard 15444-9:2005.

Mallat, S. G. (1989) A theory for multiresolution signal decomposition: the wavelet representation. *IEEE Transactions on Pattern Analysis and Machine Intelligence*, **11**, 674–693.

Ortiz, J., Ruiz, V. and Garcia, I. (2004) Improving the remote browsing of JPEG 2000 images on the Web, in *Fourth IASTED International Conference on Visualization, Imaging, and Image Processing*, Marbella, Spain.

Prandolini, R., Au, T. A., Lui, A. K., Owen, M. J. and Grigg, M. W. (2000) Use of UDP for efficient imagery dissemination, in *SPIE Proceedings on Visual Communications and Image Processing*, Perth, Australia.

Taubman, D. S. and Marcellin, M. W. (2002) *JPEG2000: Image Compression Fundamentals, Standards and Practice*, Kluwer Academic Publishers, Boston, MA.

Further Reading

ISO/IEC (2003) Information Technology – JPEG 2000 Image Coding System – Part 6: Compound, ISO/IEC International Standard 15444-6:2003.

ISO/IEC (2004) Information Technology – JPEG 2000 Image Coding System – Part 1: Core Coding System, ISO/IEC International Standard 15444-1:2004.

ISO/IEC (2004) Information Technology – JPEG 2000 Image Coding System – Part 2: Extensions, ISO/IEC International Standard 15444-2:2004.

ISO/IEC (2007) Information Technology – JPEG 2000 Image Coding System – Part 3: Motion JPEG 2000, ISO/IEC International Standard 15444-3:2007.

IETF RFC 768 (1980) User Datagram Protocol. Available from World Wide Web: <http://www.ietf.org/rfc/rfc0768.txt>.

IETF RFC 793 (1981) Transmission Control Protocol. Available from World Wide Web: <http://www.ietf.org/rfc/rfc0793.txt>.

IETF RFC 2046 (1996) Multipurpose Internet Mail Extensions (MIME) Part Two: Media Types. Available from World Wide Web: <http://www.ietf.org/rfc/rfc2046.txt>.

IETF RFC 2234 (1997) Augmented BNF for Syntax Specifications: ABNF. Available from World Wide Web: <http://www.ietf.org/rfc/rfc2234.txt>.

IETF RFC 2396 (1998) Uniform Resource Identifiers (URI): Generic Syntax. Available from World Wide Web: <http://www.ietf.org/rfc/rfc2396.txt>.

IETF RFC 2616 (1999) Hypertext Transfer Protocol – HTTP/1.1. Available from World Wide Web: <http://www.ietf.org/rfc/rfc2616.txt>.

7

JP3D – Extensions for Three-Dimensional Data (Part 10)

Tim Bruylants, Peter Schelkens, and Alexis Tzannes

7.1 Introduction

Today, techniques such as computed tomography (CT), positron emission tomography (PET), magnetic resonance imaging (MRI), and virtual confocal microscopy have led to the creation of a large amount of volumetric image data sets. Moreover, as scanner technology advances, the amount of data per individual data set produced with these devices is continuously increasing. Thus, an efficient representation technology becomes crucial to allow for compact and optimal storage and transmission of these volumetric data sets. With this, other features, like random accessibility, region-of-interest (ROI) support, and resolution/quality scalability also become desirable. Many modern applications require that these large data sets can be accessed with ease on various platforms with different display and processing characteristics. As shown earlier in this book, JPEG 2000 addresses all aforementioned requirements for two-dimensional imagery in Part 1 and Part 2.

However, Parts 1 and 2 do not provide unified support for three-dimensional data sets. As a consequence, when DICOM adopted JPEG 2000 as a compression standard (see Chapter 13), many implementers took a creative approach and chose to fix the lack of three-dimensional (3-D) compression support by using the multiple component transformation (MCT) extension of Part 2 (see Section 2.9). This enabled the use of the wavelet transform along the third dimension Z or slice axis, i.e. converting slices into virtual components. This approach, though, incorporates a couple of important drawbacks. First of all, the component and spatial Z-axis data are no longer distinguishable from each other for volumetric image data sets containing multiple components. Having multiple

The JPEG 2000 Suite Edited by Peter Schelkens, Athanassios Skodras and Touradj Ebrahimi
© 2009 John Wiley & Sons, Ltd

semantic interpretation possibilities of such decoded data sets, and lacking a standardized approach, causes potential confusion, mistakes, and incompatibilities. Second, the codec does not treat all dimensions in an isotropic fashion. This is an important drawback as it negatively affects the rate-distortion optimization of the EBCOT coder in JPEG 2000. Moreover, generic region-of-interest (ROI) support, efficient random access capabilities and many other features of JPEG 2000 are also reduced in functionality. Finally, using the MCT approach limits the applicable wavelet decomposition patterns because it enforces full wavelet decomposition along the slice axis before allowing the regular X and Y decompositions. Hence, it does not allow full packet decomposition support (see Section 2.7, ARB), resulting in a lower compression performance.

In order to add improved support for other image data types, like volumetric data sets, the JPEG committee decided in 2001 to create Part 10 of JPEG 2000, also referred to as JP3D. This volumetric extension for JPEG 2000 provides isotropic support for handling three-dimensional data with multiple components and no time component. It offers the same functionality and efficiency for 3-D data sets as exists for its 2-D counterparts.

Initially, Part 10 was intended to add also support for floating-point data sets. Hence, the nature of the data, originally addressed with this part, was twofold. However, due to a lack of potential use cases and mature technology, support for the floating-point data type was dropped in an early phase of the development of the Part 10 standard. Consequently, because Part 10 only describes the extension to three-dimensional data sets, it received the name JP3D. It should be noted, though, that since 2008 interest has grown again due to the advent of high dynamic range (HDR) imaging techniques for applications such as digital photography and digital cinema. Hence, at the time this book was being finalized the JPEG committee was probing again for interest in the topic of compression of floating-point data sets (Schelkens, 2008).

As mentioned above, JP3D is entirely based on the techniques that are also used in JPEG 2000 Part 1. Because the theory behind all the mechanisms employed in Part 1 is also valid for this part, many sections in this chapter will contain references to Part 1. Nonetheless, the most important technology parts will also be briefly explained in this chapter.

This chapter is organized as follows. Section 7.2 describes JP3D and presents comparisons with JPEG 2000 Part 1. In this section the logical transition from two dimensions to three dimensions is explained, without introducing new technology. Section 7.3 describes the compressed JP3D bit-stream organization. Section 7.4 explains how region-of-interest functionality, provided by Part 9 of JPEG 2000, was modified or extended to support three dimensions. Section 7.5 discusses the compression performance of JP3D in comparison with JPEG 2000 Part 1. Finally, Section 7.6 discusses some implications to other parts of the JPEG 2000 standard as the consequence of creating Part 10 of JPEG 2000.

7.2   JP3D: Going Volumetric

Figure 7.1 shows the fundamental building blocks of a JP3D encoder, depicting the same architecture as a Part 1 encoder (see Section 1.2). Obviously, the blocks differ internally from their two-dimensional (2-D) counterparts.

The building blocks, as given in Figure 7.1, include preprocessing, discrete wavelet transform (DWT), quantization, tier-1 coding, and tier-2 coding. The three-dimensional input image can be observed as a set of two-dimensional image slices. Each slice is

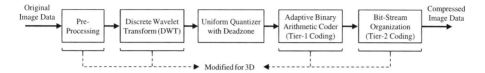

Figure 7.1 Encoder block schema

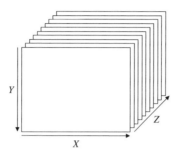

Figure 7.2 Three-dimensional input image data

then simply a two-dimensional image, having one or more components. In each slice, the horizontally aligned X axis defines the column index, while the vertically aligned Y axis defines the row index in a slice. Along the slices, the Z axis, orthogonal to the X and Y axes, defines the slice index (see Figure 7.2). While the X axis is associated with the horizontal direction and the Y axis with the vertical direction, the Z axis is associated with what is called the axial direction.

As with two-dimensional images, the three-dimensional input image data for JP3D may contain one or more components, with a maximum of 16 384 (2^{14}) components. The sample values within each component are always integer typed with a bit-depth in the range of 1 to 38 bits. Signed samples with bit-depth B would fall in the discrete interval $[-2^{B-1}, 2^{B-1} - 1]$, while unsigned samples would be in the interval $[0, 2^B - 1]$. The bit-depth and sign specifications of the samples can vary for each component. Thus, to summarize, the sample data typing for three-dimensional input image data is exactly the same as for two-dimensional input image data.

7.2.1 Preprocessing

With JPEG 2000 Part 1, the first preprocessing step is to partition the input image data into equally sized rectangular and nonoverlapping tiles. Only tiles on image borders can have different sizes. Tile sizes can range from one single pixel to a full-image dimension (see Chapter 1). In JP3D, tiling is simply extended to three dimensions. Tiles are thus no longer rectangular but instead become cuboidal subvolumes of the original input image. As in Part 1, from that point on, tiles are handled independently from each other both by the encoder and decoder.

The preprocessing steps following the first step when using JP3D are identical to those steps performed for 2-D images. Having more than two dimensions does not change the DC level shift step for unsigned samples and the point-wise intercomponent transformation step. The last step is also known as the multiple component transformation (MCT) and, for example, allows for a conversion between RGB and YUV color spaces. The restriction that the components must have identical bit-depths remains also for three-dimensional images.

After the preprocessing, each tile component (i.e. each component of each tile) is handled independently. Therefore, the subsequent sections will only be in reference to a single tile component, unless stated otherwise.

7.2.2 The 3-D Discrete Wavelet Transform (3-D DWT)

By default, the same two wavelet kernels are provided in JP3D as in JPEG 2000 Part 1, i.e. the reversible (5, 3) wavelet transformation and the irreversible (9, 7) wavelet transformation. Using the arbitrary transform kernel extension of JPEG 2000 Part 2, it is possible to specify custom transformation kernels in JP3D. For detailed information on the one-dimensional discrete wavelet transform, we refer to Section 1.2.2.1 of this book. As outlined in that chapter, a one-dimensional decomposition splits a signal into two frequency bands, i.e. one lowpass band and one highpass band. Due to the separability property of the wavelet transform kernels, a one-dimensional (1-D) discrete wavelet transformation (DWT) can be easily extended to two or more dimensions. For JP3D, the three-dimensional discrete wavelet transformation is constructed from independent one-dimensional (1-D) discrete wavelet transform (DWT) steps along the three spatial directions, X, Y, and Z. Thus, the 3-D DWT is simply a 2-D DWT with an additional 1-D DWT along the Z axis.

In two dimensions, the word 'column' describes a collection of all points that have a common X coordinate, while 'row' specifies all points that have a common Y coordinate. In three dimensions, however, the meaning of the words 'row' and 'column' become less obvious. In the text that follows, a 'column' defines a set of all points with a common X and Z coordinate and a 'row' defines a set of all points with a common Y and Z coordinate. Due to a lack of a proper naming, let a 'z-row' define a set of all points that have a common X and Y coordinate. Thus, in three dimensions, rows, columns, and z-rows each define lines in the horizontal, vertical, and axial directions respectively.

The 3-D wavelet decomposition of a volume starts by applying a 1-D analysis filter bank (or forward DWT) to each z-row of the original volume, effectively decomposing the input along the Z axis (see Figure 7.3, steps a to b). The original volume is now filtered into two subbands. Next, the same filter bank is applied to each column of the filtered and subsampled data, decomposing the volume along the Y axis (see Figure 7.3, steps b to c). Consequently, the original volume is decomposed into four subbands. Finally, the filter bank is applied a third time to each row of the filtered and subsampled data, effectively decomposing along the X axis (see Figure 7.3, steps c to d) into eight subbands. The order in which the one-dimensional wavelet transforms are applied should not matter in theory, but due to possible rounding errors it is recommended that the specified order of transforms is respected, especially if lossless decoding is desired. The one-level decomposition creates eight subbands with eight corresponding sets of wavelet coefficients. Each subband represents a filtered and subsampled version of the original volume. Similar to JPEG 2000 Part 1, JP3D has a dyadic decomposition pattern. This means that further decomposition of the image volume is possible by applying the same wavelet

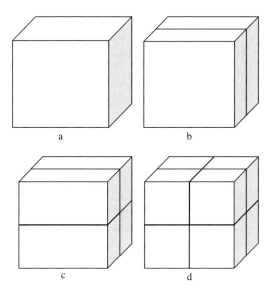

Figure 7.3 3-D DWT decomposition

filtering steps, as described above, over and over, to the lowest frequency subband, until this subband becomes too small. Each decomposition level takes the lowest frequency subband as input and decomposes it into eight smaller subbands.

The lowest frequency subband is called the LLL band, which denotes that lowpass filtering was applied in each spatial direction (X, Y, and Z respectively). The character H would indicate that highpass filtering was used in a particular direction. Each subband receives, apart from the three-character label, also a number that denotes the decomposition level by which it was created. Figure 7.4 shows a typical two-level decomposition of a volumetric image with all subbands labeled as just described.

Unlike JPEG 2000 Part 1, the number of decompositions in JP3D for the X and Y directions do not necessarily have to be equal. Logically, with 3-D, the number of decompositions in the Z direction can also differ from those in the X or Y directions. As a consequence, it is possible that for a specific decomposition level less than eight subbands

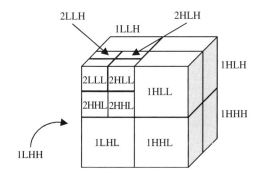

Figure 7.4 Two-level decomposition

are created. The character X in a subband label indicates that no filtering was applied in a particular direction. The flexibility of having a variable number of decompositions in each of the three spatial directions was added during the design of JP3D, because, despite the fact that JP3D targets isotropic data behavior, most volumetric data sets have different imaging properties along the slice axis compared to those in the slice plane. This is most often caused by the physical and technical limitations of image scanners, where, for example, the image resolution within slices is much higher than the resolution across slices. The flexibility in the decomposition pattern of JP3D allows the user to compensate for these kinds of discrepancies. In addition, in order to preserve the isotropic behavior of the JP3D codec, the option of variable decomposition levels was made available for all spatial directions.

Figure 7.5 shows the irregular (1, 2, 2) decomposition pattern, where a single decomposition is performed in the X direction, while two decompositions are done in Y and Z directions. The decomposition process starts by performing wavelet transformations in all three spatial directions, effectively creating the first eight subbands at decomposition level 1. Then the 1LLL subband is used for further decomposition, but now the wavelet transform is only performed in the Y and Z directions. This creates four new subbands at decomposition level 2, i.e. 2XLL, 2XHL, 2XLH, and 2XHH. It is important to note that after the requested number of decomposition levels is reached in a specific direction, only then will it no longer be considered for wavelet transformation.

Similar to JPEG 2000 Part 1, the wavelet decomposition process automatically introduces a multiresolution representation of the original image volume. Thus, resolution levels contain those subbands that are needed to reconstruct a volume at a specific resolution. The lowest resolution version of the original volume can be reconstructed using only subbands from resolution level 0. In practice, resolution level 0 is the only resolution level that contains exactly one subband. Adding subbands from the next resolution level allows a higher resolution version of the original image to be reconstructed. With all subbands from all resolution levels, the original volume can be reconstructed at full-image resolution. When a volume is decomposed using an irregular pattern (i.e. the number of decomposition levels is not equal in all spatial directions), resolution levels might become somewhat confusing to the reader. However, the number of resolution levels in a decomposed volume is simply determined by the spatial direction with the highest amount of decomposition levels.

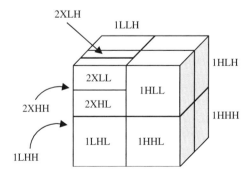

Figure 7.5   A (1, 2, 2) irregular decomposition pattern

Finally, JP3D also adds the ability of allowing different wavelet kernels in each spatial direction. However, this feature does not change the definition of resolution levels and subbands or the wavelet decomposition process.

7.2.3   Quantization

Wavelet coefficient quantization is not dependent on the dimensionality of the data. Therefore, JP3D employs exactly the same quantization strategy as used in JPEG 2000 Part 1. Thus, after the 3-D DWT stage, the encoder quantizes the wavelet coefficients per subband just like in JPEG 2000 Part 1. Quantization allows the encoder to emphasize the importance of specific subbands, or to add an extra but easy-to-implement mechanism for bit-rate control. For more detailed information on the quantization in JPEG 2000, see Section 1.2.3.

7.2.4   Bit Modeling and Entropy Coding

Subsequently, the quantized wavelet coefficients are entropy encoded to create the compressed bit-stream. As in JPEG 2000 Part 1, each subband is encoded independently from the other subbands. This allows the encoder and decoder to deliver better error resilience and higher flexibility in arranging the progression order of the bit-stream. The process of entropy coding is also referred to as tier-1 coding. JP3D uses a three-dimensional version of the EBCOT (embedded block coding with optimized truncation) algorithm for the entropy coding step. EBCOT partitions each subband into smaller units, called code-blocks, and then encodes each of these code-blocks independently. In JPEG 2000 Part 1 code-blocks are two-dimensional rectangular entities, while in JP3D code-blocks are three-dimensional cuboidal volumes. Alternatively, a 3-D code-block can be seen as a set of 2-D code-block slices, with XY plane oriented slices and the slice index in the Z direction. The dimensions of a 3-D code-block are defined at the encoder side and can be freely chosen with some constraints. The width, height, and depth of the code-blocks must be an integer power of two, and one code-block can only contain a maximum of 2^{18} and a minimum of 4 wavelet coefficients.

7.2.4.1   3-D EBCOT

As previously mentioned, the designers of JP3D wanted to use as many techniques as possible from JPEG 2000 Parts 1 and 2 without any modifications. 3-D EBCOT differs, for this reason, only slightly from its 2-D counterpart (Section 1.2.4). Each 3-D code-block, containing wavelet coefficients, is independently entropy encoded into an embedded code-stream. The 3-D EBCOT algorithm scans the coefficients in a bit-plane by bit-plane order, starting with the most significant bit-plane and proceeding until the least significant bit-plane. If the coefficients have a bit-depth of N bits, then the most significant bit-plane equals bit-depth level $N - 1$. The least significant bit-plane (zero) is located at bit-depth level 0. Thus, more generically, bit-plane P represents the bits at bit-position P of all the wavelet coefficients of the code-block. Bits in a bit-plane are spatially located at the same coordinates as their respective coefficients. It is important to note that bit-planes in JP3D are in fact three-dimensional bit-volumes, which can be best described as a set of 2-D bit-plane slices. The scanning order inside each bit-plane

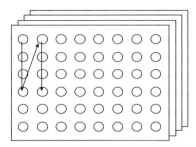

Figure 7.6   3-D Stripe-based scanning pattern

(or bit-volume for that matter) follows a bit-plane-slice by bit-plane-slice order, where the bit-plane slices are situated in the *XY* plane (see Figure 7.6). The bit-plane slice with the lowest *Z* coordinate is scanned first. Inside each bit-plane slice, the stripe-based scanning pattern of JPEG 2000 Part 1 is used.

From this point on, 3-D EBCOT is identical to 2-D EBCOT as described in the JPEG 2000 Part 1 standard. In fact, the only effective difference between the two EBCOT instantiations lies within this 3-D extended scanning pattern in bit-planes.

Starting with a code-block of wavelet coefficients, EBCOT applies a fractional bit-plane coding process by using the same three coding passes as in JPEG 2000 Part 1 for encoding each bit-plane. During each pass, the bits in the bit-plane are scanned one by one in the stripe-based scanning order, where each bit is encoded by only one of the three coding passes. Internally, these coding passes work in the exact same way as in 2-D EBCOT (see Section 1.2.4.2). The significant propagation pass encodes the bits of coefficients that are not yet significant, but that have at the same time at least one significant immediate neighbor. The magnitude refinement pass, on the other hand, encodes bits of coefficients that are already significant. Finally, the last pass or cleanup pass encodes all the bits that were not yet coded by the other two passes, and hence its name. All three coding passes make use of so-called coding primitives. These are smaller building blocks that send out bit-values, each associated with a specific context label, to the context-based adaptive binary arithmetic entropy coder, or simply the MQ-coder (see Section 1.2.4.1). The bit-values contain the actual content information that is needed to reconstruct the code-block wavelet coefficients at the decoder side and are encoded by the MQ-coder. The contexts are causal. Hence, they can be calculated at the decoder side, while not being directly signaled in the code-stream. Using contexts helps to optimize the probability models in the binary arithmetic encoder. JP3D employs the same 18 contexts and the same context models as JPEG 2000 Part 1. It is important to notice that the context models, defined in JPEG 2000 Part 1, only use state information in two dimensions, even though code-blocks in JP3D have three dimensions. In JP3D, the 2-D context models are applied on the *XY* bit-plane slices and do not (yet) exploit state information in the *Z* direction. Research showed that it is possible to improve the compression efficiency even more when applying a 3-D context model (Bruylants *et al.*, 2007) or by applying alternative bit scanning patterns (An *et al.*, 2006). Unfortunately, it has also been shown that the design of such a generic three-dimensional context model is not trivial. Moreover, most of the volumetric images today are not isotropic, inhibiting the potential benefits of advanced 3-D context models. Hence, it was decided not to include them yet in the standard. Nonetheless, the specification is written in such a way that additional context models can be added at a later point in time via an amendment procedure while maintaining interoperability with current implementations.

For each code-block, the fractional bit-plane coding process delivers a bit-stream that can be truncated at multiple positions. More specifically, each coding pass introduces a potential truncation point. This allows the encoder to be very flexible in terms of the rate/distortion (R-D) allocation of the compressed code-stream, facilitating an optimal R-D control strategy.

7.3 Bit-Stream Organization

After completing entropy coding, also referred to as tier-1 coding, JP3D continues with what is referred to as tier-2 coding. During this step, the various independent code-block bit-streams are rearranged into logical units, called packets. A packet typically contains small pieces of multiple code-block bit-streams. The packets are subsequently, depending on the type of application, combined in a particular order into a final JP3D code-stream. JP3D uses a similar bit-stream organization to that of JPEG 2000 Part 1. In fact, the few differences that do exist are only caused by the higher dimensionality of JP3D. Hence, this section explains the exact code-stream differences between JP3D and JPEG 2000 Part 1 (Section 1.3). It describes relevant structures, like components, tiles, subbands, resolution levels, and code-blocks, and how these are defined in three dimensions. These structures partition the volumetric image data into color channels (components), spatial regions (tiles), frequency regions (subbands and resolution levels), and space–frequency regions (code-blocks). Additionally, we describe how a precinct is defined in three dimensions and how the extra dimensionality of the data affects packets and layers in the code-stream.

7.3.1 The Three-Dimensional Canvas Coordinate System

In JPEG 2000 Part 1, two-dimensional images are registered with respect to a two-dimensional reference grid (see Section 1.3.1). It is a rectangular grid of points that starts at coordinates (0, 0) and extends to (Xsiz-1, Ysiz-1). An image area inside this reference grid is then defined by its origin (XOsiz, YOsiz), which gives the position of the upper-left corner and extends to (Xsiz-1, Ysiz-1). The extra dimension of JP3D forces the extension of the definition of the reference grid. Coordinates in the reference grid have now three-dimensional coordinates (Figure 7.7). The grid starts at (0, 0, 0) and extends to (Xsiz-1, Ysiz-1, Zsiz-1), with the image area starting at (XOsiz, YOsiz, ZOsiz).

As with JPEG 2000 Part 1, the spatial positioning of all resolution levels and subbands of JP3D is specified with respect to their own coordinate systems. The (three-dimensional) canvas coordinate system consists of the complete collection of coordinate systems. For a complete description of this canvas coordinate system, consult Annex B of the official Part 10 specification (ISO/IEC, 2008).

7.3.1.1 Components, Tiles, Tile Components, and Subsampling

This subsection gives a very brief overview of how components, tiles, and tile components are defined with respect to the reference grid. It also introduces the parameters that are used

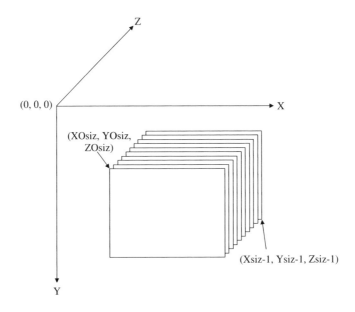

Figure 7.7   Three-dimensional reference grid

to define the component subsampling. However, this part is not critical in understanding the principles of JP3D (or JPEG 2000 Part 1 for that matter).

For a component c, samples are located at integer multiples of ($XRsiz^c$, $YRsiz^c$, $ZRsiz^c$) on the reference grid. Each component domain is then a subsampled version of the reference grid with the (0, 0, 0) coordinate as a common point. The samples of the component are mapped in a cuboid, defined by corner coordinates (cx_0, cy_0, cz_0) and ($cx_1 - 1$, $cy_1 - 1$, $cz_1 - 1$) where cx_0, cy_0, cz_0, cx_1, cy_1, and cz_1 are given as

$$cx_0 = \left\lceil \frac{XOsiz}{XRsiz^c} \right\rceil, \quad cx_1 = \left\lceil \frac{Xsiz}{XRsiz^c} \right\rceil \tag{7.1}$$

$$cy_0 = \left\lceil \frac{YOsiz}{YRsiz^c} \right\rceil, \quad cy_1 = \left\lceil \frac{Ysiz}{YRsiz^c} \right\rceil \tag{7.2}$$

$$cz_0 = \left\lceil \frac{ZOsiz}{ZRsiz^c} \right\rceil, \quad cz_1 = \left\lceil \frac{Zsiz}{ZRsiz^c} \right\rceil \tag{7.3}$$

From these coordinates it is trivial to calculate the dimensions of a given component, i.e. (width, height, depth) is given by ($cx_1 - cx_0$, $cy_1 - cy_0$, $cz_1 - cz_0$).

For a volumetric image, tiling is defined by two sets of three parameters. (XTsiz, YTsiz, ZTsiz) defines the width, height, and depth of each tile, while (XTOsiz, YTOsiz, ZTOsiz) defines the origin of the first tile. It is obvious that this origin must be located outside of the image area and that a tile cannot be empty (i.e. at least one image sample must reside in a tile). The reference grid is now partitioned into a regular-sized three-dimensional array of tiles, with the tiles being numbered in raster order (i.e. from left to right, top to bottom, front to back). Notice that all tiles have equal sizes, except possibly those tiles that are located at the borders of the image volume, since there is no need to encode the reference grid where no image samples exist. The number of tiles in each direction can be calculated as follows:

$$numXtiles = \left\lceil \frac{Xsiz - XTOsiz}{XTsiz} \right\rceil \tag{7.4}$$

$$numYtiles = \left\lceil \frac{Ysiz - YTOsiz}{YTsiz} \right\rceil \tag{7.5}$$

$$numZtiles = \left\lceil \frac{Zsiz - ZTOsiz}{ZTsiz} \right\rceil \tag{7.6}$$

With the number of tiles in each direction known, it is possible to calculate the begin and end coordinates of specific tiles. Given a tile with tile-index coordinates (p_x, p_y, p_z), the corner coordinates (tx_0, ty_0, tz_0) and (tx_1, ty_1, tz_1) with respect to the reference grid are then given as follows:

$$tx_0(p_x, p_y, p_z) = max(\text{XTOsiz} + p_x \cdot \text{XTsiz}, \text{XOsiz}),$$

$$ty_0(p_x, p_y, p_z) = max(\text{YTOsiz} + p_y \cdot \text{YTsiz}, \text{YOsiz}),$$

$$tz_0(p_x, p_y, p_z) = max(\text{ZTOsiz} + p_z \cdot \text{ZTsiz}, \text{ZOsiz}),$$

$$tx_1(p_x, p_y, p_z) = min(\text{XTOsiz} + (p_x + 1) \cdot \text{XTsiz}, \text{Xsiz}), \tag{7.7}$$

$$ty_1(p_x, p_y, p_z) = min(\text{YTOsiz} + (p_y + 1) \cdot \text{YTsiz}, \text{Ysiz}),$$

$$tz_1(p_x, p_y, p_z) = min(\text{ZTOsiz} + (p_z + 1) \cdot \text{ZTsiz}, \text{Zsiz}),$$

It is now trivial to calculate the dimensions of a specific tile, i.e. (width, height, depth) is given by $(tx_1 - tx_0, ty_1 - ty_0, tz_1 - tz_0)$.

Within the domain of image component c, a tile becomes a tile component. The samples of the tile component are mapped in a cuboid, defined by corner coordinates (tcx_0, tcy_0, tcz_0) and $(tcx_1 - 1, tcy_1 - 1, tcz_1 - 1)$, where $tcx_0, tcy_0, tcz_0, tcx_1, tcy_1,$ and tcz_1 are given as

$$tcx_0 = \left\lceil \frac{tx_0}{\text{XRsiz}^c} \right\rceil \quad tcx_1 = \left\lceil \frac{tx_1}{\text{XRsiz}^c} \right\rceil \tag{7.8}$$

$$tcy_0 = \left\lceil \frac{ty_0}{\text{YRsiz}^c} \right\rceil \quad tcy_1 = \left\lceil \frac{ty_1}{\text{YRsiz}^c} \right\rceil \tag{7.9}$$

$$tcz_0 = \left\lceil \frac{tz_0}{\text{ZRsiz}^c} \right\rceil \quad tcz_1 = \left\lceil \frac{tz_1}{\text{ZRsiz}^c} \right\rceil \tag{7.10}$$

Again, the dimensions of a specific tile component, i.e. (width, height, depth), are given by $(tcx1 - tcx0, tcy1 - tcy0, tcz1 - tcz0)$.

7.3.1.2 Resolution Levels and Subbands

The definition of resolution levels in JP3D is somewhat more complicated than in JPEG 2000 Part 1. This is because JP3D has the extra flexibility of allowing a variable number of decompositions in each spatial direction (see Section 7.2.2). For a single-component image volume that is wavelet transformed with N_{LX} decomposition levels in the horizontal direction, N_{LY} decompositions in the vertical direction, and N_{LZ} decompositions in the axial direction, the direction with the highest number of decompositions determines the number of resolution levels. Thus, with $N_{LMAX} = \max (N_{LX}, N_{LY}, N_{LZ})$, there are $N_{LMAX} + 1$ distinct resolution levels. Initially, a single-image component volume has one resolution level, i.e. the resolution of the original image volume, and zero decomposition levels. Similar to JPEG 2000 Part 1, decomposition level 0 represents the original image volume. The JP3D codec then continues with the multidimensional wavelet transformation, creating extra decomposition levels one by one and with incremental level indexes.

In the following text, a complete explanation of the relaxed decomposition pattern of JP3D is given. For this, three temporary state variables L_X, L_Y, and L_Z are introduced. They are initially set to N_{LX}, N_{LY}, and N_{LZ}, respectively, and represent the number of decompositions that are still left to be performed. For every decomposition cycle, a multidimensional wavelet transformation is executed in each spatial direction that has a nonzero state variable. During every cycle, the state variables of transformed spatial directions are decremented by one. Consequently, the complete transformation cycle is repeated $N_{LMAX} - 1$ times until all three state variables become zero. The decompositions are always performed on the lowest frequency subband at the highest or most recently created decomposition level. For every decomposition cycle, a well-defined number of subbands are created out of the previous low-frequency subband. Figure 7.8 shows the possible outcomes of a wavelet decomposition stage, depending on the values of the state variables L_X, L_Y, and L_Z. Therefore, eight subbands are generated (LLL, LLH, LHL, LHH, HLL, HLH, HHL, and HHH) if all three state variables are nonzero. However, only four subbands will be generated when exactly one of the three state variables is already equal to zero. For example, when L_X is equal to zero and both L_Y and L_Z are nonzero, then the following subbands are generated: XLL, XLH, XHL, and XHH. As previously mentioned, the character X in the subband names denotes that no decomposition took place in its respective spatial direction. Also, only two subbands will be generated in the case where two state variables of the set to N_{LX}, N_{LY}, and N_{LZ} are zero (and one is not).

With each added decomposition level, new subbands are defined where the subband with the lowest frequency information is further decomposed until the requested amount of decomposition levels is reached. The lowest frequency subband at the highest decomposition level now defines resolution level 0. The next resolution level, $r = 1$, contains the other subband(s) at the highest decomposition level. Then, if applicable, resolution level 2 is defined by the subbands of the previous decomposition level and so on. Thus, every multidimensional wavelet decomposition stage adds a new resolution level. Without any decompositions, only one resolution level ($r = 0$) exists. One way of looking at resolution levels is that they typically hold the particular subbands that are needed to increase the resolution of a reconstructed image. For each added resolution level during reconstruction, the number of reconstructed samples doubles in spatial directions that were subject to wavelet decomposition.

The true meaning of resolution levels is still identical to what is defined in JPEG 2000 Part 1. In fact, the definition of resolution levels in JP3D is perfectly compliant with that of JPEG 2000 Part 1. The differences between both definitions are only caused by

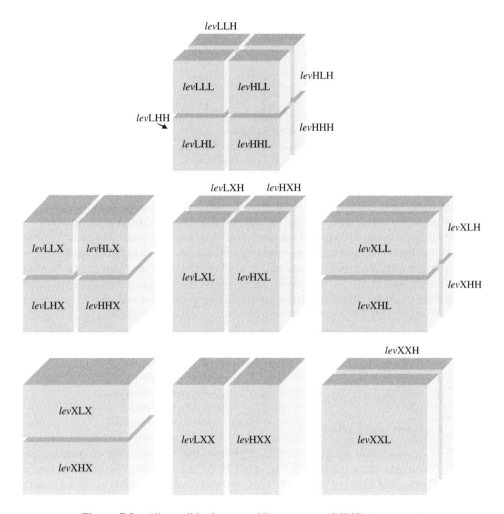

Figure 7.8 All possible decomposition patterns of JP3D (one stage)

(1) the extra dimension and (2) the flexibility of JP3D that allows for a different amount of decomposition levels in each spatial direction.

Within a tile component, a resolution level r ($0 \le r \le N_{LMAX}$) is a cuboidal volume with its corners at coordinates (trx_0, try_0, trz_0) and $(trx_1 - 1, try_1 - 1, trz_1 - 1)$, where trx_0, try_0, trz_0, trx_1, try_1, and trz_1 are given as

$$trx_0 = \left\lceil \frac{tcx_0}{2^{min(N_{Lmax}-r,N_{LX})}} \right\rceil, \quad trx_1 = \left\lceil \frac{tcx_1}{2^{min(N_{Lmax}-r,N_{LX})}} \right\rceil, \tag{7.11}$$

$$try_0 = \left\lceil \frac{tcy_0}{2^{min(N_{Lmax}-r,N_{LY})}} \right\rceil, \quad try_1 = \left\lceil \frac{tcy_1}{2^{min(N_{Lmax}-r,N_{LY})}} \right\rceil, \tag{7.12}$$

$$trz_0 = \left\lceil \frac{tcz_0}{2^{min(N_{Lmax}-r,N_{LZ})}} \right\rceil, \quad trz_1 = \left\lceil \frac{tcz_1}{2^{min(N_{Lmax}-r,N_{LZ})}} \right\rceil. \tag{7.13}$$

The tile component coordinates map in a similar manner to any particular subband b, yielding corner coordinates (tbx_0, tby_0, tbz_0) and $(tbx_1 - 1, tby_1 - 1, tbz_1 - 1)$, where $tbx_0, tby_0, tbz_0, tbx_1, tby_1,$ and tbz_1 are given as

$$tbx_0 = \left\lceil \frac{tcx_0 - (2^{xn_b - 1} \cdot xo_b)}{2^{xn_b}} \right\rceil, \quad tbx_1 = \left\lceil \frac{tcx_1 - (2^{xn_b - 1} \cdot xo_b)}{2^{xn_b}} \right\rceil \quad (7.14)$$

$$tby_0 = \left\lceil \frac{tcy_0 - (2^{yn_b - 1} \cdot yo_b)}{2^{yn_b}} \right\rceil, \quad tby_1 = \left\lceil \frac{tcy_1 - (2^{yn_b - 1} \cdot yo_b)}{2^{yn_b}} \right\rceil \quad (7.15)$$

$$tbz_0 = \left\lceil \frac{tcz_0 - (2^{zn_b - 1} \cdot zo_b)}{2^{zn_b}} \right\rceil, \quad tbz_1 = \left\lceil \frac{tcz_1 - (2^{zn_b - 1} \cdot zo_b)}{2^{zn_b}} \right\rceil \quad (7.16)$$

The quantities xn_b, yn_b, and zn_b define the number of performed decompositions in their respective spatial direction, associated with subband b. Quantities xo_b, yo_b, and zo_b indicate whether a subband is highpass or lowpass. A value of 0 is used for L- and X-type spatial directions, while a value of 1 is used for H-type spatial directions. Note that for negative exponents $(xn_b - 1)$, $(yn_b - 1)$, or $(zn_b - 1)$ the respective values xo_b, yo_b, or zo_b are always zero, since no decompositions are performed (i.e. the X-type spatial direction).

The following simple example helps to clarify the use of the formulas to calculate the coordinates of a subband. Suppose that an image volume is decomposed using a $(3, 1, 2)$ pattern into four resolution levels. Then, in order to calculate the coordinates of, for example, subband 2LXH, the values of the six quantities in the formulas should be set as follows: $xn_b = 2$, $yn_b = 1$, $zn_b = 2$, $xo_b = 0$, $yo_b = 0$, and $zo_b = 1$.

7.3.1.3 Precincts and Code-Blocks

Apart from having three dimensions, the definition and the meaning of precincts and code-blocks remains the same in JP3D as in JPEG 2000 Part 1. As mentioned before, all subbands are divided into small equally sized cuboidal units, called code-blocks, which are then independently encoded by the MQ-coder. Therefore, random accessibility of the image data becomes possible at a code-block level. Figure 7.9 shows how a wavelet transformed image volume is partitioned into code-blocks of equal sizes. Similar restrictions

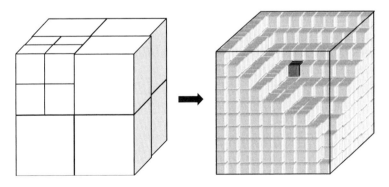

Figure 7.9 Code-block partitioning in 3-D

apply to the code-block dimensions in 3-D as in 2-D, with the most important restriction that the sizes must be a power of two. Typical code-block sizes in JP3D are $16 \times 16 \times 16$ and $32 \times 32 \times 32$ and sizes can be defined per tile component.

However, most often the wavelet coefficients from a single decoded code-block are useless when reconstructing spatial image data. In order to be able to perform an inverse wavelet transformation, various code-blocks from other subbands addressing the same spatial region are crucial. Therefore, a coarser partitioning of wavelet coefficients than code-blocks exists in JPEG 2000, called precincts. A precinct defines a group of code-blocks over all subbands at a specific resolution level, describing the same spatial region in the original image volume. Thus, a precinct typically contains multiple code-blocks from different subbands. Precincts can be defined per resolution level and are restricted to sizes that are powers of two. Both precinct and code-block partitions are anchored at (0, 0, 0) in the reference grid, causing precinct borders always to coincide with code-block borders. A precinct contains one or more code-blocks, but can never be smaller than a code-block. Despite the fact that code-blocks are equally sized in the complete tile component, code-block sizes will be automatically constrained in a particular subband in the case where the locally defined precinct sizes are smaller. For example, if a tile component has code-blocks of $32 \times 32 \times 32$ samples and precinct dimensions at a particular resolution level are defined to be only $16 \times 16 \times 16$ in size, then the code-blocks at that resolution level will be constrained to a size of $16 \times 16 \times 16$. Moreover, both precincts and code-blocks are also constrained by the subband boundaries to which they belong. Precincts determine at what precision the image data are spatially random accessible. Smaller precincts provide finer spatial granularity, but increase overhead in the code-stream. Alternatively, precincts can also be defined to contain an entire subband, effectively disabling the use of precincts and thus also minimizing the overhead at the cost of having very coarse spatial random access.

7.3.1.4 Layers, Packets, and Progression Order

In exactly the same way as with JPEG 2000 Part 1, JP3D precincts are distributed across one or more quality layers. In theory, all code-blocks contribute compressed data to each layer. In practice though, a zero contribution is possible. Remember that compressed bit-streams of code-blocks can be truncated between coding passes. This allows the encoder to use pieces of these code-block bit-streams and distribute them over various layers. Each precinct is literally chopped at the code-block bit-stream level into smaller pieces, called packets. Packets are the smallest units that build a JPEG 2000 code-stream. Each packet contains a piece of the compressed bit-stream data belonging to a specific tile, component, resolution, precinct, and layer. Because packet construction and layer creation are mostly identical with JPEG 2000 Part 1 and JP3D, the reader is advised to read Section 1.3.4. The main difference between 2-D and 3-D is that, due to the extra dimension and the more flexible decomposition pattern, a packet can now contain more than three different types of subbands at once. The order in which these subbands should appear in each packet is important and deterministically defined. The compressed image data in a packet is ordered such that the possible contributions from the LLL, XLL, LXL, LLX, LXX, XLX, XXL, HLL, HXL, HLX, HXX, LHL, XHL, LHX, XHX, HHL, HHX, LLH, XLH, LXH, XXH, HLH, HXH, LHH, XHH, and HHH subbands appear in that order, i.e. the Morton scanning order (Morton, 1966). Furthermore, each packet starts with a packet header. This header contains information about the packet, like the number of coding passes per code-block inside the packet and the amount of bytes in the packet body. Packet headers in JP3D are identical to the packet headers of JPEG 2000 Part 1 (see Section 1.3.5). However, although the packet headers are identical, there is still a minor issue with respect to their use in practice. The number of zero bit-planes and the number of coding passes per code-block are encoded with tag trees. In JPEG 2000 Part 1, tag trees are a method used to encode two-dimensional matrices of positive integer numbers. In JP3D, however, these tag trees are extended to encode three-dimensional matrices. Packets are consecutively stored in the code-stream in a predefined order, called the progression order. JP3D uses the same progression orders as defined for JPEG 2000 Part 1, with the same restrictions. One such important restriction is that coding passes of the code-blocks must appear in a causal order in the code-stream from the most significant bit-plane to the least significant bit-plane. Without this restriction, it would often be impossible to immediately use code-block contributions, making the code-stream no longer optimally ordered. Section 1.3.6 gives a detailed description of the five defined progression orders of JPEG 2000.

7.3.2 Code-Stream

JP3D maintains the original JPEG 2000 code-stream syntax. Like with JPEG 2000 Part 1, a valid JP3D code-stream starts with an SOC (start of code-stream) marker and ends with an EOC (end of code-stream) marker. Immediately following the SOC marker, a sequence of marker segments is placed, constituting the main header of the code-stream. After the main header, the actual compressed data are situated in the packet structure. However, these packets are first grouped into logical units, called tile parts, to separate packets from different tiles. In doing so, multiple (at least one) tile parts can exist per tile, and these tile parts can be interleaved with each other. Each tile part is preceded with a private tile-part header with marker segments specific to that tile part. Section 1.3.7.1 gives a more detailed overview of the various markers and marker segments that exist for JPEG 2000 Part 1, grouping them into one of six categories. In order to accommodate an extra spatial dimension, a new marker segment (NSI) is added and a number of other marker segments are modified. Table 7.1 lists all the markers and marker segments that the JP3D specification defines. The last column shows the origin of the marker and whether or not it is extended by JP3D. All marker segments listed in Table 7.1 that originate from JPEG 2000 Part 2, with the exception of the region of interest (ROI), can straightforwardly be used with JP3D (see Chapter 2).

7.3.2.1 Extending the Code-Stream (CAP and NSI)

A standard JPEG 2000 compliant code-stream always starts with the SOC marker, followed by the SIZ marker segment to signal the image and tile sizes in two dimensions. Signaling the third dimension of the image volume and tile sizes with JP3D requires extra functionalities in the code-stream. However, it is not possible to enhance the existing SIZ marker segment through modification, because this would break the JPEG 2000 standard code-stream compliance. Existing codecs will then most definitely fail with unpredictable errors. Moreover, a pure JPEG 2000 Part 1 or Part 2 decoder implementation should not be burdened with JP3D or any other future JPEG 2000 parts it does not know about.

Table 7.1 List of JP3D markers and marker segments

	Symbol	Code	Main header	Tile-part header	ITU-T Rec. T.80x | ISO/IEC 15444-x Heritage/Extended
Delimiting markers and marker segments					
Start of code-stream	SOC	0xFF4F	Required	Not allowed	ITU-T Rec. T.800 | ISO/IEC 15444-1
Start of tile part	SOT	0xFF90	Not allowed	Required	ITU-T Rec. T.801 | ISO/IEC 15444-2
Start of data	SOD	0xFF93	Not allowed	Last marker	ITU-T Rec. T.800 | ISO/IEC 15444-1
End of code-stream	EOC	0xFFD9	Not allowed	Not allowed	ITU-T Rec. T.800 | ISO/IEC 15444-1
Fixed information marker segments					
Image and tile size	SIZ	0xFF51	Required	Not allowed	ITU-T Rec. T.800 | ISO/IEC 15444-1
Additional dimension image and tile size	NSI	0xFF54	Required	Not allowed	
Functional marker segments					
Coding style default	COD	0xFF52	Required	Optional	ITU-T Rec. T.800 | ISO/IEC 15444-1, Extended
Coding style component	COC	0xFF53	Optional	Optional	ITU-T Rec. T.800 | ISO/IEC 15444-1, Extended
Region of interest	RGN	0xFF5E	Optional	Optional	ITU-T Rec. T.801 | ISO/IEC 15444-2, Extended
Quantization default	QCD	0xFF5C	Required	Optional	ITU-T Rec. T.800 | ISO/IEC 15444-1, Extended
Quantization component	QCC	0xFF5D	Optional	Optional	ITU-T Rec. T.800 | ISO/IEC 15444-1, Extended
Arbitrary transformation kernels	ATK	0xFF79	Optional	Optional	ITU-T Rec. T.801 | ISO/IEC 15444-2
Component bit-depth definition	CBD	0xFF78	Optional	Optional	ITU-T Rec. T.801 | ISO/IEC 15444-2
Multiple component transformation definition	MCT	0xFF74	Optional	Optional	ITU-T Rec. T.801 | ISO/IEC 15444-2
Multiple component transform collection	MCC	0xFF75	Optional	Optional	ITU-T Rec. T.801 | ISO/IEC 15444-2
Multiple component transform ordering	MCO	0XFF77	Optional	Optional	ITU-T Rec. T.801 | ISO/IEC 15444-2
Nonlinearity point transformation	NLT	0xFF76	Pptional	Optional	ITU-T Rec. T.801 | ISO/IEC 15444-2
Variable DC offset	DCO	0XFF70	Optional	Optional	ITU-T Rec. T.801 | ISO/IEC 15444-2

(continued overleaf)

Table 7.1 (*continued*)

	Symbol	Code	Main header	Tile-part header	ITU-T Rec. T.80x | ISO/IEC 15444-x Heritage/Extended
Pointer marker segments					
Tile-part lengths	TLM	0xFF55	Optional	Not allowed	ITU-T Rec. T.800 | ISO/IEC 15444-1
Packet length, main header	PLM	0xFF57	Optional	Not allowed	ITU-T Rec. T.800 | ISO/IEC 15444-1
Packet length, tile-part header	PLT	0xFF58	Not allowed	Optional	ITU-T Rec. T.800 | ISO/IEC 15444-1
Packed packet headers, main header	PPM	0xFF60	Optional	Not allowed	ITU-T Rec. T.800 | ISO/IEC 15444-1
Packed packet headers, tile-part header	PPT	0xFF61	Not allowed	Optional	ITU-T Rec. T.800 | ISO/IEC 15444-1
In bit-stream markers and marker segments					
Start of packet	SOP	0xFF91	Not allowed	Not allowed in tile-part header, optional in bit-stream	ITU-T Rec. T.800 | ISO/IEC 15444-1
End of packet header	EPH	0xFF92	Optional inside PPM marker segment	Optional inside PPT marker segment or in bit-stream	ITU-T Rec. T.800 | ISO/IEC 15444-1
Informational marker segments					
Component registration	CRG	0xFF63	Optional	Not allowed	ITU-T Rec. T.800 | ISO/IEC 15444-1, Extended
Comment	COM	0xFF64	Optional	Optional	ITU-T Rec. T.800 | ISO/IEC 15444-1

Therefore, JP3D uses two mechanisms to maintain code-stream compliance and to signal the extra information:

1. JP3D requires that its presence in a code-stream is signaled through the extended capabilities (CAP) marker segment, as defined in JPEG 2000 Part 2 Amendment 2. This marker segment allows existing and future JPEG 2000 parts to signal their presence and private extended capabilities, without ever breaking code-stream compliance. Apart from SOC and SIZ, JPEG 2000 Part 1 and Part 2 impose no particular ordering on the marker segments in the main header of a JPEG 2000 code-stream, which means that the CAP marker segment can appear anywhere within the main header. However, in order to make it easier and more convenient to recognize a JP3D code-stream, JP3D expects the mandatory CAP marker segment immediately to be placed after the SIZ marker segment.

2. The new marker segment *n*-size (NSI), which is similar to the SIZ marker segment, signals the image and tile sizes in the third spatial direction (*Z* axis). Currently, only

one extra dimension is supported, but the NSI marker segment is extendable and can support up to 253 additional spatial dimensions (hence the 'N' in its name). The NSI marker segment is also mandatory in the main header of a JP3D code-stream.

Thus, a JP3D code-stream starts with the SOC marker, followed by the SIZ marker segment. The JP3D code-stream can then be recognized through the presence of a CAP marker segment, which indicates usage of JPEG 2000 Part 10, and the mandatory NSI marker segment, both in the main header.

7.3.2.2 Modifying Existing Marker Segments

Using the CAP/NSI pattern, it is possible to distinguish a JP3D code-stream correctly from any other JPEG 2000 code-stream before other markers and marker segments occur. Therefore, it is possible to use modified versions of JPEG 2000 Part 1 marker segments from this point on, without breaking code-stream compliance, in order to accommodate more than two dimensions. For JP3D the following marker segments were modified: coding style (COD and COC), quantization component (QCD and QCC), region of interest (RGN) and component registration (CRG). For most of these marker segments, the modifications are straightforward and only add extra fields for the Z direction. However, the changes to the COD and COC markers also include the necessary extra fields to allow the flexible wavelet decomposition pattern for JP3D, i.e. having a different number of decompositions and different wavelet kernels per spatial direction. Aside from the minor modifications to some of the marker segments, the semantics of these code-stream elements remain the same as with JPEG 2000 Part 1 and Part 2.

7.3.2.3 N-Dimensionality

JP3D introduces a third dimension for spatial data in images, but it is designed in such a way that it can support even more dimensions for future extensions of the JP3D specification. The NSI marker segment contains a field (Ndim) that defines the dimensionality of the data set (disregarding the component dimension). Currently, this field should always contain a value of 3 to indicate that three-dimensional image data are encoded.

7.3.3 Rate Control

Two mechanisms exist in JPEG 2000 that allow the allocation of bits in the final compressed image code-stream to be controlled. One method quantizes the wavelet coefficients in the subbands to adjust the quality and bit-rate of the encoded image. However, quantization-based rate control requires multiple re-encoding iterations to achieve a specific bit-rate and is therefore not preferred. JPEG 2000 is also able to deliver excellent rate control through its embedded block coding with optimized truncation (EBCOT) approach. With EBCOT, an encoder can truncate data of the encoded code-blocks at its own discretion in order to satisfy a specific target bit-rate or target quality. By using EBCOT an encoder can perform optimal rate control in one pass (PCRD-opt), without the need to perform multiple encoding cycles. JP3D is the straightforward extension of JPEG 2000 Part 1 and thus uses the exact same methods for the bit-stream rate control. Section 1.4 gives a more detailed explanation of the JPEG 2000 rate control mechanisms. Unlike floating-point transforms, making the (lossless) integer wavelet transforms unitary is not evident. This unitary property is necessary in order to achieve good lossy coding performance. By calculating the L_2 norm of the low- and highpass filters, the normalization factors can be determined. For 2-D transforms this is not a problem, since the typical scaling factors used to obtain a unitary transform are approximately powers of two (Said and Pearlman, 1996). However, for lossless 3-D transforms, unitarity can only be achieved if the total number of decompositions influencing each individual wavelet coefficient is even. Hence, some proposals have been formulated (Kim and Pearlman, 1999; Xiong et al., 1999) that make use of a wavelet packet transform (Xiong, Ramchandran, and Orchard, 1998) to achieve this goal, while assuming that the L_2-based normalization factors for the supported kernels scale-up with $\sqrt{2}$ for the lowpass and $1/\sqrt{2}$ for the highpass kernels. Fortunately an easier solution exists for JPEG 2000, because the post-compression rate-distortion (PCRD) optimization procedure (Taubman, 2000) allows compensation for the fact that a nonunitary transform has been used. By correcting the calculated distortions for each pass with the appropriate subband weighing factor, the coding system will perform as if a unitary transform was used (or approximated when using integer powers of $\sqrt{2}$) (Schelkens et al., 2003). This correction enables support for a nonunitary transform without compromising the capability of lossless coding, a problem that does occur with classic tree-based coders in the case of volumetric encoding.

7.4 Additional Features of JP3D

7.4.1 Region of Interest

JP3D explicitly defines the volumetric extension of region-of-interest (ROI) coding. It adopts the methods from JPEG 2000 Part 1 (Max-shift) and JPEG 2000 Part 2 (scaling), thus actually extending both methods (see Section 1.6.1 and Chapter 2). The encoding and decoding mechanisms for 3-D ROI areas are identical to those of JPEG 2000 Part 1 and Part 2. The JPEG 2000 Part 1 Max-shift method requires no changes and is by design compatible with JP3D. However, in order to apply the Part 2 scaling ROI mode, it is necessary for both the encoder and decoder to generate a region-of-interest mask. The mask is a bit-plane that indicates where the scaled image samples are located. For JP3D, such a mask is a cuboid grid of scaling values. Each point in the grid matches one sample in the original image, relative to the reference grid. Therefore, JP3D supports the logical extension of the two basic ROI mask shapes from JPEG 2000 Part 2, i.e. the cuboid and the ellipsoid. Both shapes are characterized by a set of six arguments and a type value that are signaled in the extended RGN marker segment. Figures 7.10 and 7.11 show

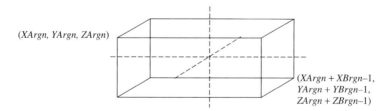

Figure 7.10   Cuboidal region of interest

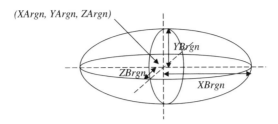

Figure 7.11   Ellipsoidal region of interest

how JP3D defines the two supported types of shapes for defining volumetric region-of-interest areas.

7.5   Compression performances: JPEG 2000 Part 1 versus JP3D

This section complements the section on the performance comparison of the JPEG 2000 encoder options in Chapter 1. Because JP3D only adds an extra dimension to the image data, most of the existing encoder options, such as tile sizes, code-block sizes, or precinct sizes, will affect the compression performances in the same way as with two-dimensional data in JPEG 2000 Part 1. Smaller code-blocks, for example, create more overhead than larger code-blocks and thus negatively affect the compression performance. The same observation is true for the type of wavelet kernels and for the various mode settings in the entropy coding step.

7.5.1   Test Setup

This section offers two sets of comparison tests that show the impact on the overall compression performance when (1) performing three-dimensional wavelet decompositions and (2) applying three-dimensional code-blocks. The comparison tests make use of various medical volumetric images of different scanner types (CT, PET, MRI, etc.). Table 7.2 shows the main properties of the volumetric images used for the comparison tests in this section.

The results obtained during the tests are generated by the official JPEG 2000 Part 10 reference software (Bruylants, 2008), which is capable of generating both Part 1 and

Table 7.2   Properties of the test images

Name	Width (pixels)	Height (pixels)	Depth (pixels)	Bit-depth (bpp)
ECHO	256	256	256	8
PET	128	128	39	15
AxialCT	512	512	100	12
NormalCT	512	512	44	12
SpiralCT	512	512	64	12
MRI	256	256	200	12
Ultrasound	500	244	201	8

Table 7.3 Generic encoder settings

Option	Value
Tiling	No tiles used
Wavelet kernel	5×3 and 9×7
Progression order	RLCP
Precincts	Maximally sized[a]
Code-block sizes	$64 \times 64 \times 1$ and $16 \times 16 \times 16$
Code-block style	0(ISO/IEC 15444-1)
EPH signaling	No
SOP signaling	No

[a]Maximally sized precincts will in practice disable the use of precincts. Each precinct contains exactly one subband.

Part 10 compliant code-streams. The codec is written in ANSI/C and does not apply any implementation-specific optimizations. It serves as a reference model for the JP3D standard and tries to match this standard as closely as possible with respect to algorithms and design choices. Moreover, by using only one codec it is guaranteed that the same (encoder implementation specific) rate/distortion optimization algorithm is used for all tests. This makes it possible to have a fair comparison between 2-D (Part 1) and 3-D (Part 10) modes of JPEG 2000. For all the tests, the set of JPEG 2000 encoder options given in Table 7.3 was fixed.

In fact, the settings that do change throughout the various test runs are the applied wavelet transformation kernel (and associated quantization values), the number of wavelet decompositions, and the code-block sizes. Table 7.4 shows the number of wavelet decomposition levels for each volumetric image in 2-D and 3-D modes.

Both the integer (5, 3) and the floating-point (9, 7) wavelet kernels are used on the full set of volumetric images. In the 2-D mode the code-block sizes are set to $64 \times 64 \times 1$, while in the 3-D mode both the $64 \times 64 \times 1$ and the $16 \times 16 \times 16$ code-block sizes are tested. Thus, code-blocks always contain 4096 samples in both 2-D and 3-D modes (with the exception of border cases). This also means that for each data set all the generated code-streams will contain roughly the same amount of packets, regardless of the applied encoder settings (i.e. 2-D versus 3-D wavelet decompositions and 2-D versus

Table 7.4 Applied number of DWT decomposition levels (for X, Y, and Z dimensions)

Name	2-D DWT	3-D DWT
ECHO	5, 5, 0	5, 5, 5
PET	5, 5, 0	5, 5, 3
AxialCT	5, 5, 0	5, 5, 5
NormalCT	5, 5, 0	5, 5, 3
SpiralCT	5, 5, 0	5, 5, 5
NormalMRIBrain	5, 5, 0	5, 5, 5
Ultrasound	5, 5, 0	5, 5, 5

3-D code-blocks). This, in turn, makes the results comparable and better highlights the impact of changing a single encoder setting.

For lossless compression[1] the generated single-layer code-stream is not truncated and the achieved bit-rate and the peak signal-to-noise ratio (PSNR) are reported. For lossy compression, both the target and the actual achieved bit-rates after code-stream truncation are reported, complemented with a PSNR value to measure the distortion. For an image with n-bit samples, the PSNR is given by

$$\text{PSNR} = 10\log_{10}\left(\frac{(2^n - 1)^2}{\text{MSE}}\right) \tag{7.17}$$

The mean square error (MSE) represents the sum of the squared differences of each sample between the original and the reconstructed image. Please note that the PSNR equation (7.17) differs slightly from the one defined in Chapter 1. Because the various volumetric images have different sample bit-depths, it defines PSNR also as a function of the sample bit-depth (n). Chapter 1 uses a bit-depth of $n = 8$ by default. Additionally, we report the obtained compression ratios since this metric is not relative to the sample bit-depth. The bit-rate and PSNR values show how the compression performance of JP3D compares with that of JPEG 2000 Part 1 (i.e. 2-D compression).

7.5.2   Lossless Compression

The first two sets of results show for all the volumetric test images the achieved bit-rates after compression without truncation of the code-stream. The latter contains a complete representation of all the code-blocks. Table 7.5 shows lossless bit-rates and compression ratios when using the integer (5, 3) wavelet kernel. The table does not show PSNR values because image reconstruction is perfect in this case. It is clear that JP3D with 3-D wavelet decompositions compresses significantly better than JPEG 2000 Part 1. Going

Table 7.5   Compression performances for encoding with the lossless 5×3 filter kernel

Data set 5×3	Bit-rate (bpp)			Compression ratio		
	3-D	3-D	2-D	3-D	3-D	2-D
	$16 \times 16 \times 16$	$64 \times 64 \times 1$		$16 \times 16 \times 16$	$64 \times 64 \times 1$	
PET	8.8135	8.8173	10.1675	1.7019	1.7012	1.4753
ECHO	3.8109	3.8519	4.2747	2.0993	2.0769	1.8715
AxialCT	3.8374	3.8730	4.0795	3.1271	3.0984	2.9416
NormalCT	5.0228	5.0462	5.1663	2.3891	2.3780	2.3228
SpiralCT	5.3761	5.4005	5.8105	2.2321	2.2220	2.0653
NormalMRIBrain	4.0795	4.1288	4.7228	2.9416	2.9064	2.5409
SpineUltraSound	4.8389	4.8348	5.0542	1.6533	1.6547	1.5828

---

[1] Lossless compression is only achievable with the integer 5×3 wavelet kernel. The floating-point 9×7 wavelet kernel introduces rounding errors and thus delivers only near-lossless compression. However, we always refer to lossless compression for clarity and ease of reading when no code-stream truncation is performed.

Table 7.6 Compression performances for encoding with the near-lossless 9 × 7 filter kernel

Data set 9 × 7	Bit-rate (bpp)			Ratio			PSNR (dB)	
	3-D	3-D	2-D	3-D	3-D	2-D	3-D	2-D
	16 × 16 × 16	64 × 64 × 1	64 × 64 × 1	16 × 16 × 16	64 × 64 × 1	64 × 64 × 1	Both	64 × 64 × 1
PET	3.4459	3.4464	5.8347	4.3531	4.3523	2.5709	68.4368	70.2936
ECHO	5.8896	5.9325	6.9364	1.3583	1.3485	1.1533	75.9700	76.3309
AxialCT	2.0516	2.0757	2.8050	5.8491	5.7811	4.2781	70.8294	74.2460
NormalCT	3.5260	3.5434	4.2083	3.4033	3.3866	2.8515	71.2385	74.1001
SpiralCT	3.3925	3.4124	4.4227	3.5372	3.5166	2.7133	69.5273	72.5529
NormalMRIBrain	2.2958	2.3287	3.5989	5.2268	5.1531	3.3343	71.0047	74.9874
SpineUltraSound	7.1214	7.1098	7.9280	1.1234	1.1252	1.0091	98.0819	97.6489

from flat 2-D code-blocks to 3-D code-blocks improves the compression performance even more.

Similar results are shown in Table 7.6 for the floating-point (9, 7) wavelet kernel. Again, 3-D wavelet decompositions and 3-D code-blocks help to improve the compression performance significantly. This table does show PSNR values, because perfect reconstruction is no longer possible due to the rounding errors introduced by the floating-point (9, 7) transform. The PSNR values of the 3-D mode are slightly lower than those of the 2-D mode because the extra wavelet decompositions introduce extra rounding errors. Please note that the PSNR values are very high and that the differences are rather imperceptible in that range. Also, the table shows only one PSNR value per data set for the 3-D mode, because changing the code-block size does not affect the introduced rounding errors.

7.5.3 Lossy Compression

To evaluate the rate-distortion performance for lossy compression, the full set of volumetric images is compressed at the following target bit-rates: 4, 2, 1, 0.5, 0.25, 0.125, and 0.0625 bits per pixel (bpp). The target bit-rate of 4 bpp is dropped for images that compress losslessly to less than 4 bpp. The rate-distortion optimization algorithm of JP3D is identical to that of JPEG 2000 Part 1. The codec terminates the code-stream when it reaches the target bit-rate, dropping the unused fragments of encoded code-blocks. The JP3D reference software will never exceed a given bit-rate limit and will try to approximate the upper bound as closely as possible.

Figure 7.12 shows the compression results for the various volumetric images when compressing with the (9, 7) wavelet kernel. The 2-D mode uses code-blocks of 64 × 64 × 1, while the 3-D mode uses code-blocks of 16 × 16 × 16. It can be seen that the compression efficiency improvement from 2-D to 3-D depends on the imaging modality. The 3-D performance is heavily dependent on the slice axis correlation. For the MRI data set an improvement of the compression efficiency with a factor of 2 is observable for the full bit-rate range. For the CT and ultrasound data sets the gain is smaller, respectively due to the lower axial correlation and the lack of motion matching. Similar results are reported for the integer (5, 3) wavelet kernel (Schelkens et al., 2003).

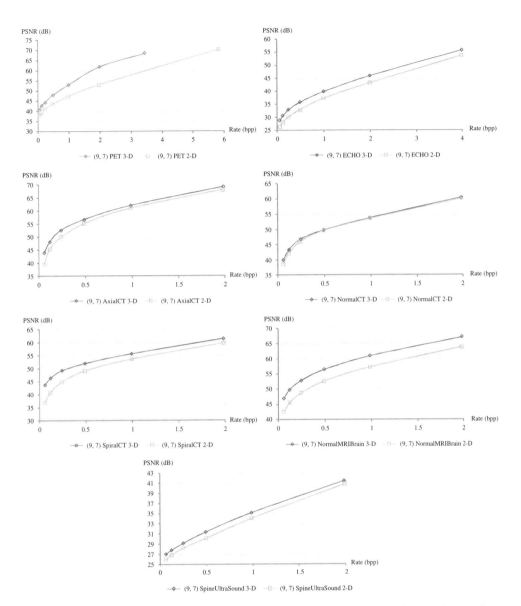

Figure 7.12 These graphs show for the test data sets the lossy 9×7 compression results at various bit-rates, with the 3-D mode using 3-D code-blocks

7.5.4 Time Complexity

To measure the algorithmic complexity of JP3D in comparison to JPEG 2000 Part 1, the processing time was measured. We measured the time spent during the wavelet decomposition step and the code-block entropy coding step separately. Doing so offers a better insight into the effect of having three-dimensional wavelet decompositions.

Table 7.7 Average times (in seconds) of multiple runs spent by the encoder. The times spent during the wavelet decomposition steps and the entropy coding steps for (near-)lossless compression are listed separately. Each data set is encoded with both the 2-D and the 3-D wavelet decomposition patterns. The last two columns show for each data set the total encoding time for the 3-D and 2-D decomposition patterns, respectively

Data set	Wavelet time (s)		Entropy time (s)		Total time (s)	
	3-D	2-D	3-D	2-D	3-D	2-D
PET 5×3	0.11	0.03	0.87	0.97	0.97	1.01
PET 9×7	0.16	0.08	0.46	0.66	0.62	0.74
ECHO 5×3	3.36	0.85	14.43	15.57	17.80	16.42
ECHO 9×7	4.79	2.05	19.78	22.16	24.57	24.21
AxialCT 5×3	2.27	1.42	26.43	28.60	28.70	30.02
AxialCT 9×7	4.51	3.16	19.31	23.68	23.82	26.84
NormalCT 5×3	0.84	0.63	13.09	14.34	13.92	14.97
NormalCT 9×7	1.89	1.40	10.66	12.76	12.55	14.16
SpiralCT 5×3	1.33	0.91	17.26	19.57	18.59	20.48
SpiralCT 9×7	2.78	2.07	13.17	16.81	15.96	18.88
NormalMRIBrain 5×3	2.17	0.65	12.78	14.25	14.95	14.89
NormalMRIBrain 9×7	3.30	1.54	9.42	12.31	12.72	13.85
SpineUltraSound 5×3	2.10	1.28	24.19	21.69	26.29	22.97
SpineUltraSound 9×7	4.21	2.91	28.41	30.60	32.63	33.50

With regards to the memory complexity of JP3D, it can easily be shown that in the worst case it is proportional to the dimensions of the input. Clever workarounds exists to lower the memory footprint drastically, even when performing a 3-D DWT. This is, however, beyond the scope of this chapter and we refer to Chapter 17 for wavelet transform implementation strategies.

To obtain the results reported in Table 7.7, the JP3D codec was slightly modified such that it measures the time spent performing the wavelet decompositions and the entropy coding of the code-blocks. The timings in Table 7.7 are actually averages over several runs of the same test in order to increase the reliability. The encoder settings for each of the volumetric images were still identical to the tests described above. Hence, the differences between the 2-D and 3-D timings are only caused by the different wavelet decomposition patterns in the third spatial dimension.

As expected, Table 7.7 shows that more time is required for the 3-D wavelet decomposition than for its 2-D equivalent. Indeed, the 3-D mode of JPEG 2000 performs extra wavelet decompositions in the Z direction and, thus, logically takes more time.

However, less expected, Table 7.7 also shows that the entropy coding step is faster for the 3-D mode of JPEG 2000 than for the 2-D mode. For both the 2-D and 3-D modes of JPEG 2000, the code-block dimensions in our tests are fixed to a size of $64 \times 64 \times 1$. This means that the number of coefficients of a code-block that needs to be encoded is identical in both modes. The differences in speed can only be explained by the fact that the extra wavelet decompositions for the 3-D mode help to decorrelate more efficiently the signal and hence decreases the signal entropy of the transform domain representations. To simplify, overall the code-block coefficients will be less significant and the higher

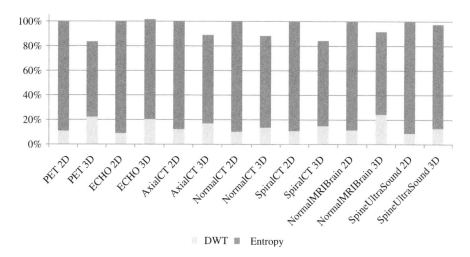

Figure 7.13 This figure shows the normalized timings for the 9×7 wavelet kernel. The bottom part of each bar in the graph represents the proportional amount of time needed for the wavelet transform, while the upper part represents the entropy coding time. The time values are relative to the total amount of time needed for the JPEG 2000 baseline mode

bit-planes will contain more zero bits. This in turn makes the entropy coding faster as fewer bits need to be encoded and often faster and less complex code-paths can be followed.

Figure 7.13 gives an alternative view on how the processing complexity changes for near-lossless coding when performing a 2-D or 3-D (9, 7) wavelet transform. The bottom part of each bar in the graph represents the proportional amount of time needed for the wavelet transform, while the upper part represents the entropy coding time. The time values are relative to the total amount of time needed for 2-D mode compression per image. Since for the JPEG 2000 baseline mode the times are always at exactly 100%, it becomes easier to compare between the different images.

7.5.5 Conclusions

The test results clearly show that JP3D offers a significant improvement for the compression of volumetric imagery. By allowing 3-D wavelet decompositions, superior compression performances are provided for a wide set of imaging modalities in the full range of bit-rates. Both lossless and lossy modes benefit from the 3-D DWT. Furthermore, the processing complexity results (see Table 7.7) indicate that often JP3D is faster than JPEG 2000 Part 1 when processing volumetric data sets. Hence, in contradiction to what would intuitively be expected, JP3D delivers a significantly improved rate-distortion performance in conjunction with a lower computational time.

7.6 Implications for Other Parts of JPEG 2000

Most parts of the JPEG 2000 family of standards are unaware of the dimensionality of the image data and are thus unaffected by the presence of JP3D. However, some parts (like Part 9) do need an update to add support for more than two dimensions and thus support for JP3D.

7.6.1 Volumetric Extension to JPIP

As explained in Chapter 6, JPEG 2000 Part 9, or JPIP, is the interactive imagery protocol of JPEG 2000. Initially, JPIP could only handle two-dimensional images, but with the development of JP3D, it was extended also to support images with more than two dimensions. Amendment 3 of the JPIP specification defines this straightforward extension to three or more dimensions. Additional fields in the request URLs allow a client application to specify more than two dimensions for the view-window region.

Acknowledgments

The authors would like to thank Chris Brislawn (Los Alamos National Labs), Gabriel Cristobal (CSIC, Madrid), Xavier Giró i Nieta (UPC, Barcelona), Adrian Munteanu (Vrije Universiteit Brussel), Joeri Barbarien (Vrije Universiteit Brussel), Mónica Díez García (University of Valladolid), So-youn An (Sejong University), Joo-hee Moon (Sejong University), Min-cheol Park (Sejong University), Jin Ri Zhu (Sejong University), and Hyoung-mee Park (Sejong University) for their fruitful contributions to the standardization process of JPEG 2000 Part 10.

References

An, S.-Y., Moon, J.-H., Park, M.-C., Zhu, J.-R. and Park, H.-M. (2006) Proposal and some results on 3D context selection for entropy coding, ISO/IEC JTC1/SC29/WG1, N3955, Perugia, Italy.

Bruylants, T. (2008) JPEG 2000 Part 10: Verification Model v1.1.1, ISO/IEC JTC1/SC29/WG1, N4808, Busan, Korea.

Bruylants, T., Alecu, A., Deklerck, R., Munteanu, A., Kimpe, T. and Schelkens, P. (2007) An optimized 3D context model for JPEG2000 Part 10, in *SPIE Proceedings on Medical Imaging*, vol. 6512-165, San Diego, CA.

Kim, Y. and Pearlman, W.A. (1999) Lossless volumetric medical image compression in *Proceedings of SPIE Conference on Applications of Digital Image Processing XXII*, vol. 3808, pp. 305–312.

Morton, G. M. (1966) *A Computer Oriented Geodetic Data Base and a New Technique in File Sequencing*, IBM Ltd, Ottawa, Canada.

Said, A. and Pearlman, W.A. (1996) An image multiresolution representation for lossless and lossy compression, *IEEE Transactions on Image Processing*, **5**, 1303–1310.

Schelkens, P. (2008) Call for information on applications and technology for floating-point datasets, ISO/IEC JTC1/SC29/WG1, N4727, Poitiers, France.

Schelkens, P., Munteanu, A., Barbarien, J., Galca, M., Giro i Nieto, X. and Cornelis, J. (2003) Wavelet coding of volumetric medical datasets, *IEEE Transactions on Medical Imaging*, **22**(3), 441–458.

Taubman, D. (2000) High performance scalable image compression with EBCOT, *IEEE Transactions on Image Processing*, **9**(7), 1158–1170.

Xiong, Z., Ramchandran, K. and Orchard, M.T. (1998) Wavelet packet image coding using space frequency quantization, *IEEE Transactions on Image Processing*, **7**, 892–898.

Xiong, Z., Wu, X., Yun, D.Y. and Pearlman, W.A. (1999) Progressive coding of medical volumetric data using three-dimensional integer wavelet packet transform, in *SPIE Conference on Visual Communications*, vol. 3653, pp. 327–335.

Further Reading

ISO/IEC (2004) Information Technology – JPEG 2000 Image Coding System – Part 1: Core Coding System, ISO/IEC International Standard 15444-1, ITU Recommendation T.800.

ISO/IEC (2004) Information Technology – JPEG 2000 Image Coding System – Part 1: Extensions, ISO/IEC International Standard 15444-2, ITU Recommendation T.801.

ISO/IEC (2006) Extended Capabilities Marker Segment, ISO/IEC International Standard 15444-2:2004/AMD 2.

ISO/IEC (2005) Information Technology – JPEG 2000 Image Coding System – Part 9: Interactivity Tools, APIs and Protocols, ISO/IEC International Standard 15444-9, ITU Recommendation T.808.

ISO/IEC (2008) JPIP Extensions to 3D Data, ISO/IEC International Standard 15444-9:2005/ FDAM 3.

ISO/IEC (2008) Information Technology – JPEG 2000 Image Coding System – Part 10: Extensions for Three-Dimensional Data, ISO/IEC International Standard 15444-10, ITU Recommendation T.809.

8

JPWL – JPEG 2000 Wireless (Part 11)

Frédéric Dufaux

8.1 Introduction

Mobile computing is now ubiquitous, as a result of the availability of high-performance mobile devices such as mobile phones, personal digital assistants, and laptops, combined with the pervasiveness of broadband wireless networks such as 3rd generation mobile telecommunication systems (3G) and wireless local area networks (WLANs) based on the IEEE 802.11 specifications (WiFi). As a result, and in conjunction with the advent of efficient imaging technologies, the transmission of video and images in mobile applications is becoming a reality. Indeed, market researches show that sales of camera phones are now exceeding those of digital cameras.

However, wireless networks are characterized by the frequent occurrence of transmission errors, putting demanding constraints on the transmission of video and images in the mobile environment. In this context, the deployment of robust and efficient wireless imaging solutions is paramount. JPEG 2000 is the latest and state-of-the-art standard for still image coding (Skodras, Christopoulos, and Ebrahimi, 2001; Taubman and Marcellin, 2002). Thanks to its high coding efficiency, JPEG 2000 is very appealing for wireless multimedia applications. Furthermore, its seamless and effective scalability makes it very suitable for novel and efficient quality-of-service policies in strenuous wireless network environments.

In order to cope with transmission errors better, the baseline JPEG 2000 specification defines several error resilience tools, as described in Moccagatta *et al.* (2000) and Taubman and Marcellin (2002). Their performances are evaluated in Santa-Cruz, Grosbois, and Ebrahimi (2002) and Bilgin, Wu, and Marcellin (2003), and a comparison with MPEG-4 for still image coding is presented in Moccagatta *et al.* (2000). A thorough video quality evaluation of Motion JPEG 2000 in mobile applications is presented in Winkler

The JPEG 2000 Suite   Edited by Peter Schelkens, Athanassios Skodras and Touradj Ebrahimi
© 2009 John Wiley & Sons, Ltd

and Dufaux (2003). A performance comparison of Motion JPEG 2000 and MPEG-4 in an error-prone environment shows the good performance of the former (Dufaux and Ebrahimi, 2004). However, these error resilience tools are not capable of correcting transmission errors, but rather of limiting their impact by detecting the occurrence of an error, concealing the corrupted data, and resynchronizing the decoder. Furthermore, they do not deal with transmission errors in the image header, despite the fact that it is the most critical part of the JPEG 2000 code-stream.

To alleviate these limitations, a number of techniques have been proposed, including forward error correction (FEC) codes to protect the image header (Nicholson *et al.*, 2003), unequal error protection (UEP) to protect the code-stream (Natu and Taubman, 2002; Nicholson *et al.*, 2003; Dufaux and Ebrahimi, 2004), data interleaving (Frescura *et al.*, 2003), selective retransmission (Grangetto, Magli, and Olmo, 2003a), unequal power distribution (Atzori, 2003), and robust arithmetic coders (Grangetto, Magli, and Olmo, 2003b; Guionnet and Guillemot, 2003).

Recognizing the importance of wireless imaging applications and the shortcomings of the JPEG 2000 baseline error resilience tools, the JPEG committee launched in 2002 a new work item referred to as JPEG 2000 Wireless or JPWL (Dufaux and Nicholson, 2004; Dufaux *et al.*, 2006). The purpose of JPWL is to extend the baseline JPEG 2000 specification in order to enable the efficient transmission of JPEG 2000 images over error-prone wireless networks. More specifically, JPWL defines a standardized framework for wireless imaging to support error correcting techniques. Furthermore, JPWL also defines means to describe the sensitivity of the code-stream to transmission errors, as well as to describe the locations of residual errors in the code-stream. JPWL is presently a Final Draft International Standard (FDIS) (ISO/IEC, 2007) and is expected to shortly become an International Standard (IS). JPWL is not linked to a specific network or transport protocol, but rather provides a generic solution for the efficient transmission of JPEG 2000 encoded images over error-prone channels.

This chapter is structured as follows. In Section 8.2, we first review the impact of transmission errors on compressed images. Techniques for error detection, resilience, concealment, and correction techniques are then discussed. Finally, we examine the error resilience tools in the baseline JPEG 2000 specification. In Section 8.3, we present an overview of JPWL, and in particular its scope, main functionalities, and system configurations. The normative aspects of the specifications are thoroughly discussed in Section 8.4, including the four marker segments defined by JPWL: error protection capability (EPC), error protection block (EPB), error sensitivity descriptor (ESD), and residual error descriptor (RED). In Section 8.5, two informative parts are presented: error resilient arithmetic coding and unequal error protection (UEP). Finally, a summary is provided in Section 8.6.

8.2   Background

8.2.1   Transmission Errors and Their Impact on Compressed Data

Delivering media content over wireless channels is challenging due to their error-prone nature. Indeed, transmission errors are the result of wireless channel characteristics such as multipath, fading, and transmission interference. These errors can be roughly classified into two types: random bit errors (e.g. bit inversion, insertion, or deletion) and erasure burst errors (e.g. packet loss). Note that transmission errors may also occur in other contexts. For instance, in wired communications, packets may get lost in packet networks due to congestion. Similarly, burst errors may occur when reading a stream from storage media due to physical defects.

When transmitting digital images or videos over a wireless network, the occurrence of transmission errors may have very different effects on the end-user visual quality depending on which parts of the code-stream are corrupted. This is especially true for JPEG 2000 due to the structure of its code-stream. The image headers are most critical, and their alteration may render the whole code-stream undecodable. Conversely, errors in some quality layers may have a minimal impact.

More generally, as compression aims to remove redundancies in the data, it makes the resulting compressed code-stream much more sensitive to transmission errors. Indeed, most coding schemes rely on a variable length code (VLC), which is very efficient at compressing data but very sensitive to errors. Specifically, the decoder often loses synchronization in the presence of errors. As a result, not only is the part of the code-stream where an error occurs lost but also all the data until the decoder is able to resynchronize, as illustrated in Figure 8.1.

In addition, if the decoding of a part of the code-stream depends on the correct decoding of previous parts of the code-stream, the effect of transmission errors will propagate. More precisely, all parts of the code-stream that depend on the corrupted data will also be degraded. For instance, in MPEG-like video coding schemes (Sikora, 1997; Wiegand *et al.*, 2003), errors in one frame may also impact subsequent frames due to motion-compensated interframe coding.

8.2.2 Error Detection, Resilience, Concealment, and Correction

Several approaches attempt to alleviate the impact of transmission errors, namely error detection, error resilience, error concealment, and error correction. They are now discussed in more detail.

Error detection is the ability to detect and locate transmission errors. This can be achieved in several ways. For instance, a cyclic redundancy check (CRC) (Castagnoli, Braeuer, and Herrman, 1993) or a forward error correction (FEC) code (Clark and Cain, 1981; Lin and Costello, 1983) can be used to detect errors explicitly. Alternatively, the decoder is capable of identifying the occurrence of illegitimate syntactic elements or codewords, signaling the likely presence of errors. Note that error detection is a prerequisite for the subsequent approaches.

Error resilience consists in making the code-stream more robust to transmission errors during encoding, in order to limit the resulting degradation. For this purpose, several techniques have been developed. Resynchronization markers are unique codes that can be judiciously inserted in the code-stream, and allow the decoder to resynchronize in the

Figure 8.1   Impact of transmission errors on a variable length code (VLC)

presence of errors (Sodagar, Chai, and Wus, 2000). A reversible variable length code (RVLC) is a special case of a VLC where the code-stream can be decoded both in the forward (i.e. from beginning to end) and backward (i.e. from end to beginning) directions (Wen and Villasenor, 1998). An RVLC allows more data to be recovered successfully in the case of erroneous transmission. Data partitioning consists in having coded units that can be independently decoded, hence avoiding the propagation of errors.

Error concealment consists in estimating or interpolating the corrupted data from the received information at the decoder (Hemami and Gray, 1997; Wang *et al.*, 2000). This step is outside the scope of standardization.

Error correction includes the further ability to identify and correct transmission errors. This is essentially achieved in two ways: automatic repeat request (ARQ) and FEC. In ARQ schemes (Lin and Costello, 1983), the receiver sends an acknowledgment message back to the transmitter for each data unit (e.g. packets) received error-free. Whenever it does not get an acknowledgment, the transmitter re-sends the data. ARQ is very efficient at error-free data transmission, but it is unfortunately not applicable in applications with low delay requirements.

With FEC techniques (Clark and Cain, 1981; Lin and Costello, 1983), the transmitter adds redundancy to the message, which allows the receiver to detect and correct errors, within some error-rate bound, without the need for additional information. FEC codes that include the original input data are referred to as systematic.

Unlike other forms of data, in a compressed multimedia code-stream various parts of the stream tend to have a very different impact on the subjective quality of the decoded content. Unequal error protection (UEP) refers to the case where different strengths of error protection are assigned to different portions of the code-stream based on their significance (Girod, Horn, and Belzer, 1996).

When a burst error occurs, many consecutive bits get corrupted, often exceeding the error correction capability of an FEC code. To circumvent this drawback, interleaving is applied to the data prior to transmission. More specifically, this consists in a reordering of the data. In this way, a burst error will, after de-interleaving, affect a correctable number of bits in each codeword, hence permitting successful detection and correction by the FEC.

8.2.3 Error Resilience Tools in JPEG 2000 Baseline

We now briefly review the error resilient tools in JPEG 2000. A more thorough description is given in Moccagatta *et al.* (2000) and Taubman and Marcellin (2002), whereas detailed performance evaluations are presented in Moccagatta *et al.* (2000), Santa-Cruz, Grosbois, and Ebrahimi (2002), and Bilgin, Wu, and Marcellin (2003). Note that, although JPEG 2000 defines error resilient tools, the procedure that the decoder shall follow in order to cope with the possible presence of errors is not standardized.

JPEG 2000 relies on both resynchronization markers and data partitioning to limit the impact of transmission errors. More specifically, the code-stream is composed of packets, with each packet corresponding to a quality layer, a resolution, a component, and a precinct. As these packets constitute independently coded units, this data partitioning limits the spread of transmission errors to a great extent. In addition, start-of-packet (SOP) resynchronization markers can be optionally inserted in front of every packet,

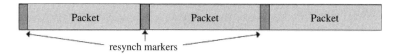

Figure 8.2 SOP resynchronization markers in JPEG 2000

as illustrated in Figure 8.2. These markers enable the decoder to resynchronize in the presence of errors.

Moreover, the quantized wavelet coefficients are partitioned into code-blocks, each code-block being independently coded using an MQ arithmetic coder. A number of options can be used to strengthen the resilience of the arithmetic coder. First, the arithmetic coder can be required to use a predictable termination procedure at the end of each coding pass. In addition, the arithmetic coder can be restarted at the beginning of each coding pass. Finally, a segmentation symbol can be encoded at the end of each bit-plane. In this case, if the segmentation symbol is not correctly decoded at the decoder side, an error is flagged in the preceding coding passes. As a direct consequence of these mechanisms, shorter coding passes will entail better error resilience. Therefore, small code-blocks tend to lead to better performance in the presence of errors, unlike the error-free case.

In the case of Motion JPEG 2000, another important feature in terms of resilience is that it relies on intraframe coding. As each frame is coded independently of the other ones, transmission errors in one frame do not propagate to subsequent frames. This is in contrast with motion compensated video coding techniques such as the ones defined in MPEG standards.

While the above tools detect where errors occur, conceal the erroneous data, and resynchronize the decoder, they do not correct transmission errors. Furthermore, these tools do not apply to the image header despite the fact that it is the most important part of the code-stream.

8.3 JPWL Overview

8.3.1 Scope

Recognizing the importance of wireless imaging applications, and given the limitations of the error resilience tools in JPEG 2000 baseline, in 2002 the JPEG committee launched a new activity referred to as JPEG 2000 Wireless or JPWL. JPWL has become an International Standard in 2007 (ITU-T Rec. T.810 (05/2006) | ISO/IEC 15444-11:2007).

JPWL provides a generic solution for the robust transmission of JPEG 2000 code-streams over error-prone networks. JPWL is not linked to a specific network or transport protocol and does not make assumptions on the characteristics of transmission errors. Therefore, while the main target of JPWL is wireless applications, the same tools can also be employed in other error-prone applications.

Whereas in this chapter we mainly consider the use of JPWL for JPEG 2000 still images, the same tools can also be applied to enhance the robustness of Motion JPEG 2000 video sequences in the presence of errors. For instance, acting at the code-stream level, JPWL tools can straightforwardly be used to protect the individual code-stream of each frame in the sequence.

8.3.2 Main Functionalities

In order to handle transmission errors more efficiently, JPWL mainly introduces three new functionalities:

- to protect the code-stream against transmission errors;
- to describe the sensitivity of different parts of the code-stream to transmission errors;
- to describe the locations of residual errors in the code-stream.

The protection of the code-stream consists of two steps: an error protection process applied at the encoder and an error correction process performed at the decoder. The error protection process modifies the code-stream such that it becomes more resilient to transmission errors. Conversely, the error correction process detects the occurrence of errors and corrects them whenever possible. Techniques to protect the code-stream include FEC (Nicholson *et al.*, 2003), UEP (Natu and Taubman, 2002; Dufaux and Ebrahimi, 2004), and data interleaving (Frescura *et al.*, 2003). While some tools are specified in a normative way, JPWL also has provision for future techniques that are registered and managed by a Registration Authority (RA).

Transmission errors may have very different impacts on the image quality depending on where they occur in the code-stream. This is especially true for JPEG 2000, given the embedded nature of the code-stream. The error sensitivity descriptor describes how different parts of the code-stream are sensitive to transmission errors. This information can easily be extracted during encoding, but it can also be derived from a JPEG 2000 code-stream. The error sensitivity information can subsequently be exploited when protecting the image. For instance, UEP can be applied where different degrees of error protection are assigned to different parts of the code-stream based on their importance (Nicholson *et al.*, 2003; Dufaux and Ebrahimi, 2004). This information also proves useful for optimizing ARQ by assigning a larger number of retransmission attempts for the most significant components of the code-stream (Grangetto, Magli, and Olmo, 2003a).

Even though JPWL error protection tools are applied to protect a code-stream, residual errors may remain in the code-stream in the case of harsh transmission conditions. The residual error descriptor specifies the occurrence and location of these residual errors in the code-stream. This information is typically generated during the error correction process and can subsequently be exploited when attempting to decode the corrupted code-stream. For instance, error concealment aims at estimating or interpolating the corrupted information (Atzori, Corona, and Giusto, 2001). If this fails, the decoder may also request retransmission of the lost data.

8.3.3 System Configuration

A JPWL system can be configured in many ways. We can first distinguish between a JPWL encoder whose input is a source image or a JPWL transcoder whose input is a JPEG 2000 code-stream, as is illustrated in Figures 8.3 and 8.4, respectively.

In Figure 8.3, a JPWL encoder consists of three modules working in concert: a JPEG 2000 baseline encoder compressing the input image, a generator of the error sensitivity description, and a tool to protect the code-stream against errors. The output of the JPWL encoder is a JPWL code-stream robust to transmission errors. The JPWL code-stream,

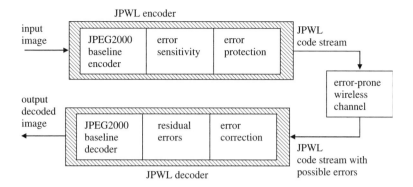

Figure 8.3   JPWL system description: JPWL encoder and decoder

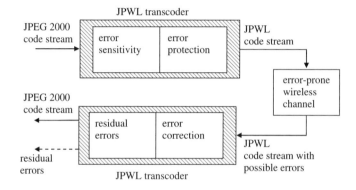

Figure 8.4   JPWL system description: JPWL transcoder

possibly including transmission errors, is then consumed by a JPWL decoder. The latter is composed of three modules: the first one corrects for transmission errors, the second one produces the residual error description, and finally the third one performs JPEG 2000 baseline decoding.

Alternatively, the input can be an already compressed JPEG 2000 code-stream. In this case, as depicted in Figure 8.4, a JPWL transcoder generates the error sensitivity description from the JPEG 2000 code-stream and applies error protection tools. Conversely, at the receiving side, a JPWL transcoder corrects the transmission errors and generates the residual error description. The output is a JPEG 2000 code-stream, which can be sent to a JPEG 2000 baseline decoder, along with residual error information.

Straightforwardly, many other configurations are also possible. Two of them are illustrated in Figures 8.5 and 8.6. Whereas in Figures 8.3 and 8.4 the generation of the error sensitivity description and the application of the error protection tool are concurrent, in Figures 8.5 and 8.6 the two operations are performed sequentially by two independent systems. Specifically, a JPWL encoder, respectively transcoder, first produces a JPWL code-stream containing error sensitivity information. Later on, a JPWL transcoder optimally applies error protection by exploiting the sensitivity information and generates another JPWL code-stream robust to transmission errors.

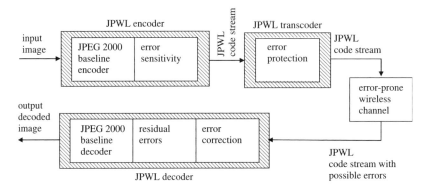

Figure 8.5 JPWL system description: another configuration

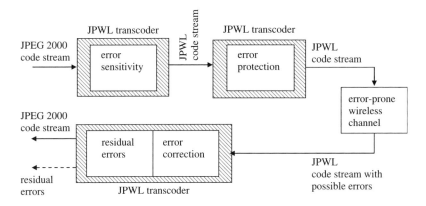

Figure 8.6 JPWL system description: another configuration

These four figures, describing various JPWL systems, are only illustrative examples. It should be clear that many more configurations are possible.

8.4 Normative Parts

For the sake of standardizing the minimum for interoperability, all encoding processes, defined as the conversion of source image data to compressed image data, are specified informatively. An encoder is an embodiment of the encoding process. In order to be JPWL compliant, the encoder shall generate compressed image data with a code-stream syntax that conforms to the syntax specified by JPWL.

Conversely, decoding processes are defined as the conversion of compressed image data to reconstructed image data. Some parts of a decoding process are normative, namely those related to extracting information contained in the JPWL-specific marker segments, as well as those that refer to the decoding of JPEG 2000 baseline features. All other aspects of the decoding process, for instance the procedure that the decoder shall follow in order to cope with the possible presence of errors and the actions it shall take to minimize their effect, are not specified as part of JPWL. However, JPWL provides guidelines for the handling of transmission errors. A decoder is an embodiment of the decoding process. In order to conform to JPWL, a decoder shall convert all, or specific parts of, compressed image data that conform to the JPWL code-stream syntax to a reconstructed image.

The normative parts of JPWL are outlined hereafter:

- Error protection capability: description of the tools that have been used to protect the code-stream and to correct the occurrence of transmission errors.
- Error protection block: tool to protect the main header and tile-part header, and to correct the occurrence of transmission errors.
- Error sensitivity descriptor: description of the sensitivity of different parts of the code-stream to transmission errors. This information is typically generated during the encoding process. It can subsequently be exploited by error protection techniques.
- Residual errors descriptor: description of the occurrence and location of residual errors in the code-stream. Residual errors are the errors that remain in the code-stream after channel decoding. This information is typically generated when decoding the code-stream.
- Code-stream syntax: definition of the code-stream syntax every JPWL code-stream must conform to.
- Registration Authority: specification of the JPWL Registration Authority (RA). Informative error protection techniques have to be registered with the RA. This mechanism allows for future rollout of new protection tools.

The marker segments specified by JPWL are summarized in Table 8.1. A more detailed description is given in the following sections.

8.4.1 Error Protection Capability (EPC)

The EPC marker segment signals the usage of JPWL normative and informative tools in order to protect the code-stream. More specifically, EPC signals the presence of the

Table 8.1 Marker segments defined by JPWL

Functionality	Name	Code	Main header	Tile-part header
Error protection block				
Signals the presence in the header of error protection data generated from a systematic code	EPB	0xFF66	Optional	Optional
Error sensitivity descriptor				
Describes the sensitivity of the code-stream to transmission errors	ESD	0xFF67	Optional	Optional
Error protection capability				
Indicates the methods applied to protect the code-stream against transmission errors	EPC	0xFF68	Required	Optional
Residual errors descriptor				
Describes the location of residual errors in the code-stream	RED	0xFF69	Optional	Optional

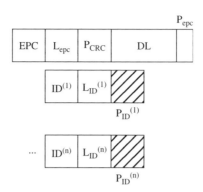

Figure 8.7 Syntax for the EPC marker segment

EPB, ESD, and RED market segments in the code-stream, as well as the optional use of informative tools previously registered with the JPWL RA. The information conveyed by EPC is useful for a decoder to identify quickly which JPWL information is available in the code-stream and whether it is capable of decoding it.

EPC is mandatory in the main header and optional in tile-part headers, but at most one EPC should appear in each header. The syntax is illustrated in Figure 8.7. The flexibility of the syntax allows for the signaling of existing tools and the future rollout of new ones registered after the publication of the specification.

The EPC marker segment is identified by the unique marker EPC (0xFF68), while L_{EPC} signals its length in bytes (as for all JPEG 2000 marker segments, this excludes the two bytes of the marker itself, but includes the two bytes of the length parameter). In order to verify that the marker segment data are not corrupted, a cyclic redundancy check (CRC) (Castagnoli, Braeuer, and Herrman, 1993) is used. More precisely, the CRC is computed on the complete marker segment, excluding P_{CRC} itself, using the 16-bit CRC-CCITT. The P_{CRC} field contains the resulting parity bits.

When the EPC marker segment is in the main header, the DL field indicates the code-stream length in bytes (from the first byte of the SOC marker to the last byte of the EOC marker). Similarly, when the EPC marker segment appears in a tile-part header, DL is the length of this tile-part in bytes (from the first byte of the SOT marker to the last byte of the tile part). DL is useful to resynchronize the decoder after a decoding failure due to transmission errors. The occurrence of the ESD, RED and EPB marker segments in the code-stream, as well as the use of informative techniques, is signaled by the P_{EPC} field.

Finally, when informative techniques are optionally used, $ID^{(i)}$ indicates the corresponding identification number issued by the RA, whereas $P_{ID}^{(i)}$ may contain associated parameters and $L_{ID}^{(i)}$ is the length of $L_{ID}^{(i)} + P_{ID}^{(i)}$ in bytes. Using this information, the decoder can query the RA about these tools and decides the most appropriate actions.

8.4.2 Error Protection Block (EPB)

The main and tile-part headers are the most important parts of the code-stream. Indeed, the occurrence of errors within the headers often results in a code-stream that is no longer decodable. The primary function of EPB is to protect the main and tile-part headers

(Nicholson *et al.*, 2003). It contains information about the error protection parameters and redundancy data used to protect the code-stream against errors. Note, however, that EPB can also be used to protect the remainder of the code-stream (Nicholson *et al.*, 2003; Dufaux and Nicholson, 2004).

EPB is optional. It may be placed in the main header and/or tile-part headers. Multiple EPB marker segments may appear in each header. In order to enable resynchronization when errors occur, the first EPB of the main header has to be placed immediately after the SIZ marker segment. Similarly, the first EPB of a tile-part header is required to be positioned right after the SOT marker. The syntax of the EPB marker segment is given in Figure 8.8.

EPB is a unique marker with value 0xFF66, and the length of the marker segment is indicated by L_{EPB}. The D_{EPB} field specifies some parameters regarding the usage of EPB. For instance, it includes a counter to index successive EPB marker segments and signals whether the present EPB is the last one in the current header. LDP_{EPB} is the length of the data to be protected by the redundant information carried within the current EPB marker segment.

Predefined error correction codes are specified to be used in EPB. For this purpose, Reed–Solomon (RS) codes (Reed and Solomon, 1960; MacWilliams and Sloane, 1977) are used, a well-known family of systematic codes. We denote them by $RS(N,K)$, where K is the source length and N is the code length. Such a code generates $N - K$ redundancy bytes and can correct up to $(N - K)/2$ errors. The predefined codes have been chosen to offer a large correction capability in order to cope with harsh transmission conditions, while simultaneously limiting padding. The following predefined codes have been selected empirically: RS(160, 64) for the first EPB of the main header, RS(80, 25) for the first EPB of the tile-part header, and finally RS(40, 13) for other EPB marker segments. Codes other than the predefined ones can also be used. This includes other RS codes as well as CRC codes. There are signaled by the P_{EPB} field. Finally, EPB data contain the data to perform error correction, typically redundancy bytes.

The use of a single EPB to protect the main header is illustrated in Figure 8.9. In this example, L2 redundancy bytes are used to protect the L1 bytes corresponding to the SOC marker, the SIZ marker segment, and EPB, L_{EPB}, D_{EPB}, LDP_{EPB}, and P_{EPB}, using

Figure 8.8   Syntax for the EPB marker segment

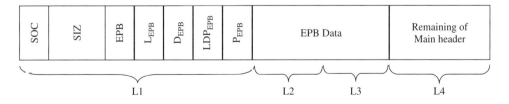

Figure 8.9   Usage of a single EPB marker segment to protect the main header

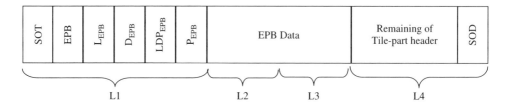

Figure 8.10 Usage of a single EPB marker segment to protect a Tile-part header

the predefined RS(160, 64) code. In turn, the L4 bytes of the remaining main header are protected by L3 redundancy bytes, using the error correction code specified by P_{EPB}. In this instance, LDP_{EPB} is equal to $L1 + L4$.

Similarly, Figure 8.10 depicts the use of a single EPB to protect a tile-part header. In this example, the L1 bytes corresponding to the SOT marker and EPB, L_{EPB}, D_{EPB}, LDP_{EPB}, and P_{EPB} are protected by L2 redundancy bytes using the predefined RS(80, 25) code. The L4 bytes of the remaining tile-part header are protected by L3 redundancy bytes, using the error correction code specified by P_{EPB}.

Note that the EPB marker segment placed in the tile-part header may protect data beyond the tile-part header itself. This property is useful in order to enable the protection of the whole code-stream with UEP using EPB (Nicholson *et al.*, 2003).

8.4.3 Error Sensitivity Descriptor (ESD)

One of the important features of JPEG 2000 is the embedded nature of the code-stream, which allows for efficient scalability. This property is also important in the framework of JPWL, as different parts of the code-stream have a very different impact on the quality of the decoded image. For this purpose, the ESD marker segment encapsulates information about the sensitivity of the code-stream to transmission errors. More specifically, it aims at quantifying the effect of corrupting parts of the code-stream on the quality of the decoded image. This information can be exploited in a number of ways, as described hereafter.

Straightforwardly, in UEP techniques (Nicholson *et al.*, 2003; Dufaux and Nicholson, 2004), more powerful codes are used to protect the most sensitive portion of the code-stream. As a consequence, this approach usually outperforms EEP. At the encoder, the ESD information can be exploited in order to select the most appropriate code for each part of the code-stream based on its sensitivity.

The ESD data also prove useful for selective retransmissions (Grangetto, Magli, and Olmo, 2003a). More specifically, an efficient ARQ is optimized by allocating a larger number of retransmission attempts for the most critical parts of the code-stream as identified by ESD.

Another potential use of ESD is for smart prefetching. In a video streaming system, the most important packets, according to ESD, can be prefetch and send in advance. Consequently, a larger number of retransmissions can be attempted in case some of these packets are lost.

Finally, the information carried in ESD can also be used for other applications not strictly related to JPWL. For instance, an efficient transcoding to adapt the data rate to

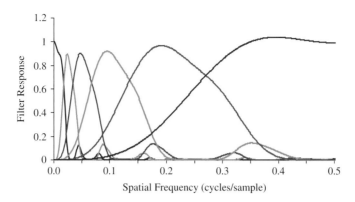

Plate 1 Frequency responses for a five-level dyadic decomposition (six frequency bands) using the (9, 7) filter pair (see page 11)

Plate 2 From left to right and top to bottom are shown bit-planes 4, 6, 8, and 11 of the JPEG 2000 encoded Lena image as described in Table 1.20. Green pixels denote the location of the wavelet coefficients that become significant during the significance propagation pass in that bit-plane, red pixels denote the coefficients that turn significant during the cleanup pass in that bit-plane, black pixels denote coefficients that are refined, and white pixels denote coefficients that remain insignificant after completion of the three coding passes (see page 62)

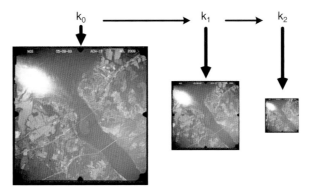

Plate 3 Multilevel access control: given one encrypted image, the user with key k_0 can access the high, medium, or low resolution images, the user with key k_1 can access the medium or low resolution images, while the user with key k_2 can only access the lowest resolution (see page 279)

(a.) Original "woman" image
(512 × 640)

(b.) Protected with lossy semi-fragile authentication,
PSNR = 38dB

(c.) Attacked image

(d.) Verification result of attacked image: unauthentic; Attacked area highlighted in red color

Plate 4 Examples of lossy embedding authentication under given LABR (see page 286)

Plate 5 Landsat 7 image of the NASA Kennedy Space Center in 1999 at a resolution of 15 meters (image courtesy of EROS Data Center) (see page 311)

Plate 6 Hyperspectral cube of Moffet Field, California, taken on August 20, 1992, using the AVIRIS sensor (image courtesy of NASA JPL) (see page 313)

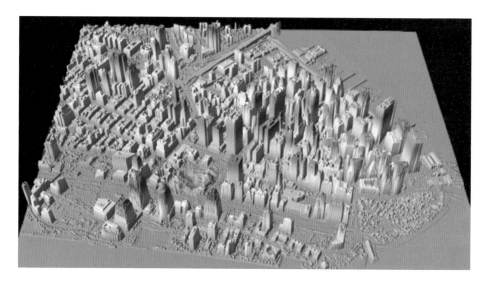

Plate 7 Terrain elevation image of lower Manhattan after the September 11, 2001 terrorist attack (image courtesy of NOAA/U.S. Army JPSD) (see page 316)

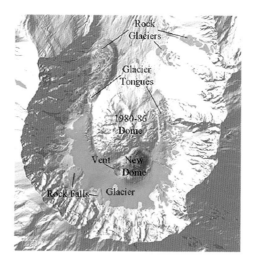

Plate 8 Airborne LIDAR data fused with an IKONOS image of Mt St Helens volcano (image courtesy USGS and NASA) (see page 316)

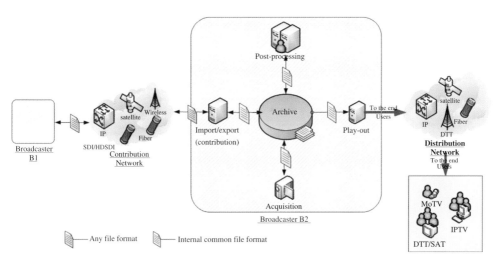

Plate 9 Overview of the broadcast chain (see page 390)

Plate 10 3-D scene with 3-D meshes wrapped up with 2-D textures (reproduced by permission of GeoID) (see page 422)

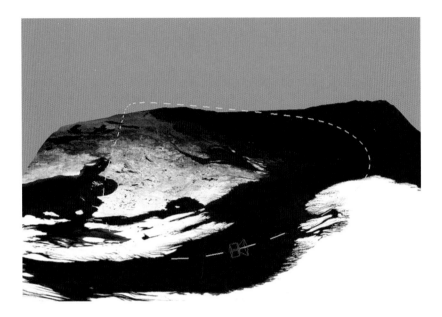

Plate 11 3-D walkthrough path (dashed) with user specified viewing directions (red camera view representation) over the 3-D scene (reproduced by permission of NASA) (see page 423)

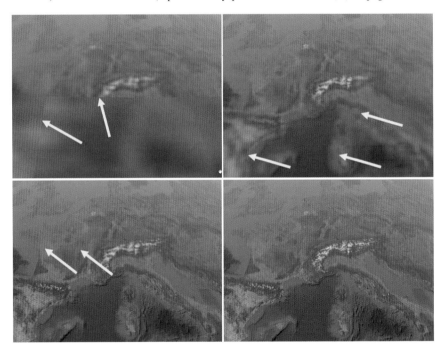

Plate 12 Gradual texture improvement with vanishing tiling effects (arrows) over texture download time. When time passes by, more texture information is transmitted, enhancing the visual quality of the rendering over low-bandwidth/high-latency networks (reproduced by permission of NASA) (see page 425)

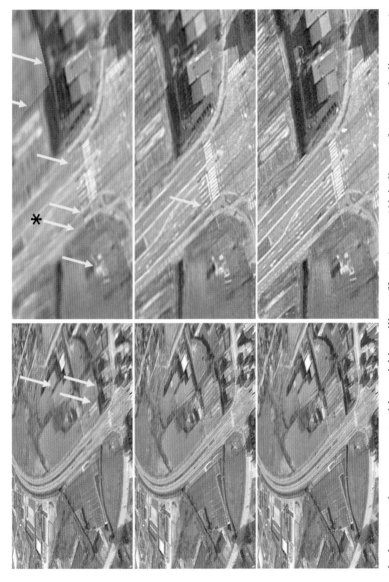

Plate 13 Gradual texture improvement with vanishing tiling effects (arrow '*' indicates feature misalignments) over texture download time. When time passes by, more texture information is transmitted, enhancing the visual quality of the rendering over low-t-andwidth/high-latency networks (reproduced by permission of Agentschap voor Geografische Informatie Vlaanderen, Belgium) (see page 426)

Plate 14 JPEG 2000 reduced tiling effect, i.e. even over very short download times a smooth texture transition is obtained (top), using a prioritized texture packet transmission protocol. Zoom-in (bottom) shows very smooth texture transitions without any abrupt tiling effect (reproduced by permission of Agentschap voor Geografische Informatie Vlaanderen, Belgium) (see page 427)

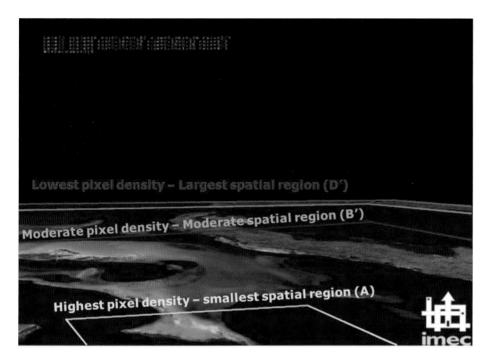

Plate 15 Texture texel/pixel densities as a function of viewing distance in the 3-D scene (reproduced by permission of NASA) (see page 429)

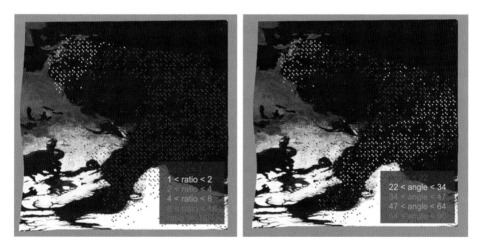

Plate 16 3-D terrain scene with distribution of the viewing distance and accompanying ratio between visualized pixels and texture texels resolution (left) and viewing angle (right) (reproduced by permission of NASA) (see page 429)

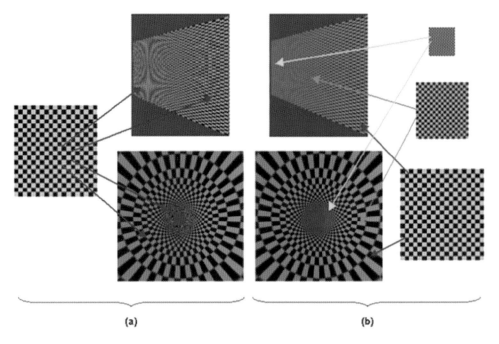

(a)　　　　　　　　　　　　　　(b)

Plate 17 Rendering of a flat wall (top) and interior of a tube (bottom) without mipmapping: (a) one texture resolution is used for any viewing conditions and with mipmapping; (b) different, antialiased texture resolutions are used for different viewing conditions (see page 430)

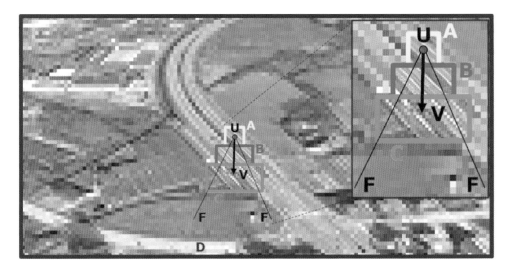

Plate 18 Mipmap texture resolutions (A, B, C, D) required for the user standing in position U and looking in the viewing direction V along the viewing cone frustum F (see page 431)

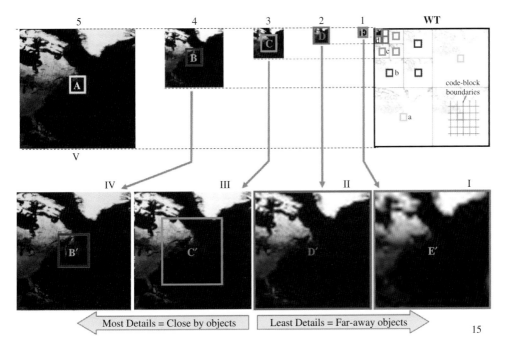

Plate 19 The relation between the WT (top-right), the mipmaps (top-left), and the clipmaps (bottom) for the terrain texture (see page 432)

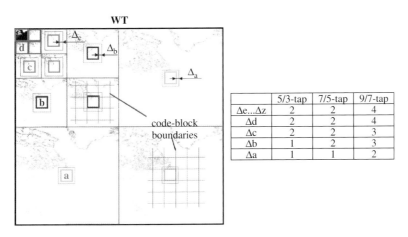

	5/3-tap	7/5-tap	9/7-tap
$\Delta e...\Delta z$	2	2	4
Δd	2	2	4
Δc	2	2	3
Δb	1	2	3
Δa	1	1	2

Plate 20 Code-blocks in JPEG 2000 used for the regions A, B, C, etc., in Figure 16.15 (left) and the additional amount of wavelet data for proper IWT filtering (right) (see page 433)

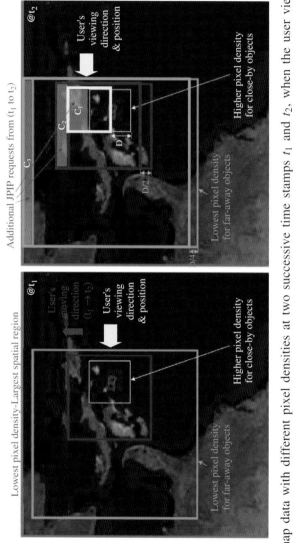

Plate 21 Clipmap data with different pixel densities at two successive time stamps t_1 and t_2, when the user viewpoint moves toward the top of the texture over distance D. Clipmap data have to be updated over respectively distances D, $D/2$, and $D/4$ with data additions C_1, C_2, and C_3 (reproduced by permission of NASA) (see page 434)

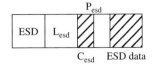

Figure 8.11   Syntax for the ESD marker segment

the network bandwidth can take advantage of the ESD information in order to truncate the code-stream optimally and to make sure of providing a reasonable image quality.

The ESD marker segment is optional and can be placed in any valid positions in the main and/or tile-part headers. Furthermore, multiple ESDs are allowed in a header. Finally, note that this information is not strictly needed for decoding a JPWL code-stream. The syntax is given in Figure 8.11.

ESD is a unique marker with the value 0xFF67 and L_{ESD} is the length of the marker segment. The ESD data refer either to a single component or an average over all components, and is specified by the C_{ESD} field. ESD supports three modes to express the portion of the code-stream that the ESD data are describing: packet, packet range, and byte range. In the packet mode, a sensitivity value is defined for each packet in the code-stream (when located in the main header) or tile part (when located in the tile-part header). In the packet-range mode, a sensitivity value is provided for a range of packets specified by the start and end packets. Similarly, in the byte-range mode, a sensitivity value is given for a segment of the code-stream defined by its start and end bytes.

Error sensitivity can be expressed in two ways: relative or absolute. Relative error sensitivity is expressed as an unsigned integer that describes the importance of a given part of the code-stream relative to other parts. In this case, high values correspond to high priority data, whereas the highest value is reserved for data units encompassing the main or tile-part headers. In turn, absolute error sensitivity values are specified by means of a quality metric. Supported quality metrics are mean square error (MSE), total square error (TSE), peak signal-to-noise ratio (PSNR) and absolute peak error (MAXERR). Their definitions are given as follows:

$$\text{MSE} = \frac{1}{N} \sum_{i=1}^{N} (x_i - r_i)^2, \tag{8.1}$$

$$\text{TSE} = \text{MSE} * \text{N}, \tag{8.2}$$

$$\text{PSNR} = \frac{M^2}{\text{MSE}}, \tag{8.3}$$

$$\text{MAXERR} = \max_i |x_i - r_i|, \tag{8.4}$$

where x_i denotes pixel values from the original image and r_i the corresponding values of the decoded image, N is the number of pixels in the image, and M is the maximum pixel value (e.g. $M = 255$ for 8-bit images). Two additional incremental metrics are supported: MSE decrease and PSNR increase. In these two cases, the improvement in MSE, respectively PSNR, resulting from decoding that data unit is specified.

The code-stream addressing mode and the error sensitivity metric, along with some usage parameters, are identified by the P_{ESD} field. Finally, ESD data contains the error sensitivity data itself.

8.4.4 Residual Error Descriptor (RED)

After channel decoding, some residual errors may still be present in the code-stream, even if JPWL tools are applied. The RED marker segment signals the occurrence and location of residual errors in the code-stream. This information can be exploited by a JPEG 2000 decoder in order to cope with residual transmission errors better. Although the behavior of the decoder in the presence of errors is not normative, a few techniques can be successfully applied to mitigate their impact. For instance, the decoder can request retransmission of the corrupted data. Another course of action is to apply error concealment (Atzori, Corona, and Giusto, 2001). This essentially consists in estimating or interpolating the lost information from the uncorrupted data. Finally, if the corrupted information is deemed of low visual importance, it can simply be discarded.

The RED marker segment is optional and can be placed in any valid positions in the Main and/or Tile-part headers. Its syntax is given in Figure 8.12.

RED is a unique marker whose value is equal to 0xFF69 and L_{RED} is the length of the marker segment. RED supports three modes to address the portion of the code-stream that the RED data are describing: packet, packet range, and byte range, similar to the ones in ESD. In the packet mode, a residual error value is specified or each packet in the code-stream (when located in the main header) or tile part (when located in the tile-part header). In the packet-range mode, a residual error value is provided for a range of packets specified by the start and end packets. Similarly, in the byte-range mode, a residual error value is given for a segment of the code-stream defined by its start and end bytes. The P_{RED} field specifies the code-stream addressing mode, along with some usage parameters. Finally, the record of residual error data itself is carried in RED data.

8.4.5 Registration Authority (RA)

JPWL error protection techniques fell into two categories. The first one encompasses well-known techniques such as RS codes. In this case, their usage is signaled by the EPB marker segment.

The second category consists of proprietary techniques. Prior to their use, they have to be registered with the JPWL RA. Upon registration, a unique identification number is assigned to the technique. In this case, its usage is signalled by the EPC marker segment, which contains the unique identification number along with private parameters. A JPWL decoder/transcoder may have to query the JPWL RA in order to get a description of

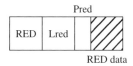

Figure 8.12 Syntax for the RED marker segment

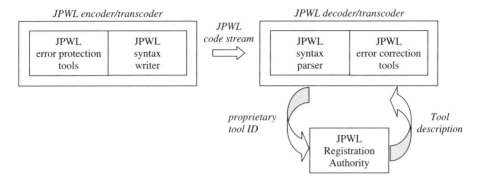

Figure 8.13   Registration Authority (RA)

the technique and be able to interpret the image data correctly, including detecting and correcting transmission errors, as illustrated in Figure 8.13. With this registration process, provision is made for future tools to be identified and registered.

The JPWL specification defines the management of the RA, including the processes to apply for the registration of a technique, to evaluate applications, and to notify identification number assignment.

8.5 Informative Parts

Besides the normative aspects previously described, JPWL also defines some informative parts as outline:

- Error resilient entropy coding: a modified arithmetic coder with improved error resilience.
- Unequal error protection (UEP): UEP applies different protections to various parts of the code-stream based on their significance.
- Encoding and decoding guidelines: indicates recommended behaviors for the encoding process to enhance robustness, and decoding processes in the presence of transmission errors.
- Interoperability: defines the interoperability of JPWL with other parts of the JPEG 2000 specifications, and more specifically Part 1 (JPEG 2000 baseline), Part 3 (Motion JPEG 2000), and Part 8 (Secured JPEG 2000 – JPSEC).

The first two items are discussed in more detail in the following sections.

8.5.1 Error Resilient Entropy Coding

Entropy coding is notoriously sensitive to transmission errors. This is especially true for arithmetic coding, as used in JPEG 2000. The main reason for this weakness is the loss of synchronization at the decoder in the presence of errors in the code-stream. Erroneous symbols may also cause unpredictable behaviors in the decoding of the coding passes.

To alleviate these problems, JPEG 2000 baseline includes some error resilient tools in the arithmetic coder: a specific termination procedure at each coding pass, restart of the coder at each coding pass, and insertion of a segmentation symbol at each bit-plane, as discussed in Section 8.2. With these tools, the arithmetic coder is able to cope with moderate error rates. However, they are not sufficient in the case of harsh conditions such as those often faced in wireless networks.

For this reason, JPWL defines a modified arithmetic coder which is able to handle transmission errors better. It extends the baseline JPEG 2000 MQ arithmetic coder by defining forbidden symbols, segmentation symbols, and soft decoding (Grangetto, Magli, and Olmo, 2003b; Guionnet and Guillemot, 2003).

8.5.1.1 Forbidden Symbols

In the JPEG 2000 baseline arithmetic coder, at each binary decision the probability interval is subdivided into two subintervals, corresponding to the most probable symbol (MPS) and the least probable symbol (LSP), as illustrated in Figure 8.14(a). Defining the interval A and the current estimate of the LPS probability Q_e, the subintervals are given by

$$\text{MPS subinterval: } A - (Q_e \cdot A) \approx A - Q_e,$$

$$\text{LPS subinterval: } Q_e \cdot A \approx Q_e,$$

where the approximations derive from the fact that the value of A is of order unity.

With the JPWL arithmetic coding with forbidden symbols (Grangetto, Magli, and Olmo, 2003b), the interval is now subdivided into three subintervals, for the MPS, the LSP, and the forbidden symbol (FS), as illustrated in Figure 8.14(b). FS is never actually encoded, but rather serves to detect errors. Defining the FS probability as Q_f, the subintervals are now given by

$$\text{MPS subinterval: } A - [(Q_e + Q_f) \cdot A] \approx A - (Q_e + Q_f),$$

$$\text{LPS subinterval: } Q_e \cdot A \approx Q_e,$$

$$\text{FS subinterval: } Q_f \cdot A \approx Q_f,$$

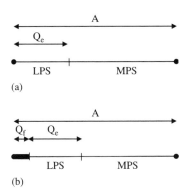

Figure 8.14 Probability intervals for (a) the arithmetic coder and (b) the arithmetic coder with forbidden symbols

At the decoder side, a transmission error is detected whenever a decoded symbol falls in the FS interval. Note that when $Q_f = 0$, the JPWL arithmetic coder with forbidden symbols is equivalent to the JPEG 2000 baseline MQ arithmetic coder.

8.5.1.2 Segmentation Symbols

In order to improve the detection of errors, three segmentation symbols are defined:

- SEGMARK: insertion of the marker 1010 at the end of each bit-plane, namely at the end of each cleanup pass (identical to the segmentation symbol defined in JPEG 2000 baseline).
- SEGMARKPASS: insertion of the marker 1010 at the end of each significance and refinement coding passes.
- SEGMARKSTRIPE: insertion of the marker 10 ($n = 1$) or the marker 1010 ($n = 2$) at the end of each stripe, where a stripe contains four rows of code-block samples.

The segmentation symbols are encoded using the arithmetic coder. The decoder is therefore expecting to find them; otherwise an error is flagged and the preceding decoded data are discarded. By having markers at the end of each coding pass or each stripe, transmission errors can be identified with a finer granularity, hence better limiting their impact to the part of the code-stream that is actually corrupted. Furthermore, these segmentation symbols allow the decoder to resynchronize during soft decoding (Guionnet and Guillemot, 2003).

8.5.1.3 Soft Decoding and Error Correction

Thanks to the forbidden symbols and the segmentation symbols, it is possible to have a JPWL decoder correcting bit errors at the code-stream level. More specifically, whenever the error detection tools identify the presence of bit errors, the JPWL coder may attempt to correct those errors by means of sequential decoding on a bit clock basis (Guionnet and Guillemot, 2003). It can be modeled as a state transition automaton, as illustrated in Figure 8.15. Let us define the input bit CD[n], where n represents the position in the code-stream. The state $\sigma[n]$ contains all internal state information, such as the arithmetic decoder states. The input bit CD[n] sets off the transition between state $\sigma[n-1]$ and $\sigma[n]$, with a variable number of output binary decisions D associated with this transition.

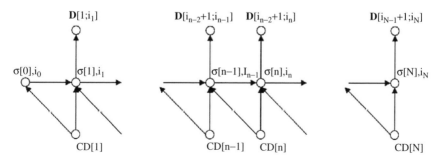

Figure 8.15 State transition automaton representation of the decoding process (ISO/IEC, 2007)

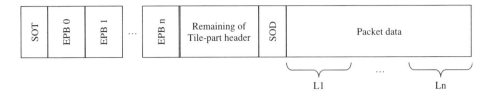

Figure 8.16   EPB for unequal error protection

At each bit depth n in the decoding automaton, a set of candidate code-streams CD_k is considered. In order to decide on the most likely candidate code-stream CD, corresponding to the correct decisions D, each code-stream candidate CD_k with its corresponding decoded decisions D_k is ranked according to a metric. Several metrics can be used, including a maximum *a posteriori* (MAP) metric, Hamming distance, or Euclidean distance.

8.5.2   Unequal Error Protection (UEP)

On top of protecting the image headers, the EPB marker segment may also be used to protect the remaining code-stream. More specifically, EPB marker segments placed in the tile-part header may protect data beyond the tile-part header itself. Furthermore, as each EPB may use a different code, this feature can be used for UEP (Nicholson *et al.*, 2003). The efficiency of this technique for the transmission of Motion JPEG 2000 video sequences over a noisy wireless channel is verified in (Dufaux and Nicholson, 2004).

More precisely, the data range to be protected is specified by the LDP_{EPB} parameter, whereas the error correction code for each segment is defined by P_{EPB}. UEP is obtained by appropriately selecting these different codes, namely by choosing their error correction capabilities depending on the significance of the respective packets that are protected.

This approach is illustrated in Figure 8.16. In this example, the tile-part header is protected by the marker segment EPB0. The subsequent EPBs protect the packet data. More precisely, EPB1 to EPBn contain redundancy information for the segments L1 to Ln of the code-stream.

In order to select the error correction codes optimally for each part of the code-stream, the error sensitivity information carried in the ESD marker segment can be exploited. In fact, the most significant parts of the code-stream are more heavily protected, i.e. using a higher redundancy, whereas the less important ones are less protected, i.e. using a lower redundancy. Finally, note that it is also possible to apply other UEP techniques not using the EPB marker segment, providing that they have been registered with the JPWL RA and that they are properly signaled using EPC.

8.6   Summary

In this chapter, we have reviewed the JPEG 2000 Wireless or JPWL standard. Its goal is to extend the baseline JPEG 2000 specification to provide a standardized framework for wireless imaging. JPWL enables the efficient transmission of JPEG 2000 images in an error-prone environment. It supports not only error correcting techniques such as forward error correction (FEC) codes, unequal error protection (UEP), and data interleaving, but also provides mechanisms to describe the error sensitivity of the code-stream as well as the locations of residual errors. The framework is open and flexible, hence ensuring a straight path for future extensions.

We have first discussed the impact of transmission errors on compressed images and techniques to overcome them, such as error detection, resilience, concealment, and correction. We have also reviewed the error resilience tools in the baseline JPEG 2000 specification. We then presented an overview of JPWL, including its scope, main functionalities, and system configurations. It was followed by a thorough presentation of the normative aspects of JPWL, including the four marker segments it defines. We finally described two informative techniques for error resilient arithmetic coding and UEP.

Acknowledgments

The author would like to thank all the individuals who have participated in the development of JPWL, and in particular Didier Nicholson, Fabrizio Frescura, Enrico Magli, and Giuseppe Baruffa.

References

Atzori, L. (2003). Transmission of JPEG2000 images over wireless channels with unequal power distribution, *IEEE Transactions on Consumer Electronics*, **49**(4), 883–888.

Atzori, L., Corona, S. and Giusto, D. D. (2001) Error recovery in JPEG 2000 image transmission, in *IEEE Proceedings of International Conference on Acoustics, Speech, and Signal Processing*.

Bilgin, A., Wu, Z. and Marcellin, M. W. (2003) Decompression of corrupt JPEG 2000 code-streams, in *Proceedings of 2003 Data Compression Conference*, Snowbird, UT.

Castagnoli, G., Braeuer, S. and Herrman, H. (1993) Optimization of cyclic redundancy-check codes with 24 and 32 parity bits, *IEEE Transactions on Communications*, **41** (6), 883–892.

Clark, G. C. and Cain, J. B. (1981). *Error-Correction Coding for Digital Communications*, Plenum Press, New York.

Dufaux, F. and Ebrahimi, T. (2004) Error-resilient video coding performance analysis of Motion JPEG 2000 and MPEG-4, in *SPIE Proceedings on Visual Communication and Image Processing*, San Jose, CA.

Dufaux, F. and Nicholson, D. (2004) JPWL: JPEG 2000 for wireless applications, in *SPIE Proceedings on Applications of Digital Image Processing XXVII*, Denver, CO.

Dufaux, F., Baruffa, G., Frescura, F. and Nicholson, D. (2006) JPWL – an Extension of JPEG 2000 for wireless imaging, in *IEEE Proceedings of International Symposium on Circuits and Systems (ISCAS)*, Island of Kos, Greece.

Frescura, F., Giorni, M., Feci, C. and Cacopardi, S. (2003) JPEG 2000 and MJPEG 2000 transmission in 802.11 wireless local area networks, *IEEE Transactions on Consumer Electronics*, **49**(4), 861–871.

Girod, B., Horn, U. and Belzer, B. (1996) Scalable video coding with multiscale motion compensation and unequal error protection, in Y. Wang, S. Panwar, S. P. Kim, and H. L. Bertoni (eds), *Multimedia Communications and Video Coding*, pp. 475–482, Plenum Press, New York.

Grangetto, M., Magli, E. and Olmo, G. (2003a) Error sensitivity data structures and retransmission strategies for robust JPEG 2000 wireless imaging, *IEEE Transactions on Consumer Electronics*, **49**(4), 872–882.

Grangetto, M., Magli, E. and Olmo, G. (2003b) Robust video transmission over error-prone channels via error correcting arithmetic codes, *IEEE Communications Letters*, **7**(12), 596–598.

Guionnet, T. and Guillemot, C. (2003) Soft decoding and synchronization of arithmetic codes: application to image transmission over noisy channels, *IEEE Transactions on Image Processing*, **12**(12), 1599–1609.

Hemami, S. S. and Gray, R. M. (1997) Subband-coded image reconstruction for lossy packet networks, *IEEE Trans. on Image Processing*, **6** (4), 523–539.

ISO/IEC (2007) JPEG 2000 Image Coding System – Part 11: Wireless, Final Draft of ISO/IEC International Standard 15444-11, ITU-T Rec. T.810 (05/2006) | ISO/IEC 15444-11:2007.

Lin, S. and Costello, D. J. (1983) *Error Control Coding: Fundamentales and Applications*, Prentice-Hall, Englewood Cliffs, NJ..

MacWilliams, F. J. and Sloane, N. J. A. (1977) *The Theory of Error-Correcting Codes*, North-Holland, New York.

Moccagatta, I., Soudagar, S., Liang, J. and Chen, H. (2000) Error-resilient coding in JPEG-2000 and MPEG-4, *IEEE Journal on Selected Areas in Communications*, **18**(6), 899–914.

Natu, A. and Taubman, D. (2002) Unequal protection of JPEG 2000 code-streams in wireless channels, *IEEE Proceedings of GLOBECOM'02*, Taipei, China.

Nicholson, D., Lamy-Bergot, C., Naturel, X. and Poulliat, C. (2003) JPEG 2000 backward compatible error protection with Reed–Solomon codes, *IEEE Transactions on Consumer Electronics*, **49**(4), 855–860.

Reed, I. S. and Solomon, G. (1960) Polynomial codes over certain finite fields *Journal of Society of Industrial and Applied Mathematics*, **8**(2), 300–304.

Santa-Cruz, D., Grosbois, R. and Ebrahimi, T. (2002) JPEG 2000 performance evaluation and assessment, *Signal Processing: Image Communication*, **17**(1), 113–130.

Sikora, T. (1997) The MPEG-4 video standard verification model, *IEEE Transactions on Circuits and Systems for Video Technology*, **7** (1), 19–31.

Skodras, A., Christopoulos, C. and Ebrahimi, T. (2001) The JPEG 2000 still image compression standard, *IEEE Signal Processing Magazine*, **18**(5), 36–58.

Sodagar, I., Chai, B.-B. and Wus, J. (2000) A new error resilience technique for image compression using arithmetic coding, in *IEEE International Conference on Acoustics, Speech, and Signal Processing (ICASSP '00)*.

Taubman, D. and Marcellin, M. (2002) *JPEG 2000: Image Compression Fundamentals, Standards and Practice*., Kluwer Academic Publishers, Boston, MA.

Wang, Y., Wenger, S., Wen, J. and Katsaggelos, A. (2000) Review of error resilient coding techniques for real-time video communications, *IEEE Signal Processing Magazine*, **17**(4).

Wen, J. and Villasenor, J. (1998) Reversible variable length codes for efficient and robust image and video coding, in *IEEE Proceedings of the Data CompressionConference (DCC)*.

Wiegand, T., Sullivan, G., Bjontegaard, G. and Luthra, A. (2003) Overview of the H.264/AVC video coding standard, *IEEE Transactions on Circuits and Systems for Video Technology*, **13**(7), 560–576.

Winkler, S. and Dufaux, F. (2003) Video quality evaluation for mobile applications, in *SPIE Proceedings on Visual Communication and Image Processing (VCIP)*, Lugano, Switzerland.

Part B

9

JPEG 2000 for Digital Cinema

Siegfried Fößel

9.1 Introduction

Digital media content is widely used today. With DVD, DVB, and the Internet a broad acceptance of digital content distribution took place. Digital cinema is one of the latest steps in the digitization of media content. It encompasses not only the replacement of film reels by digital data packages, but in a broader sense the complete digitization of the production and utilization chain for high-quality moving images. For a clear differentiation the digital cinema distribution to theatres is called in this chapter 'D-Cinema.'

In former times the theatres were the main refinancing instrument for movie producers. Nowadays, movies will be shown not only in theatres or on TV but also in many other different formats. Today the DVD is the main revenue source. Nonetheless, consuming markets are still changing rapidly. Analysts claim that this will shift in future to Internet and video-on-demand markets. The expected worldwide market for video-on-demand shall change from US$281 million in 2006 to US$4.4 billion in 2010 (Hohenauer, 2007). However, this shall only inhibit the revenues of the TV and DVD market. An example for the Western world changeover is the comparison of the market share between cinema and video in Germany in between the years 1999 and 2006. While the turnover in the theatres kept nearly constant with €808 million in 1999 and €814 million in 2006, new markets in the digital media area could be opened. In the meantime these new markets have overrun the values of the cinema market considerably (see Figure 9.1). This demonstrates that multiple digital utilization of content is the key element for digital movie distribution in the future. The process has already started. In the future, movies will be consumed not only in theatres or by TV at home, but anytime anywhere. In the meantime the quality of home entertainment has been improved by digital HDTV and Blu-Ray DVD, and so cinema came under pressure. One way out of this dilemma is D-Cinema, which allows

The JPEG 2000 Suite   Edited by Peter Schelkens, Athanassios Skodras and Touradj Ebrahimi
© 2009 John Wiley & Sons, Ltd

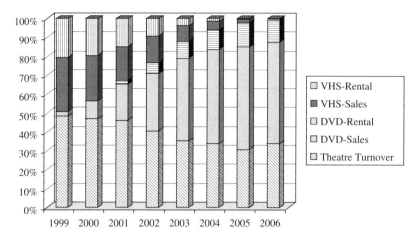

Figure 9.1   Market shares of cinema and video in Germany (from BVV (Bundesverband Audio-visuelle Medien), 2007)

better quality, reduced distribution costs, improved viewing experiences, and a seamless integration into a complete digital production chain.

For very high quality requirements JPEG 2000 with its intraframe coding and new features, like scalability, is a good option for producing, archiving, and distribution of moving images. As shown on the simplified diagram of the production chain (Figure 9.2), the only remaining market actors using film are the existing film archives. The encircled items indicate the main standardized or proposed areas for JPEG 2000. Areas that especially require the preservation of very high quality are well suited for JPEG 2000.

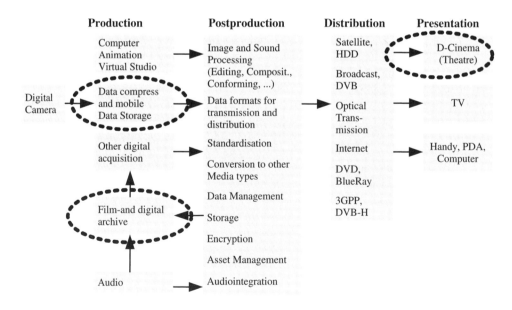

Figure 9.2   Digital movie production chain (Fraunhofer IIS)

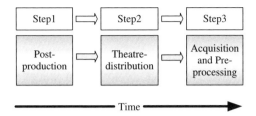

Figure 9.3 Temporal order of the transition from analog to digital technology

While moving images used for TV and the Internet have been processed and played back digitally for a long time, this was not the case for the high-end area of cinema movies, because of the huge amount of data and the very high requirements for image quality. This caused a separate film processing chain for cinema moving images. The first steps into digitization were taken in the postproduction, where special effects and animations in movies like *Star Wars* resulted in new viewing experiences.

Because of the permanently increasing processing power of computers and the increasing capacity of storage media cinema movie production is also subject to a new evolution. As in other application areas, analog technology is being replaced in the cinema production and distribution by digital processing (Figure 9.3).

9.2   General Requirements for Digital Cinema

9.2.1   General Requirements

The main benefit of a cinema presentation is – in parallel with the social aspects – the immersive viewing experience. For the industry this is the most important feature, as this brings people from home to theatres. Otherwise people would not pay a high ticket price for a single presentation. An immersive viewing experience demands that the projection should have high resolution, high contrast and dynamic range, a very good color reproduction, and outstanding sound. For this reason theatres have a big screen, low ambient light conditions, and excellent loudspeaker systems. To take full advantage of these benefits the content needs to be presented from a high-quality distribution format.

In former times the carrier for distribution was a 35 mm film print. Tests within ITU-R (ITU, 2001; Morton, Maurer, and DuMont, 2002) have demonstrated that typical 35 mm film answer prints for theatres have a vertical resolution of less than 800 lines. The 35 mm film during acquisition, however, can have a vertical resolution of up to 2000 lines. The main losses in the chain are due to the duplication and copy processing steps. Based on these results two formats were selected by the industry, one format with 2048×1080 pixels ('2k format') and the other with 4096×2160 pixels ('4k format'). To allow the use of only one distribution master, a scalable compression format was requested. This allows only one distribution format to be produced for both projection sizes, and the playback system can extract the requested size from one code-stream.

For the successive contrast in a scene, a nominal ratio of 2000:1 was requested. This allows for the perception of fine details in different scene conditions (e.g. dark scenes versus sun bright scenes). Within one single frame the contrast can be lower. This correlates with the adaptation behavior of the human eye. A bit-depth of equal or more than

12 bits per component is the consequence. This high contrast ratio is the real difference to home presentations based on 8 bits per component transmission formats and allows for immersive experiences.

Color reproduction is also an important aspect in cinema. Film has a very high color gamut, whereas TV in former times had a color gamut limited by the cathode-ray tube (CRT) phosphors. With the invention of new projection technologies like DLP®[1] and LCOS, more suitable digital projection systems are currently available. So that color reproduction in D-Cinema systems did not need to be restricted to actual technology or devices, the XYZ colorspace was chosen. This colorspace is a mathematical model that was defined in 1931 by the CIE (Commission Internationale de l'Eclairage) and can represent all human visible colors. Therefore future projection systems like laser projectors will not need to request a new colorspace (Kennel, 2006).

A movie will be shown in many countries all over the world. Because of country-specific movie rating restrictions, scenes within the movie may need to be cut off or replaced by less offensive content. Hence, country-specific reassembly is an additional requirement. A frame accurate reassembling is possible by using country-specific composition play lists. These play lists define in- and out-points for the movie, so that only one distribution package has to be used. The reassembling is effected during playback in real-time. A typical movie with a playtime of 100 minutes has an uncompressed size of about 1.5 TB in the 2k format and 6 TB in the 4k format. Even with today's increasing storage capacities this is too large for distribution. Therefore compression is still necessary. As discussed before, the special requirements for a compression format in D-Cinema are scalability, high-resolution capabilities of at least 4k, high bit-depths of at least 12 bits per component, the use of special colorspaces, and frame accurate reassembling of play lists. The visibility of compression artifacts is not acceptable. Hence, even with compression the data rates remain high. Based on these requirements JPEG 2000 as an intraframe compression format was chosen for D-Cinema and special profiles were created. With enough flexibility and headroom JPEG 2000 is not solely suitable for D-Cinema but also for the complete Digital Cinema chain.

9.2.2 Additional Requirements in the Acquisition Area

Image quality in the acquisition of movies is the most important requirement since, as with the analog film process, all following processing steps will reduce the image quality. Hence, the compression has to be lossless or near lossless. However, the required processing power for compression will be even higher than for D-Cinema, as variable frame rates up to 150 fps for slow motion capturing will be recorded and the compression has to be carried out in real-time before storing the data. On the other hand, the use of scalable compression formats would allow an easy extraction of 'dailies' (previews for quality control at the end of the day). An additional copy and processing step for the dailies after recording would require a multiple of the recording time.

9.2.3 Additional Requirements in the Postproduction Area

The huge amount of data in cinema movie production requires special processing systems. Even with today's technology most 4k productions will be processed first, with

---

[1] DLP® is a trademark of Texas Instruments.

smaller representation copies of the movie (proxies). This allows faster editing and color correction. With JPEG 2000 these proxies can be extracted directly from a lower resolution level of the bit-stream. When producing the final master copy, all processing steps will be reapplied to the original data at maximal resolution. The processing steps will be stored, for example, in metadata like EDL (Editing Decision List) and CDL (Color Decision List) (ASC, 2008).

9.2.4 Additional Requirements in the Archive Area

For archiving of movies two main, contradictory requirements exist. On the one hand, the image shall be preserved in its original quality, which implies lossless compression or uncompressed storage with huge amounts of data. This requirement is necessary to give future audiences an authentic impression of the artistic intention of the director. It preserves the cultural heritage of movie makers. Furthermore, as soon as better distribution media are available (e.g. BlueRay DVD), the movie can be remastered and released in a better image quality. On the other hand, fast browsing and retrieval of content in a large archive system is desired. With uncompressed or lossless compressed images one movie will exceed a data amount of one or more TB. To read the data from media alone requires too much time and is not appropriate for browsing and searching. A higher compression ratio is therefore necessary for such applications.

9.2.5 Summary

Digital cinema imposes the handling of big images with very high quality requirements, leading to a huge amount of data to be processed and stored. Because of the aforementioned quality and accessibility constraints, compression is, however, limited to intraframe processing and very low compression ratios. Hence, for browsing, editing, and retrieval, a scalable compression algorithm with the option of extracting lower resolution representations efficiently is helpful in reducing processing power. JPEG 2000 fulfils these requirements. Figure 9.4 gives an overview of the main interesting features of JPEG 2000 for the digital cinema chain.

9.3 Distribution of Digital Cinema Content

Whereas digital movie postproduction can and will be done in many different ways and with a multiplicity of different systems, the output of these systems has to be standardized, such that the content can be presented on all screens without further processing. The film reel with its standardized 35 mm film fulfilled this requirement in a perfect way. Film could be shown on projectors in America, Europe, or Asia. To allow this worldwide interchangeability a large effort was made to standardize digital cinema distribution. The Digital Cinema Initiative was the result of this effort and was supported additionally by the Society of Motion Picture and Television Engineers (SMPTE) and ISO/IEC (and ITU-T).

9.3.1 Digital Cinema Initiatives, LLC (DCI)

When it became obvious that cinema had to take the digital highway, the big Hollywood studios started in 2002 with the DCI project to define specifications for digital cinema

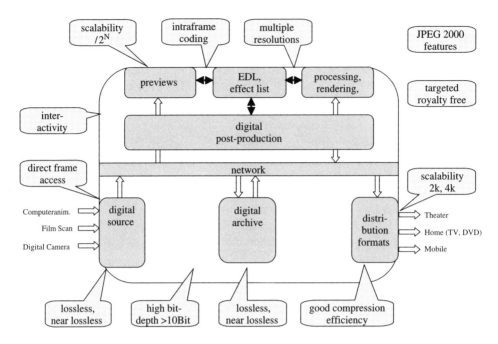

Figure 9.4   Features of JPEG 2000 useful for digital cinema (Edwards and Foessel, 2001)

distribution. The mission statement on the DCI website (http://www.dcimovies.com) states:

> DCI's primary purpose is to establish and document voluntary specifications for an open architecture for digital cinema that ensures a uniform and high level of technical performance, reliability and quality control.

During this still ongoing effort many evaluation and comparison tests were performed to ensure the highest quality and security. A mini-movie (StEM = standard evaluation material) was produced to allow for quality tests. A system architecture was developed that enables the replacement of film reels by digital movies in the form of digital cinema packages (DCP), which were encrypted and can be unlocked by a key contained by a key delivery message (KDM).

After extensive compression tests at the Digital Cinema Lab of the Entertainment Technology Center (http://www.etcenter.org/digital-cinema-lab), JPEG 2000 was chosen in 2004 as the compression algorithm, because of its scalability, the quality, and some other features. It has to be noted that quality comparisons were done between JPEG 2000 and other coding standards. For high-quality applications, where the noise level of the image or the grain in scanned films becomes relevant, JPEG 2000 is a very good solution (see also Section 1.5). In these circumstances, interframe coding algorithms like MPEG cannot benefit from the temporal redundancy of successive images. In addition, JPEG 2000 can be used to integrate several resolution levels inside the code-stream. For D-Cinema distribution the quality scalability in JPEG 2000 was not used, which means that only one quality layer is present.

The main results of the DCI are:

- StEM Mini-Movie (DCI, 2006);
- Digital Cinema System Specification V1.2 (DCI, 2008);
- Digital Cinema System Specification Compliance Test Plan V1.0 (DCI, 2007);
- Reference DCP and KDM.

Since the industry immediately supported this effort by developing devices, the DCI Digital Cinema System Specification soon became a de facto standard.

9.3.2 System Concepts and Processing Steps

The principal workflow of the D-Cinema content distribution is as follows (see Figure 9.5):

- During postproduction a D-Cinema distribution master (DCDM) will be created. The DCDM has specific format constraints to enable efficient standardized processing for the distribution format packaging (see the SMPTE standard family 428-x, Section 1.3.4). The packaging format enables the embedding of image, audio, subtitle, and caption content.
- The Digital cinema package (DCP) is derived from the DCDM. It consists of MXF and XML files. The MXF files are created by wrapping audio, subtitles, captions, and JPEG 2000 compressed images into these files. The images are typically encrypted

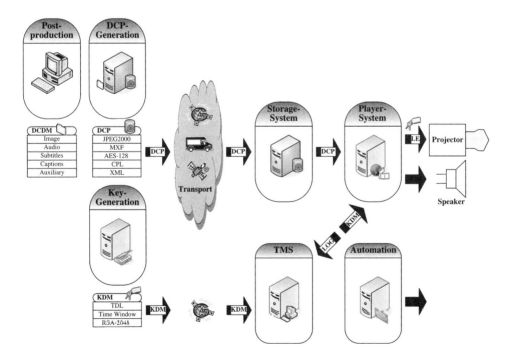

Figure 9.5 Distribution workflow for D-Cinema (Foessel *et al*., 2008)

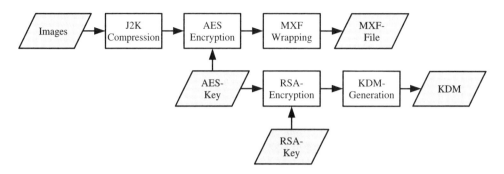

Figure 9.6   Encryption process for images

with AES128 before wrapping, so that unauthorized access is prevented. Additional information for the package (a packaging list and asset map) and the playback procedure (a composition package list) are included in the XML files.

- To enable the presentation of the movie the AES128 key is enclosed in a key delivery message (KDM). The AES128 key is again encrypted by a public–private key pair encryption technology, called RSA-2048 (see Figure 9.6). The KDM uses a public key to encrypt parts of the message and only the corresponding private key of the player/presentation device can decrypt the KDM. In addition, messages within the theatre and to the outside are defined (see the SMPTE standard family 430-x, Section 1.3.4, for further information; see also Bilgin and Marcellin, 2006, and Bloom, 2006).
- Both DCP and KDM are required for playback of a movie. The DCP contains the movie, whereas the KDM is the activation key for the playback. The DCP has a size of 100 to 300 Gbytes. The KDM has a size of only a few kbytes. Today DCP distribution is done by mailing hard disk drives. At a later stage satellite or Internet distribution is planned. The KDMs are send by email, USB-Sticks, or can be acquired over the Internet.
- In the theatre, the DCP will be ingested into a central storage system or directly into a player system. The KDM is valid for only one specific player system. A media block inside the player system can decrypt the KDM, access the AES key, and decrypt the DCP. If the media block is not in the projector a separate encryption has to be used between the media block and the projector. A theater management system (TMS) can control multiple player and projector systems.

9.3.2.1   Requirements and Derived Processing Steps

Several requirements were imposed for the introduction of D-Cinema distribution. The first set of requirements considers the replacement of an old proven technology by a new one:

- Traditional film reels shall be replaced by digital data packages with all the advantages of film and digital.
- The movie shall be able to play back worldwide and the systems shall be interoperable.
- The same organizational distribution channels shall be used for a seamless exchange.
- The visual quality shall be the same or even higher than 35 mm film.

- The digital playback system shall be as robust as film projectors with the same amount of low number of failures.

For IT equipment, the latter requirement is certainly a challenge. The second set of requirements considers the new advantages of a digital system:

- The access to the market shall be faster, because hybrid systems (film and digital) are no longer required.
- The digital data distribution shall be much cheaper than the film distribution, mainly because of the lower cost of the digital copy in relationship to the film print.
- The production of digital cinema movies shall fit seamlessly to the rest of the digital movie production (DVD, TV, etc.).
- What is very important, especially in the last few years, is that new technologies like digital 3-D movies shall be possible now.

Of course, some disadvantages also exist. Piracy of a digital copy is much easier than for a film print. Therefore more than half of the DCI specification covers security and watermarking items.

The distribution of D-Cinema content requires many new processing steps, which are derived from other postproduction, coding, and encryption technologies. The concatenation of processing steps can be shown in a simplified way, as in Figure 9.7. Subtitles and captions are neglected.

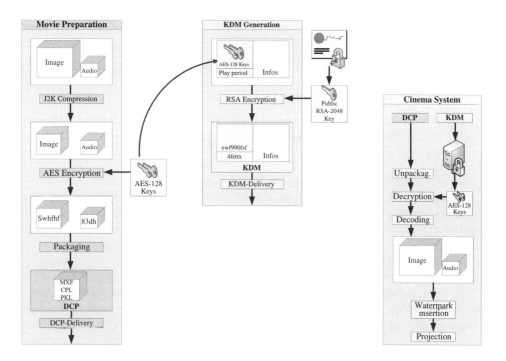

Figure 9.7   Processing steps for D-Cinema (Foessel *et al.*, 2008)

This led to the following specifications reflecting the aforementioned requirements:

- To reduce the amount of data JPEG 2000 compression for the images was chosen. Each frame is compressed individually. Only Part 1 of the JPEG 2000 standard 15444-1 is used. DCI did not adopt JPEG 2000 Part 3, i.e. the Motion JPEG 2000 file format, but preferred to adopt the more suitable MXF specification of SMPTE. The JPEG 2000 profiles for digital cinema distribution defined the maximum bit-rate for 2k/4k distribution at 250 Mbit per second. This reduces the amount of data for 2k images by a factor of more than 8 and for 4k images by a factor of more than 30.
- Two resolutions are selectable from the code-stream, one for big screens and one for smaller. This resulted in two image containers: 2048×1080 pixels and 4096×2160 pixels. The scalability feature of JPEG 2000 was used to create only one data-stream for both screen sizes. For feature films with an aspect ratio of 2.39:1 this results in a resolution of 2048×858 pixels or 4096×1716 pixels.
- The selected bit-depth is 12 bits per color component to allow enough headroom for dynamic range and contrast. An XYZ colorspace was introduced, which allows for a device-independent coding. Future projectors can benefit from this colorspace without additional color grading.
- For security purposes the JPEG 2000 code-streams are AES-128 encrypted. The key is only known during the postproduction or inside the playback devices in a specially secured environment (Media Block). No decrypted image data are available outside these devices.
- The audio information is uncompressed (PCM format). The DCP allows the integration of up to 16 audio channels.
- The frame rate is currently 24 frames per second. This is sufficient to avoid flicker problems under the low ambient light conditions of the theatres. An extension to other frame rates is now under consideration.

9.3.3 Digital Cinema Package (DCP)

The digital cinema package replaces the traditional film reel and contains all movie essence, such as images, soundtrack, and subtitles. It is a set of files with the following structure (see Figure 9.8):

- The most basic element is a track file, which contains essence and/or metadata. It is an MXF wrapped file format and holds JPEG 2000 compressed images, audio, and/or subtitles. Analog to film the digital track files will be split into reels. This allows the use of different encryption keys throughout the movie. Typically three reels are used for D-Cinema.
- A composition playlist (CPL) is created in the digital cinema mastering process to assemble a complete composition. This composition consists of all of the essence and metadata required for a single presentation of a movie, a trailer, an advertisement, or a logo. A single composition playlist contains all of the information on how the files are to be played, at the time of a presentation, along with the information required to synchronize the track files.
- The packing list (PKL) contains information and identification about each of the individual files that will be delivered in a digital cinema package (DCP).

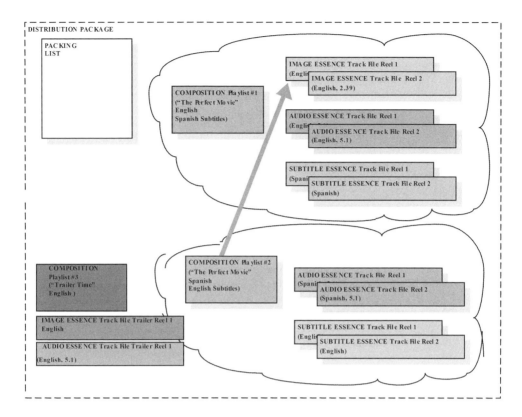

Figure 9.8   Structure of the distribution package (DCI, 2008)

9.3.4   Standardization of D-Cinema Distribution

9.3.4.1   Society of Motion Pictures and Television Engineers (SMPTE)

To achieve a broader acceptance the DCI specifications were fed to SMPTE and ISO/IEC for standardization. Based on the specifications a complete set of new standards is under development. Table 9.1 is an extract of the most important standards

9.3.4.2   ISO/IEC SC29 WG1

The JPEG 2000 profiles, which are referenced in the SMPTE 429-4 standard and used for D-Cinema distribution, are defined in ISO/IEC 15444-1:2004/Amd 1:2006, Profiles for Digital Cinema Applications (ISO/IEC, 2006).

9.3.4.3   ISO TC36 Cinematography

For a worldwide standardization of the D-Cinema distribution the SMPTE standards are now under standardization within ISO TC36. The standards are in a so-called fast-track procedure, which allows a faster standardization. By mid 2008 about 27 documents of Table 1 were in the ISO standardization process.

Table 9.1   SMPTE D-Cinema standards (SMPTE, 2007)

427		Link Encryption for 1.5 Gb/s Serial Digital Interface
428	428-1	D-Cinema Distribution Master: Image Characteristics
	428-2	D-Cinema Distribution Master: Audio Characteristics
	428-3	D-Cinema Distribution Master: Audio Channel Mapping and Channel Labeling
	428-4	D-Cinema Distribution Master: Audio File Format/Constraints
	428-5	D-Cinema Distribution Master: Image Mapping into TIFF
	428-6	D-Cinema Distribution Master: PCM Format
	428-7	D-Cinema Distribution Master: Subtitle
	428-8	D-Cinema Distribution Master: Image Metadata
	428-9	D-Cinema Distribution Master: SDI Mapping
429	429-1	D-Cinema Packaging: Packaging Guideline
	429-2	D-Cinema Packaging: Operational Constraints
	429-3	D-Cinema Packaging: Sound and Picture Track File
	429-4	D-Cinema Packaging: MXF JPEG 2000 Application
	429-5	D-Cinema Packaging: Timed-Text Track File
	429-6	D-Cinema Packaging: MXF Track File Essence Encryption
	429-7	D-Cinema Packaging: Composition Playlist
	429-8	D-Cinema Packaging: Packing List
	429-9	D-Cinema Packaging: Asset Mapping and File Segmentation
	429-10	D-Cinema Packaging: Stereoscopic Picture Track File
	429-11	D-Cinema Packaging: Auxiliary Sound for CPL
430	430-1	D-Cinema Operations: Key Delivery Message
	430-2	D-Cinema Operations: Digital Certificate
	430-3	D-Cinema Operations: Generic Extra-Theater Message Format
	430-4	D-Cinema Operations: Log Record Format Specification
	430-5	D-Cinema Operations: Security Log Event Class and Constraints
	430-6	D-Cinema Operations: Auditorium Security Messages for Intra-Theater Communications
	430-7	D-Cinema Operations: Facility List
	430-8	D-Cinema Operations: Show PlayList
	430-9	D-Cinema Operations: Key Delivery Bundle
431	431-1	D-Cinema Quality: Screen Luminance Level, Chromaticity and Uniformity
	431-2	D-Cinema Quality: Reference Projector
433		XML Data Types

9.3.5   JPEG 2000 D-Cinema Distribution Profiles

For D-Cinema distribution code-stream restrictions, two new profiles were introduced. The two profiles correspond to a 2k (2048 × 1080) pixel profile and a 4k (4096 × 2160) pixel profile. The most important parameters are given in Table 9.2. A complete set can be found in ISO/IEC 15444-1:2004/Amd 1:2006, Profiles for Digital Cinema Applications (ISO/IEC, 2006).

The main reason for the code-stream restrictions was to constrain the codec flexibility such that manufacturers can create systems more easily. A detailed explanation of the parameters can be found in Bilgin and Marcellin (2006). The main restrictions are: defining a 2k and 4k format, using maximally up to five or six resolution levels, using only one

Table 9.2 Important parameters of JPEG 2000 profiles for D-Cinema distribution

	2K digital cinema profile	4K digital cinema profile
SIZ marker segment		
Profile indication	Rsiz = 3	Rsiz = 4
Image size	Xsiz <= 2048, Ysiz <= 1080	Xsiz <= 4096, Ysiz <= 2160
Tiles	One tile for the whole image	Same
Subsampling	No subsampling	Same
Number of components	3	Same
Bit-depth	12 bits unsigned	Same
Marker locations		
Packed headers (PPM, PPT)	Disallowed	Same
COD, COC, QCD, QCC	Main header only	Same
COD/COC marker segments		
Number of decomposition levels	$N_L <= 5$	$1 <= N_L <= 6$
Number of layers	Shall be exactly 1	Same
Code-block size	32 × 32	Same
Code-block style	SPcod, SPcoc = 000 000 00	Same
Precinct size	PPx = PPy = 7 for N_L LL band, else 8	Same
Progression order	CPRL, POC marker disallowed	There shall be exactly one POC marker segment in the main header
Tile parts	Each compressed image shall have exactly 3 tile parts. Each tile part shall contain all data from one color component	Each compressed image shall have exactly 6 tile parts. Each of the first 3 tile parts shall contain all data necessary to decompress one 2k color component. Each of the next 3 tile parts shall contain all additional data necessary to decompress one 4k color component
Application-specific restrictions		
Max. compressed bytes for any image frame (aggregate of all 3 color components)	1 302 083 bytes for 24 fps 651 041 bytes for 48 fps	1 302 083 bytes (for 24 fps)
Max. compressed bytes for any single color component of an image frame	1 041 666 bytes for 24 fps 520 833 bytes for 48 fps	1 041 666 bytes for 2k portion of each component (for 24 fps)

quality layer, using a fixed code-block size of 32 × 32, using 12 bits per color component, and restricting the bit-rate to 250 Mbit/s.

9.4 Archiving of Digital Movies

With the introduction of D-Cinema distribution and the continued growth of digital post-production and digital movie acquisition, archiving of digital cinema movies in its original format is requested by the industry. Film is certainly today the best physical long-term archiving media for movies. Nonetheless, since the film substrate is still prone to long-term deterioration, other carriers need to be looked at in order to provide redundant storage architectures and standardized archive formats, such that long-term digital preservation becomes acceptable. Additionally, as more and more content is digitally born, the transfer back to film (e.g. the YCM Separation Archival Master) costs much in terms of effort, money, and potential quality.

Traditional film archives therefore have to look carefully at the way in which the preservation of movies as cultural heritage can be done in the digital age. On the one hand, the content has to be preserved for the long term; on the other hand, with the change in utilization of movies (see the introduction) direct short-term access for reusing digital movie data is necessary in order to help refinance the archival costs.

9.4.1 EDCine

To address the problem of digital movie archiving an initiative was started in 2006 by the Cinémathèque Royale de Belgique (CRB), the Fraunhofer IIS, and MOG Solutions, as part of the European funded project called EDCine (EDCine, 2008). The project started with the collection of requirements for digital movie archives and intends to meet all requirements by presenting a two-tier storage model that provides both a long-term preservation file format with the best quality, called the Master Archive Package (MAP), and a simple access format, called the Intermediate Access Package (IAP) (Read and Mazzanti, 2008). As the ideal physical storage media for digital long-term data storage today has still not been identified, EDCine concentrates its effort in defining digital storage formats independently from the physical media.

A digital preservation strategy data migration has been recommended by AMPAS (2007):

> Data migration involves the transfer of data from old physical media to new physical media, a process that often (but not always) includes updating file formats for currency with the latest-generation operating system and/or software application.

The main goal in the project is to define JPEG 2000 and MXF-based file formats as open standards with a long-term lifetime, so that the migration process is limited to the data transfer from old to new physical media. With the scalability features of JPEG 2000 a high flexibility for data access in browsing, searching, and retrieval applications shall be achieved.

9.4.2 Requirements for Digital Movie Archives

All moving image film, video, and digital archives need a procedure that stores their images (and related sound) in their highest quality and provides access to them in an increasing variety of formats and qualities. For search and retrieval purposes additional metadata for categorizing and indexing the archive items are necessary.

With the transition of film movies to D-Cinema distribution packages the storage of digital movie data has to be taken into account. To avoid quality loss and additional processing steps archiving of DCI/SMPTE-compliant DCPs has to be supported as well.

Film for cinema has always generated its own unique experience in the audience, which has varied with time, location, and technology for over 100 years. Archives, movie libraries, and specialist cinemas require the cinema projection of heritage films (best defined as films shot and released prior to digital projection becoming used or standardized) to be authentic and bring as faithful as possible a representation of the projected original film (a requirement already observed in the restoration of old films). Hence, it should respect the rendered intent.

The following list (an excerpt of the details provided by the Cinémathèque Royale de Belgique in consultation with the FIAF Technical Commission to research partners) records some of the characters that provide this authenticity, but is not exhaustive. A complete list and a white paper can be downloaded from the EDCine Website (www.edcine.org):

- The resolution of the projected screen image should not be visually lower than that of the original film/movie image (it is anticipated that this requirement is difficult to quantify).
- The frame rate of a digital cinema projection should be the same as that of the original film.
- The aspect ratio of the image should be that of the original film.

In brief, the basic requirements of a digital storage system for film are as proposed:

- Ingest should be from all available essence: film images, film sound, and any analog, digital video, or data version of a programme, images, or sound.
- For purposes of access, the intermediate access package IAP must be capable of producing a wide range of current formats and media output (from D-Cinema quality to HD and standard definition television, to Internet-browsable versions).
- The MAP, the long-term digital storage format, must store data in a lossless format.
- D-Cinema output versions, film output versions, and any other high-quality output versions of the future that are generated from long-term digital storage formats should be as close to the original film or digital version as possible.

9.4.3 System Concepts and Archival Workflows

The EDCine Digital Film Archive System Architecture presents two different file formats to store images, audio and metadata (see Figure 9.9). Each is able to store archived items in the best possible quality (at this time) and to facilitate access to the archived items into many different distribution formats. The proposed architecture consists of two packages, the Master Archive Package (MAP) for long-term preservation and an Intermediate Access Package (IAP), designed to make access to the stored items faster and simpler. Both MAP and IAP are designed as information packages where the content (image, sound, texts, etc.) is stored jointly with its cataloguing information and technical metadata to ensure that the content is correctly displayed when accessed. This is in accordance with the Open Archival Information System (OAIS) Reference Model standardized as ISO 14721:2003. Archived material may be stored in either one of the two formats or in both formats simultaneously, depending on the usage scenarios.

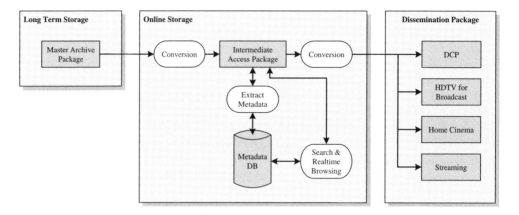

Figure 9.9 Two-tier approach for digital movie archives (Fraunhofer IIS)

The main design goal for the MAP is to store the source images and sound without any loss of information in order to preserve the original viewing experience. This practice usually requires lossless compression and results in large amounts of data. For the MAP a JPEG 2000 lossless profile is recommended.

The main design goal for the IAP is to allow frequent access to archived items. This usually does not require the original lossless quality of the source images and data. Instead the focus lies on easy and standardized access through local and remote channels. This results in the requirement for a significant (and often extreme) reduction of the original amount of data and the restriction of certain coding parameters to ensure maximum compatibility with a wide range of decoding and playback equipment.

9.4.4 Archive Packages MAP and IAP

The key specifications of the *Master Archive Package* are within the limitations of JPEG 2000 and AES (Audio Engineering Society) recommended practices:

- image resolutions up to 16k (16384 × 8640 pixels);
- any image frame aspect ratio;
- any image bit-depth (only limited by the JPEG 2000 maximum bit-depth, which is for practical implementations 32 bits per component);
- any colorspace;
- any frame rate,
- image components up to 8;
- mathematically lossless compression for image content (can also be lossy if required, e.g. when archiving an already compressed DCP);
- audio data in RAW format (optional MPEG-4 SLS);
- no audio sampling restrictions (at least support of recommended values as in AES-5);
- no restrictions for word length in audio (at least 16 bits or 24 bits);
- discrete (i.e. no matrix encoding) audio channels (unrestricted number);

- MXF wrapper (implementation currently limited to operational patterns Op1a and Op1b or Op2a); for more information on MXF format specifications see Devlin *et al.* (2006).

Within the limitations of JPEG 2000 and AES recommended practices, the key specifications of the *Intermediate Access Package* are:

- image resolution up to 2k (2048 × 1080) or 4k (4096 × 2160) (depending on the adopted profile);
- any image frame aspect ratio;
- image bit-depth up to 12 bits;
- any frame rate;
- image components up to 8;
- compression aimed at producing a bit-rate up to 250 Mbits/s for 2k images with 24 fps, 500 Mbit/s for 2k images with 48 fps, or 500 Mbits/s for 4k images (this level is set at a current practical level that allows real-time presentation with current technology);
- audio data in RAW format (optional MPEG-4 SLS);
- 48 kHz or 96 kHz audio sampling frequencies;
- 16-bit or 24-bit word length for audio;
- discrete (i.e. no matrix encoding) audio channels (up to 16 channels);
- MXF wrapper (implementation currently limited to operational patterns Op1a and Op1b or Op2a).

9.4.5   Standardization of Archive Profiles

The EDCine consortium started a standardization effort within ISO SC29 WG1 in liaison with the Association des Cinémathèques Européennes (ACE), Fédération Internationale des Archives du Film (FIAF), and SMPTE to define and standardize the new profiles for JPEG 2000. It is known as ISO 15444-1: Amd 2, Extended Profiles for Digital Cinema Applications (ISO/IEC, 2007). Further recommendations are planned for the integration of the JPEG 2000 code-streams into the MAP and IAP data packages, similar to the D-Cinema distribution package (DCP).

9.4.6   JPEG 2000 Archive Profiles

For digital cinema archive applications three new profiles are proposed within ISO: one profile for lossless long-term archiving and two other profiles as intermediate access formats with higher flexibility than the D-Cinema distribution profiles: scalable 2k digital cinema and scalable 4k digital cinema (Table 9.3). The profiles are being standardized as ISO 15444-1: Amd 2, Extended Profiles for Digital Cinema Applications (ISO/IEC, 2007).

The two scalable profiles contain DCP-compliant tile parts, meaning that by only properly parsing the code-stream can a D-Cinema distribution code-stream (15444-1:Amd 1) be created without decompression and recompression. With the additional layer 1 a bit-rate up to 500 Mbit/s can be achieved for 4k image movies and 2k image movies at 48 fps. The long-term storage profile has very flexible parameter ranges, such that the original format can be preserved.

Table 9.3 Important parameters of JPEG 2000 profiles for digital movie archives

	Scalable 2k digital cinema profile	Scalable 4k digital cinema profile	Long-term storage profile
SIZ marker segment			
Profile indication	Rsiz = 5	Rsiz = 6	Rsiz = 7
Image size	Xsiz <= 2048, Ysiz <= 1080	Xsiz <= 4096, Ysiz <= 2160	Xsiz <= 16 384, Ysiz <= 8640
Tiles	One tile for the whole image	One tile for the whole image	One tile for the whole image or minimum tile size YTsiz + YTOsiz >= 1024 XTsiz + XTOsiz >= 1024
Subsampling	No subsampling	No subsampling	No restriction
Number of components	Csiz = 3	Csiz = 3	Csiz <= 8
Bit-depth	Ssiz^i = 11 (i.e. 12-bit unsigned)	Ssiz^i = 11 (i.e. 12-bit unsigned)	No restriction
Progression order	CPRL	CPRL	CPRL
Number of layers	L = 2	L = 2	L <= 5
Multiple-component transform	No restriction	No restriction	No restriction
Number of decomposition levels	N_L <= 5. Every component of every image of a distribution shall have the same number of wavelet transform levels	$1 <= N_L <= 6$. Every component of a distribution shall have the same number of wavelet transform levels	No restriction, with respect to: (Xsiz-XOsiz)/D(I) <= 64 (Ysiz-YOsiz)/D(I) <= 64 and D(I) = pow (2, N_L) for each component I
Code-block size	xcb = ycb = 5	xcb = ycb = 5	xcb <= 6, ycb <= 6
Transformation	9-7 irreversible filter	9-7 irreversible filter	No restriction
Application-specific restrictions			
Max. compressed bytes for any image frame (aggregate of all 3 color components)	1 302 083 bytes	2 604 166 bytes	No restrictions
Max. compressed bytes for any single color component of an image frame	1 041 666 bytes	2 083 332 bytes	No restrictions
Max. compressed bytes for quality layer 0 of any image frame (aggregate of all 3 color components)	1 302 083 bytes for 24 fps 651 041 bytes for 48 fps	1 302 083 bytes (for 24 fps)	No restrictions
Max. compressed bytes for layer 0 of any single color component of an image frame	1 041 666 bytes for 24 fps 520 833 bytes for 48 fps	1 041 666 bytes for 24 fps for 2k portion of each component	No restrictions
Purpose	Used as IAP	Used as IAP	Used as MAP

9.5   Future Use of JPEG 2000 within Digital Cinema

9.5.1   Acquisition

During the acquisition of digital feature movies preservation of high quality is required. Whereas in HDTV applications broadcast cameras compress the images down to 100 to 200 MBit/s, this is not acceptable for feature movies. Consequently, either an uncompressed recording is used or a lighter compression, which results in data rates of 400 to 800 Mbit/s. Next-generation digital cameras will have higher resolutions (up to 4k) and higher frame rates (up to 150 frames per second); hence the necessary data rate will also increase. Feature movies run through a complex postproduction process with many image processing steps (pan and scan, zoom, contrast enhancement, etc.), where compression artifacts can become visible. Tests within the project WorldScreen (WorldScreen, 2003) have demonstrated that compression factors of 4:1 to 8:1 are acceptable as a maximum; otherwise postproduction is limited to specific operations.

First cameras based on JPEG 2000 algorithms have been implemented. Cameras developed for movie and medical purposes that use this technique are, for example, the Thomson Infinity camera series or the RED camera. In the future more cameras are expected. However, high quality and high bit-depth are the basic requirements for its use.

9.5.2   Postproduction

Postproduction is a very data-intensive process. Today 2k real-time postproduction is possible. However, in many cases, especially for movies with 4k images, the standard process is to generate proxies, meaning a copy of the captured images with a smaller resolution, and to work with these proxies. Hence, editing can be done faster and more easily. Typically CDLs (color decision lists) and EDLs (editing decision lists) will be generated as metadata and later used in the conforming process. In this process the decisions will be applied to the original captured data to generate the final version.

JPEG 2000 would ideally eliminate the need for proxies, as several resolutions are inherently present inside the JPEG 2000 code-stream. Several projects like WMP (Tools for Media Production) (WMP, 2008) demonstrated that the use of JPEG 2000 from acquisition to distribution is a good approach for future postproduction systems. Today, it is hampered by the high processing complexity of the JPEG 2000 algorithm. In the future, however, with higher computational power and/or better JPEG 2000 ASICs, this seems to be a good option.

9.5.3   Mastering with JPEG 2000

In the past a scanned film master was the starting point for different distribution formats. However, the number of distribution formats has increased in the last few years (more than 30) and will increase more, so that the handling of the original uncompressed high data volumes has become a bottleneck in the creation of these formats. With the DCP as the highest quality distribution format, a new workflow is possible (see Figure 9.10). Tests have demonstrated that the creation of lower quality distribution formats directly from a DCP have no quality loss in comparison to the original workflow, where the distribution formats are processed directly from the uncompressed master. The only workflow

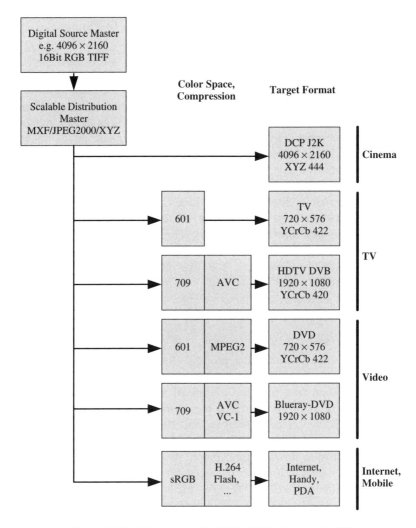

Figure 9.10 Mastering with JPEG 2000 (Fraunhofer IIS)

restriction is the limited scalability and metadata information of the DCP. Metadata in this case means information about the processing steps from one distribution format to the other. This problem can be solved by defining a more flexible JPEG 2000 package called a DVP (digital video package) (Lukk, 2008; Foessel, 2008) or to use the scalable 2k or 4k profiles of the digital cinema archive profiles (see Section 1.4.6) in conjunction with XML metadata files.

9.5.4 Enhanced Distribution Profiles

During visual tests it could be demonstrated that extremely high quality feature movies cannot be presented in their original acquisition quality when limiting the data rate to 250 MBit/s, as specified in the D-Cinema distribution profile (DCP). To reach this quality, it is necessary to use exclusive lenses and a scanned 65 mm film or computer-generated images. For scanned 35 mm film or today's digital cameras this is not the case. Therefore DCI and SMPTE have restricted the parameters in the DCP to avoid higher costs due to the complexity of the systems.

However, it can be expected that future digital cameras will be improved and that 3-D movies will receive a higher acceptance. In this case, profiles with higher data rates are useful. The scalable 2k and 4k profiles in the archive section seem suitable for these applications. Further investigations are needed to verify this and a guideline for the use of JPEG 2000 in digital cinema applications is in progress. The guideline will be published in the future as ISO/IEC 15444-1:2004/Amd 3, Guidelines for Digital Cinema Applications (ISO/IEC, 2008). It will contain information about visual weighting factors for frequency bands, wireless transmission, and enhanced profile coding for digital cinema.

9.6 Conclusions

With JPEG 2000 an excellent solution for high-quality digital cinema movie compression was found. The intraframe codec is ideal for editing and for frame accurate access during playback and browsing. Bit-depths of 12 or more bits per component allow for coding of high dynamic range images, which are typical for digital cinema environments. The scalability of JPEG 2000 enables the generation of one compressed image track, which can be used simultaneously for 2k and 4k screens. In addition, the scalability of JPEG 2000 offers future possibilities for fast editing and browsing in postproduction and archive applications.

Acknowledgments

This contribution is dedicated to my beloved wife Angela for her patience and her continuous support, even if the time for writing papers and contributions is limited to weekends and vacation time.

- The text and tables from amendments to ISO/IEC 15444-1:2004, Information Technology – JPEG 2000 Image Coding System – Part 1: Core Coding System, are reproduced with the permission of the International Organization for Standardization, ISO. This standard can be obtained from any ISO member and from the Website of ISO Central Secretariat at the following address: www.iso.org. Copyright remains with ISO.
- The author thanks DCI and the partners of EDCine for permission to reuse contributions and excerpts from different papers and specifications. This chapter contains figures, quotations, and excerpts from EDCine/Archives – A Summary of the EDCine Project – Archival Applications and DCI System Specification, which were produced with permission of the EDC and DCI.
- Research on Digital Cinema is done within the projects WorldScreen and EDCine. The author wishes to thank the European Commission for the continuous support in this research area. The WorldScreen project – Layered Compression Technologies for Digital Cinematography and Cross Media Conversion – was funded by the European Commission within the 6th Framework Programme FP6/2003/IST2, Contract No.

511333 WorldScreen. The EDCine project – Enhanced Digital Cinema – is funded by the European Commission within the 6th Framework Programme FP6/2004/IST/4.1, Contract No. 038454 EDCine.

- This chapter contains quotations and drawings from "DCI System Specification" which were produced with permission of Digital Cinema Initiatives, LCC.

References

AMPAS (The Science and Technology Council of the Academy of Motion Pictures Arts and Sciences) (2007) *The Digital Dilemma – Strategic Issues in Archiving and Accessing Digital Motion Picture Materials*, AMPAS.

ASC (American Society of Cinematographers) (2008) ASC color decision list workflow presentation, in *Digital Cinema Summit NAB 2008*, Las Vegas, NV, April 21–24, 2008. Available at: http://www.theasc.com/clubhouse/committee_tech.html#.

Bilgin, A. and Marcellin, M. W. (2006) JPEG 2000 for digital cinema, in *IEEE International Symposium on Circuits and Systems (ISCAS)*, May 21–24, 2006.

Bloom, J. A. (2006) Digital cinema content security and the DCI, in *40th Annual Conference on Information Sciences and Systems*, March 22–24, 2006, pp. 11761181.

BVV (Bundesverband Audiovisuelle Medien e.V.) (2007) Der Deutsche Videomarkt, BVV Business Report 2006/2007, Hamburg, April 2007.

DCI (2006) StEM Access Procedures, http://www.dcimovies.com, October 1, 2006.

DCI (2007) Digital Cinema System Specification Compliance Test Plan, Version 1.0, http://www.dcimovies.com, October 16, 2007.

DCI (2008) Digital Cinema System Specification, Version 1.2, http://www.dcimovies.com, March 7, 2008.

Devlin, B., Wilkinson, J., Beard, M. and Tudor, Ph. (2006) *The MXF Book: An Introduction to the Material eXchange Format*, Focal Press, ISBN-10: 024080693X.

EDCine (2008) Project Overview and Deliverables of Project, www.edcine.org.

Edwards, E. and Foessel, S. (2001) JPEG 2000 for Digital Cinema Applications, http://www.jpeg.org//public/DCINEMA-v2.pdf.

Foessel, S. (2008) Digital film archival using JPEG 2000 and MXF, in *SMPTE Technical Conference*, Hollywood, CA, 2008.

Foessel, S. *et al.* (2008) System Specifications for D-Cinema in Germany, Commissioned by FFA, http://www.ffa.de/, February 2008.

Hohenauer, F. (2007) Marktanalyse: Strategy Analytics prognostiziert dramatischen Wandel der Film- und Fernsehindustrie, *PresseBox*, München, Pressemitteilung BoxID 133566, October 29, 2007.

ISO/IEC (2006) Profiles for Digital Cinema Applications, ISO/IEC 15444-1:2004/Amd 1:2006.

ISO/IEC (2007) Extended Profiles for Digital Cinema Applications, ISO/IEC 15444-1:2004/PDAmd 2.

ISO/IEC (2008) Guidelines for Digital Cinema Applications, ISO/IEC 15444-1:2004/PDAmd 3.

ITU (2001), 35mm Cinema Film Resolution Test Report, ITU-R Doc-Nr. 6/14920.09.2001.

Kennel, G. (2006) *Color and Mastering for Digital Cinema*, Focal Press.

Lukk, H. (2008) Digital video package (DVP), in *SMPTE Technical Conference*, Hollywood, CA, 2008.

Morton, R. R. A., Maurer, M. A. and DuMont (2002) Assessing the quality of motion picture systems from scene-to-digital data, *SMPTE Journal*, **111**(2).

Read, P. and Mazzanti, N. (2008) A summary of the EDCine Project – Archival Applications (Version 1), *EDCine Archives*, March 2008.

SMPTE (Society of Motion Pictures and Television Engineers) (2007) SMPTE D-Cinema Standards, http://www.smpte.org.

WMP (2008) Werkzeuge für die Medienproduktion, http://www.wmp-bayern.de.

WorldScreen (2003) Layered Compression Technologies for Digital Cinematography and Cross Media Conversion, EU-Project, FP6/2003/IST2, Contract No. 511333, http://www.worldscreen.org.

10

Security Applications for JPEG 2000 Imagery

John Apostolopoulos, Frédéric Dufaux, and Qibin Sun

10.1 Introduction

The range of imaging applications continues to grow, as do their popularity. Many of these applications either require or their usefulness can be enhanced through the application of security services. For example, in surveillance applications it is important to ensure that the captured images have not been tampered with. For professional content, such as with digital cinema, it is important to prevent piracy. It is also useful to be able to adapt protected image content for delivery without compromising the security. This chapter describes some of the large number of security applications that involve JPEG 2000 coded images.

There are two general classes of approaches for securing JPEG 2000 coded images. In the first class of approaches the compressed image is treated as generic data and conventional data security techniques are applied. This class of approach does not exploit knowledge that the data to be secured correspond to a JPEG 2000 coded image. Alternatively, in the second class of approaches the security techniques explicitly account for the fact that it is a JPEG 2000 coded image that is being secured. A key example of this second class of approaches is the recent JPEG 2000 Security (JPSEC) standard, covered in Chapter 5, which is the media security standard for protecting JPEG 2000 images. We next discuss this issue of media security versus data security in more depth, and then identify some of the different types of security services that may be applied to JPEG 2000 coded imagery.

10.1.1 Media Security versus Data Security

Conventional approaches for securing images or general media are media unaware in that they treat the media as if it were a block of data. For example, consider the case

The JPEG 2000 Suite   Edited by Peter Schelkens, Athanassios Skodras and Touradj Ebrahimi
© 2009 John Wiley & Sons, Ltd

of protecting the confidentiality of an image by encrypting it to prevent eavesdropping or piracy. The media can be encrypted at the application layer as a file, then packetized and transported over the network, or the media can be packetized, then encrypted at the network packet layer using network-layer encryption. While these approaches do provide confidentiality, they also lose all of the beneficial properties that are inherent to the media or its coded representation. For example, unlike the typical case for data where one requires the entire data in order for the data to be useful, for media it is often the case that only a portion of it is either necessary (based on the user's goals) or available (because of packet loss in delivery). In addition, these file or packet-based security approaches prevent the flexible processing of the media because all of the structure of the coded media is lost. For example, it is often desirable to adapt or transcode the media, e.g. to be better matched to the receiving client's capabilities or for delivery over the available networks. The above characteristics motivated early work on jointly designing the security, compression, and delivery to preserve the valuable media characteristics even when the media is protected.

10.1.2 Different Types of Security Services

There are a number of basic security services that may be applied to JPEG 2000 coded images. The JPEG 2000 Security (JPSEC) standard, which is covered in Chapter 5, is a media security standard for protecting JPEG 2000 digital images. The JPSEC standard was designed to support the following six security services (ISO/IEC, 2007):

1. Confidentiality via encryption and selective encryption.
2. Integrity verification (including data and image content integrity, using fragile and semi-fragile verification).
3. Source authentication.
4. Conditional access and general access control.
5. Registered content identification.
6. Secure scalable streaming and secure transcoding.

The 6th service is a nontraditional security service that arises for media, and does not have an analogous service in the context of data security.

   The above security services may be achieved using the JPSEC standard to provide these security services. They may also be achieved by non-tandard (proprietary) approaches. In this chapter we will largely focus on security applications that can be created by building on the JPSEC standard.

   An important related area is digital rights management (DRM), which relates to the management of one's rights to digital content. This is a very important topic which is outside the scope of this chapter. Interested readers are referred to recent overviews available in Zeng and Lin (2006) and Kundar et al. (2004).

10.1.3 Chapter Overview

This chapter continues by examining a number of important digital imagery security applications. Section 10.2 begins by discussing secure transcoding and secure streaming. Media adaptation, referred to as transcoding, is often required to adapt JPEG 2000 coded content for clients with diverse device capabilities (e.g. small display sizes or low bit-rate network connections) and for time-varying network conditions. We describe how to perform transcoding of protected content in a secure manner, that is without unprotecting the content, e.g. without requiring decryption. This functionality is referred to as secure transcoding to stress that the transcoding or media adaptation is performed without compromising the security. The related topics of multilevel access control and selective or partial encryption of image content are briefly highlighted in Sections 10.3 and 10.4, respectively.

The important area of image authentication is examined in Section 10.5. In many applications authentication is the most important security service. We stress this because the lay person often considers encryption, to provide confidentiality, as the most important security service. Image authentication is typically provided by coupling conventional digital signature-based authentication schemes with a representation of the image content. For example, image authentication may support complete authentication, which detects if even a single bit is changed, to semi-fragile authentication, where minor modifications that do not change the perceived content are acceptable. This large range of possible types of image authentication is described in Section 10.5.

The emerging area of region-based scrambling for privacy is discussed in Chapter 11. This technology can be used in video surveillance systems to show what is occurring in a scene under surveillance while preserving the privacy of the individuals in the scene. This may also be used in a variety of other settings, such as to hide the identity of a source in TV news reporting or to anonymize the individuals communicating over a video conference. Detailed discussion about this new application area is given in Chapter 11.

10.2   Secure Transcoding and Secure Streaming

10.2.1   Motivation

A media delivery system may need to deliver media to different clients with diverse device capabilities and connection qualities, e.g. different display sizes and low bit-rate network connections. This may require the media to be adapted, or transcoded, in order to be best matched to the downstream client capabilities and time-varying network conditions. The transcoding may need to be performed by mid-network nodes, or proxies, that have knowledge about local and downstream network conditions and downstream client capabilities, and then act as control points by transcoding compressed streams according to these statistics.

Another important property is security to protect content from eavesdroppers, pirates, or malicious attackers. For example, to protect the confidentiality of the content makes it necessary to transport the content in encrypted form. The problem of adapting to downstream conditions while also providing security is particularly acute for mobile, wireless clients, since the available bandwidth can be highly dynamic and the wireless communication makes the transmission highly susceptible to eavesdropping.

In this context, conventional mid-network transcoding poses a serious security threat because it requires decrypting the stream, transcoding the decrypted stream, and re-encrypting the result. Since every transcoder must decrypt the stream, each network transcoding node presents a possible breach to the security of the entire system. Thus, this is not an acceptable solution in situations that require end-to-end security.

Therefore it is desirable to have the ability to transcode or adapt at a mid-network node or proxy in a manner that does not require decrypting the content. This capability is called *secure transcoding* to stress that the transcoding operation does not unprotect the content and thereby preserves the end-to-end security of the system (Dufaux *et al.*, 2004).

10.2.2 Secure Transcoding for JPEG 2000 and JPSEC

This section now examines how secure transcoding can be supported for JPEG 2000 content secured using JPSEC. Consider the use case for privacy protection of JPEG 2000 coded content. A sender encrypts the JPEG 2000 content into a JPSEC code-stream using a prescribed key. An allowed receiver may decode the JPSEC code-stream using its prescribed key. If the client capabilities and network conditions permit, the content can be received and rendered in its full, original quality. Otherwise, the content may need to be adapted to match the client capabilities, e.g. matching a low-resolution display by lowering the resolution, and the network conditions, e.g. by lowering the bit-rate to match the available bandwidth better.

When sharing conventional (unprotected) JPEG 2000 streams in this scenario, the solution is quite straightforward because of the inherent scalability properties that JPEG 2000 was designed to provide. Specifically, the JPEG 2000 standard uses structures such as tiles, precincts, resolution levels, quality layers, and color components and provides a syntax that allows easy access to these various components. Thus, reducing the resolution simply requires extracting a subset of resolution levels. Likewise, reducing the bit-rate simply requires extracting a subset of quality layers. Generally speaking, adapting or transcoding JPEG 2000 code-streams is a relatively straightforward operation and enables low-complexity implementations of mid-network transcoders. The scalability of the JPEG 2000 code-stream is one of the key differentiation features provided by the standard.

However, straightforward encryption of the JPEG 2000 code-stream to provide confidentiality would result in the loss of the scalability properties of the JPEG 2000 code-stream. Therefore, this motivates designing the protection method so that it provides mechanisms for retaining various levels of scalability within the protected code-stream. This should be done in a way that allows protected streams to be transcoded without decrypting the content. Furthermore, the receiver should be able to authenticate or verify that the protected content was modified or transcoded in only a valid and permissible manner, and that no unintentional (e.g. from an error) or malicious (e.g. from an attacker) modification has occurred.

An example use scenario is shown in Figure 10.1, where a sender transmits protected JPEG 2000 content to a mid-network node or proxy, which performs a secure transcoding operation to adapt the received protected data for each of three clients: a low-bandwidth client, a medium-bandwidth client, and a high-bandwidth client. The mid-network node performs a secure transcoding, that is a transcoding without decryption, and therefore the end-to-end security is preserved – the content is JPSEC protected throughout the delivery chain from the server to each receiving client. Note that the encryption keys are only available to the server and the receiving clients, and not to the mid-network node.

Another example use scenario is shown in Figure 10.2, which illustrates the streaming of privacy-protected JPSEC code-streams stored on a server. In this case the server may

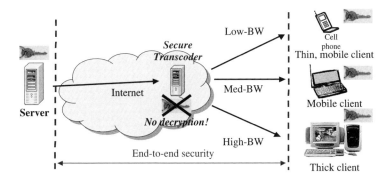

Figure 10.1 Example use case of end-to-end security and mid-network secure transcoding. The media is encrypted at the sender and decrypted only at the receiver; it remains encrypted at all points in-between. The mid-network node or proxy performs a secure transcoding to adapt the protected content for each receiving client. It performs this secure transcoding without unprotecting the content (without decryption) and therefore preserves the end-to-end security

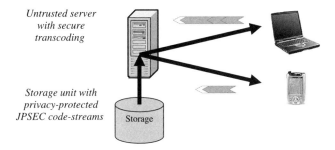

Figure 10.2 Example use case of remote browsing of JPSEC protected images. The server can perform secure transcoding to adapt the protected content for the different devices; specifically, it can perform this adaptation without decrypting the protected content

be untrusted; e.g. it does not have the encryption keys and is undesirable to provide the server with the encryption keys. Secure transcoding enables the server to adapt the protected content for streaming to different clients with different capabilities.

Further information and example system designs are given in Wee and Apostolopoulos (2001a, 2001b, 2003, 2004).

10.2.3 Security Properties

A key design principle within the secure transcoding framework has been not to design new cryptographic primitives, as generally they are vulnerable to flaws. Instead the goal is to use known and well-studied cryptographic primitives. As a result, the security of secure scalable streaming and secure transcoding depends on the security of the applied cryptographic primitives, such as the encryption cipher and message authentication code or digital signature. The illustrative examples described above have been implemented using the JPSEC-compliant encryption cipher AES, in CBC and CTR modes, and HMAC using SHA-1 for authentication. The AES cipher and HMAC using SHA-1 are believed to be secure as of the date of this writing. Of course, other JPSEC-compliant ciphers and MACs may be used in addition to those illustrated in our examples. Furthermore, this approach may be used with other JPSEC digital signature, access control, and key management tools.

10.2.4 Summary

This section describes secure scalable streaming and secure transcoding with JPSEC, which enables the two seemingly conflicting properties of end-to-end security with secure transcoding at mid-network nodes. This allows the JPSEC code-stream to be transcoded without requiring decryption, that is without requiring unprotection of the code-stream. Furthermore, this method also provides authentication that the transcoding was performed only in a valid and permissible manner. This enables a (potentially untrusted) server or mid-network node (e.g. proxy) to perform secure transcoding and allow a JPSEC consumer to give authentication that the received content was transcoded in a valid and permissible manner.

10.3 Multilevel Access Control

A related functionality to secure transcoding is the ability to provide multilevel access control. Security protection tools, such as encryption, may be applied to an image using one key or multiple keys. The advantage of using multiple keys is that they can provide multiple levels of access control. For example, different access rights may be provided to different individuals by providing each individual with a different key. These access rights may correspond to different qualities of the image, e.g. given JPEG 2000's various forms of scalability, one can provide access to different resolutions, quality levels (pixel fidelity), spatial regions or regions of interest, etc. These capabilities may be achieved by encrypting, for example, the different layers in the compressed bit-stream using different encryption keys. In this manner the rich scalability built into the JPEG 2000 standard may be exploited to provide different levels of access control.

The multiple keys may be independent of each other. However, the distribution and other logistics issues related to having multiple keys instead of a single key can lead to complications. Another common approach, which provides the benefits of multiple keys but with the distribution benefits of a single key, is for the multiple keys to be related in a structured manner from a single key. For example, the multiple keys may be recursively computed from a master key using a hash chain. Given a master key k, a sequence of keys may be computed by applying a one-way hash function $H(\ )$, where $k_{i+1} = H(k_i)$. For example, with a two-level wavelet decomposition three resolution levels are available {Low, Med, High}. By encrypting these three levels with the three keys $\{k_2, k_1, k_0\}$, where $k_1 = H(k_0)$ and $k_2 = H(k_1) = H(H(k_0))$, a user with k_0 can generate k_1 and k_2 and thereby decrypt all three resolution layers to get the High resolution image. Similarly, a user with k_1 can generate k_2 and thereby decrypt two resolution layers to get the Med resolution, and a user with k_2 can only decrypt one resolution layer to get the Low resolution version of the image, as illustrated in Figure 10.3 from (Apostolopoulos *et al.* 2006). This brief discussion focused on a one-dimensional (1-D) hash chain for

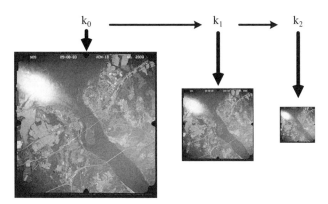

Figure 10.3 Multilevel access control: given one encrypted image, the user with key k_0 can access the high, medium, or low resolution images, the user with key k_1 can access the medium or low resolution images, while the user with key k_2 can only access the lowest resolution (see Plate 3)

key generation as it is one of the conceptually simplest approaches. However, there are a variety of straightforward extensions including general tree structures (including nonbinary, unbalanced, and M-D trees), which can provide very valuable flexibility and richness for access control.

In summary, there are a variety of applications where it is desirable to provide multiple levels of access control. In the approach described above, which exploits JPEG 2000 scalability features and JPSEC protection tools, it is possible for one copy of encrypted media content to provide multiple levels of access control where the access depends on the key available to the accessor.

10.4   Selective or Partial Encryption of Image Content

Much of the prior discussion has considered the case where the entire image is encrypted. There are also interesting applications where it may be beneficial to encrypt only a selected portion of the image. For example, JPSEC may be used to encrypt selectively different semantically meaningful portions of an image. This is illustrated via the following two examples. In Figure 10.4 portions of a JPEG 2000 image are encrypted to place a marking on the image. An end user without the key can still understand what is in the image and therefore decide whether to purchase the unmarked image. On the other hand, the image with the marking is clearly much less desirable than the unmarked image, and the marking cannot be eliminated without the loss of valuable information. If the end user purchases the image they will receive the key and can then decrypt the marked portion of the image to get the unmarked image.

In the second example, in Figure 10.5 from Apostolopoulos *et al.* (2006), a portion of the JPEG 2000 coded image is selectively left unencrypted, while the remaining portions are encrypted. An end user without the key can still see part of the image contents and therefore decide whether to purchase it or not. Note that once the end user can see the image on the left they have access to all of the compressed bit-stream and only need the key to gain access to the decoded image, i.e. only a small amount of additional information is required (determined by the key length).

Figure 10.4 The compressed bits corresponding to the image details at the spatial locations of the pattern (right) are encrypted. The marked image (left) is sufficient for understanding the image content and making the decision on whether to purchase the unmarked image

Another interesting note here is that these encrypted JPEG 2000 bit-streams were decoded using a JPEG 2000 decoder, and not a JPSEC decoder, i.e. the encrypted bit-streams in these examples were designed to be decoded usefully by a JPEG 2000 decoder, which does not have the key or have knowledge about what was encrypted. On the other hand, given the key a JPSEC decoder can recover the intended final image. This functionality is achieved by ensuring that the encrypted JPEG 2000 bit-stream corresponds to a valid (and hence decodable) JPEG 2000 bit-stream.

10.5 Image Authentication

Authentication is one of the basic security services, and this section begins by discussing some basic requirements of authentication for JPEG 2000 coded images. We then briefly

Figure 10.5 Selected spatial regions are left unencrypted while the remaining regions are encrypted. An end user can tell who is in the image and then determine if they would like to purchase the image. Note that a JPEG 2000 decoder (without the key) can decode the left image while a JPSEC decoder with the key can recover the right image

describe a unified digital signature based authentication system (ISO/IEC, 2007; Zhang *et al.*, 2004), which is able to protect JPEG 2000 images in different flavours, including complete authentication and semi-fragile authentication. The former is to protect the image at its code-stream level, while the latter is to protect the image at its content level. At the code-stream level, any modification in code-stream data will make the image unauthentic, while at the content level, only modifications that result in a change of perceivable content can make the image unauthentic. The content authentication can be further classified into lossy and lossless authentication. The original image can be recovered after verification when using lossless authentication, while the original image cannot be recovered after verification when using lossy authentication. In addition, we also describe some application scenarios where this system can be used to protect JPEG 2000 images. The whole system is compliant with the public key infrastructure (PKI) and can be incorporated into the current network security framework (Zhang *et al.*, 2004).

10.5.1 Motivation

Figure 10.6 shows a typical surveillance image. However, when we take a close look at pictures (a) and (b), we cannot directly tell which one is the true (unmanipulated) image if no security measures are applied at the time when the image is taken. Furthermore, with advances in intelligent surveillance, and to satisfy network bandwidth constraints and abide by privacy policies (Hata *et al.*, 2005), these systems may choose to transmit only a portion or processed version of the image or video instead of the entire image or video. For example, the systems may only transmit to the monitoring center the detected events (e.g. segment of the video in the time domain), or resized image frames (e.g. scaled down video in the spatial domain, as shown in Figure 10.7(a)), or the detected regions of interests (ROI, shown in Figure 10.7(b)). The application of these forms of intelligent processing in surveillance systems usually requires some robustness when designing a surveillance image authentication system. Specifically, as long as the image content (i.e. semantic meaning) is not changed, the authentication verification should be successful.

(a)   (b)

Figure 10.6 Examples of surveillance image: (a) true image and (b) attacked image by adding a walking person

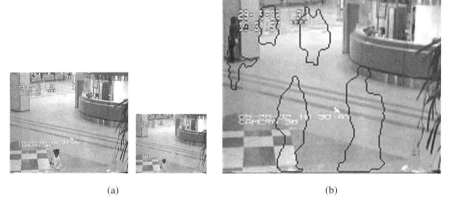

(a)   (b)

Figure 10.7 Transmitting surveillance images in different ways: (a) resizing and (b) ROI segmentation

In practice, the authentication should be robust to a number of operations or effects including: (1) image format conversion, e.g. between JPEG 2000 and JPEG; (2) image resizing and frame dropping due to bandwidth constraints; (3) segmentation of objects and background; and (4) resilience to packet loss due to unreliable network connections.

In the next subsection, we continue by introducing a robust and flexible JPEG 2000 image authentication scheme to meet most of the above-mentioned requirements.

10.5.2 A Unified Digital Signature Scheme for JPEG 2000

This subsection presents a unified authentication system for JPEG 2000. This authentication system integrates both complete and semi-fragile authentication. These two different authentication procedures complement each other, and hence this system can satisfy users with different requirements. If the JPEG 2000 image is exchanged in an environment where no incidental distortion is introduced, then complete authentication may be used. Otherwise, semi-fragile authentication may be preferred since it is more robust to incidental distortions.

In this system, we use a new parameter called the 'lowest authentication bit-rate' (LABR) to define a threshold quantitatively for acceptable transcoding operations (Sun *et al.*, 2002). For instance, if a JPEG 2000 image is protected at the LABR of 2 bpp (bits per pixel), then any transcoded version of the image shall be rendered as authentic as long as the bit-rate after transcoding is not lower than 2 bpp. Obviously, this parameter, LABR, is easier than the reference quantization step size for users to specify the largest acceptable lossy compression ratio (equivalently, the smallest acceptable bit-rate after compression).

In the authentication system, digital signature related data are embedded into the image when semi-fragile authentication is used. There are two methods for data embedding: lossy and lossless. With the lossy method, once data are embedded into an image, the original image can never be recovered. On the other hand, with the lossless method the data are embedded in a reversible way; i.e. the original image can be recovered if the protected image has not been altered. With both embedding methods, the semi-fragile authentication can support lossy-to-lossless authentication, which is consistent with lossy-to-lossless compression of the JPEG 2000 standard. Moreover, if the protected image has been modified then the semi-fragile authentication is able to identify the attacked locations.

10.5.2.1   General Overview

Figure 10.8 illustrates the unified authentication system for JPEG 2000 images. The left part is the encoder, while the right part is the decoder. The encoder accepts three sets of parameters: (1) encoding parameters (e.g. compression bit-rate), (2) original image to be encoded, and (3) authentication parameters (such as LABR, protected locations, and authentication methods). Depending on the specified authentication method, different authentication modules will be invoked while the image is being encoded. Specifically, if the complete authentication procedure is specified, the *complete sign* module is invoked to generate the digital signature from the code-stream. If the lossy semi-fragile authentication procedure is specified, the *semi-fragile (lossy) sign* module is invoked such that a digital signature is generated and the signature-related information is embedded into the image as a watermark in a lossy way. If the lossless semi-fragile authentication procedure is specified, the *semi-fragile (lossless) sign* module is invoked such that a digital signature is generated and signature-related information is embedded into the image in an invertible manner. The final outputs of the encoder are a JPSEC code-stream that consists of protected image data, digital signature, public key, and all necessary parameters for verification and decoding.

On the other end, a decoder accepts the JPSEC code-stream generated by the encoder. According to the specified authentication method signaled in the JPSEC code-stream, different verification modules (*complete verify, semi-fragile (lossy) verify*, or *semi-fragile*

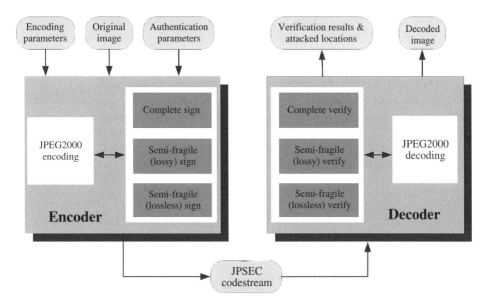

Figure 10.8   Unified JPEG 2000 authentication scheme

(lossless) verify) will be invoked. The final outputs of the decoder are the decoded image, verification results, and if the image has been attacked then the locations of the detected attacks. Furthermore, if the image is protected with the lossless authentication method, the decoded image will be exactly the same as the encoded image.

10.5.2.2 Complete Authentication

As defined in JPSEC, the complete authentication protects images at the code-stream level by directly applying a crypto/digital signature technique (e.g. DSA) on a specified JPEG 2000 data set. In the authentication system presented here, the complete authentication is able to exploit the rich structure of the JPEG 2000 code-stream; i.e. it can selectively protect any tile(s), layers(s), resolution(s), precinct(s), code-block(s), or arbitrary combinations of the above. According to JPEG 2000's rate-control structure (EBCOT), a bit-rate specified by the LABR is associated with a number of most important layers; i.e. we can use the LABR to specify the most important layers to be protected. In the extreme case when the LABR is equal to the full rate, all layers are protected. The protected image is able to pass the verification as long as the protected part of the code-stream remains intact.

To demonstrate the use of complete authentication, the JPEG 2000 standard test image 'woman' (512 × 640) is encoded with six resolutions and 4 bpp. Assume we want to protect only the resolutions from 0 to 3 of the image (and not resolutions 4 and 5); the LABR is set to 2 bpp, SHA-1 is used for hashing, and DSA is used for signature generation and verification (ISO/IEC, 2007). The first test case is to transcode the protected code-stream based on resolutions, and the second test case is to transcode the protected image based on bit-rates. Figure 10.9 shows the verification results of the transcoded code-stream. Since traditional cryptographic algorithms are applied directly to the protected code-stream, any bit change of the protected code-stream will make the image unauthentic. As shown in Figure 10.9(a), the verification results are authentic when the protected image was transcoded by discarding resolutions 4 and 5, which did not change any bit in the protected part, i.e. resolutions 0, 1, 2, or 3. If resolution 3 is discarded, as shown in Figure 10.9(b), the verification result becomes unauthentic. Similarly, Figures 10.9(c) and (d) show the verification result when the image is transcoded to different bit-rates. So long as the resultant bit rate is not less than the LABR, the verification result will be authentic. Otherwise, it is unauthentic. Note that because complete authentication directly applies cryptographic techniques on JPEG 2000 data, it does not allow even one bit change in the specified protection areas.

10.5.2.3 Semi-fragile Authentication

The semi-fragile authentication portion of the system consists of four parts: feature extraction, error correction coding (ECC) of features, signature generation, and data embedding. Semi-fragile authentication uses a set of features to characterize the content of the image, instead of using the code-stream as in the complete authentication. These features are extracted from the most significant bit-planes of wavelet coefficients. With the specified LABR value, the rate control process based on the EBCOT scheme of JPEG 2000 decides the number of most significant bit-planes from which features are extracted. Therefore, the extracted features closely representing the image content will be preserved if the image is transcoded to a bit-rate that is not lower than the LABR.

(a) Discarding resolution 4 and 5
leads to authentic

(b) Discarding resolution 3, 4 and
5 leads to unauthentic

(c) Transcoding to 2.5 bpp leads
to authentic

(d) Transcoding to 0.5 bpp leads
to unauthentic

Figure 10.9   Examples for complete authentication

To improve the robustness, i.e. to allow possible minor changes of the selected features caused by the transcoding operation, error correction coding (ECC) (Sun *et al.*, 2002) is used to protect the features, generating a set of codewords. All these codewords are concatenated into one bit-string, which is then used to generate a global signature using the traditional cryptographic algorithms such as RSA or DSA (ISO/IEC, 2007). As a result, the generated signature is robust against certain transcoding operations, as indicated by the specified LABR value.

In order to increase the robustness of the generated signature further, the parity check bits (PCBs) of each codeword are embedded into the local blocks of the image using lossy or lossless embedding methods. In ECC, the PCBs correspond to redundant bits that can help to correct a certain number of bit-errors. Therefore, embedding PCBs into the image will help the receiver to recover the original features, in case some incidental distortion is introduced. Furthermore, since PCBs are embedded into local blocks, embedding PCBs into the image can also help to locate the attacked areas, which correspond to those codewords with uncorrectable bit errors.

As mentioned before, two data embedding methods are used for semi-fragile authentication: lossy and lossless embedding (Zhang *et al.*, 2004). For lossy embedding, the selected bit-planes are used to embed the PCBs into those bit-planes that will remain in the bit-stream after applying transcoding, whose resultant bit-rate is equal to the LABR. Since PCBs embed data by replacing the lower bit-planes, it minimizes the perceived visual artifacts. In addition, the embedded data are robust to transcoding operations as long as the resultant bit-rate is not lower than the LABR. For lossless embedding, a subband is selected for data embedding according to the specified LABR. A larger LABR will lead us to choose a subband at the higher resolution level, and vice versa. After that, it divides the selected subband into 8×8 blocks, and embeds one bit in each block by slightly modifying half of the 64 coefficients. During the data extraction, all the modifications can be reversed to obtain the original coefficients.

The same test image is also used for semi-fragile authentication with lossy data embedding. The image is encoded using the (9, 7) DWT filter (ISO/IEC, 2000), LABR is set to 2 bpp, SHA-1 is used for hashing, and DSA is used for signature generation. Figure 10.10(a) and (b) show the original and the protected images, respectively. The PSNR of the protected image is 38 dB, and the human eye cannot perceive any artifact in the protected image. In addition, we also attacked the protected image as shown in

(a) Original "woman" image
(512 × 640)

(b) Protected with lossy semi-fragile authentication,
PSNR = 38 dB

(c) Attacked image

(d) Verification result of attacked image: unauthentic; Attacked area highlighted in red color

Figure 10.10 Examples of lossy embedding authentication under given LABR (see Plate 4)

(a) Image protected with lossless semi-fragile authentication. PSNR = 49 dB

(b) Recovered image after verification; PSNR = ∞

Figure 10.11 Examples of lossless embedding authentication under given LABR

Figure 10.10(c), where one of the beads on the woman's head is changed from blue color to green color, the number '0' is removed at the top-right corner, and the rubber band on her left wrist is changed from blue color to red color. Figure 10.10(d) illustrates the verification results, which declare the image unauthentic and approximately point out the attacked areas, as indicated by the highlighted areas.

For semi-fragile authentication with lossless data embedding, the 'woman image is encoded using the (5, 3) DWT filter, the LABR is set to 2 bpp, SHA-1 is used for hashing, and DSA is used for signature generation. Figure 10.11(a) shows the protected image (with PSNR = 49 dB) and Figure 10.11(b) shows the recovered image after verification (with PSNR = infinite).

10.5.3 Image Authentication for Surveillance

In the previous subsection, we have shown that the described JPEG 2000 image authentication is robust to format conversion and resizing (i.e. transcoding) with many flexible features. In this subsection, we briefly describe how the system also works seamlessly for object/ROI based systems.

The advanced intelligent video surveillance system can identify and segment the interested objects and then only transmit the segmented objects to the information center. In such cases, the typical attacks on object-based authentication is to change the background associated with the objects or substitute the segmented object(s) for another object(s), as shown in Figure 10.12. Therefore, building a secure link between the transmitting objects and its associated background stored at the center becomes an interesting topic in media authentication. Encoding the region of interest (ROI) is a new feature in the JPEG 2000 standard (ISO/IEC, 2000). The JPEG 2000 standard adopts the method of coefficient scaling, called the *Max-shift* algorithm for ROI coding. The basic idea is to identify and scale-up the magnitudes of those wavelet coefficients that belong to the ROI and those that belong to the background will be scaled down, before quantization and bit-plane

(a)   (b)

Figure 10.12   Object-based attacks on surveillance image

(a)   (b)   (c)   (d)

Figure 10.13   Authenticating ROI: (a) decoded with ROI image: 0.6 bpp, and (b) authentication result: 0.6 bpp with ROI (LABR: 0.5 bpp). Since there is no change within ROI, the image still passes verification: (c) decoded without ROI: 0.6 bpp and (d) authentication result: 0.6 bpp without ROI (LABR: 0.5 bpp)

entropy coding. Scaling-up the magnitudes results in putting more ROI-related coding bits into the lower bit-rate stream. As the previously described authentication scheme is controlled by LABR, it can naturally be applied for ROI-based JPEG 2000 coding.

Figure 10.13 illustrates authentication results for ROI-based authentication (Sun and Chang, 2005; Sun *et al.*, 2002). We define one rectangular ROI (left: 150, top: 240, width: 350, height: 400) and encode with the Max-shift method (ISO/IEC, 2000), as shown in Figure 10.13(a). Its target CBR is around 2.7 bpp while the LABR is set to 0.6 bpp. We then decode the ROI-embedded JPEG 2000 image in full rate (i.e. 2.7 bpp) and manipulate the image content as shown in Figure 10.13(c). Figure 10.13(a) shows the decoded image with a watermarking ROI model in 0.6 bpp. The authentication result is shown in Figure 10.13(b), where the image passes the authentication because the whole ROI is protected at 0.6 bpp. In comparison the no-ROI case (the CBR is still around 2.7 bpp and the LABR is set to 0.6 bpp but with no embedded ROI) in Figure 10.13(c) shows the decoded image with the watermarking non-ROI model at 0.6 bpp. Its corresponding authentication result is shown in Figure 10.13(d).

10.5.4   Summary for Image Authentication

This section described the need for image authentication and introduced a unified digital signature scheme for JPEG 2000 image authentication. The scheme can flexibly authenticate JPEG 2000 images in different modes: complete authentication, semi-fragile lossy authentication, and semi-fragile lossless authentication. This approach provides a significant amount of flexibility, as illustrated in the examples, which can be used to meet a range of requirements as needed by different image authentication applications and surveillance systems.

10.6   Summary

This chapter provided an overview of a number of important security applications for digital imagery, namely secure transcoding and secure streaming, and image authentication. As can be seen in this chapter, a wide range of potential image security applications exists that are very different from conventional data security applications, and future applications may lead to other creative combinations of JPEG 2000 compression and security. A further example of such applications is given in Chapter 11.

Acknowledgments

The authors would like to thank Susie Wee, Touradj Ebrahimi, and Zhishou Zhang for their valuable collaborations, without which the research that led to this chapter would not have occurred.

References

Apostolopoulos, J., Wee, S., Dufaux, F., Ebrahimi, T., Sun, Q. and Zhang, Z. (2006) The emerging JPEG-2000 security (JPSEC) standard, in *IEEE ISCAS*.

Dufaux, F., Wee, S., Apostolopoulos, J. and Ebrahimi, T. (2004). JPSEC for secure imaging in JPEG 2000, in SPIE Proc. Applications of Digital Image Processing XXVII, Denver, CO.

Hata, T., Kuwahara, N., Nozawa, T., Schwenke, D. L. and Vetro, A. (2005) Surveillance system with object-aware video transcoder, Technical Report TR-2005-115, MERL, www.merl.com, October 2005.

ISO/IEC (2000) Information Technology – JPEG 2000 Image Coding System – Part 1, ISO/IEC International Standard 15444-1, ITU Recommendation T.800.

ISO/IEC (2007) Information Technology – JPEG 2000 Security – Part 8, ISO/IEC International Standard 15444-8, ITU Recommendation T.800.

Java (2002) Cryptography Architecture API Specification and Reference, http://java.sun.com/j2se/1.4.2/docs/guide/security/CryptoSpec.html.

Kundar, D., Lin, C.-Y., Macq, B. and Yu, H. (eds) (2004) Special issue on enabling security technologies for digital rights management, in *Proceedings of the IEEE*.

Sun, Q. and Chang, S.-F. (2005) A secure and robust digital signature scheme for JPEG 2000 image authentication, IEEE Transactions on Multimedia, **7**(3), June, 480–494.

Sun, Q., Chang, S.-F., Kurato, M. and Suto, M. (2002) A quantitive semi-fragile JPEG 2000 image authentication system, in *IEEE International Conference on Image Processing*.

Wee, S. J. and Apostolopoulos, J. G. (2001a) Secure scalable video streaming for wireless networks, in *IEEE ICASSP*.

Wee, S. J. and Apostolopoulos, J. G. (2001b) Secure scalable streaming enabling transcoding without decryption, in *IEEE International Conference on Image Processing*.

Wee, S. J. and Apostolopoulos, J. G. (2003) Secure scalable streaming and secure transcoding with JPEG-2000, in *IEEE International Conference on Image Processing (ICIP)*.

Wee, S. J. and Apostolopoulos, J. G. (2004) Secure transcoding with JPSEC confidentiality and authentication, in *IEEE International Conference on Image Processing (ICIP)*.

Zeng, W., Yu, H. and Lin, C.-Y. (eds) (2006) *Multimedia Security Technologies for Digital Rights Management*, Academic Press, New York.

Zhang, Z., Qiu, G., Sun, Q., Lin, X., Ni, Z. and Shi, Y. Q. (2004) A unified authentication framework for JPEG 2000, in *IEEE International Conference on Multimedia and Expo*.

11

Video Surveillance and Defense Imaging

Touradj Ebrahimi and Frédéric Dufaux

11.1 Introduction

Video surveillance systems are becoming ubiquitous. They are widely deployed in many strategic places such as airports, banks, public transportation, or busy city centers. While people usually appreciate the sense of increased security brought by video surveillance, they often fear the loss of privacy that accompanies it. This legitimate concern often slows down the deployment of video surveillance. Another expected evolution is toward smart video surveillance systems. Indeed, video understanding by means of analysis techniques allows the detection of events in the scene and identification of regions of interest.

This information can then be successfully exploited, for instance to decrease the reliance on a security guard monitoring control screens and triggering an alarm in case of suspect situations, which is notoriously inefficient and unreliable.

In this chapter, we discuss the use of JPEG 2000 in a smart video surveillance system for security and defense applications. It can be said that one of the most important novelties of surveillance systems based on JPEG 2000 is that regions corresponding to privacy-sensitive information can be scrambled after detection. In particular, people under surveillance cannot be recognized, hence successfully addressing the loss of privacy issues, which are present in many applications and of concern in the deployment of surveillance systems in security and even defense applications. More specifically, a video analysis procedure identifies regions of interest in the scene by means of computer vision algorithms, such as change detection or face detection techniques. Scrambling is then applied to the corresponding regions of the video compressed by JPEG 2000. As a consequence, the scene remains understandable, but privacy-sensitive content such as people are unidentifiable. Furthermore, the amount of distortions in the protected video can be controlled

The JPEG 2000 Suite   Edited by Peter Schelkens, Athanassios Skodras and Touradj Ebrahimi
© 2009 John Wiley & Sons, Ltd

by applying the scrambling to some resolution levels only. In other words, a light scrambling (e.g. limited to high-resolution levels) will result in privacy-sensitive objects such as people appearing visible but sufficiently fuzzy to be unrecognizable, whereas a heavy scrambling (e.g. applied to all resolution levels) will result in replacing those regions by noise. The scrambling process depends on one or more encryption keys, which can be put in the possession of law-enforcement authorities or various trusted third parties. They are therefore the only ones able to unlock and view the whole scene clearly.

Privacy being a very important concern in video surveillance, this issue has been addressed by many researchers in various contexts. The use of JPEG 2000 and JPSEC facilitates an international standards compliant approach to protect privacy-sensitive information in video surveillance and defense applications. Such an approach is more efficient and practical when compared to alternatives such as systems that are based on an object-oriented representation of the scene, in which the system re-renders a modified video based on the end-user access control authorizations. During re-rendering, areas of the image are blanked out or replaced by computer graphics. The relevant information in the scene is therefore preserved, but privacy-sensitive details are not transmitted. Similarly, another alternative is to use a privacy buffer operating on incoming sensor data to prevent access to sensitive information or transform data to remove private information. These privacy filters are expressed using a privacy grammar, which is tedious to define and even more difficult to implement. Nevertheless, the concerns about the use of video surveillance are growing, and several reports have warned about the consequences and the threat to privacy brought by advanced video analytics such as face recognition techniques, which can automatically identify people in a video surveillance scene. Despite the fact that other algorithms have been proposed to de-identify faces, these approaches remain impractical and often undermine the reasons for which video surveillance is deployed in the first place.

The use of scrambling in video surveillance, however, offers a number of advantages and is therefore more appealing. Following such approaches, the camera outputs a single protected code-stream. This very same code-stream is transmitted to all clients regardless of their access control credentials. On the one hand, unauthorized clients who do not possess the private key(s) required for unscrambling the content can only view distorted versions of the content where private information is not identifiable. On the other hand, authorized clients, e.g. law-enforcement authorities, can unscramble the code-stream and recover the complete undistorted scene when duly justified. In addition, such scrambling approaches remain very flexible. They can be restricted to an arbitrary-shaped region of interest in the video, for instance the people in the scene. They can also be restricted to some resolution levels, hence controlling the amount of distortion introduced from merely fuzzy to completely noisy. Finally, compliance with JPEG 2000 and Secured JPEG 2000 (JPSEC) standards allows interoperability.

11.2 Scrambling

In this section we discuss the problem of scrambling regions of interest in images and video sequences. Scrambling is closely linked to the scheme used to compress image and video. Most coding schemes are based on transform coding; namely, images are transformed using an energy compaction transform such as the discrete cosine transform

(DCT) or wavelet transform. The resulting coefficients are then entropy coded using techniques such as Huffman or arithmetic coding. Basically, scrambling can be applied at three different stages: in the image domain prior to coding, in the transform domain during coding, or in the code-stream domain after coding. These approaches will be discussed in more detail hereafter.

The following features are important for an efficient solution. The scrambling should not entail lower coding performance or a significant increase in complexity. It should cope with arbitrary-shaped regions. Finally, it should be flexible, allowing for the adjustment of the amount of distortion introduced.

11.2.1 Image-Domain Scrambling

The first approach is to perform scrambling in the original image prior to encoding, as illustrated in Figure 11.1. This can be achieved, for instance, by randomly flipping the bits in one or more bit-planes of the pixels belonging to the regions of interest.

This approach has the advantage of being very simple and independent from the encoding scheme subsequently used. However, it has the disadvantage of significantly altering the statistics of the video signal, hence making the ensuing compression less efficient.

Note that the same effect could be achieved to some extent by masking the pixels corresponding to the regions of interest (e.g. replacing them by a solid color) or by applying a lowpass filter (e.g. making the region sufficiently blurred). However, these two approaches have the drawback that they preclude the possibility of ever unscrambling the video. In addition, the masking approach provides an all-or-nothing solution without flexibility to control the amount of distortion introduced.

11.2.2 Transform-Domain Scrambling

A second approach is to apply scrambling during encoding, as shown in Figure 11.2. Scrambling takes place after the DCT or wavelet transform and before entropy coding. More specifically, this can be done by randomly flipping the sign of the transform coefficients corresponding to the regions to be scrambled. Besides its simplicity, this approach does not adversely affect the subsequent entropy coding. Furthermore, thanks to the frequency analysis property of the transform, the strength of the scrambling can be controlled by restricting the scrambling to certain frequencies.

Another benefit of this approach is that it preserves the syntax of the code-stream, e.g. maintaining standard compliance. This enables content adaptation or transcoding at mid-network nodes or proxies, as is often required in a multimedia delivery system.

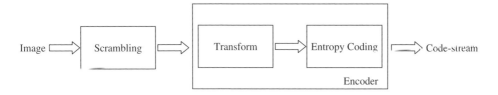

Figure 11.1 Image-domain scrambling

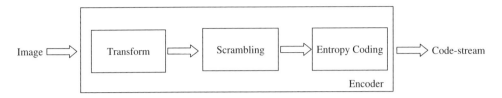

Figure 11.2   Transform-domain scrambling

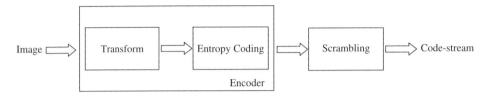

Figure 11.3   Code-stream-domain scrambling

11.2.3   Code-Stream-Domain Scrambling

In the third approach, scrambling is applied after encoding, as illustrated in Figure 11.3. More specifically, the compressed code-stream is directly scrambled. Again, this can be efficiently done by randomly flipping bits in the stream.

One of the drawbacks of this approach is that the code-stream has to be parsed in order to identify which parts correspond to the regions to be scrambled, hence entailing a larger computational complexity. Furthermore, another severe drawback is that it may be difficult to guarantee that the scrambled code-stream will not crash a decoder. Finally, the strength of the scrambling cannot be easily adjusted.

11.3   Overview of a Typical Video Surveillance System

A high-level description of a typical video surveillance system is illustrated in Figure 11.4. A number of smart surveillance cameras are positioned so as to cover the area under surveillance. Each camera processes the captured video sequence in order to analyze, encode, and secure it. The resulting code-stream is then transmitted over an IP network. The network can either be wired or wireless. The latter case is especially appealing as it makes it very easy to deploy and relocate cameras. The central server stores the code-streams received from the various cameras, along with corresponding metadata information from an eventual video analysis module (e.g. events detection). Based on this metadata information, the server can decide to trigger alarms and to archive the sequences corresponding to events. Heterogeneous clients can access the server in order to monitor the live or archived video surveillance sequences. In the case of compression by JPEG 2000, as the code-stream is scalable, the server is able to adapt the resolution and bandwidth of the delivered video depending on the performance and characteristics of the client and its network connection. If the video compression used is not scalable, e.g. if it is based on MPEG-4 or AVC/H.264, transcoding is required to adapt the compressed bit-stream. Again, delivery can be through a wired or wireless network. The second case

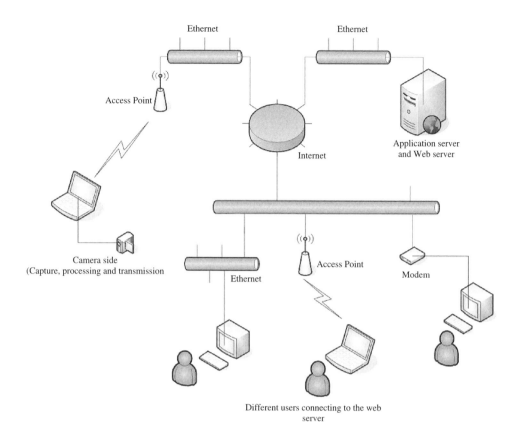

Figure 11.4   Typical video surveillance system architecture

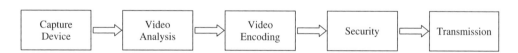

Figure 11.5   Processing steps in a typical smart camera

allows for mobile clients to access the system. For instance, policemen or security guards can be equipped with laptops or PDAs while on patrol.

The typical video processing taking place in the smart surveillance camera is depicted in Figure 11.5. The video content is first analyzed. This step detects the occurrence of events in the scene (e.g. intrusion, presence of people). Typical algorithms used are based on change detection and face detection methods. In most situations, and for the sake of low complexity, the change detection is a simple frame difference algorithm using a previously stored background. In order to handle changes of illumination in the scene, the background is gradually updated in some situations. Moreover, in order to smooth and clean up the resulting segmentation mask, various postprocessing operations can be applied. The detection of the presence of people in the scene is one of the most relevant pieces of information a video surveillance system can convey.

Unless specified otherwise, in the rest of this chapter we assume that the video sequence is compressed using JPEG 2000. Furthermore, if fully or partially encrypted, we assume that such an operation is performed in compliance with JPSEC specifications. Indeed, JPEG 2000 is well suited for video surveillance applications, and several solutions are already offered in products and services based on this standard. First, even though JPEG 2000 may lead to inferior coding performance compared to an interframe coding scheme such as MPEG-4 or AVC/H.264, intraframe coding allows for easy browsing and random access in the encoded video sequence, requires lower complexity in the encoder, and is more robust to transmission errors in an error-prone network environment. Moreover, JPEG 2000 intraframe coding outperforms previous intraframe coding schemes such as JPEG, and achieves a sufficient quality for a video surveillance system. JPEG 2000 also offers two very useful functionalities. It supports regions of interest coding, enabling significant foreground objects in the scene to be encoded with higher quality. Furthermore, the resulting code-stream is fully scalable, both in resolution and quality. This property is very essential in video surveillance applications when clients with different computational resources and characteristics have to access the system.

In addition, two extensions of the baseline JPEG 2000 are also particularly interesting for video surveillance and defense applications, namely Secured JPEG 2000 (JPSEC) and Wireless JPEG 2000 (JPWL), as discussed in other chapters in this book. JPSEC defines an open framework for secure imaging supporting techniques, such as conditional access control, source authentication, or data integrity verification. The JPSEC conditional access control technique can be used to scramble the whole video or sensitive regions corresponding to people and goods in the scene, as discussed in more detail in the previous section. JPWL defines additional mechanisms to achieve the efficient transmission of JPEG 2000 imagery over an error-prone network. When surveillance cameras are connected through IP networks, and especially when in the presence of a wireless network, JPWL can be used in order to make the code-stream more robust to transmission errors.

11.4 Overview of a Video Surveillance System Based on JPEG 2000 and ROI Scrambling

In this section, an example of a smart video surveillance using JPEG 2000 is described. In particular, the system relies on JPEG 2000 compression to allow for compression of captured video, and employs a scrambling technique based on the region-of-interest (ROI) feature of JPEG 2000 in order to ensure privacy.

11.4.1 Explicit Region of Interest (Max-shift)

The Max-shift method is an explicit approach for the ROI. More precisely, the ROI mask is specified in the wavelet domain. At the encoder side, a scale factor of 2^s is determined such as to be larger than the magnitude of any background wavelet coefficients. All coefficients belonging to the background are then scaled down by this factor, which is equivalent to shifting them down by s bits. As a result, all nonzero ROI coefficients are guaranteed to be larger than the largest background coefficient, as illustrated in Figure 11.6. All

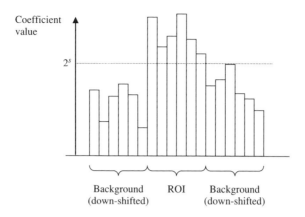

Figure 11.6   Max-shift ROI coding

the wavelet coefficients are then entropy coded and the value s is also included in the code-stream. At the decoder side, the wavelet coefficients are entropy decoded, and those with a value smaller than 2^s are shifted up by s bits.

The Max-shift method is therefore an efficient way to convey the shape of the foreground region without having to actually transmit additional shape information. Note also that this method supports multiple arbitrary-shaped ROIs. Another consequence of the method is that coefficients corresponding to ROI are prioritized in the code-stream so that they are received before the background at the decoder side. This is an interesting feature for video surveillance situations, as the ROI often correspond to the most interesting and salient part of the scene. A drawback of the approach is that the transmission of any background information is delayed, resulting in a sometimes undesirable all-or-nothing behavior at the low bit-rate. However, this drawback is less critical in video surveillance as the background corresponds to less interesting portions of the scene under surveillance.

11.4.2   Implicit Region of Interest

Another implicit approach can also be used for ROI coding. The JPEG 2000 code-stream is composed of a number of quality layers, where every layer can include a contribution from each code-block. This contribution is usually determined during rate control based on the distortion estimates associated with each code-block.

An ROI can therefore be implicitly defined by upscaling the distortion estimate of the code-blocks corresponding to this region. As a result, a larger contribution will be included from these respective code-blocks.

Note that, in this approach, the code-stream does not contain explicit ROI information. The decoder merely decodes the code-stream and is not even aware that a ROI has been used. One disadvantage of this approach is that the ROI is defined on a code-block basis, as opposed to the approach presented in the previous section. However, this drawback is again less critical in video surveillance applications, as a very precise outline of the ROI is less critical.

11.4.3 Scrambling of ROIs

In this section we discuss how the scrambling of a ROI can be achieved in order to pre-serve privacy. The approach is compatible with both mechanisms of maxshift ROI and implicit ROI. In the discussion below, we consider a video sequence with an associated segmentation mask, produced by a given video analysis component present in the video surveillance system. The specific algorithm to generate the mask is not critical in the following steps, and can be a change detection, skin detection, face detection, or license plate detection method, among others. We assume that the foreground objects (the result of the video analysis) contain privacy-sensitive information and have to be scrambled. Con-versely, the background is of lower relevance and can be transmitted without scrambling and eventually with an inferior quality.

A block diagram illustrating the encoding and scrambling process is shown in Figure 11.7. Basically, this technique adds a pseudo-random noise in parts of the JPEG 2000 code-stream corresponding to the regions to be scrambled. Authorized users who know the pseudo-random sequence can easily remove the noise. Conversely, unautho-rized users do not know how to remove this noise and have only access to a distorted image.

In order for the decoder side to receive a low-resolution version of the background with-out delay, the implicit ROI method is used to prioritize all the code-blocks from lower resolution levels. In particular, the purpose of this stage is to circumvent the all-or-nothing

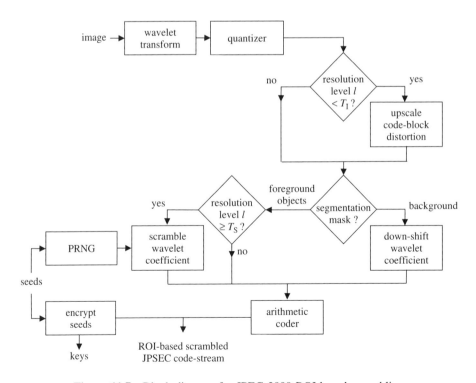

Figure 11.7   Block diagram for JPEG 2000 ROI-based scrambling

behavior characteristic of the maxshift method, as explained earlier. For this purpose, a threshold T_I (with $T_I = 0, 1, 2, \ldots$) is defined so that code-blocks belonging to the resolution level l are incorporated in the ROI if $l < T_I$. This is achieved by upscaling the distortion estimate for these code-blocks.

The segmentation mask is then used to classify wavelet coefficients to the background or foreground. Also, a second threshold T_S (with $T_S = 0, 1, 2, \ldots$) is defined in order to control the strength of the scrambling. At this stage, the maxshift ROI method is used to convey the background/foreground segmentation information. Accordingly, coefficients belonging to the background are downshifted by s bits, where s is determined such that the scale factor 2^s is larger than the magnitude of any background wavelet coefficients. Conversely, coefficients corresponding to the foreground and belonging to resolution level l are scrambled if $l \geq T_S$. Remaining foreground coefficients are unchanged.

The scrambling relies on a pseudo-random number generator (PRNG) driven by a seed value. For the sake of simplicity and low complexity, the scrambling consists in pseudo-randomly inverting the sign of selected coefficients. Note that this method modifies only the most significant bit-plane of the coefficients. Hence, it does not change the magnitude of the coefficients, therefore preserving the maxshift ROI information. The sign flipping takes place as follows. For each coefficient, a new pseudo-random value is generated and compared with a density threshold. If the pseudo-random value is greater than the threshold, the sign is inverted; otherwise the sign is unchanged. In order to improve the security of the system, the seed can be changed frequently. To communicate the seed values to authorized users, they are encrypted and inserted in the JPEG 2000 code-stream based on JPSEC specifications. Hence, the resulting code-stream can be decoded by any JPEG 2000 compliant decoder to visualize the scrambled video, and can be unscrambled with any JPSEC compliant decoder.

11.4.4   Unscrambling of ROIs

At the decoder side, the following operations are carried out, as illustrated in Figure 11.8. The decoder receives the ROI-based scrambled JPSEC code-stream, including the value s used for maxshift, the encrypted seeds for PRNG, and the threshold T_S. The wavelet coefficients are first entropy decoded. The coefficients with a value smaller than 2^s are classified as background. As they have not been scrambled, it is sufficient to simply shift them up by s bits in order to recover their correct values.

The remaining coefficients correspond to the foreground and those belonging to resolution level $l \geq T_S$, which are scrambled. On the one hand, unauthorized users do not have possession of the keys. Therefore, they cannot decrypt the seeds nor reproduce the sequence of pseudo-random numbers and consequently are unable to unscramble these coefficients. To them, the decoded image will appear distorted. On the other hand, authorized users can reproduce the same sequence of pseudo-random numbers as used during encoding. They are therefore able to unscramble these coefficients and to see the unprotected image.

Note that the use of the implicit ROI to prioritize code-blocks corresponding to the background and belonging to low-resolution levels is transparent to the decoder. Therefore any JPEG 2000 compliant decoder will be able to decode the scrambled video, whereas

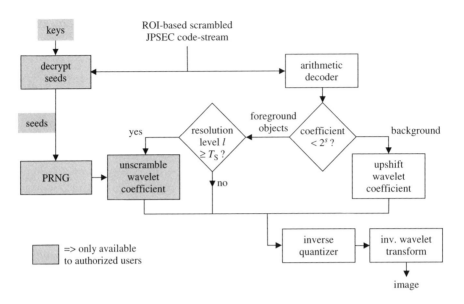

Figure 11.8   Block diagram for decoding

only JPSEC compliant decoders can proceed with unscrambling if the encryption key for scrambling is provided.

11.4.5   Experimental Results

In this section, we present some results obtained with by the ROI-based scrambling described above, in order to allow readers to assess its perfomance better. The hall monitor video sequence in CIF format is used along with a ground-truth segmentation mask, as shown in Figure 11.9.

Figure 11.10 shows the flexibility of the scrambling technique to adjust the amount of distortion introduced by varying T_S. More specifically, with a heavy scrambling ($T_S = 0$),

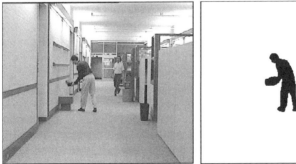

Figure 11.9   Hall monitor: original frame and corresponding segmentation mask

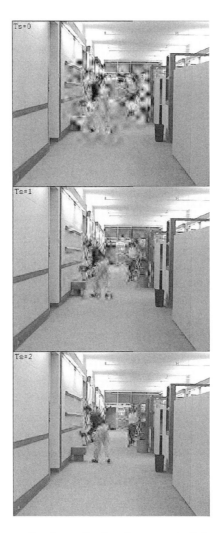

Figure 11.10 Scrambling results when adjusting the amount of distortion: from top to bottom, $T_S = 0$, 1, 2 ($T_I = 0$ and rate $= 4$ bpp)

the foreground is replaced by noise, whereas with a medium or light scrambling ($T_S = 1$ or 2), the people in the scene are still visible but are too fuzzy to be recognizable.

Figure 11.11 shows the importance of simultaneously considering both the explicit (maxshift) and implicit ROI mechanisms, as discussed before. When using solely the maxshift method ($T_I = 0$), the foreground objects are completely transmitted before the decoder receives background information. At a low bit-rate, this results in an all-or-nothing behavior. By allowing for implicit ROI ($T_I = 1$ or 2), all the code-blocks from lower resolution levels (level 0 for $T_I = 1$, levels 0 and 1 for $T_I = 2$) are included in the ROI even though those belonging to the background are not scrambled. Consequently, a low-resolution version of the background is received without any delay.

Figure 11.11 Scrambling results when including a low-resolution background in ROI: from top to bottom, $T_I = 0$, 1, 2 ($T_S = 0$ and rate = 0.75 bpp)

In Figure 11.12, one can see comparison results between ROI-based scrambling with a technique performing scrambling on a code-block basis. In the latter case, the shape of the scrambled region is restricted to match code-block boundaries. This becomes a significant drawback in the case of small arbitrary-shaped regions, as can be observed, but may be still acceptable for video surveillance situations when *a priori* one expects only large foreground objects. Indeed, with 32×32 code-blocks, the scrambled region is significantly larger than the foreground mask. This drawback is slightly alleviated with smaller 16×16 or 8×8 code-blocks. However, the use of a smaller code-block size is detrimental to both compression performance and computational complexity. In contrast, with the ROI-based scrambling technique discussed previously, the scrambled region matches fairly well the foreground mask, independently from the code-block size.

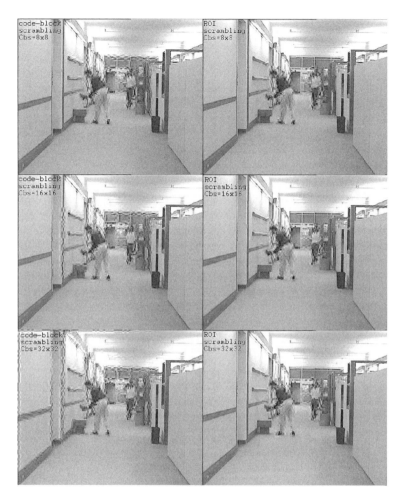

Figure 11.12 Comparison between scrambling on a code-block basis and an arbitrary ROI-based scrambling with code-block size of 8×8, 16×16, or 32×32 ($T_I = 0$, $T_S = 2$ and rate = 4 bpp)

Figure 11.13 Typical results for heavy scrambling at a rate of 4 bpp or 0.75 bpp ($T_I = 2$, $T_S = 0$)

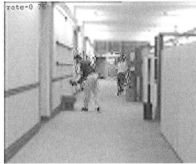

Figure 11.14   Typical results for light scrambling at a rate of 4 bpp or 0.75 bpp ($T_I = 2$, $T_S = 2$)

In the light of the above results and to summarize our observation, it appears that $T_I = 2$ is a suitable threshold to include low-resolution background information in the ROI, $T_S = 0$ leads to heavy scrambling, and $T_S = 2$ is suitable for light scrambling. Typical heavy and light scrambling results at high and low bit-rates can be seen in Figures 11.13 and 11.14.

Further Reading

Dufaux, F. and Ebrahimi, T. (2004) Video surveillance using JPEG 2000, in *Proceedings of SPIE on Applications of Digital Image Processing XXVII*, Denver, CO, August 2004.

Dufaux, F. and Nicholson, D. (2004) JPWL: JPEG 2000 for wireless applications, in *Proceedings of SPIE on Applications of Digital Image Processing XXVII*, Denver, CO, August 2004.

Dufaux, F., Wee, S., Apostolopoulos, J. and Ebrahimi, T. (2004) JPSEC for secure imaging in JPEG 2000, in *Proceedings of SPIE on Applications of Digital Image Processing XXVII*, Denver, CO, August 2004.

Fidaleo, D. A., Nguyen, H.-A. and Trivedi, M. (2004) The networked sensor tapestry (NeST): a privacy enhanced software architecture for interactive analysis of data in video-sensor networks, in *Proceedings of the ACM 2nd International Workshop on Video Surveillance and Sensor Networks*, New York, 2004.

Hampapur, A., Brown, L., Connell, J., Pankanti, S., Senior, A. W. and Tian, Y.-L. (2003) Smart surveillance: applications, technologies and implications, in *IEEE Pacific-Rim Conference on Multimedia*, Singapore, December 2003.

ISO/IEC (2004) JPEG 2000 Part 8 (JPSEC) FCD, ISO/IEC JTC1/SC29 WG1N3480, November 2004.

Lienhart, R., Kuranov, A. and Pisarevski, V. (2002) Empirical Analysis of Detection Cascades of Boosted Classifiers for Rapid Object Detection, MRL Technical Report, Intel Labs, December 2002.

Newton, E., Sweeney, L. and Malin, B. (2003) Preserving Privacy by De-identifying Facial Images, Technical Report CMU-CS-03-119, Carnegie-Mellon University, Pittsburgh, PA.

Rivest, R. L., Shamir, A. and Adleman, L. M. (1978) A method for obtaining digital signatures and public-key cryptosystems, *Communications of the ACM*, **21**(2), 120–126.

Senior, A. W., Pankanti, S., Hampapur, A., Brown, L., Tian, Y.-L. and Ekin, A. (2003) Blinkering Surveillance: Enabling Video Privacy through Computer Vision, IBM Technical Report RC22886.

Skodras, A., Christopoulos, C. and Ebrahimi, T. (2001) The JPEG 2000 still image compression standard, *IEEE Signal Processing Magazine*, **18**(5), September, 36–58.

Taubman, D. and Marcellin, M. (2002) *JPEG 2000: Image Compression Fundamentals, Standards and Practice*, Kluwer Academic Publishers, Boston, MA.

Viola, P. and Jones, M. (2001) Rapid object detection using a boosted cascade of simple features, in *IEEE Proceedings of CVPR*, Hawaii, December 2001.

12

JPEG 2000 Application in GIS and Remote Sensing

Bernard Brower, Robert Fiete, and Roddy Shuler

12.1 Introduction

The Geographic Information Systems (GIS) and remote sensing community has been an early adopter of JPEG 2000. This chapter reviews the history of GIS and remote sensing, makes recommendations for encoding remote sensing data using the JPEG 2000 Part 1 standard, and finally describes GIS and remote sensing applications for other parts of the JPEG 2000 family of standards.

12.2 Geographic Information Systems

Geographic Information Systems (GIS) is more than map making; it is the ability to organize and quickly visualize information about a geographical location. The very first GIS applications involved just mapping the location of shores, cities, and roads. Mapping, in some form, has been around for more than two thousand years. Mapping was traditionally performed though physically surveying the land and water. It was not until the 1800s that aerial photography and remote sensing began to impact mapping. The use of remote sensing enabled information to be assembled more quickly and with greater accuracy than ever before.

GIS applications evolved with the use of remote sensing to focus on providing information beyond merely geographical location, for applications such as military (troop strengths and movements), natural resources, agriculture, and game and wildlife management. The digital revolution for GIS started in the 1960s with the development of the Synagraphic Mapping System (SYMAP) by Howard Fisher of the Harvard Laboratory

The JPEG 2000 Suite   Edited by Peter Schelkens, Athanassios Skodras and Touradj Ebrahimi
© 2009 John Wiley & Sons, Ltd

for Computer Graphics and Spatial Analysis and the start of the Environmental Science Research Institute by Jack and Laura Dangermond (Gupta, 2008).

The combination of the digital information in graphics forums and the digital remote sensing imagery GIS applications has become more powerful than ever. Government organizations were the first to adopt this technology, for applications like city planning, infrastructure planning, emergencies and disaster relief, crime mapping, agriculture, natural resource, and game/wildlife management. Next, the business sector began to use GIS information for applications such as store location planning, cellular tower site selection, target marketing, population trends, and locating manufacturing and distribution centers. In 1996, MapQuest® brought GIS applications to the masses with the first consumer-based interactive mapping application. Internet mapping applications continue to grow with the addition of Yahoo!® Maps, Google Maps™, Google Earth™, and Microsoft Live Search Maps™.

These systems continue to evolve so that individual users can add information to geographical locations. The information grows exponentially from street names to street number locations, to business locations, to information about that business. For example, these systems have evolved from finding a street to mapping the direction from the user's house to a given restaurant, to the restaurant's menu, to reviews from other users, to personal information limited to private users. These systems not only use the remote sensing imagery to produce information but now enable all users to view and interact with the remote sensing data.

Figure 12.1 Earliest existing aerial photograph, taken in 1860 by J. Wallace Black from a balloon above Boston

12.2.1   What Is Remote Sensing?

Remote sensing is generally defined as a method for collecting information about an object without making direct contact with that object. We humans are 'remote sensors' with our ability to sense acoustic (sound), chemical (smell), and radiometric (sight) information about objects without touching them. The most common use of the term today is for aerial and space-based systems that collect images of the Earth, specifically by detecting and measuring electromagnetic radiation.

12.2.2   History of Remote Sensing

After the invention of photography by Nicéphore Niépce and Louis Jacques Mandé Daguerre in the early 1800s, the first known overhead images were taken by Gaspard Felix Tournachon, also known as 'Nadar,' from a hot air balloon 1200 feet (370 m) over Paris in 1859. Unfortunately, none of his original images survived. The earliest surviving overhead image was taken by J. Wallace Black in 1860 (Figure 12.1).

The development of lighter weight cameras using roll film by George Eastman in 1888 allowed further experimentation with collecting overhead images, including the use of kites and even pigeons (Figure 12.2). The invention of the airplane in 1903 made

Figure 12.2   Carrier pigeons were used to collect overhead images in Germany in 1909

Figure 12.3 Photoreconnaissance proved useful for mapping trench networks during World War I, such as this image over Fort Douaumont near Verdun, France

overhead imaging more practical than with kites, pigeons, or hot air balloons. During World War I, France and Germany used airplanes for overhead reconnaissance but quickly realized that verbal reports were often wrong or exaggerated. Capturing overhead images with cameras not only provided accurate information on enemy troops but also proved valuable for making maps of the trenches (Figure 12.3).

Although Alfred Maul patented a gyroscopically stabilized camera for rockets in 1904, rockets were not capable of taking cameras into orbit until the advancement of rocket technology in the 1950s. The first image of the Earth from space was transmitted from Explorer VI, launched on August 7, 1959. Figure 12.4 shows an image of the northern Pacific Ocean taken by Explorer VI on August 14, 1959, from an altitude of 27 000 km. Although crude, it demonstrated the capability of photographing the Earth's cloud cover using a camera in space.

NASA developed the TIROS Program (Television and Infrared Observation Satellite) as the world's first meteorological satellite. TIROS-1 was launched on April 4, 1960, and consisted of two television cameras. Although it collected data for only 78 days, it successfully demonstrated the utility of meteorological satellites. The first television picture taken from space by TIROS-1 is shown in Figure 12.5.

Figure 12.4   First image of the Earth from space, taken by Explorer VI on August 14, 1959 (image courtesy USGS and NASA)

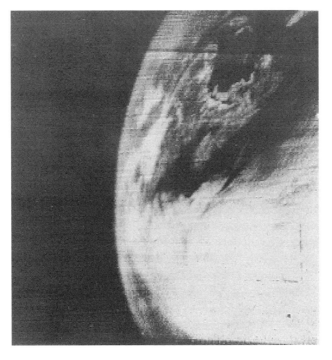

Figure 12.5   First TIROS-1 image and the first television picture taken from space on April 1, 1960 (image courtesy of NASA)

Figure 12.6   First RBV image from ERTS 1 (Landsat 1) over Dallas, Texas, in 1972 at a resolution of 80 meters (image courtesy of NASA)

With the success of the weather satellites, NASA developed a series of satellites to collect scientific data on the Earth's resources. The Earth Resources Survey (ERS) Program was developed in 1965 to develop space-based methods for monitoring Earth resources on a systematic, repetitive basis. The first satellite, the Earth Resources Technology Satellite (ERTS), was launched on July 23, 1972 and placed in a near-polar orbit at an altitude of 970 km. With the launch of ERTS 1, remote sensing as we know it today was born. Images were collected using a return beam vidicon (RBV) and a multispectral scanner (MSS) to collect more precise spectral data for agricultural studies. Figure 12.6 shows the first RBV image from ERTS 1. In 1975 NASA changed the name of the ERTS satellite to Landsat. Landsat 7, launched in 1999 to an altitude of approximately 705 km, has the capability to collect images at a 15 meter resolution. Figure 12.7 shows a Landsat 7 image of NASA's Kennedy Space Center in 1999.

Space Imaging's IKONOS satellite, launched on September 24, 1999, was the first commercial satellite to capture imagery of the Earth at a resolution of one meter. Figure 12.8 shows an IKONOS image of Washington, DC, acquired on September 30, 1999, from an altitude of 680 km. DigitalGlobe's QuickBird satellite, launched on October 18, 2001, captures images at 0.6 meter resolution, and their WorldView-1 satellite, launched on September 18, 2007, captures images at 0.5 meter resolution. GeoEye's GeoEye-1 satellite, launched on September 6, 2008, captures images at 0.41 meter resolution.

12.2.3   Sensor Types

Remote sensing systems are generally categorized by the wavelength region of the electromagnetic spectrum that they image. The earliest cameras used black and white film

Figure 12.7 Landsat 7 image of the NASA Kennedy Space Center in 1999 at a resolution of 15 meters (image courtesy of EROS Data Center) (see Plate 5)

that captured panchromatic images, i.e. images corresponding to the visible portion of the electromagnetic spectrum, approximately 0.4 to 0.7 μm. The highest resolution images collected are typically black and white images over the visible spectrum, because the light is captured over a broad spectrum of wavelengths, thus allowing a high enough signal to collect at high resolution. Visible imaging systems generally acquire images using the sun as the illumination source.

Multispectral images collect several images within different narrow bands of the electromagnetic spectrum, but collectively span a broader portion of the spectrum. The most common multispectral image is a true-color image, where three images are collected with one imaging the red portion of the visible spectrum, another imaging the green portion of the spectrum, and the other imaging the blue portion of the spectrum. When these three images are viewed together on a color monitor, a true-color image is produced.

Hyperspectral images are collected with a much finer spectral resolution than multispectral images. A multispectral collection may divide up a spectral region into a few bands, whereas a hyperspectral collection may divide up the same spectral region into tens to hundreds of bands. Each pixel in a hyperspectral image contains spectral characteristics about the surface material imaged within that pixel. Hyperspectral imagery

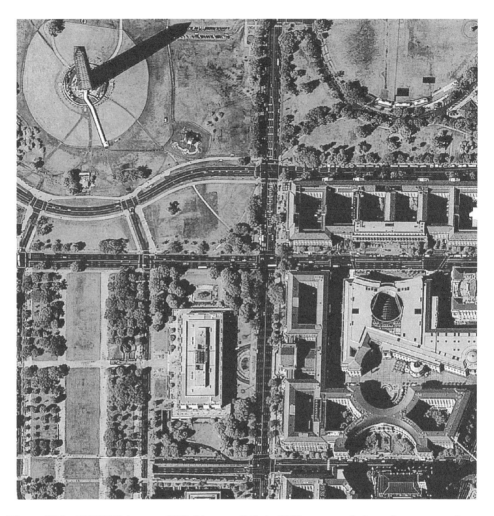

Figure 12.8 IKONOS image of Washington, DC, in 1999 at a resolution of one meter (image courtesy of GeoEye)

is collected to help identify and discriminate different materials in the scene and is generally used with machine exploitation algorithms. Figure 12.9 shows a hyperspectral cube of Moffet Field, California, where the spectral images are stacked in order of the spectral wavelength so that the z axis shows the spectrum of each pixel. The hyperspectral data were collected using NASA's AVIRIS (airborne visible/infrared imaging spectrometer) sensor, which has 224 channels from 0.4 μm to 2.5 μm, each with a spectral bandwidth of approximately 10 nm.

Infrared images are collected using sensors that detect the electromagnetic spectrum between 0.7 μm and 100 μm, outside the range visible to the human eye. The infrared band is further divided into the NIR (near-infrared) band, spanning between 0.7 μm and 1.4 μm; the SWIR (short-wave infrared) band, spanning between 1.4 μm and 3 μm; the MWIR (mid-wave infrared) band, spanning between 3 μm and 8 μm; the LWIR

Figure 12.9   Hyperspectral cube of Moffet Field, California, taken on August 20, 1992, using the AVIRIS sensor (image courtesy of NASA JPL) (see Plate 6)

(long-wave infrared) band, spanning between 8 μm and 15 μm; and the FIR (far infrared) band, spanning between 15 μm and 1000 μm. The Earth emits radiation primarily at wavelengths between 3 μm and 20 μm; this region is typically referred to as thermal infrared and can be used to acquire images that give information regarding temperature changes on the Earth. Figure 12.10 shows an example of a thermal IR image from NASA's HCMM (Heat Capacity Mapping Mission) satellite. Note that the water is brighter due to higher heat capacitance of water compared to land.

SAR (synthetic aperture radar) imaging emits high-frequency radio signals and converts the returned signal into imagery. The motion of the platform is used to synthesize a larger antenna and produce higher resolution imagery. Since SAR supplies its own illumination source and the radio waves are not scattered by clouds, SAR systems can acquire images during any time of the day or night, and under cloudy conditions. This makes SAR systems ideal for mapping the surface of a planet; hence a SAR system was used by the Magellan spacecraft to map the surface of Venus from 1990 to 1994. Figure 12.11 shows an SAR image of Washington, DC, taken from the space shuttle using the SIR-C/X-SAR system.

LIDAR (light detection and ranging) systems have been developed recently to provide detailed topographic information. A LIDAR system typically uses a scanning mirror to

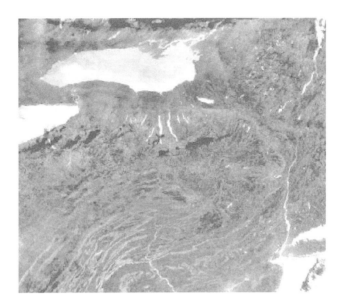

Figure 12.10   Thermal IR image of New York from the HCMM satellite (image courtesy of NASA)

Figure 12.11   Image of Washington, DC, taken with the SIR-C/X-SAR system from the space shuttle (image courtesy of NASA JPL)

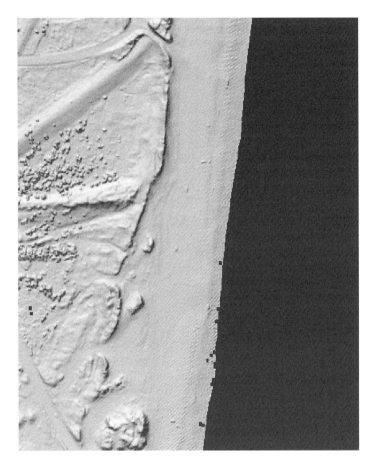

Figure 12.12   LIDAR image of the Outer Banks, North Carolina, at 1 meter resolution (image courtesy of NOAA)

project a pulse of laser light on to the ground. A receiver measures the time for the transmission and the return of the pulse, giving an accurate measure of the range between the instrument and the ground. LIDAR systems require instruments, such as a global positioning system (GPS) and an inertial measurement unit (IMU), to determine the absolute position and orientation of the LIDAR sensor at all times during the data collection. Figure 12.12 shows an example of a three-dimensional terrain image generated from LIDAR data.

12.2.4   Applications of Remote Sensing Data

Remote sensing has become critical for many different disciplines, such as meteorology, archeology, forestry, geology, monitoring the environment, and city planning. Of growing importance is the ability of remote sensing data to aid in disaster relief. Figure 12.13 shows a three-dimensional (3-D) image of the lower Manhattan generated by LIDAR collections after the September 11, 2001 terrorist attacks. These data showed crews where

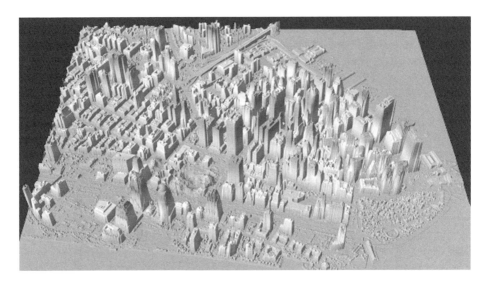

Figure 12.13 Terrain elevation image of lower Manhattan after the September 11, 2001 terrorist attack (image courtesy of NOAA/U.S. Army JPSD) (see Plate 7)

to pinpoint their recovery efforts by locating foundation support structures, elevator shafts, and basement storage areas.

Fusing data from different remote collection systems can provide information that may be unobtainable or difficult to interpret with each sensor alone. Figure 12.14 shows an example of combining topographical information derived from an LIDAR sensor with the visual information provided by a panchromatic sensor.

Figure 12.14 Airborne LIDAR data fused with an IKONOS image of Mt St Helens volcano (see Plate 8)

Two challenges facing the remote sensing community are accurate georeferencing and the ability to handle large amounts of data. The ability to fuse or extract geospatial information from remote sensing data of the Earth requires accurate georeferencing. Ever-increasing amounts of remote sensing data are available today from current airborne and satellite platforms and from historical collections as well. By the mid-1990s, over 100×10^{12} bits per day were being collected from Earth-observing satellite systems. Since then, more systems are flying and collecting more data than ever before. For example, the QuickBird satellite alone, launched in 2001, collects approximately 2×10^{12} bits per day, while the WorldView-1 satellite, launched in 2007, collects over 5×10^{12} bits per day. The current and archived data volumes associated with remote sensing systems are growing dramatically each year, and the industry faces growing challenges with handling and processing this vast amount of data.

12.2.5   JPEG 2000 for GIS and Remote Sensing Applications

The GIS and remote sensing community was an early adopter of the JPEG 2000 standards. Several different proprietary compression algorithms were emerging before the adoption of JPEG 2000. These proprietary products were primarily developed to meet the requirements of the image sizes, bit-depths and number of components (bands) that were not satisfied by baseline JPEG DCT, and they gained popularity because of their functionality. Most of these techniques were based on wavelet compression, which enabled the resolution scalability that is very important for the very large images of remote sensing. The capability of JPEG 2000 to compress very large, multiple-component, greater bit-depth images makes it an obvious open standard solution. While these capabilities meet the minimum requirements to handle the remote sensing images, it is the functionality that has made JPEG 2000 successful in the GIS and remote sensing markets. The scalability in resolution and quality and the region of interest accessibility are key enablers for users to exploit the GIS and remote sensing information.

12.2.6   JPEG 2000 Scalability and Access

The true power of JPEG 2000 is the flexibility and scalability of the code-stream. Scalability in JPEG 2000 comes in two main forms: resolution scalability and image quality scalability. The building blocks of JPEG 2000 are the wavelet transform, the bit-plane arithmetic encoding, and the embedded bit-stream. The pyramidal nature of the wavelet transform eliminates the need for secondary reduced resolution data set generation; i.e. direct access to reduced resolutions is achieved without the overhead to manage and store multiple representations of the image. The bit-plane arithmetic encoder of JPEG 2000 enables image quality scalability (also referred to as layering). A single JPEG 2000 file can provide direct access to various quality increments ranging from lossless (approximately 2:1 compression) to 100:1 (or greater) lossy compression. The JPEG 2000 code-stream syntax and embedded bit-stream allow access to both the resolution and quality scalability while enabling fast region of interest (ROI) access at a given quality and resolution.

Resolution scalability is important when viewing very large remote sensing images. High-quality displays are typically limited to four megapixels or less, while the remote sensing images are commonly greater than 100 megapixels. The only efficient way to review these images is to access a subset of the data at the desired resolution and region of interest. With a JPEG 2000 code-stream, the built-in reduced resolution data could be used for quick searching or preview of a scene (contextual overview). When needed, high-resolution data at a given ROI can be retrieved and displayed. A client application can simply (and smartly) choose which resolution should be used given a compressed JPEG 2000 image or Motion JPEG 2000 video sequence, without any additional processing costs.

Quality scalability within JPEG 2000 allows remote sensing images to be disseminated to multiple users with different quality requirements or bandwidth limitations. For example, a user performing algorithmic evaluation of data may require full 'numerically lossless' quality, while another user may require only 'visually lossless' quality at 4:1 compression or 'high-quality' data at 10:1 compression.

Fast region-of-interest access is useful for transmission and decoding purposes. JPEG 2000 enables access to regions of an image at a desired resolution and quality. Region-of-interest access reduces memory, bandwidth, and processing requirements of users, since only the data that fulfills the display requirements of the users must be transmitted and decoded.

12.3 Recommendations for JPEG 2000 Encoding

The following sections discuss critical parameter selection and recommendations that should be considered when encoding remote sensing imagery.

12.3.1 Wavelet Filters

Two different wavelet transforms are described in ISO/IEC 15444-1 Annex F. One is a reversible 5-3 (5-3R) integer wavelet transform that allows for lossless encoding/decoding whenever all bits in a compressed file are received. The other wavelet transform is an irreversible 9-7 (9-7I) transform. As shown in Section 1.2.2, the 9-7I is the more efficient filter and produces better image quality versus bit-rate than the 5-3R filter. Therefore, the 9-7I filter is recommended for remote sensing applications that are based on visual analysis of the compressed imagery.

There are many remote sensing applications, however, that are based on radiometric accuracy. Multispectral and hyperspectral data often require numerically lossless compression, because these data types are commonly used for algorithm and machine exploitation. Algorithmic applications like band ratios, classification, and spectral matching all require radiometric accurate date. Infrared data are commonly used to measure the temperature of objects (e.g. water temperature), and any inaccuracy of the compressed data directly affects the accuracy of the temperature measurements. Similarly, inaccuracy in LIDAR data may reduce the accuracy of height measurements. On remote sensing imagery types such as these that are used for radiometric exploitation, the requirement for extreme numerical accuracy dictates the use of numerically lossless compression that can only be achieved with the 5-3R filter.

Table 12.1 illustrates the impact of filter selection on image quality for various compression ratios, using a representative remote sensing image (the 8-bit test image 'AERIAL2.tif' from the JPEG committee, Figure 12.15). Numerical quality is reported in PSNR, where $PSNR = 20 \log_{10}(RMSE/255)$.

Table 12.1 Impact of the wavelet filter on image quality for AERIAL2.tif

Compression ratio	5-3R filter PSNR (dB)	9-7I filter PSNR (dB)
2:1 (4.0 bpp)	46.72	48.96
4:1 (2.0 bpp)	37.48	38.13
8:1 (1.0 bpp)	32.77	33.27
16:1 (0.5 bpp)	30.08	30.64
32:1 (0.25 bpp)	28.06	28.59
64:1 (0.125 bpp)	26.09	26.51
128:1 (0.0625 bpp)	24.30	24.62
256:1 (0.03125 bpp)	22.71	22.99

Figure 12.15 JPEG committee test image 'AERIAL2.tif'

12.3.2 Resolution Levels

The wavelet decompositions enable access to the data at reduced resolution by powers of two: the original, 2× reduction in each dimension, 4× reduction, 8× reduction, and so forth, where the total number of available resolutions is one more than the number of stages of wavelet decomposition. For example, consider a one gigapixel square image (32 768 pixels × 32 768 pixels). Such an image would require at least five decomposition

Table 12.2 The reduced resolution image size as a function of decomposition level

Decomposition level (reduction)	Lines (pixels)	Samples (pixels)	Image resolution (megapixels)
R0 (original)	32 768	32 768	1024
R1 (2×)	16 384	16 384	256
R2 (4×)	8192	8192	64
R3 (8×)	4096	4096	16
R4 (16×)	2048	2048	4
R5 (32×)	1024	1024	1
R6 (64×)	512	512	0.25
R7 (128×)	256	256	0.0625
R8 (256×)	128	128	0.015625

levels in order for the full extents of the lowest resolution representation to fit on a one megapixel display screen. This image would require three additional decomposition levels to produce a 128 × 128 overview thumbnail. For this example with eight decomposition levels, the resulting reduced resolution image sizes for each decomposition level are shown in Table 12.2. (Note that in this table, an upper-case 'R' refers to the decomposition level, where R0 is the original resolution, using nomenclature that is common in remote sensing applications. This is in contrast to the lower-case 'r' that refers to the resolution level in the JPEG 2000 standard, where r0 is the lowest resolution.)

As shown in Section 1.5.1.4, in general the compression ratio increases with more decomposition levels. Typically, the optimal compression efficiency is achieved when the size of the lowest resolution representation of a single tile is approximately the size of a code-block. For example, when encoding an image with a tile size of 1024 × 1024 and a code-block size of 64 × 64, the optimal compression efficiency is expected with four decomposition levels. After the first three levels of wavelet decomposition, however, the impact on compression efficiency for each additional level is minimal. Thus, selection of the number of decomposition levels is typically driven by the requirements for access to the lower resolution images. For large remote sensing images, it is common to use eight or nine (or more) decomposition levels to enable access to a relatively small overview thumbnail image.

12.3.3 Compression Ratio, Quality Layers, and Rate Control

Remote sensing, like medical imagery, was developed for the information that is gathered from the imagery. The scientists and image analysts are mainly focused on the exploitation quality or radiometric accuracy of the data and less on the aesthetic appearance of the imagery. With this in mind, the data should always be archived with enough quality to ensure that even the most discerning user will be satisfied with the image quality. Imagery types from which users expect to extract accurate radiometric information (e.g. multispectral, hyperspectral, and infrared imagery) should be compressed in such a manner that the original image may be exactly reconstructed (numerically lossless). Imagery that is used mainly for visual inspection (e.g. high-resolution panchromatic, color, and pan-sharpened imagery) should be stored at a compression ratio that results in quality that

Table 12.3 Impact of quality layers on the lossless compression ratio

Number of quality layers	Lossless bit-rate (bpp)	Compression ratio
1	5.4411	1.4703:1
2	5.4453	1.4692:1
5	5.4536	1.4669:1
10	5.4588	1.4655:1
20	5.4608	1.4650:1
50	5.4621	1.4646:1

is visually indistinguishable from the original (visually lossless). The archival quality of remote sensing data is important for long-term studies that evaluate scene changes over time, such as global warming and deforestation.

With the increased access to GIS and remote sensing data through Internet applications like Google Earth™ and ImageAtlas™, the most prevalent use of these data is no longer scientific and does not have the strict quality requirements. Given the limited bandwidth available for this mass access of the very large remote sensing images, the data may be delivered to these less discerning users at higher compression ratios. Each application and user may have different quality requirements or limits that warrant different compression ratios. Both the high-quality requirements for the scientists and the different compression ratio requirements for the common users may be achieved in a single JPEG 2000 compressed representation through the use of multiple quality layers. These layers should include the very highest quality required for the scientific users (either numerically lossless or visually lossless) and several layers at different visual quality to meet the various needs of the common users. The number of quality layers does not significantly impact the overall compression ratio, especially with larger images. Table 12.3 illustrates the impact of the number of quality layers on the lossless compression ratio for the 2048 × 2048 test image AERIAL2.tif:

The compression ratios/bit-rates for the quality layers can be allocated based on the requirements of the users, and the following recommendations can help guide this selection:

- First, quality layers should be considered to meet specific system quality requirements and/or bit-rate limitations. For instance, if the network bandwidth is limited, there may be a specific bit rate that will meet a user's quality requirements without exceeding some bandwidth allocation. Alternatively, based on the available memory or processing power, a user may be unable to effectively use data beyond some specific bit-rate. The use of quality layers enables data providers or libraries to access quickly the data that is required or requested by a user. For example, if a user only requires 1.0 bpp, the data provider may parse and deliver the data by stripping all of the quality layers above 1.0 bpp, without the need to decode and re-encode the data for that user. For the LRCP progression order (as described in Section 12.3.6), this is accomplished through a simple truncation of each tile's data at the desired quality layer.

- Second, the layers should be more finely spaced at the higher compression ratios (lower bit-rates), because the slope of the compression ratio to quality is steeper at the lower bit-rates; i.e. at lower bit-rates, a small increment in bit-rate has more impact on quality

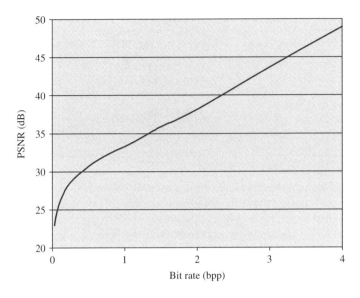

Figure 12.16 Graph of image quality versus bit-rate

than the same incremental change at higher bit-rates. For example, consider the plot of image quality versus bit-rate for the AERIAL2.tif test image (Figure 12.16). This shows a 1.47 dB difference in PSNR between 0.5 bpp and 0.75 bpp, and a larger PSNR difference of 2.06 dB between 0.25 bpp and 0.5 bpp. In order to achieve a fine progression in quality, the layers must be spaced closer together at the lower bit-rates.

- Third, very low quality layers should be considered that are correlated to the delivery of reduced resolution imagery. The compaction of the energy in the compression process ensures that most of the data required to produce a high quality reconstruction of a lower resolution is in a relatively small number of bits. Table 12.4 illustrates the difference in reduced resolution reconstruction quality at various quality layers versus the lossless reconstruction of all layers (using the AERIAL2.tif test image with the 5-3R wavelet). The difference in quality from one layer to the next is not very significant,

Table 12.4 RMSE of a lossless image of R5, R4, and R3 images at reduced quality versus all of the layers

Layer bit rate (bpp)	RMSE at R3	RMSE at R4	RMSE at R5
5.45 (lossless)	0.0000	0.0000	0.0000
4.0	0.5946	0.0000	0.0000
2.0	1.2000	0.4644	0.0000
1.0	1.6376	0.9165	0.0000
0.5	2.5543	1.4688	0.6984
0.25	4.5193	2.4515	1.1952
0.125	7.4521	3.6834	2.2811
0.0625	12.7801	5.4955	2.6616

especially at the lowest resolution. For the LRCP progression order, the use of reduced quality layers enables faster access to the low-resolution/overview imagery, because the server/decoder has less data to sift through when seeking the desired resolution level.[1]

- Finally, for interactive client-server applications, even if the user requests a high-quality representation, intermediate layers help optimize the order in which data are delivered to the client, such that higher quality is reconstructed sooner when only a subset of the requested data has been received. This is especially important over lower band-width connections. For this purpose, a typical scheme is to provide layers with roughly logarithmically spaced bit-rates with two to three layers per octave. For example, a numerically lossless image with 15 quality layers may include intermediate layers at the following target bit rates (in addition to the lossless layer): 4.0, 2.8, 2.0, 1.4, 1.0, 0.71, 0.50, 0.35, 0.25, 0.18, 0.13, 0.088, 0.063, and 0.044 bpp.

The mechanism by which the compressed bytes are allocated to each quality layer is known as rate control. The JPEG 2000 standard leaves this matter up to the implementer, but it is recommended that a postcompression rate-distortion optimization method should be used to optimize the quality achieved at each layer based on an analysis of the rate (number of bytes) versus distortion (image quality) contribution of each coding pass from each code-block. For tiled imagery, this rate control optimization may be performed independently for each tile or globally across all tiles of the image (or somewhere in between). Individual tile rate control allocates the same number of bytes to each tile regardless of that tile's scene content. This allows every tile to be layered and processed independently. If the scene content between neighboring tiles varies dramatically, there may be noticeable image quality differences (such as visible tile boundaries) between adjacent tiles within a given layer, especially at very high compression ratios. Full image rate control, on the other hand, produces a more uniform image quality by allowing busy tiles (those with fine detail) to contribute more bytes to the compressed file at the expense of less busy tiles. The aggregate bit-rate for the entire image is still maintained, but the tile-by-tile bit-rate may vary. In general, full image rate control consumes more memory during compression since all the compressed data from each tile must be retained until the final layer is formed. If overall image quality is the main goal, rate control using the full image (or as much of the image as can be processed given the compressor's memory constraints) is preferred. However, individual tile rate control may be preferred if the goal is that any chipped tile (or layer therein) will have a predetermined size.

Typically, the goal of rate control is to assign the best possible quality at a predetermined bit-rate for each layer. An alternate paradigm for rate control may be employed for applications where constant quality across images is more important than a constant bit-rate. In this case, the target for each layer is a slope on the rate-distortion curve for each code-block rather than a number of bytes. This approach can greatly simplify processing requirements, as it allows each code-block to be independently analyzed for layering decisions, and full image rate control is achieved without retaining the entire image in memory. It also simplifies the development of heuristics for early termination of the bit-plane encoding, as discussed in Section 12.3.8.8, and may lead to higher processing throughput. This approach may be especially useful for video applications, where the target

---

[1] This does not apply to layer-last progression orders such as RPCL that are recommended for interactive applications.

slopes could be modified from one frame to another if too much or too little data is being generated. The downside of slope-based allocation is that the resulting tile and file sizes are indeterminate, which may not be acceptable for some remote sensing applications.

12.3.4   Quantization

When compressing with the 9-7I wavelet transform, there is another mechanism that impacts the achieved bit-rate and quality of the resulting image. After the wavelet transform, the noninteger wavelet coefficients must be quantized into the integer values that are bit-plane encoded. For each subband, a quantization step size controls how finely the quantized bins are spaced within the original coefficient space. The wavelet coefficients are effectively divided by the quantization step size. The larger the step size, the coarser is the quantization and thus the fewer bit-planes and less data generated by the MQ encoder. The quantization step sizes used during compression are signaled in the codestream headers so that a decoder can undo the effects of quantization.

Most compressor implementations provide a means of specifying a base step size that drives the calculation of quantization step sizes for each subband. This offers a simple control mechanism to adjust the number of bytes generated by the bit-plane encoder via a single value. Depending on how this base step size is specified, it may be necessary to adjust it based on the sample bit-depth of the imagery (e.g. by a factor of ($1/(2^{\wedge}$ bit_depth))) in order to achieve the full quality for higher bit-depths. When specifying just a single base step size, the implementation typically will scale this value for each subband in order to maximize numerical image quality. For some applications such as visual weighting (see Section 12.3.8.3), it may be preferable to specify quantization step sizes independently for each subband.

A common option is to choose a relatively fine quantization step size and rely on postcompression rate-distortion optimization to achieve control over the target bit-rate, as described in Section 12.3.3. Another option is to vary the quantization step size based on the expected image statistics to control the amount of data generated by the bit-plane encoding. This approach can lead to constant quality across an entire image or series of images. Furthermore, it typically results in slightly better image quality than that achieved by finer quantization with truncation to the same bit-rate. It also has the benefit that excess bit-planes are not arithmetically encoded only to be discarded during the rate control step, and thus can optimize processing throughput. However, if the actual image statistics are significantly different than expected, the resulting file sizes may be significantly larger or smaller than desired. In many situations, the best approach may be to combine quantization step size selection based on image statistics with postcompression rate-distortion optimization to generate quality layers (and optionally enforce an upper limit on the achieved bit-rate).

12.3.5   Tiles and Precincts

JPEG 2000 allows for two fundamental methods of encoding the data to enable region-of-interest (ROI) access to the data. The first is a classic preprocessing method that spatially segments the image into separate smaller images or tiles. These tiles are then independently transformed and encoded, which enables fast access to the region of interest based on access to the tile or tiles that are included in the given ROI. The second method is to segment the compressed data into partitions known as precincts. Precincts are similar to tiles in that they provide segmentation of the image into regions that can be accessed by the user. The main difference is that segmentation into precincts is performed in the wavelet transformed domain, whereas tiling is performed in the original pixel domain. Tiles and precincts are not mutually exclusive; i.e. images can be partitioned first into tiles, and then each tile can be partitioned into multiple precincts per resolution after performing the wavelet transform. For example, an image compressed with a relatively large tile size (such as 4096 × 4096) may be further broken down into precincts to achieve faster and more efficient access to regions of interest. There are advantages and disadvantages to each of these methods.

Breaking the image into independent tiles enables easy and efficient parallel processing for compression and decompression. Because the tiles are transformed and encoded independently, the tiles can be pulled out (or 'chipped') to form a new image that is a spatial subset of the original image, without any need to decode and re-encode. For example, if a user would like to keep a 4096 × 2096 image that is within a much larger image (e.g. 16 384 × 16 384), a set of eight 1024 × 1024 tiles may be extracted without decoding and re-encoding the data. A new image header and updated tile headers can represent the result as a standalone 4096 × 2096 image (rather than a 16 384 × 16 384 image with several missing tiles). The new header may represent the new image offsets so as to retain its location relative to the original image grid, or it may represent the image at the upper-left origin of a new image grid (subject to certain restrictions such as nominal tile dimensions that are integer powers of two).

For a given image, all tiles have the same nominal size. Tiles may be rectangular in dimension and spatially offset relative to the top-left corner of the image. The image size need not be an integer multiple of the tile width and tile height. This leaves open the possibility that the tiles on the image border may be of reduced size in the vertical or horizontal dimension and possibly both. These incomplete tiles are not padded, but rather they are clipped to the image border. In order to maximize the possibilities for parallel processing and chipping, both the nominal tile width and height must be integer powers of 2 and no smaller than 2 (number of decompositions). As shown in Section 1.5.1.1, tiles need to be at least 256 × 256 so as not to impact compression efficiency. For practical considerations, however, tile sizes should typically be 1024 × 1024 or greater. Otherwise, the overhead of managing tiles can become a heavy burden on processing, especially when decoding low-resolution thumbnail images. Table 12.5 shows the impact of tile size upon processing a 1024 × 1024 overview from an original 32 768 × 32 768 image.

Table 12.5 Impact of tile size on a 1024 × 1024 overview from a 32 768 × 32 768 image

Tile size	Number of tiles	Pixels per tile at reduced resolution
8192 × 8192	16	65 536
4096 × 4096	64	16 384
2048 × 2048	256	4096
1024 × 1024	1024	1024
512 × 512	4096	256
256 × 256	16 384	64

This table highlights how the number of tiles that need to be accessed for a low-resolution overview increases exponentially as the tile size decreases. Although the time spent strictly decoding pixels is relatively unchanged, when the tile size is small enough the decoding time becomes overshadowed by the overhead required to access the tiles, process their header information, and set up the tile decoding machinery. This added overhead can lead to a significant reduction in the speed of displaying these data. Thus, for remote sensing images, it is common to consider even larger tile sizes like 4096×4096 or 8192×8192 (or even a single tile for the entire image, using just precincts to partition the data into regions of interest).

Precincts have been shown to be better for browsing large images versus tiles because they offer more accurate selection and dissemination of a given region of interest. As discussed, tile size strictly decreases by a power of two for each successively lower resolution level. Efficient access to lower resolutions thus requires very large tiles, but this limits the usefulness of tiles to provide ROI access at higher resolutions. Precincts, on the other hand, do not suffer from this limitation. The precinct size may be specified independently for each resolution, avoiding the need to access an excessive number of precincts in generating a low-resolution overview. Typically, optimal interactive browsing may be achieved by using relatively small precincts of the same size across all resolutions. For very large images, however, one may consider larger precincts for the highest resolutions in order to reduce the total number of precincts that need to be managed. For example, one could use 256×256 precincts for the highest two resolutions and 128×128 precincts for all lower resolutions.

Even if an image was not originally compressed using relatively small precincts, some dissemination tools may transcode these data 'on the fly' into smaller precincts to more efficiently disseminate the data to the users. Such a transcoding step could be used, for instance, to deliver small regions of interest from a tiled image with maximal precincts. It might also be used, however, to break existing precincts into even smaller partitions, operating 'just in time' on only those original precincts that intersect with a region of interest without the overhead required to manage the large number of precincts that may be present if the image were initially compressed with the smaller precincts. By choosing precinct sizes that are at least twice the height and width of the code-block size,[2] this transcoding step can be relatively quickly performed without any MQ decoding/encoding or wavelet transform (though packet headers do need to be decoded and re-encoded). Since the code-block size is constrained by the precinct boundaries, if a smaller precinct were chosen the code-block dimensions would need to be modified, requiring arithmetic decoding and re-encoding (though still no wavelet transform would be necessary).

The parameters of tiles versus precincts and quality layers can be selected or optimized for several different applications. As discussed in the next section, it is important to match these parameters with the progression order to ensure the achievement of goals. For most GIS and remote sensing applications, the data are tiled so that parallel processing and chipping can be achieved. As the processing power and memory capabilities of computers

---

[2] Or at least as large as the code-block size for the lowest resolution LL subband. Recall that precinct dimensions are specified with regard to the resolution, whereas code-block sizes are defined in the subband space. Thus, for all subbands other than the LL, a precinct maps to a region in the subband space that is half the precinct width and half the precinct height.

continue to increase, tile sizes are likely to increase, and the use of precincts will become the predominant mechanism to provide access to regions of interest.

12.3.6 Code-Stream Organization

Much of the utility of JPEG 2000 derives from the manner in which the compressed code-stream is organized to facilitate incremental access to the data necessary to perform specific tasks. The JPEG 2000 standard defines five choices of progression order for organizing packets of data within a tile: LRCP, RLCP, RPCL, PCRL, and CPRL (where L = layer, R = resolution, C = component, and P = position), as discussed in Section 1.3.6. The choice of progression order can have profound impacts on the efficiency of access to regions of interest.

For many applications, layer-last progressions (RPCL, PCRL, and CPRL) are preferred. These progressions keep the code-stream contributions to each precinct in a single contiguous segment. For the decompressor, this facilitates the efficient management of code-block data that are fed into the MQ decoder. With either of the non-layer-last progressions (LRCP and RLCP), the contributions to a code-block from various layers are scattered and need to be reassembled as they are input to the MQ decoder. Furthermore, when used in conjunction with PLT (or PLM) marker segments (see Section 12.3.7), an application can easily index a layer-last code-stream by maintaining a single pointer per precinct into the original data representation. Although PLT-based indexing is possible with non-layer-last progressions, it is less practical due to the multiplication of the amount of indexing information needed by the number of layers.

Of these layer-last progressions, RPCL is typically the preferred order for interactive viewing of large images. This assumes that the typical scenario is to start with a low-resolution overview display of the entire image extents, with the user progressively zooming in to one or more regions of interest. The RPCL progression places all the low resolution information together at the front of the tile for fast access to the overview image. Next, within a resolution, precincts are contiguous, facilitating access to the higher resolution regions of interest. Individual components (spectral bands) are not contiguous across the tile; instead, all components corresponding to a given region are collocated by precinct, which is appropriate for viewing visual images where all the color components (e.g. red, green, and blue) are required to reconstruct each pixel of the region.

Other progressions may be preferred for specific applications. For instance, an application that merely chips out tiles of the original image at a specific quality layer may prefer LRCP progression, where the highest order index within the tile is the layer. Multispectral applications that process a single component (spectral band) at a time may prefer CPRL progression (although RPCL may be preferred if using the Part 2 multicomponent transform, which performs a third dimension of transformation across multiple components at each pixel location). Finally, though perhaps the least intuitive of the available progression orders, PCRL may be useful for compressing huge untiled images. Its position-major ordering enables an incremental flushing of the compressed data to file as full precincts are completed, while processing the input data in raster order from top to bottom. To take full advantage of this incremental flushing, the precinct sizes need to decrease in height for each successively lower resolution. The resulting code-stream may not be as efficient for interactive viewing as RPCL, but it may be necessary to avoid excessive memory usage problems on the encoder. (Note that this memory usage problem can also be avoided by using tiles and flushing the compressed data to file as each row of tiles is completed.)

With very large, heavily tiled imagery, choosing an appropriate progression order such as RPCL alone may not be sufficient to enable efficient interactive viewing. Even if data within a tile are ordered by resolution, viewing an overview of the entire image will require reading a small amount of data from the start of each tile. If the tiles are organized sequentially in the file, reconstruction of an overview image will require reading a little bit of data from the first tile, followed by a disk seek and reading a little bit of data from the second tile, etc. File systems are very inefficient at reading many small scattered pieces of information, and thus there is a significant latency involved in reading the relatively small amount of data used in reconstructing the overview image. This problem can often be solved by breaking each tile into multiple tile parts and interleaving the tile parts from all tiles in a fashion that is optimal for a particular application. For example, progression by resolution across the entire file may be achieved using the following progression of tile parts: $T_0R_0, T_1R_0, T_2R_0, \ldots, T_{N-1}R_0, T_0R_1, T_1R_1, T_2R_1, \ldots, T_{N-1}R_1$, etc. In this case, the tradeoff with introducing multiple tile parts is that accessing an individual tile of the full-resolution image requires seeking to discrete segments of the file for each resolution. Since very large images may have many, many tiles but typically only a few resolutions, however, the penalty for access to the full-resolution tiles is trivial compared to the savings for access to the low-resolution overview.

The good news is that a JPEG 2000 code-stream can be easily reordered from one progression order to another. Since this does not require any arithmetic decoding/encoding or wavelet transform, there is absolutely no loss of image fidelity, and the operation may be implemented with very low computational complexity. An excellent example of this is illustrated by the two compression profiles recommended by the Basic Image Interchange Format (BIIF) profile for JPEG 2000 (BPJ2K). The North Atlantic Treaty Organization (NATO) Secondary Imagery Format (NSIF) Preferred JPEG 2000 Encoding (NPJE) profile recommends that original image providers compress with a single tile part per tile in LRCP progression. This involves a straightforward implementation of the compressor, whereby individual tiles can easily be processed in parallel and simply concatenated into the full contiguous code-stream. However, the NPJE profile is a very poor choice for interactively viewing large images. In particular, access to the low-resolution overview is very inefficient, since the lowest resolution contributions are scattered throughout the entire image file – both within each tile and across all the tiles of the image. To work around this problem, the BIIF profile prescribes use of an Exploitation Preferred JPEG 2000 Encoding (EPJE) for images that are submitted to their libraries for dissemination to interactive viewers. The EPJE profile organizes the data in a resolution-first progression (RLCP),[3] with one tile part per resolution per tile, such that all the lowest resolution contributions from all tiles are at the front of the file, followed by the next resolution from all tiles in the image, etc. With this profile, reconstruction of the overview image requires reading only a relatively small contiguous segment of data at the start of the file and is very efficient.

Typically, system requirements can be met using one of these five basic progression orders (possibly with multiple tile parts per tile), but the JPEG 2000 standard offers

---

[3] Although RLCP was selected for the EPJE profile, later experiments have demonstrated that RPCL would offer even better performance, at least for viewers who fully exploit the benefits of layer-last progression orders.

the option of changing the progression order for specific resolutions, components, and/or layers. These changes are signaled via POC (progression order change) marker segments in either the main or tile-part headers. For instance, a progression order change could be used to present the lowest resolution in RPCL progression, followed by all higher resolutions in LRCP progression, to provide a compromise of fast access to the lowest resolution overview with easy extraction of specific quality layers for the higher resolution data.

12.3.7   Pointer Marker Segments

Suppose that an optimum progression order has been chosen for a certain application – does this imply that all data accesses will be fast and efficient? Not necessarily. As a simple example, consider an application that merely chips out full tiles from an original JPEG 2000 file with one tile part per tile presented in raster order from top-left to bottom-right. If the tile of interest happens to be the upper-left tile, we can simply parse the main header and then our file pointer will be right at the SOT marker segment for this first tile in the file. From the SOT, we can easily parse the length of the tile and copy it to the output target. However what if the tile of interest happens to be in the middle of the image or, worse yet, the lower-right corner of the image? The JPEG 2000 standard allows for an optional TLM marker segment that describes the length of each tile part in the main header. With this information, an application can quickly determine the offset to the tile of interest and seek directly to the appropriate location in the file. Without the TLM marker segment, the application would need to read the SOT for the first tile, parse its length, and seek to the SOT of the second tile; this process of read a few bytes and then seek would be repeated for every tile until the desired one is reached. Although very little processing is involved, this is very inefficient for the file system and would lead to significant latency in accessing later tiles in heavily tiled images.

On the other hand, consider a large single-tile image that makes judicious use of precincts and is organized in the RPCL progression order. Accessing the low-resolution overview from this file would be fairly straightforward, but what if the user zooms in to a high-resolution region of interest? To address this concern, the JPEG 2000 standard allows for optional PLM or PLT marker segments that describe the length of each packet in either the main header (PLM) or tile-part headers (PLT). With this information, an application can quickly determine the offset to each packet that contributes to the region of interest and seek directly to the appropriate locations in the file. Without the PLM or PLT marker segments, the application would need to decode each packet header sequentially in order to determine the sum of the lengths of the compressed image data contributed by each code-block to this packet; after fully decoding the packet header, the application can seek over the compressed image data and process the next packet header, repeating the process until the desired packet is reached. For large images, this is very inefficient both in terms of the file access/seeking and processing of the packet headers.

For remote sensing images, it is highly recommended to use both TLM and PLT marker segments so that applications can easily index directly to tiles and/or packets that correspond to regions of interest. (Of course, the TLM does not provide any value for single-tile images and may be omitted in this case.) With this information, an interactive viewer or server application can save both memory and disk access time by reading in only those tiles and/or packets that are needed and discarding them from memory when they are no longer needed (e.g. once the region has been fully decoded or a different region of interest is selected). Typically, PLTs are preferred over PLMs, because they facilitate parsing only those packet lengths that correspond to the tile parts of interest. In fact, for extremely large single-tile images, it is recommended that multiple tile parts be introduced to exploit this concept. For instance, with the recommended RPCL progression order, place each resolution in its own tile part so that an interactive application can easily delay parsing the great number of packet lengths for the higher resolutions until they are needed.

12.3.8 Other Parameters and Implementation Considerations

Although the preceding sections have outlined the most fundamental choices that need to be considered when compressing remote sensing images with JPEG 2000, the standard offers many additional encoding options. Some might even argue that there is too much flexibility, making it difficult to implement compliant decoders. With this in mind, it is recommended that any parameters chosen comply with the code-stream restrictions defined in ISO/IEC 15444-1 Annex A.10. By advertising compliance to one of these restricted profiles in the SIZ marker segment, decoders that implement a subset of the standard can easily determine whether or not they have implemented all the features necessary to process a given JPEG 2000 code-stream. If using a single tile (with appropriate precincts), it should be reasonable to conform to the most restrictive Profile-0. For multiple tiles, Profile-0 only supports tile sizes of 128×128, which is too small to be practical with remote sensing images, so the goal would be to conform to the slightly less restrictive Profile-1. Of course, for certain applications, there may be a strong motivation to consider parameters that do not conform to these restricted profiles, but one must take into account that this may impact which decoders will be able to handle the resulting JPEG 2000 image.

The following sections summarize some of the other JPEG 2000 parameters and implementation considerations that may be of interest for remote sensing applications.

12.3.8.1   File Formats

Depending on the arena, remote sensing JPEG 2000 code-streams are typically wrapped in one of two file formats, which include metadata such as geographic coordinates in a defined manner. For commercial applications, the JP2 file format (described in ISO/IEC 1544401 Annex I) is commonly used with metadata embedded either in a UUID box from the GeoTIFF box specification from LizardTech, Inc.[4] or the rich Geography Markup Language (GML) model from the OpenGIS GML in JPEG 2000 for Geographic Imagery (GMLJP2) encoding specification from the Open Geospatical Consortium, Inc. (OGC). For government applications, JPEG 2000 code-streams are typically embedded as image segments in a National Imagery Transmission Format (NITF) 2.1 file per MIL-STD-2500C (or the equivalent NATO Secondary Imagery Format (NSIF) 1.0 file per ISO/IEC NSIF01.01), using the BIIF (Basic Image Interchange Format) profile for JPEG 2000 per

---

[4] The GeoTIFF metadata for JPEG 2000 was originally developed by Mapping Science, Inc. (MSI) under the name GeoJP2™, and was subject to certain licensing restrictions. LizardTech acquired the assets of MSI in 2004 and prefers that the metadata format be publicly available and thus not referred by the trademark of the original MSI encoder now owned by LizardTech.

ISO/IEC BPJ2K01.00. One advantage of the JP2 file format with either GeoTIFF or GML metadata is that any compliant JPEG 2000 decoder can support decoding and viewing the image data, even if it does not include any intelligent handling of the geographic metadata. Proper decoding of an NITF file, on the other hand, requires an application that understands how to parse the metadata to find the JPEG 2000 code-stream(s) in the file. The advantage of the NITF file format is the extensive metadata support that it offers, using the existing paradigm for non-JPEG 2000 image segment types that is well known throughout international government imaging organizations.

12.3.8.2 Color Transformation

With typical color images, the compression efficiency (i.e. quality versus bit-rate) may be substantially improved by decorrelating the spectral bands via an intercomponent transformation prior to the wavelet transformation. For compression with the 9-7I wavelet filter, Part 1 of the JPEG 2000 standard defines an irreversible multiple-component transformation that converts an RGB image into the YCbCr (luminance/chrominance) representation. For compression with the 5-3R wavelet filter, a reversible multiple-component transformation is used that approximates the RGB to YCbCr conversion. While this transformation alone improves numerical accuracy, perceptual quality can typically be increased further by applying visual weights that emphasize the luminance component over the chrominance components, as described in Section 12.3.8.3. Note that there is no need to subsample the chrominance components spatially in order to perform such a weighting.

For multispectral and hyperspectral imagery, it may be advantageous to extend this concept by applying a multicomponent transform across multiple components (i.e. spectral bands) as described in Part 2 of the JPEG 2000 standard. If the spectral bands are highly correlated, this transformation may significantly improve compression efficiency. This improved compression efficiency, however, comes at the cost of loss of direct access to individual spectral bands. In order to decode a single spectral band of a region, all spectral bands involved in the multicomponent transform must be arithmetically decoded and inverse wavelet transformed prior to applying the inverse multicomponent transform. Depending on the application, it may be preferred to treat each spectral band independently without any intercomponent transformation.

12.3.8.3 Visual Weights

Visual weighting is a mechanism by which the compression can be optimized for a specific viewing condition. There are two steps in the compression process at which visual weights may be applied: during quantization and during rate control. During quantization, the step size of each subband may be weighted by some relative factor above and beyond the nominal energy-based step sizes; any such weightings applied during quantization must be signaled in the code-stream as scalar expounded step sizes in the QCD (or QCC) marker segment. During postcompression rate-distortion optimization, the distortion contribution of each coding pass may be weighted, based on its subband, in such a way that affects the order in which coding passes are allocated into quality layers; these rate control decisions are made strictly on the encoder side and are not signaled in the code-stream. Typically, it makes sense to apply the same relative weighting at both steps, but if the quantization step size is fine enough (or if there is no quantization due to use of the 5-3R wavelet) it may be sufficient to apply visual weights only during rate control. Conversely, if no postcompression rate-distortion optimization is performed, any desired visual weighting must be applied during quantization.

The most fundamental question to ask when considering visual weighting is which is more important – numerical accuracy or perceptual quality? If numerical accuracy is more important for an application, no visual weights should be applied. If the intended use of the imagery is for viewing by human observers, appropriate weights may be chosen based on a model of the frequency response of the human visual system. In this case, the trick is to understand (at compression time) the conditions under which the images will be viewed. Factors such as the selected resolution level, the resolution of the display, postprocessing enhancements, and the viewer's distance from the display all impact the spatial frequency incident upon the retina. Although any given image is likely to be viewed under various conditions, it may be possible to characterize the most likely and/or most important viewing conditions. For the consumer photography market, for instance, the target viewing condition may an 8 inch × 10 inch print of the full image extents at a viewing distance of 20 inches. For remote sensing, on the other hand, a likely viewing condition may be zoomed in to a full-resolution region of interest on a 24 inch LCD monitor with 1920 × 1200 resolution at a viewing distance of 30 inches. By modeling the viewing conditions and the human visual system response versus the effective spatial frequency of each subband and wavelength of each component, appropriate weighting factors can be chosen for each subband of each component.

12.3.8.4 Sample Bit-Depth

It is important to match the sample bit depth to the imagery source properly. For instance, if the imagery is from a 12-bit source (pixel counts from 0 to 4095), this should be encoded with a sample bit-depth of 12. A common mistake would be to encode these data with a sample bit-depth of 16 if the input imagery were stored in 16-bit words. There are a few problems with using the incorrect bit-depth. First, for unsigned data, prior to the wavelet transform, the pixel values are subtracted by half the dynamic range – using an overly high bit-depth will place large negative values in the LL subband that are slightly harder to compress than the correct bit-depth values. Second, the number of bit-planes required for proper decoding increases with the sample bit-depth. In some cases, this may lead to a decoder using higher precision 32-bit arithmetic instead of a more efficient 16-bit processing path. Alternatively, in more extreme cases, it may lead to a requirement for more bit-planes than the decoder supports. Finally, upon decoding, most viewer applications will by default display the imagery scaled to a dynamic range based on the bit-depth – if the signaled bit-depth is too high, the decompressed image will appear dark (unless the viewer provides automatic dynamic range adjustment).

12.3.8.5 Code-Blocks

As described in Section 1.2.4, code-blocks are the smallest partition of wavelet-transformed coefficients that can be independently encoded. Larger code-blocks are able to take advantage of a larger neighborhood of statistics and thus lead to higher compression efficiency. Smaller code-blocks, on the other hand, offer finer control over region-of-interest selection. For remote sensing applications, the maximum allowed square code-block size of 64×64 typically offers sufficient control over the ROI, so this code-block size is recommended to optimize image quality and minimize the overhead of managing the number of code-blocks. For hardware encoder implementations with limited memory availability, however, it may be desired to reduce the code-block height so that less input pixel data are associated with each row of code-blocks, in which case there may be benefit in considering a rectangular code-block size of 128×32.

12.3.8.6 Guard Bits

Guard bits are additional bits beyond the nominal 'growth' of the wavelet transform that are used to prevent overflow in the magnitude bit-plane representation of the quantized coefficients. Most 'natural' images will compress easily without any guard bits, but with the right combination of pixel values that hit the 'resonant frequency' of the wavelet filter there may be a potential for overflow at one or more coefficients. Typically, one guard bit is sufficient to avoid overflow for natural images and two guard bits are sufficient for all but the most contrived synthetic images. There is a tradeoff, however, in choosing too many guard bits. With the 5-3R wavelet transform, increasing the number of guard bits may limit the sample bit-depth that may be reversibly compressed with a given encoder implementation. With the 9-7I wavelet transform, increasing the number of guard bits may limit how many fractional bit-planes are available for distortion estimations, or with even higher sample bit-depths (or finer quantization) may limit the number of bit-planes that can be encoded. A conservative approach is to always use two guard bits, unless the number of bit-planes is tightly limited (e.g. if using a hardware encoder implementation). If insufficient guard bits are chosen, a well-designed encoder would identify this problem – otherwise, it may inadvertently discard the most significant bit-plane of one or more wavelet coefficients.

12.3.8.7 Region-of-Interest Encoding

When performing layer allocation during rate control, the encoder is free to include coding passes from one code-block before those from another. This can be used to emphasize a region of interest so that it will be delivered and decoded first. However, what if finer control over the region of interest is desired? Part 1 of the JPEG 2000 standard provides a clever method for signaling a region of interest by shifting the coefficient bit-planes of the ROI above those of the background. With this method, any arbitrary ROI mask can be used, e.g. to highlight individual objects in the scene. The amount by which the ROI coefficients are shifted is signaled in an RGN marker segment, but the location of the ROI is implied by detecting the shifted coefficients and is not explicitly signaled in the code-stream. This implicit ROI shift must be large enough such that there is no overlap of the ROI bit-planes with the background bit-planes. The main drawback of this 'Max-shift' technique is that it requires significantly more bit-planes in both the encoder and the decoder in order to maintain the full quality of the downshifted background coefficients. For this reason, if the requirements of an application can be achieved by merely reordering the layering of one code-block relative to another, such a method is preferred to encoding with an implicit ROI shift.

Another option to consider is to encode an arbitrary rectangular or elliptical ROI using the extended RGN marker segment defined in Part 2 of the JPEG 2000 standard. In this case, since the position of the ROI is explicitly signaled, any shift can be used. The ROI does not need to be shifted completely above the background, and thus it may be possible to maintain a higher fidelity from the background for a given sample number of bit-planes in the encoder or decoder implementation. Of course, whenever considering using features from Part 2, one must consider that many decoders will not support these features and therefore the intellectual property rights and patents that may cover these features must be reviewed.

12.3.8.8 Early Termination of Coding Passes

When using postcompression rate-distortion optimization, some of the compressed data that was generated by the MQ encoder will typically be discarded to achieve a desired target compression ratio. If the quantization step size is extremely fine, the amount of data truncated may be significant, causing both excess memory usage and excess MQ encoder processing for the bit-planes that are eventually discarded. For improved efficiency, an encoder implementation may employ heuristics to identify which coding passes are not likely to contribute to the target compression ratio and to stop encoding of the associated code-blocks early. This can significantly improve throughput when using fine quantization step sizes, but there exists the possibility that a poor decision could be made and the final output quality may not be quite optimal. Such heuristics are particularly problematic if the image statistics are highly nonuniform. For most natural images, however, an appropriate heuristic can be applied to terminate early with minimal impact on the resulting image quality. Of course, if the quantization step size has been carefully tuned for the expected image statistics, this performance optimization would not be necessary, and it would be preferable to disable any such early termination in order to optimize image quality.

12.4   Other JPEG 2000 Parts to Consider

In addition to the baseline ISO/IEC 15444-1, other JPEG 2000 parts are applicable to the GIS and remote sensing market, and several of these have been adopted. This section will review some of the technology associated with these extensions, the advantages of this technology, and its application within the GIS and remote sensing community.

12.4.1   ISO/IEC 15444-2: Extensions

This standard describes optional features and value-added extensions that were not considered to be critical to the mainstream users. While there are several technologies described in the extensions, this section is focused on three technologies that have the most impact on the GIS and remote sensing community.

12.4.1.1   Trellis-Coded Quantization

Trellis-coded quantization (TCQ), as described in ITU-T Rec T.801 | ISO/IEC 15444-2 Annex D (Section 2.4), has been shown to improve the visual quality of several image types when compressed to the same bit-rate as the standard quantization of Part 1.

The computation of the TCQ parameters is slightly more complex than the standard quantization and does not enable the same scalability features. While the quality improvement can be significant, the increased computation and decreased scalability have limited this technology to applications where the visual quality at a given bit-rate is the most critical requirement.

12.4.1.2 Multiple-Component Transform

The multiple-component transform, as described in ITU-T Rec T.801 | ISO/IEC 15444-2 Annex J (Section 2.9), includes both three-dimensional wavelet transforms and component linear transforms that increase the compression efficiency for multiple-component data. Common to the remote sensing applications are multispectral, hyperspectral, and ultraspectral data. These imaging systems, which collect up to hundreds of bands, present an even greater challenge to storage and transmission and require significantly more compression. Lossless compression ratios for multiple-component imagery can increase from 2:1 for standard JPEG 2000 lossless compression to 3:1 with the proper use of the reversible 5-3 wavelet transform in the component direction. The Karhunen–Loeve transform (KLT) produces the greatest compaction of spectral information but is computationally complex. Fourfold increases in the compression ratio can be achieved by using the KLT and other linear transforms while maintaining the same quality.

Users in the hyperspectral community are always concerned about the impact of compression on the radiometric accuracy and exploitation results of these data. Studies have shown that significant compression can be achieved before impacting the exploitation quality of the data (Shen, Kasner, and Wilkinson, 2001). The increase in compression efficiency from a component transform enables quicker access to the full data set, but this comes with an increase in computation complexity and reduced access to individual bands. Properly decoding an individual band that is part of a component collection (either wavelet or linear) requires accessing all the bands that result from the component transform. For example, a user may want to access and display three bands that are all in different component collections and therefore must access all of the bands in each component collection simply to view the three selected bands. Optimization of both the compression efficiency and access to individual bands may be achieved through appropriate partition into component collections based on system requirements. For example, standard 'display' bands can be encoded in one component collection compressed band-by-band for quick access, while the rest of the bands can be encoded in a separate component collection using a multiple-component transformation to achieve greater compression.

12.4.1.3 Region-of-Interest Coding and Extraction

Region-of-interest (ROI) coding and extraction, as described in ITU-T Rec T.801 | ISO/IEC 15444-2 Annex L (Section 2.13), can enable fast access to regions of a remote sensing image that may be more important than other regions. For example, in an aerial photograph of a harbor scene the water may not be of interest to many users, while the ships, docks, or crates may be of significant interest and therefore may be encoded as a region of interest. Both the Max-shift ROI encoding included in Part 1 and the ROI encoding extensions of Part 2 can achieve improved compression efficiency by reducing the quality in the surrounding areas while maintaining high quality for the ROI. The ROI encoding in Part 2 enables more flexibility at the cost of added complexity, while the ROI in Part 1 can be decoded by any compliant Part 1 decoder and does not need explicit signaling of the ROI mask.

12.4.2 ISO/IEC 15444-3: Motion JPEG 2000 and ISO/IEC 15444-12: ISO Base Media File Format

ISO/IEC 15444-3: Motion JPEG 2000 and ISO/IEC 15444-12: Base Media File Format define the file format and metadata required for Motion JPEG 2000. Currently these standards are not used in the GIS and remote sensing community but are expected to become more common as digital video cameras used for remote sensing continue to increase in resolution. Currently, video cameras used for remote sensing collect imagery at resolutions associated with standard-definition television or high-definition television (e.g. 1920 × 1080), which are then encoded with standard MPEG-2, MPEG-4, or H.264 (MPEG-4 Part 10). As these systems grow to include new sensors developed for digital cinema (e.g. 4096 × 2160) and beyond, Motion JPEG 2000 may become a more important standard in this arena. While the compression efficiency gained by the interframe coding of H.264 may continue to be preferred for lower resolution streaming, the quality and resolution scalability of JPEG 2000 will offer advantages for access to regions of interest within the larger frame sizes.

12.4.3 ISO/IEC 15444-4: Conformance Testing

ISO/IEC 15444-4: Conformance Testing is a useful standard in testing an application for compliance to the standard and profiles within the standard (section 16.2). Understand that many of the remote sensing images are larger than the parameters covered by the baseline compliance classes (Part 4) and profiles (Part 1). Because of the increased image size, bit-depth, and number of components typical in remote sensing images, it is important to test applications with a representative data set to assure interoperability and compliance. Applications for remote sensing use will be expected to meet the increased memory and computational requirements to handle these large images properly.

12.4.4 ISO/IEC 15444-5: Reference Software

ISO/IEC 15444-5: Reference Software can be used to help develop JPEG 2000 baseline software or in experiments, but is not recommended for applications. Commercial software products have shown to be significantly more efficient and capable than the reference software included in this international standard.

12.4.5 ISO/IEC 15444-6: Compound Documents

ISO/IEC 15444-6: Compound Documents compresses the different components of compound documents (text, graphics, and images) with different encoders using the mixed raster content (MRC) defined therein. While this standard was originally developed for scanned documents that include imagery, text, and graphics, the use for delivery of combined GIS information and remote sensing imagery is an ideal application. The combination of GIS information layers with remote sensing images is typically performed at the final stage of viewing, pulling data from two different databases and then fusing the data for display. The compound document format would enable a user to fuse this information at the server and then stream it to a user. This process would enable the users to display these data without having custom applications for reading and displaying the GIS information and trying to fuse the data with remote sensing data.

12.4.6 ISO/IEC 15444-8: JPSEC

JPEG 2000 Part 8: Secure JPEG 2000 provides the tools that allow applications to generate, consume, and exchange Secure JPEG 2000 code-streams. The standard enables users to protect the remote sensing imagery and metadata with such tools as digital watermarking, digital signature, encryption, and authentication. The producers of remote sensing data are commonly commercial companies or government agencies that need to protect their data from unauthorized use. Similar to how unauthorized copying and use of digital music and movies impacts the profits for the entertainment industry, the remote sensing industry is even more susceptible because the cost of a full satellite image may cost thousands of dollars (in contrast to a ten to twenty dollar DVD or music CD). Digital watermarking and encryption are two techniques that can be used to protect the remote sensing data. The unique aspect of this standard is that these protection techniques can be used within the full functionality and scalability of the JPEG 2000 standard. For example, each resolution level or each quality layer can be encrypted differently, which enables the owner to limit users to a given resolution or quality until they have purchased a key. Similarly, a given tile or region of interest may be encrypted because it is considered a sensitive area (e.g. a military base, nuclear power plant, or government building). The protection of a given ROI could allow common users to view the data at only a given resolution, while people who are authorized will have the key to view the data at full resolution. The digital watermarking enables the data provider to include information in the image that can track any unauthorized use of the data back to the responsible party. Users can be assured that the information in the image is correct through authentication of the data from the original source and verification that the image has not been modified.

12.4.7 ISO/IEC 15444-9: Interactivity Tools, APIs, and Protocols (JPIP)

JPEG 2000 Part 9: JPIP is the key to accessing the scalability and functionality of JPEG 2000 Part 1 within a client/server environment. JPIP, which typically operates over standard network transports such as HTTP and TCP, is a client–server protocol that provides the capability for a client to interact with a large image or video sequence on a server. The JPIP client sends requests to a JPIP-enabled server to deliver a given region of interest in a streaming fashion to the client. The common scenario in which a user is interested in localized regions of a large image elevates JPIP as the most important standard for remote sensing beyond the baseline JPEG 2000 Part 1. This protocol enables a client to request only the portions of an image (by region, quality, and/or resolution level) that meet the client's needs. A user may quickly review an image and focus into a region of interest at full quality and resolution without having to download the entire image, or the user may directly request the server to send a 'chip' from the image. For example, a remote sensing server may have an image that covers a 17 kilometer by 17 kilometer square area on the Earth at 0.5 meter resolution in four bands (at 16 bpp per band). Downloading and locally storing this full image (nearly 10 gigabytes uncompressed) would be prohibitive, whereas interactively streaming these data to a user who may only be interested in their house, a historical site, or some other attraction may take only seconds. This protocol is especially critical when smaller devices such as cellular phones and navigation systems access these large remote sensing images over wireless networks with limited bandwidth.

Part 9 offers both tile-based ('JPT-stream') and precinct-based ('JPP-stream') media types. JPT-streams deliver data as contiguous tiles and enable the client to store a 'streamed' set of data directly to a JPEG 2000 file in the same organization as the original file on the server. JPP-streams, on the other hand, offer random access to precincts and are typically more efficient in the delivery of regions of interest. The recommendation is to use precinct-based streaming for most GIS and remote sensing applications, regardless of whether or not the imagery is tiled.

Servers for GIS and remote sensing applications should also support both 'stateful' sessions and stateless requests. If a session is initiated between the client and server, the server will mirror the state of the client's code-stream cache to track what information the client has already received. Thus, the server need not continually resend data over and over as the user roams around an image or changes resolution (i.e. zooms in or out). It is expected that many clients would include the ability to cache most of the data in an interactive session, but stateless requests may be preferred for smaller client devices that do not include significant data caching.

12.4.8 ISO/IEC 15444-10: 3-D Volumetric Data

JPEG 2000 Part 10: 3-D Volumetric Data is a continuation of the multiple-component compression described in Part 2: Extensions. This part provides the means to encode a three-dimensional set of original or transformed data directly. This standard is slightly more complex and requires more memory management but produces greater compression efficiency. Users should consider testing the multiple-component compression in Part 2 versus the multiple-component compression in Part 10 to evaluate which performs better for a given data set.

12.4.9 ISO/IEC 15444-11: Wireless

JPEG 2000 Part 11: Wireless is used to protect JPEG 2000 streaming data against errors that may occur during the transmission over wireless networks. As the access to GIS and remote sensing data continues to grow, one would expect that the remote sensing data will be available to users on their cellular phones and the wireless navigation tools for automobiles. Currently, many of these applications are simply GIS maps and do not include remote sensing imagery, but in the future these applications will include full quality streaming of remote sensing data.

References

Gupta, R. (ed.) (2008) Mapping GIS milestones: 1960–1970, http://www.gisdevelopment.net/history/1960-1970.htm.

Shen, S. S., Kasner, J. H. and Wilkinson, T. S. (2001) New hyperspectral compression options in JPEG-2000 and their effects on exploitation, in *Imaging Spectrometry VII*, paper 4480-18, July 30–August 3, 2001.

Further Reading

BPJ2K01.00 (2004) Information Technology – Computer Graphics and Image Processing – Registered Graphical Item – Class: BIIF Profile – BIIF Profile for JPEG 2000 Version 01.00.

ISO/IEC (2007) BIIF PROFILE NSIF01.00 – Information Technology – Computer Graphics and Image Processing – Registered Graphical Item, Class: BIIF Profile – NATO Secondary Imagery Format Version 01.00.

Janza, F. A. (ed.) (1975) *Manual of Remote Sensing*, vol. I, American Society of Photogrammetry, Falls Church, VA.

Lillesand, T. and Kiefer, R. (2000) *Remote Sensing and Image Interpretation*; John Wiley & Sons, Inc., New York.

MIL-STD-2500C (2006) National Imagery Transmission Format Version 2.1 for the National Imagery Transmission Format Standard.

Open Geospatial Consortium Inc. (2006) OGC 05-047r3, GML in JPEG 2000 for Geographic Imagery (GMLJP2) Encoding Specification.

Schott, J. R. (1997) *Remote Sensing, The Image Chain Approach*; Oxford University Press, New York.

13

Medical Imaging

Alexis Tzannes and Ron Gut

13.1 Introduction

In this chapter, after a brief historical review of the development of the medical imaging field, we describe how the different parts of the JPEG 2000 standard have been adopted in the medical imaging community. Specifically, we discuss how Part 1 and Part 2 of JPEG 2000 are used in the compression of still and volumetric medical imagery. In addition, we describe how Part 9 of JPEG 2000 (JPIP) is being used to improve remote access of medical imagery over bandwidth-constrained networks. The advantages and disadvantages of the various parts of JPEG 2000 are discussed and the need for future improvements is outlined.

13.2 Background

Medical imaging began with William Conrad Roentgen's discovery of X-rays in 1895, which allowed doctors to peer inside the body, noninvasively, for the first time. Since then the number of technologies used to acquire images has increased: ultrasound imaging was introduced in the 1960s, computed tomography (CT) in 1972, and magnetic resonance imaging (MRI) in 1980.

The introduction of computerized archives to store the various radiology images created the need for the different acquisition and storage devices to interoperate. The National Electrical Manufacturing Association (NEMA), representing the manufacturers of the devices, and the American College of Radiology (ACR), representing the users of the devices, jointly published the Digital Imaging and Communications in Medicine standard, better known as DICOM, to fill this need (NEMA, 2008). The DICOM standard was first published in 1985 and has since undergone several revisions, most notably in 1993 when the DICOM 3.0 standard was published. The current version of the standard is DICOM 3.0 into which many changes and supplements have been incorporated.

The JPEG 2000 Suite   Edited by Peter Schelkens, Athanassios Skodras and Touradj Ebrahimi
© 2009 John Wiley & Sons, Ltd

The DICOM standard is a protocol for the communication between different devices. The standard defines services that DICOM devices can provide or consume, queries that can be sent to and responses received from such devices, and the syntax for the information contained in the queries and responses (queries and responses collectively will be referred to as messages).

For image data transfers, the initial version of the DICOM standard only permitted raw pixel data to be included in messages. The need for compression was recognized, however, and DICOM Working Group 4 (WG4) was established to study the options and recommend which compression algorithms to include and how. In 1989, WG4 published its results: its conclusion was that compression should be allowed, as it does provide value, but that a single compression algorithm will not meet every need. Instead, the standard defined several compression blocks and a mechanism to specify which blocks were applied to the image data, and in what order. The blocks included various lossless coding methods, such as Huffman (1952) and Lempel-Ziv (Welch, 1984) coding, as well as modeling and ordering techniques, such as DPCM (Habibi, 1971), DCT (Ahmed, Natarajan, and Rao, 1974), and pyramidal coding (Burt and Adelson, 1983). The image data could, for example, be encoded using a DCT followed by Huffman coding.

This use of compression proved to be cumbersome. Too many combinations of transforms, modelers, and encoders were available, and few implementations supported all combinations. Since all implementations were required to support the use of uncompressed data, compression was rarely used during the interchange of data, thus losing the communication-bandwidth conservation advantage that compression offers. In an attempt to solve this problem, the DICOM 3.0 standard, published in 1993, standardized the use of the original JPEG standard (ISO/IEC, 1994) for the lossless and lossy compression of image data, as well as the run-length encoder (RLE) (Apple Computer, Inc., 1994), known as the pack bits algorithm, which is used in TIFF as well. DICOM 3.0 also allowed implementers to include proprietary compression schemes; DICOM devices can negotiate which compression schemes, if any, they support prior to the transmission of the image data.

The reduced overhead for the support of compression, along with the inclusion into DICOM of standardized compression for which software is readily available, helped spur the adoption of compression by device makers. Ultrasound device makers started offering RLE compression, and the makers of CT and MRI scanners added lossless JPEG compression to their systems.

Since the DICOM standard enabled devices to use proprietary compression, device makers began to investigate and then adopt new compression algorithms. In the mid-1990s wavelet-based compression algorithms became very popular. Wavelet-based compression offers several advantages over JPEG and RLE. Most notably, wavelet-based compression offers resolution scalability, which enables devices to create thumbnails easily and to adapt images to the desired display size. In addition, wavelet-based compression offers higher quality images at the same bit-rates as JPEG (or smaller compressed images at the same quality). A disadvantage of these wavelet-based techniques was their proprietary nature, which hindered interoperation between devices from different manufacturers. Two devices from the same manufacturer could take advantage of wavelet-based compression, but two devices from different manufacturers would have to resort to the default, uncompressed, storage formats.

13.3   DICOM and JPEG 2000 Part 1

During the late 1990s, the DICOM committee's WG4, which was charged with overseeing the use of compression in the DICOM standard, met several times to discuss the creation of a standardized wavelet-based image compression algorithm for inclusion in the DICOM standard. However, lack of cooperation and participation from vendors of compression software and lack of interest by the medical device makers delayed action in WG4. At the same time, the JPEG committee was working on the JPEG 2000 standard. In the late 1990s WG4 made the decision to wait until the JPEG 2000 work was completed and to adopt it whole, as it had with the earlier JPEG standards.

Shortly after Part 1 of the JPEG 2000 standard was finalized in December 2000, a new DICOM supplement titled 'JPEG 2000 Transfer Syntaxes' was proposed in WG4. The new supplement, officially known as Supplement 61, was completed about one year later. In January 2002, the DICOM committee approved the final text of DICOM Supplement 61, marking the inclusion of Part 1 of JPEG 2000 in DICOM.

13.3.1   Supplement 61

Supplement 61 defines new transfer syntaxes for the use of Part 1 of JPEG 2000 in DICOM. All features of Part 1 of JPEG 2000 are included. In fact, Supplement 61 states that compliant DICOM implementations are required to support all features of JPEG 2000 Part 1. The supplement also adds new photometric interpretations, the DICOM colorspace identifiers, to describe the colorspace rotations that may be performed as part of the JPEG2000 compression.

Although Supplement 61 defines the mechanism for using JPEG 2000 as part of DICOM, it does not attempt to define the situations where the use of lossy compression of medical images is clinically acceptable. In addition, Supplement 61 does not specify any information about the selection of the appropriate compression parameters (e.g. compression ratio) for JPEG 2000 compressed images. In fact, the supplement explicitly states that any such definitions are beyond the scope of the DICOM standard.

Supplement 61 defines two new transfer syntaxes: one specifically for lossless compression using Part 1 of JPEG 2000 and one that can be used for either lossless or lossy compression using Part 1 of JPEG 2000. Although all compression options and features of Part 1 of JPEG 2000 are allowed, the optional JP2 file format (discussed in Section 1.6.3) is not allowed. All images compressed with JPEG 2000 that are included in DICOM are to be in the code-stream form only, without the file format. This decision was made because the functionality of the JP2 file format (metadata, colorspace information, etc.) is redundant in the presence of the DICOM file structure. In addition, this is consistent with how other JPEG standards are included in DICOM. The original JPEG standard (ISO/IEC, 1994) does not include the JFIF header and JPEG-LS (ISO/IEC, 1999) (which was added to DICOM through a Change Proposal in September 2000) does not include the SPIFF header.

13.3.2   Color Transformations and Photometric Interpretations

Part 1 of JPEG 2000 allows the use of two different color transformations. These color transformations are optionally applied to three-channel RGB images to decorrelate the color channels by transforming the RGB data into the YCbCr colorspace, leading to improved compression efficiency (see Section 1.2.1). Two different versions of the RGB to YCbCr transformations are used in JPEG 2000. One is a lossy transformation that is used for lossy compression and the other is an integer-based lossless transformation that is used for lossless compression.

Supplement 61 specifies that both of these RGB to YCbCr transformations are allowed in DICOM. There is also a DICOM field that specifies the photometric interpretation of the image. This value describes the internal representation of the pixel data in terms of its color content. Before the introduction of Supplement 61, DICOM included several different photometric interpretation values, including two variants of MONOCHROME (for single-component images), PALETTE COLOR, RGB, etc. Supplement 61 defined the following two new photometric interpretations to be used in conjunction with JPEG 2000:

1. YBR_ICT: used for the situation when the source image is an RGB image with three color components, 8 bits per component, and an irreversible RGB to YCbCr trans-formation is applied (as part of the JPEG 2000 compression algorithm) to the pixel data.
2. YBR_RCT: used for the situation when the source image is an RGB image with three color components, 8 bits per component, and a reversible RGB to YCbCr trans-formation is applied (as part of the JPEG 2000 compression algorithm) to the pixel data.

Three-channel JPEG 2000 compressed images that have not had the color transformations applied are tagged with the RGB photometric interpretation, while single-channel JPEG 2000 images are tagged with the MONOCHROME photometric interpretation.

13.3.3   Multiframe Imagery and Fragments

The DICOM standard allows for multiple images (frames) to be included in a single DICOM file (object). Supplement 61 specifies that if the DICOM file includes multiple frames, the JPEG 2000 compression algorithm should be applied on a per frame basis. Other multiframe formats defined as part of the JPEG 2000 family of standards, such as the Motion JPEG 2000 file format (see Chapter 3), are not allowed.

The DICOM standard also allows for the pixel data (the compressed bit-stream) to be segmented in one or more fragments. Supplement 61 specifies that for multiframe DICOM files compressed using JPEG 2000, each fragment shall contain pixel data from a single frame. It is allowed, however, to split the pixel data from a single image into multiple fragments.

13.4   DICOM and JPEG 2000 Part 2

After the publication of Supplement 61, multiframe and three-dimensional imagery started to become more popular and volumetric image sizes continued to increase. In addition, the rising popularity of remote diagnosis and consultations required the ability to store, transfer, and view these data sets over bandwidth-constrained networks. This introduced a need for improved compression techniques that exploit the correlation between sets of consecutive frames.

In early 2003, motivated by the need for improved compression of volumetric data sets, DICOM WG4 started investigating the use of decorrelating transformations in the third dimension. These types of transformations are defined in Part 2 of the JPEG 2000 standard (see Chapter 2). They are given the name multiple-component transformations (MCTs) since they are treated as extensions to the RGB to YCbCr transformations described in the previous section. Members of WG4 performed studies testing the performance of these multiple-component transformations on volumetric medical imagery. Results showed that these transformations can improve compression efficiency, for both lossless and lossy compression. For lossless compression, the improvement was shown to be about 5–25%, depending on the type of images. More significant improvements have been observed for lossy compression (Tzannes, 2003).

Driven by these results, a new DICOM supplement was drafted in late 2003 to include the use of these JPEG 2000 multiple-component transformations in DICOM. The new supplement, officially known as Supplement 105, was completed about a year and a half later. The DICOM committee officially approved the final text of DICOM Supplement 105 in June 2005.

13.4.1 Supplement 105

Supplement 105 is officially titled 'JPEG 2000 Part 2: Multi-component Transfer Syntaxes.' The supplement defines new transfer syntaxes for the use of the multiple-component transformations (described in Part 2 of JPEG 2000) in DICOM. Both lossy and lossless multiple-component transforms are supported. Supplement 105 states that all compliant DICOM implementations must support all the multicomponent extensions as described in Annex J of JPEG 2000 Part 2 (see Section 2.9).

Similarly to Supplement 61, Supplement 105 defines two new transfer syntaxes: one specifically for lossless compression using the multiple-component transformations and one that can be used for either lossless or lossy compression using the multiple-component transformations. Consistently with Supplement 61 (which did not allow the optional JP2 file format), Supplement 105 does not allow the use of the optional JPX file format that is defined in Part 2 of JPEG 2000 (see Section 2.14).

As mentioned above, all the transformations defined in Annex J of Part 2 of JPEG 2000 are supported. This includes the following two types of transformations:

1. Array-based multiple-component transforms that form linear combinations of components to reduce the correlation between them. Array-based transforms include prediction-based transformations such as differential pulse code modulation (DPCM) (Habibi, 1971) as well as more complicated transformations such as the Karhunen–Loeve transform (Hoteling, 1933). These array-based transformations can be implemented reversibly or irreversibly.
2. Wavelet-based multiple-component transformations using the same two wavelet filters as used in Part 1 of JPEG 2000 (see Section 1.2.2).

Annex J of Part 2 of JPEG 2000 also defines a mechanism and syntax for reordering the components before the multiple-component transformations are applied. This feature is supported in Supplement 105. In addition, Supplement 105 allows for the use of component collections (see Section 2.9.2.2). In DICOM, component collections are groupings of frames. The multiple-component transformations are then applied to each component collection independently. Component collections can be used to reduce computational complexity and to improve access to specific frames on the decoder. Supplement 105 specifies that if component collections are used, each DICOM fragment shall contain encoded data from a single-component collection. It has been shown that the use of component collections does not significantly affect the compression efficiency (for details, see the last section of this chapter and Tzannes, 2003).

Part 2 of JPEG 2000 also allows arbitrary wavelet transformations with arbitrary decomposition trees. These features (which are defined in Annex H of Part 2 of JPEG 2000, see Sections 2.7 and 2.8 for details) are explicitly excluded from being used in DICOM in Supplement 105.

Some results of using the compression techniques approved in DICOM Supplements 61 and 105, on typical medical image data, are shown in the next section.

13.5   Example Results

This section presents results of applying the DICOM approved Part 1 and 2 JPEG 2000 lossless and lossy compression techniques to sample sets of medical imagery. A comparison between using Part 1 and 2 is presented along with some results of using component collections of different sizes. Results of lossless and lossy compression, using component collections of 20, 40, 80, and N frames, where N is the total number of frames in a volumetric sequence, are presented and compared.

The compression techniques were tested on three sample sets of volumetric medical data. Characteristics of the image sequences are given in Table 13.1.

13.5.1   Lossless Compression Results

Each of the three sample sequences was compressed losslessly using Part 1 JPEG 2000 and Part 2 JPEG 2000 with components collections of 20, 40, 80, and N, where N is the total number of images in the sequence. The resulting compressed image sizes are shown in Table 13.2.

Reviewing the lossless results presented in the Table 13.2, the compressed file size is significantly smaller for the three-dimensional (3-D) case compared to simple Part 1 JPEG 2000 compression. The best compression is achieved when all the components in the sequence are grouped into a single collection (the N column in the table), although the loss in compression efficiency (compared to the N column) is small even with a component collection as small as 20. In the right-most column is the percent reduction in file size

Table 13.1   Test sequence information

Sequence name	Image size	Number of images (N)	Bit-depth	Uncompressed sequence size (Mbytes)
Sequence 1	256×256	127	8	7.9
Sequence 2	512×512	449	16	224
Sequence 3	512×512	620	16	310

Table 13.2 Lossless compression results. Note that Part 2 JPEG 2000 compression reduces the compressed sequence size by 15–18%

Sequence name	Uncompressed sequence size (Mbytes)	Part 1 JPEG 2000 compressed size (Mbytes)	Part 2 JPEG 2000 compressed size, at different component collection sizes (Mbytes)				Improvement with 3-D compression collection size N (%)
			20	40	80	N	
Sequence 1	7.9	3.81	3.27	3.25	3.24	3.23	15.2
Sequence 2	224	75.8	62.7	62.3	62.1	61.8	18.5
Sequence 3	310	120	105	104	101	100	16.7

between Part 1 JPEG 2000 compression and Part 2 JPEG 2000 with a single-component collection (of all the images in the sequence).

Several other sequences of medical images were compressed using both Part 1 JPEG 2000 and Part 2 JPEG 2000, with a single-component collection. The results were similar to the ones presented in the table for the three sample sequences. For the sequences that were used in these tests, Part 2 JPEG 2000 increased the compression efficiency by 5–20%.

13.5.2 *Lossy Compression Results*

Each of the three sequences was irreversibly compressed using both Part 1 JPEG 2000 and Part 2 JPEG 2000 compression. Compression ratios in the range of 10:1 to 50:1 were used and the resulting compressed sequences were compared using the peak signal-to-noise ratio (PSNR) metric averaged over all the images in the sequence. PSNR is a commonly used metric for image fidelity; higher PSNR is equivalent to lower distortion in the image.

To test the effect of the component collection size on compressed sequence size and resulting image PSNR, the first sequence was compressed using components collections of 20, 80, and 127 images (all the images in the sequence). The resulting average PSNR over the entire image sequence is shown in Figure 13.1, at compression ratios ranging from 10 to 50:1. The three Part 2 curves represent the different component collection sizes. As in the lossless case, the best results are obtained with a single-component collection for the entire sequence. The average PSNR is only slightly lower for the other component collection sizes, but still considerably higher than that for Part 1 JPEG 2000 compression.

Results of lossy compression of Sequences 2 and 3 are shown in Figures 13.2 and 13.3. These figures show the average PSNR of the compressed sequences at various compression ratios, using a single-component collection for the entire sequence. Note that the average PSNR is consistently higher than the JPEG 2000 Part 1 compression case.

Also note that at a specific PSNR level, Part 2 achieves more than twice the compression efficiency of Part 1. For example, for Sequence 2 in Figure 13.2, Part 1 compression results in an average PSNR of 77 dB, at a 10:1 compression ratio. The same average PSNR can be achieved with Part 2 at a compression ratio of 25:1. Similarly, for Sequence 3, Part 1 achieves 67 dB at a 15:1 compression ratio. Part 2 achieves the same PSNR value at a

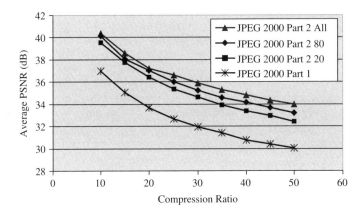

Figure 13.1 Sequence 1: average PSNR over all images in the sequence versus compression ratio. The three JPEG 2000 Part 2 curves represent the different component collection sizes

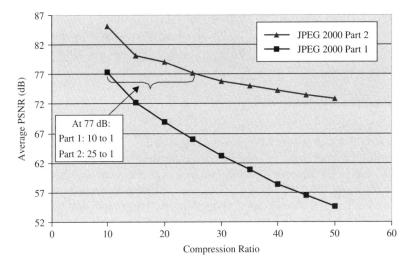

Figure 13.2 Sequence 2: average PSNR over all images in the sequence versus compression ratio. Note that Part 2 JPEG 2000 outperforms Part 1 JPEG 2000 by 8 to 18 dB. The Part 2 JPEG 2000 data in this figure was compressed using a single-component collection for the entire sequence (449 images)

40:1 compression ratio. At higher compression ratios the gains are even larger, as seen in both figures.

13.6 Image Streaming, DICOM, and JPIP

A typical radiology practice or hospital department includes one or more modalities (image-generating medical devices such as X-ray machines, CT scanners, and so on),

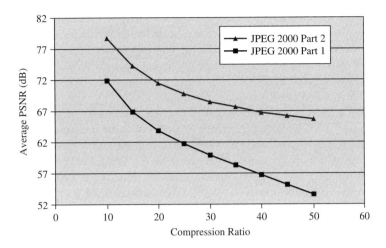

Figure 13.3 Sequence 3: average PSNR over all images in the sequence versus compression ratio. Note that Part 2 JPEG 2000 outperforms Part 1 JPEG 2000 by 6 to 13 dB. The JPEG 2000 Part 2 data in this figure were compressed using a single-component collection for the entire sequence (620 images)

an image archive, and review workstations. The radiologists use the review workstations to retrieve images from the modalities and the archives, read the images, and write a report detailing the diagnosis. Since the display of the workstation cannot show every image at full resolution simultaneously, thumbnails of the images are displayed allowing the radiologist to choose which series of images (or individual images) will be displayed at full resolution.

Prior to 2006, the DICOM standard only allowed for the transmission of complete images. In the scenario described above, the workstation would need to request and receive the full image data for each of the images for which it needs to display a thumbnail, though only low-resolution data are required. In order to speed up the display of the thumbnails, and to speed up the display of images in general, workstation manufacturers started adopting image-streaming techniques.

Image streaming denotes three innovations. The first is the idea that in a system consisting of a server that supplies images and a client that receives them, the client can display the image as it is being received over the network connecting the client to the server. When the images are transmitted uncompressed the client can build the display of the image pixel-by-pixel, row-by-row, as the image is received. When the images are transmitted compressed using JPEG, the client can build the image on the screen block-by-block from a baseline JPEG process, or progressively from low quality to full quality from a progressive JPEG process.

Wavelet compression algorithms enable another mode of streaming: progressive by resolution. In wavelet-based image compression the image is divided in successively lower resolution representations, which can be transmitted serially. This allows the client to decode and display low-resolution versions of the image before the entire image has been received.

The second innovation encompassed by image streaming is the client's ability to request the specific portions of the images it needs. As in the example above, the client needs multiple images, but at reduced resolution, in order to display thumbnails for the user. The streaming client can request from the server that only the lower resolution portions of the relevant images be sent, shortening the amount of time the user needs to wait for the images to arrive and be displayed.

The third innovation is the client's ability to keep the already-received portions of an image and to use them when a higher resolution version of the same image is needed. Where a traditional DICOM client would have to request and receive the entire image, a streaming client can make incremental requests for images adapting the amount of information it requests and receives to the user's actions. In a system where the network connection is a limiting factor in determining the speed of transferring images, the streaming client can adapt to the user's actions more quickly.

The providers of streaming-capable servers and workstations created their own proprietary communication protocols for these systems. This generally meant that the users had to purchase complete systems from a single vendor, since no interoperability between the different systems was guaranteed. In 2004, the DICOM committee sought to standardize such an image streaming protocol to enable system interoperation. This work resulted in DICOM Supplement 106, which formalized the inclusion of Part 9 of the JPEG 2000 standard (JPIP) (for more details on JPIP, see Chapter 6) in DICOM.

13.6.1 Supplement 106

Supplement 106 was drafted in late 2004 and finalized in January 2006. This supplement describes a mechanism for using JPIP to transmit partial or full image data as part of a DICOM interaction between a server and a client. According to Supplement 106, a DICOM client may request images from the server using the DICOM protocol while noting that it requires the use of JPIP. The server will respond to such requests with a DICOM response that replaces the image pixel data with a JPIP URL reference. The client will then use this JPIP URL to request the image data.

Since all of the image metadata are sent in the initial DICOM response, Supplement 106 limits the use of the JPIP protocol to image pixel data only; no image metadata is sent using JPIP. Supplement 106 further limits the use of JPIP to include only image data in the JPEG 2000 compressed format, either wholly (the entire image sent at once noninteractively) or incrementally. Supplement 106 further requires that the client support all three of the defined JPIP transmission modes: by tile part, by precinct, and by the full image return mode (see Section 6.3). Finally, Supplement 106 restricts the use of JPIP to the HTTP and HTTPS transport protocols. Other transport protocols, such as UDP, are explicitly forbidden.

13.6.2 JPIP and DICOM Use Cases

This section describes four typical usage scenarios of JPIP in a medical setting. They are examples of situations where a client wants to access a portion of the DICOM image pixel data without the need to wait for receipt of the entire image or sequence of images. The following usage scenarios are described:

1. *Large single-image navigation.* JPIP was designed for efficient browsing of images that are significantly larger than the available display, allowing the user to view regions of the image at different resolutions, without the need to receive the entire image. This situation arises in medical fields such as microscopy, pathology, and full body scans, all of which produce images of very large sizes. The client will typically start with a subresolution version of the entire image and then zoom and pan to view the area of interest at a higher resolution.

2. *Navigation of a stack of images.* CT, MRI, and similar acquisition devices often produce large stacks of uniformly sized images. In these cases, it is desirable to initially scroll through this large set of data at a lower resolution. This can be accomplished with JPIP by transmitting the low-resolution version of each image in the stack. Once a specific image or set of images is identified as being of interest, the full-resolution data for these images is transmitted. For the situation where the series of images has been compressed using Part 2 JPEG 2000, component collections can be used to access the images of interest without the need to transmit the compressed data for the entire series of images.

3. *Navigation among a series of images from varying acquisition devices.* Medical image workstations typically present thumbnails for each series of images, to aid the user in identifying the type of imagery in each series and determining which series are of interest. This can be accomplished with JPIP by transmitting a thumbnail-sized resolution version of a representative image in each series.

4. *Navigation of 3-D volume.* Stacks of CT and MRI images are also commonly viewed as rendered volumes rather than sets of 2-D slices. In these cases the combination of Part 2 MCT and JPIP allows the transmission of volumes that are reduced in resolution in all dimensions. For example, a volume of 1000 512×512 CT images could be compressed using JPEG 2000 Part 2 with a wavelet transform multiple-component transformation. A JPIP client could initially request to browse this volume at a spatial subresolution of 128×128. Since a wavelet transform MCT was utilized during compression of the volume, the client could also request to view the volume at quarter resolution in the third direction, resulting in a rendered volume of $250 \times 128 \times 128$.

References

Ahmed, N., Natarajan, T. and Rao, K. R. (1974) Discrete cosine transform, *IEEE Transactions on Computation*, **23**, 90–93.

Apple Computer, Inc. (1994) *Inside Macintosh: Operating System Utilities*, Addison Wesley Publishing Company.

Burt, P. J. and Adelson, E. H. (1983) The Laplacian pyramid as a compact image code, *IEEE Transactions on Communications*, **31**, 532–540.

Habibi, A. (1971) Comparison of nth-order DPCM encoder with linear transformations and block quantization techniques, *IEEE Transactions on Communications Technology*, **COM-19**, 948–956.

Hotelling, H. (1933) Analysis of a complex of statistical variables into principal components, *Journal of Educational Psychology*, **24**, 417–441 and 498 520.

Huffman, D. A. (1952) A method for the construction of minimum-redundancy codes, in *Proceedings of the IRE*, September 1952, pp. 1098–1102.

ISO/IEC (1994) Information Technology – Digital Compression and Coding of Continuous-Tone Still Images: Requirements and Guidelines, ISO/IEC 10918-1:1994, International Standards Organization, Geneva, Switzerland.

ISO/IEC (1999) Information Technology – Lossless and Near-Lossless Compression of Continuous-Tone Still Images: Baseline, ISO/IEC 14495-1:1999, International Standards Organization, Geneva, Switzerland.

NEMA (National Electrical Manufacturers Association) (2008) *Digital Imaging and Communications in Medicine (DICOM)*, Rosslyn, VA.

Tzannes, A. (2003) *Compression of 3-Dimensional Medical Image Data Using Part 2 of JPEG 2000* [internet], DICOM: Minutes of Working Group 4, February 2004. Available at: http://medical.nema.org/Dicom/minutes/WG-04/2004/2004-02-18/3D_compression_RSNA_2003_ver2.pdf (accessed 31 July 31, 2008).

Welch, T. A. (1984) A technique for high performance data compression, *IEEE Computer*, **17**(6), 819.

14

Digital Culture Imaging

Greg Colyer, Robert Buckley, and Athanassios Skodras

14.1 Introduction

Cultural heritage institutions, such as Museums and Art Galleries, as well as Libraries and Archives, are in possession of very extensive collections of valuable assets. Many of them are not visible to the public because of display capacity or other reasons. These materials are gradually being digitized, thus giving the possibility for them to be accessed by all interested people from anywhere at any time. Dealing with the digital surrogates of the cultural assets is demanding and continuous work. It starts with the proper digitization of the asset and continues with its preservation. *Digital preservation* or *digital curation* is the managed activities necessary for ensuring both the long-term maintenance of digital information and the continued accessibility of its contents. The digital cultural information can be of all types of media, namely text, photographs, prints, plans and maps, drawings, documentaries and films, sound and video recordings, music, philately, microfilms, periodicals, newspapers, manuscripts, or even three-dimensional (3-D) objects (BCR, 2008; Desrochers and Thurgood, 2007; IST Cultural Heritage, 2005).[1]

Digital imagery is likely to be the majority of the digital cultural heritage material. Huge image repositories, along with the associated metadata, constitute the Archival Information System. Such a system should flexibly and reliably allow the producers to ingest a digital asset and the managers to manage it, while it is distributed to the consumers (Figure 14.1).

For digital cultural imagery, it is critically important to use widely adopted standards that increase the chance of longevity in the face of technological change (MacDonald,

---

[1] 'European libraries and archives contain a wealth of material – including books, newspapers, films, photographs and maps – representing the richness of Europe's history, and its cultural and linguistic diversity. The online presence of this material from different cultures and in different languages will make it easier for citizens to appreciate their own cultural heritage as well as the heritage of other European countries, and use it for study, work or leisure' (COM, 2005).

The JPEG 2000 Suite   Edited by Peter Schelkens, Athanassios Skodras and Touradj Ebrahimi
© 2009 John Wiley & Sons, Ltd

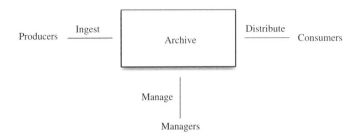

Figure 14.1 A typical Archival Information System

2006; JPEG, 2008). Long-term preservation of digital information is plagued by short media life, obsolete hardware and software, slow read times of old media, and defunct websites (Chen, 2001). Thus the use of recently developed standards that address many of the relevant issues in combination would contribute to the partial resolution of the digital preservation paradox above, and would greatly simplify the whole system, taking into consideration that the majority of the cultural institutions were using at least six digital formats in their collections (Yeung, 2004). The JPEG 2000 standard addresses many of the requirements of the cultural heritage sector. Specifically, it includes: efficient lossless compression with full color management; handling of moving and 3-D images with the same advantages; royalty- and license fee-free Core for widescale deployment; support for protection of moral and copyrights; unrestricted thumbnail viewing and encrypted high-resolution viewing from the same image file; comprehensive client/server architecture for users to be able to zoom in and out on images, or to request regions of interest to be served ahead of background material; extensive metadata possibilities, including the incorporation of Dublin Core metadata; adherence to well-accepted and defined standards, including the use of XML, HTTP, and others within the defined architecture (JPEG, 2008). Large-scale digital cultural heritage repositories have to face the preservation–access–performance issues, all at the same time. The JPEG 2000 standard possesses features that significantly contribute toward their treatment in a unified way, and this is what is going to be presented in the following sections.

14.2 The Digital Culture Context

14.2.1 Requirements

Long-term preservation, guaranteeing continued access to collections, is an important issue both motivating and constraining digitization. On the one hand, digitization enables the preservation in robust and incorruptible form of information originating from fragile or perishable objects such as ancient manuscripts. On the other hand, the rapid pace of technology development and the known difficulty, which has already been experienced, of accessing computer media whose physical or virtual formats are now out of date, calls into question the long-term viability of digital storage.

Increased access to collections for which there is insufficient gallery space to display in physical form is another motivation for digitization. Because of the international character of the Internet, online access also directly enables a wider audience. Multilingual issues therefore arise, e.g. the requirement to support international character sets.

Sophisticated kinds of online access include the ability to zoom in to examine detail as well as to see the entire object on screen at once. Access to additional data linked with the object (e.g. stories about its history) could also be included here; the primary data may be a digital image, but additional data may have a variety of other formats and properties.

Digitization must be cost-effective. This is linked to long-term preservation in an obvious way – future costs will be saved if digitization is longer-lasting – but is also a factor driving the requirement to be able to repurpose information flexibly after it has once been digitized. We will mention JPEG 2000's scalability features on many occasions below.

14.2.2 Processes and File Formats

A typical digital culture imaging workflow, depicted in Figure 14.2, consists of (Colyer and Clark, 2003):

- a digitization process whereby a clean, high-quality master image is first produced from the original object by scanning or photography;
- a management system, with associated database, responsible for linking the master image to other data about the object, about the digitization, and to other versions of the image used for other purposes;
- a delivery system, such as a web server, which makes available to users some subset of the full data (e.g. the master image may not be made available).

A typical workflow may utilize a variety of file formats, such as uncompressed TIFF for the master image, proprietary or nonimage formats managed in the central database, and compressed JPEG at various sizes for web delivery (HUL, 2008).

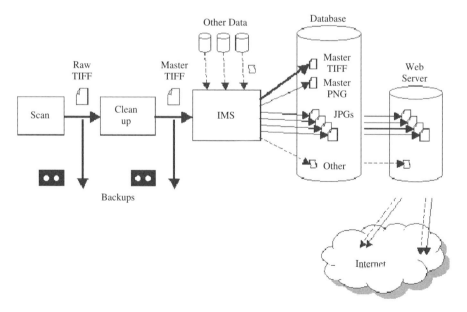

Figure 14.2 Typical digital culture imaging workflow. Permission to reproduce extracts from *PD 6777:2003, Guide to the practical implementation of JPEG 2000* is granted by BSI

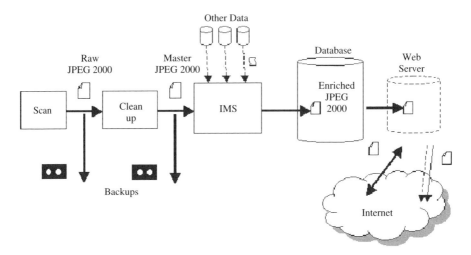

Figure 14.3 A JPEG 2000-based digital culture imaging workflow. Permission to reproduce extracts from *PD 6777:2003, Guide to the practical implementation of JPEG 2000* is granted by BSI

The above workflow could be highly simplified by the introduction of JPEG 2000, resulting in the single-stored file system of Figure 14.3 (Colyer and Clark, 2003). The scalability offered by JPEG 2000 affects the on-demand web delivery in a positive way. The delivery of different sized JPEG 2000 images can be done faster than the current transcoding from TIFF to JPEG. The image database system can potentially also be simplified, facilitating the combination of database and the web-server. As discussed later and in Chapter 6, fully interactive sessions (i.e. zoom and pan to access the image data at full quality and resolution if desired) on suitably equipped clients are also feasible.

The details may be highly application dependent. For example, if the digitized objects are multipage documents, it may be convenient to deliver them as multipage PDF files. A PDF file is not itself an image format, but has the capability of holding inside it a number of different formats for imagery that appears on its pages.

14.2.3 Technological Developments

We mention two particular aspects of developments in computer networking that strongly affect the deployment of collections online, because of the implications that they have for 'scalability' (dealt with in more detail later).

First, wired networks are becoming available with increasingly high speeds and across all sectors of society. Even home users now have access to broadband connection speeds previously associated only with direct connections to the Internet backbone. This means that online service providers can begin to deploy applications taking advantage of this bandwidth. The need for raw compression power, which was the driving force behind the adoption of JPEG, becomes less important for still imagery in this context, but advanced features such as the ability to zoom and pan efficiently become more so.

Second, interactive access across wireless networks is also becoming ubiquitous, and although their speeds are also increasing it is nevertheless the case that for the foreseeable future there will be a need to support connections with bandwidths below 10 Mbps. (Most wired networks already exceed this speed, but cellular wireless networks are considerably slower both now and in the projected next generation.) So there will also be a need to support online users for whom raw compression power *is* still a factor, along with other factors specific to wireless or mobile usage.

The human users of both types of network are of course often the same people. For this reason, and also to gain productivity on the supply side, it is desirable to use scalable systems that support both slow and fast connections, and support devices with both full-size and handheld displays. We could envisage an online art gallery, for example, that is usable on both a broadband-connected PC and a mobile phone (Politou, Pavlidis, and Chamzas, 2004).

14.2.4 Digitization

14.2.4.1 Transcoding Existing Image Repositories

In many digital culture contexts, JPEG 2000 will be considered for systems involving large existing repositories of images. The question arises as to whether they should be converted to JPEG 2000 or whether only new images should use the format. To a large extent this question is answerable only on an individual project basis, as it depends on so many external factors (such as the capability of various parts of the system to handle multiple image types). We wish to address just two aspects of the question here.

First, is there a sense in which JPEG 2000 is an 'upgraded' version of JPEG, such that JPEG tools may naturally be expected to support the new format, either immediately or soon? The answer is that there is such a sense, but that in this context it is not a very useful one. JPEG and JPEG 2000 share the low-level syntax that is used to structure the code-stream; however, this is of little more than historical interest, because what is built up with that syntax is almost completely different. Furthermore, a JPEG 2000 code-stream is surrounded by another layer of encapsulation, the JPEG 2000 file format, which is different again. (JPEG has no equivalent: in a JPEG file, any additional information is stored *within* the code-stream.) A small amount of code could probably be shared between a JPEG codec and a JPEG 2000 codec. This may help slightly those vendors whose applications already support multiple formats to add JPEG 2000 to the list; it will make a correspondingly small difference to the amount of work required to convert a JPEG-specific tool to JPEG 2000. For practical purposes, JPEG 2000 should be regarded as a completely new format.

The second aspect of the question is one on which more research could be done. In comparing the performance of different algorithms, testing has usually involved compression directly from uncompressed or lossless originals. Performance may differ when significantly compressed images of one type (e.g. JPEG) are converted to significantly compressed images of another type (e.g. JPEG 2000). In a limited small-scale test, it was found that when the size of the final JPEG 2000 code-stream was the same or not much smaller than the size of the initial JPEG code-stream, the PSNR measure of image quality fell by no more than a fraction of a decibel, and in cases of strong JPEG compression could even increase, as shown in Figure 14.4 (Colyer and Clark, 2003).[2] As these results

---

[2] An increase in PSNR sounds paradoxical: where does the new image recover the extra quality from? The improvement can, however, be thought of as due to the partial cancellation of 'sharpening' artifacts that occur

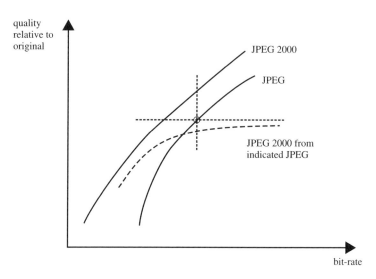

Figure 14.4 Compression efficiency of JPEG, JPEG 2000, and JPEG to JPEG 2000 transcoded images. Permission to reproduce extracts from *PD 6777:2003, Guide to the practical implementation of JPEG 2000* is granted by BSI

should not be taken as universally relevant, it would be sensible for users to conduct testing on their own data sets if there is a lossily compressed image at both ends of the conversion process.

Note that converting a highly compressed JPEG to a lossless JPEG 2000, while not introducing any new compression artifacts, will lead to a large increase in file size. If it is desired to preserve any existing compression artifacts, perhaps regarded as part of the cultural context of the image, then it may be more sensible to leave such images in the original format.

14.3   Digital Culture and JPEG 2000

In this section we first describe some of the features of JPEG 2000 that match the requirements of archivists and curators separately. We then address its ability to integrate both preservation and delivery aspects simultaneously. Finally, we consider some technical barriers to, and other aspects of, their exploitation.

14.3.1   Some Features that Match the Requirements of Archivists

14.3.1.1   Longevity, Openness, and Availability

JPEG 2000 is an international standard published as ISO/IEC 15444 in several parts, many of which are also published with identical text as ITU-T Recommendations T.800 onward. The standard is publicly available in printed and electronic form, and is likely to

---

with high JPEG compression and 'blurring' artifacts that occur with high JPEG 2000 compression, as shown in Figure 14.5. In general, the possibility also exists that completely new kinds of 'transcoding artifact' could arise from the interaction of two different compression algorithms, but this was not investigated.

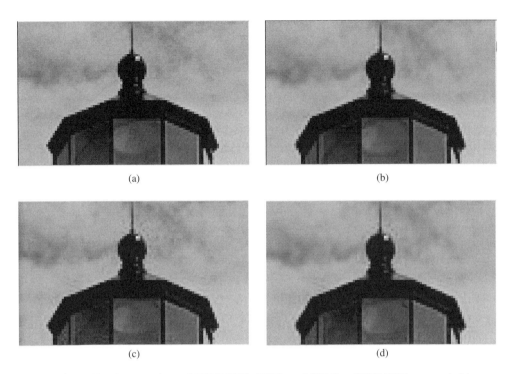

(a)   (b)

(c)   (d)

Figure 14.5   Visual comparison of JPEG 2000, JPEG, and JPEG to JPEG 2000 transcoded image: (a) original, (b) JPEG 2000 at 2 bpp, (c) JPEG at 2 bpp, (d) JPEG to JPEG 2000 transcoded at 2 bpp. All images refer to part of the color *lighthouse.tif* original of size 512×768

remain so for a long time. Some parts of the standard are available free of charge: these include the three file-format-related annexes from Parts 1 and 2, and the whole of Parts 4, 5, and 12.

This provenance also means that JPEG 2000 was developed cooperatively by experts from many different organizations and countries, working together within the open framework of the ISO/IEC standardization process. No single member organization controlled or dominated the outcome and therefore the standard does not serve only a single interest. More importantly, perhaps, future work on JPEG 2000 – corrigenda, amendments, new parts, related initiatives – is also open to new participants through their national standards bodies.

In these respects international standards differ not only from proprietary technologies that have not been published openly but even from published technologies over which the originator retains control. It is in fact quite common for the membership profile of standards committees to change as the technology changes from early development to later deployment, and the JPEG committee welcomes participation from consumer organizations as well as vendor companies and from users as well as tool creators.

Of course, it may also be true that other venues are more appropriate for certain initiatives related to JPEG 2000 that are specific to the digital culture context. Various aspects of JPEG 2000 were designed to be not only flexible but also amenable to domain-specific 'profiling.' For example, the presence or absence of certain elements of the file format structure could be recommended. The use of a certain number of wavelet decomposition levels or certain color spaces could be required, and so on. Other ISO/IEC committees and other external organizations often liaise with the JPEG committee when both JPEG 2000 and domain-specific issues are involved.

14.3.1.2 Color Management

A JPEG 2000 file includes precise information about the color space used to represent the image data, which is needed to ensure the accurate rendering of the image. sRGB colorspace and the grayscale and luminance–chrominance versions of it are the only color spaces that may be signaled directly by name in JP2 files (see Chapter 1). Two classes of ICC input profile are also supported in JP2 files; they can be used to encode a wider range of grayscale and RGB color spaces. JPEG 2000 supports image data with high dynamic range (more than 8 bits per color component), including scene-referred data. JPX files support input and output ICC profiles more generally than JP2. Output profiles enable the specification of rendering intents for gamut mapping.

14.3.2 Some Features that Match the Requirements of Curators

14.3.2.1 Points or Regions of Interest

JPEG 2000 provides two mechanisms for enhancing the coding of specific regions of interest within an image. One mechanism, specified in Part 1 of the standard, is applicable to regions of any shape and has the effect of coding those regions' data first (so that it is decoded first for display) in a layer-progressive ordering, and potentially to a higher quality than the rest of the image (see Chapter 1). The other, an extension specified in Part 2 of the standard, is applicable only to rectangular and elliptical image regions, but provides more flexibility with respect to data ordering and quality enhancement (see Chapter 2). The JPEG 2000 file format also enables regions of interest to be labeled and to have metadata associated with them. Other mechanisms can also be used to attach metadata to specific locations within an image. For example, a satellite image or aerial photograph can be associated with a mapping to geospatial coordinates.

14.3.2.2 JPIP

JPIP, the JPEG 2000 Interactivity Protocol, is defined in Part 9 of the standard, which was published in 2005. It provides a mechanism for a client system to request from a server only the image data that are appropriate to its particular current view-window. Irrelevant information (e.g. about other regions of the image or corresponding to a higher display resolution) is not sent over the network, thereby reducing the bandwidth required for a given level of performance. JPIP includes mechanisms for requesting metadata appropriate to the view-window or to the image as a whole. JPIP is particularly applicable to the interactive exploration of large image files (see Chapter 6).

14.3.3   Integration of Preservation and Delivery

14.3.3.1   Uncompressed, Losslessly Compressed, and Lossy Compressed Images

Different types of file format may be used for storing digital images:

- uncompressed, in which the digitized image data are stored in their original form or in an encoding that requires at least as much space as the original form;
- losslessly compressed, from which the original data can still be recovered exactly; and
- lossy compressed, after some information has been discarded to achieve even greater compression.

TIFF can be an example of the first type, PNG is an example of the second type, and JPEG is an example of the third type. Obviously, for a given image the third type can be produced from the first or second, but the first or second cannot be regenerated from the third. The third type is therefore generally smaller in size than the second, and this can lead to a misconception that 'compressed' necessarily means lossy or that original master files are necessarily 'uncompressed.' However, this is not the case, because lossless compression is possible and may be suitable for storing originals. (In another context, the ZIP format is a type of lossless compression suitable, for example, for text.)

JPEG 2000 incorporates both lossless and lossy image compression. Thus, it may be used both for preservation masters and for smaller, more highly compressed versions of the same image. A single set of software tools may be adequate for both these sides of digital culture usage.

Lossless compression is achieved in JPEG 2000 by choosing the reversible (integer) rather than the irreversible (floating-point) wavelet and component transforms, by omitting quantization, and by retaining all the coded data. In fact, most stages of the JPEG 2000 encoding process are lossless by design – exactly so if only the reversible transforms are used, but very nearly so if an irreversible transform is used. The discarding of information that is necessary to achieve greater, lossy compression is performed after most of the rest of the coding, and essentially consists of choosing where to truncate the lossless code-stream in a fine-grained layer-progressive order (although the code-stream may actually be written out in some other progression order and with fewer layers). Thus, lossy compression is an additional, and relatively simple, step beyond lossless compression. There is no 'boundary' in JPEG 2000 between lossless and lossy: lossless is simply the trivial choice to truncate the code-stream at the end, or in other words not to truncate it.

Thus JPEG 2000 not only supports both lossless and lossy compression, but integrates the two by relating them with a conceptually simple operation. The step from lossless to lossy compression makes use of only a part of the JPEG 2000 codec and requires correspondingly fewer computational resources. In some cases, it may be feasible to perform the transcoding from one degree of compression to another on demand, thereby eliminating the need for permanent intermediate image files of lower quality.

The same applies not only to reduced-quality versions of the image at the same size but also to reduced-resolution versions such as thumbnails: this is JPEG 2000's resolution scalability. If a master file also serves as the source for real-time delivery, the concept of 'master' may become redundant if the master is in fact the *only* file that is stored. This is the case, for example, with the National Digital Newspaper Program and their Chronicling America website, where JPEG 2000 is used for the production master. (A separate archival master is also retained offline.) The JPEG files that are delivered to clients are dynamically created on demand from these JPEG 2000 production masters, which in fact contain lossy but visually lossless tiled grayscale images with multiple resolution levels and multiple quality layers (Buckley and Sam, 2006). The use of JPEG 2000 by the National Digital Newspaper Program is described later in this chapter.

14.3.3.2 Metadata

Color space information, mentioned above, is only one type of metadata that may be associated with image data. The JPEG 2000 file format supports other specific and general types, including:

- capture resolution and intended display resolution;
- intellectual property rights;
- arbitrary XML documents;
- links to external references (URL boxes);
- arbitrary custom metadata (UUID boxes).

In the digital culture context, the user is therefore not restricted by a predefined choice of metadata fields. The DIG35 metadata schema has been incorporated into the standard, but its use is not mandated: an XML box may contain *any* XML 1.0 document and a UUID box may contain data of any format at all. The UUID box is so called because it contains a unique identifier that identifies the rest of the content. These UUIDs are easily generated for private use; there is also an official mechanism for registering them globally to achieve greater interoperability between systems. Because a system encountering an unknown UUID will simply ignore the box, a UUID box must not contain any information necessary for decoding and rendering the image. The same applies to an XML box.

14.3.3.3 Image Management

The use of JPEG 2000 for both preservation and delivery, and the integration of metadata within the JPEG 2000 file, may transform the requirements for image management. In some situations, it may eliminate the need for separate management systems or databases altogether. In others, there may still be reasons for additional management systems, but their requirements may be greatly simplified.

14.3.3.4 JPM

Part 6 of the JPEG 2000 standard defines a file format called JPM based on the Mixed Raster Content model of ISO/IEC 16485. A JPM file represents a multipage image document, where each page can be a compound image, consisting of text and image areas. In a heterogeneous document, the images of text areas and image areas, for example, may be coded using different parameters or even different algorithms according to their different characteristics and rendering requirements – not only JPEG 2000, but a variety of other codecs, including ones for black-and-white images, is supported by JPM.

JPM files may be made JP2-compatible, such that a JP2-only image reader would display a sensible image from the file. This could be chosen to be a rendering of the first page of the entire document, for example.

14.3.4 Some Technological Barriers

14.3.4.1 Market Adoption

The take-up of JPEG 2000 since Part 1 was published at the beginning of 2001 has been slow compared to the original JPEG standard, which was supported by some of the first web browsers only a year or so after its publication in 1994. Partly this is because of the relative complexity of the JPEG 2000 codec. Probably a more significant factor is that the explosion of the Internet was the 'killer app' for JPEG and the same situation has not repeated itself. Rather, JPEG 2000 is a development in a now more mature field of technology. It was designed looking ahead to more predictable growth patterns that are still in the process of happening. For example, JPEG 2000 was designed with high-compression performance over low-bandwidth wireless channels in mind, but the ubiquity of cameras in mobile phones and the consequent growth in image traffic over cellular networks is a phenomenon of only the last couple of years or so.

However, although the take-up of JPEG 2000 has not been rapid, it has been steady. Support has gradually been incorporated into widespread software tools such as Apple QuickTime, Adobe Reader, and Photoshop. More and more examples of deployment on the server side of online image delivery systems are apparent. Because of the continuing lack of native support in most popular web browsers, these systems use some form of transcoding (typically to JPEG) for delivery to the client system, and this loses some of the benefits of using JPEG 2000 throughout. However, there is an obvious upgrade path when direct delivery to the browser becomes possible. The lack of JPEG 2000 digital cameras is probably less of an immediate issue in the digital culture context, as most original digitization will be done at very high quality, and transcoding to a JPEG 2000 master would only occur once.

A number of other standardization efforts have incorporated JPEG 2000 into their own work, among them DICOM for medical imaging and the Digital Cinema Initiative (see Chapter 9). Metadata schemas for use with JPEG 2000 have also been published, e.g. GMLJP2 (which shows how to incorporate GIS metadata in a JPX file) by the Open Geospatial Consortium.

It remains to be seen how market adoption will continue, but at the time of writing there is no reason to suppose that it will reverse and be abandoned. In fact, there has been a steady growth in the use of JPEG 2000 as cultural heritage institutions deal with the increasing demands for storage of and access to their content, both born digital and digitized (Gillesse, Rog, and Verheusen, 2008).

14.3.4.2 Patents

A number of patents have been identified that may cover aspects of JPEG 2000 Part 1 (the core coding system). The JPEG committee sought a commitment from the licensors of these patents to license implementations of Part 1 on a nondiscriminatory, royalty- and fee-free basis. They are listed in an Annex to Part 1 and online at the ISO and ITU websites (ISO, 2008; ITU, 2008). A similar process is now underway for Part 13, the reference encoder.

The mere existence of these patents is sometimes perceived as an impediment to the adoption of JPEG 2000. However, with any new technology the possibility exists that a relevant patent may come to light at a later stage; the committee's efforts in fact constitute an extensive attempt at mitigation. In the context of very long-term preservation it should also be noted that a patent is a monopoly with only a limited life-span and is also another form of open publication of the patented technology. Even in the short term, a patent has limited geographical scope.

Of course, many widely used technologies involve patent licensing. Typically, a vendor will negotiate a license with any identified patent holder to cover all of the vendor's customers and may also indemnify the customers against any infringement that comes to light later due to unidentified patents. The customer of such a vendor need not worry about patent infringement in the context of JPEG 2000 any more than in other contexts. While technologies thought likely to infringe patents for which free licenses are not available, even only in one country, tend to be less favored by free software developers, open-source implementations of JPEG 2000 do already exist (two of them constituting the reference software in Part 5).

14.3.5   Exploitability of Designed-In Features

A problem often faced by software users is that, even though the software may conform to certain standards, and even though those standards were designed to enable certain benefits, some of those benefits are unavailable to end users because they are not fully supported by the software. For example, a standard may allow much flexibility in the choice of certain parameters, but early implementations may fix those parameters for the convenience of the application developers, and consequently restrict the flexibility of end users. Significant restrictions of this kind tend to decrease over time, as applications become more fully featured and as their designers respond to feedback from customers.

Convenience is not the only reason why application developers may choose to fix parameters that are left open by a standard. It may be that hardware performance or other system capabilities force them to do so. Certain choices may determine other choices. This kind of restriction may require technological improvements in other areas before it can be lifted or it may be that one of the possible choices becomes adopted by the market to the exclusion of the others. This happened with the JPEG standard, which defines many different code-stream profiles, almost all of which would be unreadable by today's JPEG software. Interoperability is thereby achieved more easily but at the cost of reduced flexibility.

How does JPEG 2000 fare in this respect? Because it is a relatively new standard, software toolkits and applications are at a relatively early stage of development compared to legacy formats and in a given case may not yet effectively enable some of the ideals that are described in this chapter. It is also an early stage at which to evaluate the relative performance of different possible coding choices, which may subsequently determine market restriction effects of the kind just described.

The intention of this section is to indicate some areas where it may be particularly beneficial to focus development efforts. Fortunately, many projects in the digital culture context involve large-scale digitization programs, which are in a strong position to influence the direction of software development by vendors. Smaller projects with similar requirements will also benefit as a result.

14.3.5.1 Scalability to Lossless – Reversible and Irreversible Transforms

The combination of lossless and lossy compression within a single codec was emphasized above. Truly lossless compression, without even the small loss of information due to floating-point rounding errors, requires the use of only reversible wavelet and component transforms: these are based on integer arithmetic. Thus, a lossy code-stream obtained from such a lossless code-stream by minimal transcoding would also use only reversible transforms. This is the scenario in which true 'scalability to lossless' is possible and in which a truly lossless archival master could potentially also serve as the delivery master.

The irreversible transforms typically achieve better compression performance; i.e. for a given quality (other than truly lossless), a smaller code-stream will typically be obtained if the irreversible transforms are used. In situations where true losslessness will never be an issue, or where compression performance is absolutely paramount, the irreversible transforms may therefore be the best choice, but the potential future workflow benefits of complete scalability should be borne in mind even if the current system architecture requires separate archival and delivery files. To require the use of reversible transforms for one purpose and irreversible transforms for the other would perpetuate a barrier between them that JPEG 2000 has the potential to render unnecessary.

Two important points should be noted here. First, the compression performance of the reversible transforms, even if slightly worse than the irreversible ones, may still be significantly better than alternatives to JPEG 2000, and this may be the most relevant comparison to make when the other benefits of scalability are also at stake.

Second, any digital image is inherently 'lossy' in the sense that the original digitization is of limited dynamic range and resolution. (A rectangular array of integer pixel values contains much less information than the real object or scene that it depicts.) Although use of the irreversible transforms is not 'truly lossless,' the loss thereby introduced may be negligible when the effect of the original digitization is considered. If this is the case, then the irreversible transforms could be used on both archival and delivery sides in the same manner as the reversible transforms. This case has already started to be applied by some institutions. As an example, Library and Archives Canada, for production or access masters, has recommended a compression ratio of 24:1 for color images, which included photographs, prints, drawings and maps, and 8:1 for grayscale images, which included newspapers, microfilm, and textual materials (Buckley, 2008; Desrochers and Thurgood, 2007).

14.3.5.2 Tiles and Precincts

Two different code-stream structures, called tiles and precincts, facilitate access to spatial subregions of JPEG 2000 images. Both are tessellations of the image data into grids of rectangles, but they occur at different stages of the coding process and consequently have different scalability properties. Tiling occurs at the beginning of the coding process, 'in the image domain.' A tile is a rectangular region of the original image and each tile proceeds independently through the subsequent coding stages before the tile data are recombined in the image code-stream. Tile size is therefore a property of the entire coded image. A precinct, on the other hand, is a rectangular region of subband data 'in the wavelet domain;' the precinct size may be different at every resolution level (see Chapter 1).

Resolution scalability is diminished when accessing regions that contain a large number of tiles; therefore it is undesirable for the tile size to be enormously smaller than the largest portion of the image that would ever be viewed at one time. For many applications, the ideal would be to encode the entire image as a single tile. (This may be described as 'no tiling.') However, a larger tile size potentially increases the burden on the wavelet transform stage of the algorithm. Performance considerations may therefore place a maximum limit on the tile size. Library and Archives Canada has recommended tile sizes of 512×512 and 1024×1024 (Desrochers and Thurgood, 2007).

A precinct is the smallest region that is independently spatially accessible in JPEG 2000. Precincts enable spatial accessibility at the sub-tile level; they may, for example, be chosen to have the same size (in the wavelet domain) at every resolution level, so that a view-window of a given pixel size corresponds to a similar number of precincts regardless of the zoom factor. In this case the precincts would have *different* sizes in the image domain. If only one precinct is used per tile at all resolution levels, then spatial accessibility is lost within the tile. (This may be described as 'no precincts.') However, if the tile size is not much larger than the smallest portion of the image that would ever be viewed on its own, then using multiple precincts may provide insufficient benefit to justify the additional coding cost.

The JPIP delivery protocol supports two kinds of data-stream, one for tiles (called JPT-stream) and one for precincts (called JPP-stream). The protocol was not really designed to exploit both structures at the same time and it is possible that one or the other approach may eventually win out in the market.

For completeness, we note here that because of differences in the coding algorithms neither tiles nor precincts are likely to give rise to artifacts analogous to the 8×8 pixel 'blocking' associated with aggressive JPEG compression. Both tiles and precincts are typically much larger than 8 pixels in size; precincts, moreover, have sharp boundaries in the wavelet domain but overlap in the image domain.

14.3.5.3 Progression Orders and Transcoding

Part 1 of the JPEG 2000 standard supports five code-stream progression orders. For simplicity we will consider only LRCP (see Section 1.3.6.1) and RLCP (see Section 1.3.6.2), furthermore suppressing consideration of the C and P dimensions, to make our point here. The four letters stand for the four dimensions of scalability: L for layer (corresponding to an increment of image quality), R for resolution, C for (color) component, and P for (spatial) position, or precinct.

LRCP is called layer-progressive, meaning that at the outer level of packet ordering within the code-stream, the lowest quality layer comes first, then the next layer, and so on up to the highest quality layer. Within the first group of packets, constituting the lowest layer, the resolution levels each occur in order: first the lowest resolution level, then the next resolution level, and so on up to the highest resolution level. The resolution levels all occur again in this order within each of the subsequent layers. Given such a code-stream, to produce a more highly compressed one essentially involves no more than truncating it earlier, discarding some number of layers at the upper end. It is a trivial operation with which there should be no performance problem. On the other hand, to extract a low-resolution version of the image involves removing the highest resolution levels from *each* layer; although this is still a conceptually simple operation – in particular, no packet decoding is required – it is clearly not so trivial as the first operation, involving a greater degree of parsing and reconstruction of the code-stream.

RLCP is called resolution-progressive, and has exactly the opposite properties. Producing a lower resolution image is essentially just truncation, whereas to increase compression requires layer removal from within *each* resolution level.

What if a highly compressed, low-resolution image is required? In this case, it may be necessary to remove both layers and resolution levels, which means that the operation is not merely truncation, whichever progression order is used. A thumbnail image used for selection rather than examination could be an example of such a requirement. The point is that even if one or other progression order is preferred for overall performance, to gain the maximum benefit from JPEG 2000's scalability it is necessary to be able to perform these other kinds of transcoding efficiently. Note, however, that even such operations do not require packet decoding: they are still just a pruning of certain packets from the code-stream. For some purposes, such as JPIP streaming, reordering of the packets may also be necessary, but this still does not require any packet decoding. Focusing development efforts on removing any performance barriers here would help to ensure that scalability benefits are maximized.

14.4 Application – National Digital Newspaper Program

Large-scale digitization projects are common now as many cultural heritage institutions and organizations look to improve access to printed materials by turning them into digital files that then can be made available online. Among them are several newspaper digitization projects, many of which use JPEG 2000. One of the first to do so was the National Digital Newspaper Program in the US.

The National Digital Newspaper Program (NDNP) is a partnership between the National Endowment for the Humanities (NEH) and the Library of Congress. It aims to enhance access to historical American newspapers by creating an Internet-accessible, digital resource of historically significant newspapers. The goals of the approach taken include:

- convenient accessibility over the World Wide Web for the general public;
- page images of sufficient spatial and tonal resolution to support effective OCR performance;
- use of digital formats with a high probability of sustainability;
- attention to the cost of digital conversion and maintenance.

The first phase began in 2005 and a prototype website was launched in May 2007 (http://www.loc.gov/chroniclingamerica/), with digitized newspaper pages from 1880 to 1910. The digitization of newspapers is performed by institutions that apply for and are awarded grants. As a rule, the awardees digitize newspapers from the microfilm at hand on which the newspaper pages were previously imaged for preservation. Figure 14.6 shows the image of a sample NDNP newspaper page that was scanned from microfilm.

Among the deliverables are two raster images for each newspaper page: a grayscale TIFF file with uncompressed image data and a JP2 file derived from it and containing JPEG 2000 compressed data encoded with parameters specifically designed for use with NDNP (Buckley and Sam, 2006; NDNP, 2008). The grayscale TIFF files, usually with a spatial resolution of 400 dpi with respect to the original material, serve as archival masters. The JP2 files derived from them serve as the production masters. The JP2 file format is specified in Part 1 of the JPEG 2000 standard (see Chapter 1 and Houchin and Singer, 2002). The use of JPEG 2000 offers multiple views and interactive zoom, pan, and rotate, which are features more easily implemented with JPEG 2000 than in other image formats.

As production masters, the JP2 files are the files that users interact with online. However, they are usually unaware that they are dealing with JPEG 2000 files since the data for the requested view is transcoded at the server to JPEG for delivery to a standard browser at the client. This extra step means that users do not need to download a plug-in to view JPEG 2000 data directly, although some institutions, such as the National Archives of Japan on their Digital Gallery website, offer users the option of installing a JPEG 2000 plug-in that uses JPIP for client–server interactions (see Chapter 6).

To meet the needs of Phase 1 of the NDNP, the recommended format for the production master is a visually lossless, tiled, JPEG 2000 compressed grayscale image, with multiple resolution levels and multiple quality layers, encapsulated in a JP2 file with Dublin Core-compliant metadata. The encoding parameters for the JPEG 2000 code-stream were selected with performance in general and decompression performance in particular in mind since the NDNP paradigm is 'Compress once, decompress many times.'

14.4.1 JPEG 2000 Code-Stream

To begin with, the decision was made to use visually lossless JPEG 2000 compression (see Section 14.3.3.1). Although the compression is lossy in the production master, an uncompressed archival master is retained as an exact record from which the original scan can be recovered. While lossy compression saves space, the differences it creates from the original should not be visible or noticeable when the images are viewed.

In tests that focused on the quality of text presented on the screen, only the most experienced observers could locate compression artifacts in page images with a compression ratio of 8:1. In most cases, the images were acceptable at higher compression ratios.[3] Also, there was no indication that lossy compression adversely affected the quality of OCR results. While, in general, the uniform compression ratio does not give uniform visual quality with respect to the uncompressed original, the divergence between the two was reduced due to the homogeneity of the material in this case.[4]

---

[3] These tests were performed using decompressed JPEG 2000 images before JPEG transcoding, as the user has the option of downloading the production masters. Artifacts in the images delivered to the client browser would be the result of applying JPEG compression.

[4] Some JPEG 2000 implementations use the PSNR instead of the compression ratio (or equivalently bit-rate) as a parameter to set the amount of compression.

Figure 14.6 Sample newspaper page (courtesy of the Library of Congress)

After the compression ratio, the next significant design choice was the number of decomposition or resolution levels. Compression improves with up to three or four resolutions levels or applications of the wavelet transform, but little if at all with additional resolution levels. Instead, additional resolution levels are used so that the lowest resolution subband image in the code-stream has a desired image size. This makes it possible, for example, to obtain a thumbnail image from a JP2 file by simply decompressing the lowest resolution level.

Consider five resolution levels applied to an 8832×6144 original image, which is about the size of a broadsheet newspaper page scanned at 400 dpi. In this case, the lowest resolution subband image will be 276×192, slightly smaller than QVGA size, which is 320×240. To obtain QVGA-sized or smaller images and in anticipation of larger originals, NDNP uses six resolution levels. To obtain QVGA-sized lowest resolution images across the broadest range of original sizes, the number of resolution levels would vary with the original image size.

Specifying a resolution-major progression order organizes the JPEG 2000 code-steam so that the compressed image data for the lowest resolution level occurs at the beginning of the code-stream, where it can be quickly found, read, and decompressed. In the case of NDNP, the progression order is RLCP, which is resolution level–layer–component–position. A resolution-major progression order fits well when the initial presentation is a low-resolution view of the entire image, followed by zooming in and panning around (see Section 14.3.5.3).

After resolution, the next element in the progression is quality layer. It was noted earlier that in some cases image quality is acceptable at compression ratios higher than 8:1. The use of multiple layers enables a single JPEG 2000 code-stream to offer multiple decompression options. Decompressing all the layers is equivalent to decompressing an image with the lowest compression ratio (and highest quality); decompressing all layers but one to an image with the next lowest decompression ratio; and decompressing only one layer to an image with the highest compression ratio (but lowest quality).

Which option is invoked can vary with the application and the view. At maximum resolution, it is usual to decompress all layers to maximize quality and minimize visible artifacts. However, at lower resolution views, such as thumbnail images, most artifacts are hardly, if at all, visible and higher compression ratios are tolerable. Decompressing fewer layers at lower resolution views means less data are accessed and decompressed. As a result, overall decompression can be faster and avoids decompressing layers that do not improve the visual quality of the decompressed image at reduced resolution, as illustrated in Figure 14.7 (see Buckley, Stumbo, and Reid, 2007). The magnitude of the effect depends on the distribution of layer data across resolution levels.

In the case of NDNP, the minimum compression ratio is 8:1 and the maximum is 512:1. (For a grayscale image, these correspond to bit-rates of 1 and 0.015625 bits per pixel.) This means that decompressing all layers is equivalent to decompressing an image with a compression ratio of 8:1; decompressing one, to an image with a compression ratio of 512:1.

This range is covered by 25 layers, selected so that the logarithms of the compression ratios corresponding to the layers are uniformly distributed between the minimum and maximum values. What is significant about the number and distribution of the layers

(a)   (b)   (c)

Figure 14.7   Reduced resolution views decompressed with variable numbers of layers to (a) 8:1, (b) 32:1, and (c) 512:1 to show quality scalability. In this presentation, (a) and (b) are indistinguishable while (c) is noticeably different (image courtesy of the Library of Congress)

is that they provide an adequate set to choose from when decompressing the image at reduced resolutions and for different applications.

These many layers create some overhead in the code-stream. When the file size is fixed, as it is when the overall compression (actually the compressed file size) ratio is set to 8:1, this overhead takes away from the number of bits available for the compressed data. As a result, with fewer compressed image bits, the image quality is lower when multiple layers are used. However, tests on the sample images used in the development of the NDNP profile showed that this reduction was imperceptible. Compressing with 25 layers instead of one was equivalent to using a compression ratio of about 8.1:1 instead of 8:1, so the cost of quality scalability is small.

Besides zooming, panning is the other main way that users interact with large images. While JPEG 2000 offers resolution and image scalability to support zooming by factors of 2 (using resolution levels), it offers spatial addressability for panning. The two mechanisms for addressing localized regions of the image in the code-stream are tiles and precincts (see Section 14.3.5.2). When creating a JPEG 2000 code-stream, one or other or both can be explicitly specified.

Besides being a mechanism for spatial addressability, tiles are also used for memory management when compressing and decompressing images. Because different JPEG 2000 implementations take different approaches to memory management, some perform noticeably better when images are tiled and some when images have precincts.

With NDNP both approaches were tested. It was found that, with the server implementation that the Library of Congress uses for NDNP, images were decompressed significantly faster when they were tiled than when they had precincts. Accordingly, NDNP uses JPEG 2000 compressed images with tiles. Smaller tile sizes mean finer addressability but also more tiles and more overhead, and thus lower image quality, if only slightly. In general, tile size would scale with image size. For the image sizes typically encountered when scanning newspapers, tiles of size 1024×1024 were judged a reasonable choice for tiling in NDNP production masters.

Choices such as the number of layers and tile size affect performance and coding efficiency. Coding efficiency is a measure of how well the coder compresses an image. The more efficient the coder, the better is the image quality (the closer the decompressed image is to the original) for a given compressed file size, or the smaller the compressed file size for a given image quality. While the coding efficiency is higher with larger tiles and fewer layers, spatial addressability is higher with smaller tiles and quality scalability is greater with more layers. As noted earlier, the choice for tile size and number of layers is thus a compromise between performance needs and coder efficiency.

Other compression parameters for which there are similar tradeoffs are code-block size, coder bypass, and wavelet filter. These parameters are described in Chapter 1. In the case of NDNP, 64-square code-blocks are specified, along with coder bypass. With coder bypass, the coder leaves untouched the least significant bit-planes of the wavelet-transformed image. As a result, some of the data are not compressed; in return for a loss in coding efficiency, the decoding (and encoding) is speeded up.

Part 1 of the JPEG 2000 standard defines two wavelet filters: a 9-7 irreversible and a 5-3 reversible (see Chapter 1). The 5-3 filter is intended for applications that need to be able to recover the original image data from the compressed data. This is not a requirement for the NDNP production masters, which use the 9-7 filter. Tests showed that the 9-7 filter was significantly more efficient than the 5-3 filter and, although the 5-3 filter was designed so that it could be implemented with integer arithmetic, there was no significant difference in decompression times between the two filters. From the beginning, the 9-7 filter was used in the tests to develop the NDNP JPEG 2000 profile. Table 14.1 gives the design choices for the main JPEG 2000 parameters in the NDNP production master.

These choices are consistent with Profile 1, one of the code-stream restrictions defined in Part 1 of the JPEG 2000 standard. Most current cultural heritage applications can use the code-stream defined in Part 1 of the standard and do not need the extensions of Part 2.

14.4.2 JP2 File Format

The JPEG 2000 code-stream conforming to the NDNP profile is embedded in a JP2 file (Houchin and Singer, 2002). A major purpose of the file format is to associate metadata with the image. In this regard, the two most significant boxes in a JP2 file are the

Table 14.1 NDNP JPEG 2000 Profile

Parameter	Value
Number of components	1 (grayscale)
Number of bits per component	8
Compression	Visually lossless
Compression ratio	8:1
Number of resolution levels	6
Progression order	RLCP
Number of layers	25
Tile size	1024×1024
Precincts	No
Code-block size	64×64
Coder bypass	Yes
Wavelet transform	9-7 irreversible

JP2 header box, which contains technical metadata, and an XML box, which contains descriptive and adminstrative metadata.

Image parameters such as image height and width are necessary for decoding the compressed image; they occur in both the code-stream and the JP2 header box. Two other technical metadata elements necessary for interpreting the image are color (in this case grayscale) specification, and resolution, both of which occur only in the JP2 header box. Since the operation that creates a JPEG 2000 production master does not alter the resolution or tone scale of the image, the color specification and resolution in the production master are the same as those in the TIFF file from which it was derived.

The two methods for specifying the gray content of the image both assume a calibrated process for capturing and digitizing the image. The first method uses a preset value, which indicates that the color encoding of the image is the grayscale equivalent of the sRGB colorspace, which means it uses the same nonlinearity or gamma correction as sRGB. The second method inserts in the colour specification box of the JP2 header an ICC monochrome input profile, which allows the use of gamma-corrected values with a different nonlinearity than sRGB.

The resolution of the newspaper page image is usually 400 dpi with respect to the original page. For a broadsheet newspaper, this gives page images with a height and width of about 9000 lines and 6000 pixels. If the reduction ratio when scanning from microfilm is too high ($20\times$ or below is preferred), then lower resolutions, typically down to 300 dpi, are permissible provided they do not adversely affect OCR performance (NDNP, 2008). The image resolution given as pixels/inch in the TIFF file is represented as pixels/meter in the JP2 file.

Meeting the NDNP aim of creating an Internet-accessible, digital resource for historical American newspapers requires providing metadata with each digitized newspaper. NDNP specifies the required metadata elements at the title, issue/edition, and page level. At the page level the metadata identifies:

- file format: image/jp2 in this case;
- newspaper title;
- publication date;
- edition order and page sequence number;
- page number;
- publication location;
- description, including LCCN (Library of Congress control number).

For each page, the combination of LCCN, publication date, edition order, and page sequence number is unique.

These metadata are encoded using RDF, following the template given in NDNP (2008), and placed in an XML box in the production master. The metadata encoding for the page shown in Figure 14.6 is:

```
<?xml version="1.0"?>
<rdf:RDF xmlns:rdf="http://www.w3.org/1999/02/22-rdf-syntax-ns#">
 <rdf:Description
 rdf:about="urn:library-of-congress:ndnp:mets:newspaper:page://sn
 84026749/1905-07-17/1/1"
```

```
xmlns:dc="http://purl.org/dc/elements/1.1/">
  <dc:format>image/jp2</dc:format>
  <dc:title>
    <rdf:Alt>
      <rdf:li xml:lang="en">Washington Evening Times. (Washington,
DC) 1905-07-17 [p 3].</rdf:li>
    </rdf:Alt>
  </dc:title>
  <dc:description>
  <rdf:Alt>
      <rdf:li xml:lang="en">Page from Washington Evening Times
(newspaper). [See LCCN: sn 84026749 for catalog record.]. Prepared
on behalf of Library of Congress.</rdf:li>
    </rdf:Alt>
  </dc:description>
  <dc:date>
    <rdf:Seq>
      <rdf:li xml:lang="x-default">1905-07-17</rdf:li>
    </rdf:Seq>
  </dc:date>
  <dc:type>
    <rdf:Bag>
      <rdf:li xml:lang="en">text</rdf:li>
      <rdf:li xml:lang="en">newspaper</rdf:li>
    </rdf:Bag>
  </dc:type>
  <dc:identifier>
    <rdf:Alt>
      <rdf:li xml:lang="en">Reel number 100493159. Sequence
number 9.</rdf:li>
    </rdf:Alt>
  </dc:identifier>
  </rdf:Description>
</rdf:RDF>
```

Since the Chronicling America website allows users to download production masters, the files can exist apart from the website. Including these metadata in the file makes it possible to link the file to the Library of Congress record.

The JPEG 2000 production masters with these metadata are the image files that are accessed as users zoom into and out of the image and pan by click and drag on the user interface. The view that is delivered to the user as a JPEG file is created dynamically on demand from JPEG 2000 production masters by accessing and decoding only as much of the code-stream as is needed in an example of demand-side imaging and smart decoding.

In March 2009, two years after the Chronicling America website was launched, more than one million newspaper pages were available. During the next 20 years, NDNP will continue digitizing newspapers with the aim of creating a 'national digital resource of historically significant newspapers published between 1836 and 1922'.

Acknowledgments

The authors wish to thank Michael Ridley of the University of Guelph, Ontario, Ann Borda of JISC, UK, and Brian Thurgood of Library and Archives Canada for helpful discussions.

References

BCR (2008) CDP Digital Imaging Best Practices Working Group, CDP Digital Imaging Best Practices, Version 2.0, June 2008. Available at: http://www.bcr.org/cdp/best/digital-imaging-bp.pdf.

Buckley, R. (2008) JPEG 2000 – A Practical Digital Preservation Standard? DPC Technology Watch Series Report 08-01. Available at: http://www.dpconline.org/docs/reports/dpctw08-01.pdf.

Buckley, R. and Sam, R. (2006) JPEG 2000 Profile for the National Digital Newspaper Program, Xerox Global Services. Available at: http://www.loc.gov/ndnp/pdf/NDNP_JP2HistNewsProfile.pdf.

Buckley, R., Stumbo, W. and Reid, J. (2007) The use of JPEG 2000 in the information packages of the OAIS reference model, in *IS&T Archiving 2007*, Arlington, VA, May 2007, pp. 24–28.

Chen, S.-S. (2001) The paradox of digital preservation, *IEEE Computer*, March, 1–6.

Colyer, G. and Clark, R. (2003) *Guide to the Practical Implementation of JPEG 2000*, PD 6777: 2003, British Standards Institution, London.

COM (2005) i2010: Digital Libraries. Available at: http://ec.europa.eu/information_society/activities/digital_libraries/doc/communication/en_comm_digital_libraries.pdf.

Desrochers, P. and Thurgood, B. (2007) JPEG 2000 implementation at Library and Archives Canada, in J. Trant and D. Bearman (eds), *Museums and the Web 2007: Proceedings*, Archives and Museum Informatics, Toronto, Canada, March 31, 2007. Available at: http://www.archimuse.com/mw2007/papers/desrochers/desrochers.html.

Gillesse, R., Rog, J. and Verheusen, A. (2008) Alternative file formats for storing master images of digitisation projects. Available at: http://www.kb.nl/hrd/dd/dd_links_en_publicaties/links_en_publicaties_intro-en.html.

Houchin, S. and Singer, D. (2002) File format technology in JPEG 2000 enables flexible use of still and motion sequences, *Signal Processing: Image Communication*, **17**(1), January, 131–144.

HUL (2008) Harvard University Library Open Collections Program: Final Report on the Open Collections Program Pilot *Women Working, 1870–1930*. Available at: http://ocp.hul.harvard.edu/report/final/.

IST Cultural Heritage (2005) Coordinating digitisation of European cultural heritage. Available at: http://cordis.europa.eu/ist/digicult/eeurope.htm.

ISO (2008) ISO patent database. Available at: http://isotc.iso.org/livelink/livelink/fetch/2000/2122/3770791/JTC1_Patents_database.html?nodeid=3777806&vernum=0.

ITU (2008) ITU-T patent database. Available at: http://www.itu.int/ITU-T/dbase/patent/.

JPEG (2008) Cultural Heritage. Available at: http://www.jpeg.org/apps/culture.html.

MacDonald, L. (ed.) (2006) *Digital Heritage: Applying Digital Imaging to Cultural Heritage*, Elsevier, Oxford.

NDNP (The National Digital Newspaper Program) (2008) Technical guidelines for applicants. Available at: http://www.loc.gov/ndnp/pdf/NDNP_200911TechNotes.pdf.

Politou, E. A., Pavlidis, G. P. and Chamzas, C. (2004) JPEG 2000 and dissemination of cultural heritage over the Internet, *IEEE Transactions on Image Processing*, **13**(3), 293–301.

Yeung, T.-A. (2004) Digital preservation: best practice for museums. Available at: http://www.chin.gc.ca/English/Pdf/Digital_Content/Digital_Preservation/digital_preservation.pdf.

Related Links

Buonora, P. and Liberati, F. (2008) A format for digital preservation of images: a study on JPEG 2000 file robustness, *D-Lib Magazine*, **14**(7/8), ISSN 1082-9873. Available at: http://www.dlib.org/dlib/july08/buonora/07buonora.html.

Janosky, J. S. and Witthus, R. W. (2004) Using JPEG 2000 for enhanced preservation and web access of digital archives – a case study, in *IS&T 2004 Archiving Conference,* San Antonio, TX, pp. 145–149. Available at: http://charlesolson.uconn.edu/Works_in_the_Collection/Melville_Project/IST_Paper3.pdf.

NINCH (2003) *The NINCH Guide to Good Practice in the Digital Representation and Management of Cultural Heritage Materials*, Humanities Advanced Technology and Information Institute and National Initiative for a Networked Cultural Heritage. Available at: http://www.nyu.edu/its/humanities/ninchguide/.

UOCL (2008) University of Connecticut Libraries, Archives and Special Collections of the Thomas J. Dodd Research Center: Charles Olson's Melville Project. Available at: http://charlesolson.uconn.edu/Works_in_the_Collection/Melville_Project/index.htm.

15

Broadcast Applications

Hans Hoffman, Adi Kouadio, and Luk Overmeire

15.1   Introduction – From Tape-Based to File-Based Production

The digitalization in the television production environment was fundamentally possible by the worldwide agreement on a common standard for the encoding parameters of analogue 625-line and 525-line Standard Definition Television. The standard, ITU-R Recommendation BT.601, defined a color difference or digital component system with one luminance (Y) and two color difference signals (C_r, C_b), with a luminance sampling frequency of 13.5 MHz and 720 luminance samples per line. The bit-depth was defined with either 8 or 10 bits (more popular today is 10 bits). In addition to the encoding standard, a bit serial digital interface (SDI) with 270 Mbit/s (8-bit or 10-bit) was developed and standardized in ITU-R BT.656 in order to interconnect different production equipment devices. Quite soon, however, it was recognized that uncompressed standard-definition television (SDTV) signals could not be handled in an economic way by video tape recorders and disk or memory-based storage systems. Different digital compression technologies were consequently developed and introduced; first in digital video tape recorders and later in nonlinear editing systems. Prominent examples were video tape recorders (VTRs) such as the Digital Betacam (Sony), IMX (Sony), DVCPRO (Panasonic), or D9 (JVC). In these applications compressed video data were stored on the tapes, but the VTRs still needed to use the uncompressed SDI interface for signal exchange within the studio. As a consequence, each copy or transfer required a decoding process back to the baseband with quality loss. In addition, early nonlinear editing systems used the SDI interface for interconnection – even if they stored internally the video and audio data in files. A paradigm change occurred in the late 1990s when the Society of Motion Pictures (SMPTE) and the European Broadcasting Union (EBU) formed a joint Task Force to encourage vendors to open and standardize their compression algorithms and to provide the compressed

The JPEG 2000 Suite   Edited by Peter Schelkens, Athanassios Skodras and Touradj Ebrahimi
© 2009 John Wiley & Sons, Ltd

video bit-stream data on a standardized interfaces point (serial data transport interface, SDTI) on the VTRs. Unfortunately, it was not possible to agree on a single compression algorithm, so two compression algorithms (MPEG-2, DV-based) for standard-definition television (SDTV) production environments came into use. As a quality criterion for the compression algorithms, the EBU defined that the perceived quality in expert viewings, ITU-R BT.500-11 (ITU, 2003), shall be transparent to the original even after seven multi-generations including pixel shift after each generation. Quite soon also nonlinear editing vendors utilized the compressed bit-stream in their products and mapping into file formats, and the Internet protocol (IP) stack allowed the first file transfers of compressed and even uncompressed video/audio between systems. A crucial point, however, was that the available file formats only partly fulfilled the requirements of professional broadcast applications. Again, several organizations such as the SMPTE, EBU, and the Pro-MPEG forum worked together to define a professional file format, the Material Exchange Format (MXF). This file format, along with the Advanced Authoring Format (AAF) used in postproduction, represents the most common exchange file format in the professional TV production environment today. Useful literature on the MXF format, including information on the relevant standards can be found in Gilmer (2004), Wells (2006), and SMPTE standard 377M (SMPTE, 2004).

15.1.1   The Advent of HDTV and Consequences for TV Production

The main driver for the increasing demand of high-definition television (HDTV) programs is the popularity of flat panel matrix displays (FPDs) in the home, which have gradually replaced cathode ray tube (CRT)-based television sets. Two important factors are to be mentioned in this context: FPDs are usually much larger in size (e.g. 42 inches to 65 inches diagonal) than CRT-based TVs and offer spatial resolutions such as 1920 pixel $\times$1080 lines (HD-Ready 1080p (EICTA, 2007)), which are beyond standard definition television. When low bit-rate (e.g. 3 Mbit/s MPEG-2) standard definition signals are presented by modern FPDs they often provide a dissatisfying image quality. This is due to the increased visibility of compression artifacts on large size displays, the need for spatial upconversion from the SDTV resolution to the display resolution, and deinterlacing required by FPD. Content providers are therefore required to provide HDTV broadcasts as soon as the penetration of the large FPDs in homes has reached a 'critical mass.' While the USA and major countries in Asia have already started HDTV broadcasts, Europe has, to date, only a few HDTV broadcasts. However, all major European broadcasters announced concrete plans to launch HDTV channels between 2009 and 2012. Further factors supporting the need for HDTV content are the availability of Blu-ray and HDTV capable game consoles on the mass market and HDTV consumer camcorders. For broadcasting HDTV, the state-of-the-art compression technology is H.264/AVC either utilizing the 720p/50-60 or 1080i/25-30 HDTV format. Recent tests of the EBU (2008) have indicated that with state-of-the-art broadcast encoders 720p/50 would require between 10 Mbit/s and 12 Mbit/s video data-rate and 1080i/25 about 20% more bit-rate for satisfactory image quality.

For content providers such as broadcasters the demand for HDTV content has led to the following consequences in production:

• Demand for SDTV material is decreasing on a worldwide basis.

- The substantial worldwide demand for HDTV content has led to a clear HDTV content gap for many broadcasters and content providers. Thus substantial investments in HDTV production equipment have been made.
- Multipurposing of material for television (SDTV and HDTV), optical disks (SD-DVD and Blu-ray), for download/streaming applications, mobile and digital cinema applications require HDTV content production.
- Some content providers plan to create archive stocks of selected content in HD in preparation for a full HDTV environment in the future (this concerns having both the rights and quality in HD).
- Using HDTV equipment that is potentially able to replace film at lower production costs is attractive to film makers.
- Film material can be stored in archives at HDTV quality.

In contrast to the situation with SDTV mainstream production where broadcasters had to deal with two major compression algorithms (DV-based and MPEG-2), the situation for HDTV is more complicated. First, there is the very high baseband data rate of HDTV, which can reach 1.48 Gbps on the serial digital interface for 720p/50-60, 1080i/25-30, and 1080p/24-25 and for future applications of 1080p/50-60 even 2.98 Gbps (called '3G interface'). Secondly, it is likely that broadcasters will have to decide for one of the currently available HDTV images formats (720p/50-60, 1080i/25-30), although the production and emission format does not need to be identical, and, thirdly, they need to take note of the fact that the industry provides a number of different compression technologies for mainstream HDTV production environments. Prominent examples are the I-frame only based compression algorithms (Sony HDCAM, Panasonic DVCPROHD 100, AVID DNxHD, Apple ProRes, and Panasonic AVC-I) or the MPEG-2 Long GOP algorithm (Sony XDCAMHD422), as well as JPEG 2000-based compression (Thomson/GVG Infinity series). These compression systems are used for storage of video on traditional tape-based storage, on optical disk, memory cards, as well as server systems, and (with the exception of HDCAM) are also available as network exchange formats in file form (usually via MXF). A further format to be mentioned is the HDCAM-SR tape format for high-quality HDTV applications with a 440–880 Mbps compressed bit-stream, but this bit-stream is currently not available outside the VTR.

The challenge of handling a multiplicity of studio compression algorithms for HDTV and the technological and business-related issue has been leading to a clear demand from broadcasters for a single compression algorithm that could be used as a master format in, for example, fully IT-based broadcast archival applications. The requirements on such a single compression algorithm can be summarized as:

- support of SDTV and HDTV image formats and future HDTV image formats;
- ability to adapt in bit-rate in order to maintain the image quality of the source format (i.e. from 50 Mbps up to lossless HDTV);
- ability to create low-resolution browsing;
- ability to extract different quality levels;
- open and fully standardized and IPR free;
- broad support from industry;
- cost-effective implementations.

Taking into account these requirements, JPEG 2000 is a serious candidate to meet the requirements above and ISO/IEC JTC1/SC29/WG1 in liaison with EBU and SMPTE is working on the definition of a particular studio broadcast profile for JPEG 2000 (ISO/IEC, 2004, 2007).

15.2 Broadcast Production Chain Reference Model

The file-based media production paradigm has established an innovative, content-centric pull-based workflow model in broadcast production, by replacing the old concept of linear push tape-based production. Production systems are no longer self-sufficient, independent islands that are rigidly interconnected with each other in a sequential production chain. Instead, they can now be highly integrated with each other, such that the same media can be accessed and processed in parallel by multiple clients. Media and related information can be transparently interchanged between the different systems. Multiple production steps can happen simultaneously without interfering with each other. File-based production brings about unprecedented advantages such as concurrent engineering, edit- and transfer-while-ingest capability, faster than real-time processing, and enriched integrations. In addition, it enables genuine cross-media production, i.e. the merge of television, radio, web, and mobile in a unified and integrated architecture. Obviously, this has introduced a complete overhaul of the media production work processes. In the next sections, the main principles and system building blocks of a file-based broadcast production chain are outlined.

15.2.1 System Building Blocks in File-Based Media Production

15.2.1.1 General Architecture

A typical file-based media production architecture is depicted in Figure 15.1. The television production infrastructure usually includes file-based cameras, studio cameras, file-based ingest system of feeds and tapes, browse and craft editing suites, newsroom computer system, file-based multichannel final control room and play-out system, play-out automation, full integration on data and metadata level of all systems, and a media asset management system. It is conveniently complemented by a production and play-out work centre for radio and online.

Central in the architecture are the concepts of a *work centre* and *central media asset management*. A work centre is an autonomous environment optimized for a specific, craft operation in the production chain. Examples are audio editing, subtitling, video editing, etc. These work centres are typically connected to a central media asset management (MAM) system that serves as a media-aware hub and media repository for production and archive material. It is the main access point to media essence for a large media production user community (e.g. journalists, program assistants), with the exception of the smaller groups of specialized craft users (e.g. video and audio editors, planners) who have the specialized work centres available.

15.2.1.2 Media Asset Management

The MAM system centralizes a number of functionalities, mainly related to the storage and access of media material. It is a core component of the architecture, managing the storage

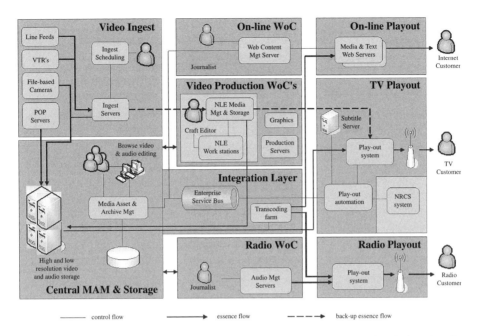

Figure 15.1   Outline reference model for file-based cross-media production in broadcasting

resources for the high-resolution and browse media essence, user access, and life-cycle of media-related information. The user can search and access a large amount of media material through the browse copy from a desktop environment, add annotations, and even perform basic cutting operations to select fragments for later editing and generation of items. The preparatory media items can then be transferred to the applicable work centre, accompanied by its related metadata. In some cases, simple browse-based edits already provide sufficient quality to be broadcasted directly (e.g. WEB TV). The media asset management enables far-reaching interwork centre and cross-media integration, by acting as a central hub and repository for both work-in-progress material with a longer life-span and longer-term archive video and audio material. The life-cycle of the media typically exceeds the short life-cycle of intermediate work. To some extent, the MAM system also handles workflow management, taking care of core essence processing (transcoding, rewrapping, conforming, and so forth) and internal and external data transfers. It is of vital importance that new compression (e.g. for high-definition television) and file formats can be easily and flexibly fitted in the media asset management.

15.2.1.3   Specialized Work Centers

The three main areas where work centers can be defined and implemented are ingest, editing, and play-out. New media material (feeds, tapes, camera material) is imported directly by dedicated ingest facilities – typically including ingest scheduling and automation functionality – in the central system or in a work center. Craft video editing, audio editing, and online multichannel editing and publishing take place in specialized, professional editing cells and work environments. The play-out environment is automated and tightly integrated with the media asset management and for news with the newsroom computer system, which is used for news content planning and rundown scheduling. In most cases, work centers have their own local media asset management functionality.

15.2.1.4 Integration Layer

Efficient integration between work centers and the media asset management system is essential to exploit fully the added value of a tapeless workflow. As essence is transported back and forth between work centers and the media asset management system throughout the production life-cycle, the production-related descriptive metadata are correspondingly enriched in each step of the workflow. Metadata integration primarily occurs on the level of the central and local media asset management. The *integration layer* takes care of synchronizing the metadata between systems and of orchestrating the various essence transfers. Technically, the integration layer can be accomplished as the core integration element in, for example, a service oriented architecture (SOA), where applications are accessed through services that are processed within the enterprise service bus (ESB).

15.2.1.5 Production Models

The specific role and interrelation of the central MAM and the different work centre strongly depend on the production model. Roughly, two main models of media production can be distinguished: project-based and item-based media production. In item-based production, the emphasis lies on simultaneous, short-term, high-volume media production and reuse by a large number of users in a heterogeneous, cross-media environment. Typical examples are news, sports, magazines, and cultural programs. The central MAM system plays a prominent role and the integration demands regarding work centres are less complex (e.g. exchange of simple timeline information). Project-based production rather relies on smaller teams of people and is generally less time-critical. Typical examples include drama and documentary production. Although cross-media reuse remains important, the focus rather lies on a specific medium (e.g. television in this case). Complex timeline-based integrations between the different system components throughout the production workflow are imposed. As a consequence, the role of the work center becomes more important, when multiple-production steps (e.g. editing, sonorization, colour correction, graphics) are combined before handing over to the central MAM (Figure 15.2).

15.2.1.6 Layered Architecture

A fully integrated end-to-end production chain is realized, from acquisition to play-out and archive, based on an open and layered approach. The following independent layers can be identified: storage and network infrastructure, production, information, and business services. Each of these layers represents an essential building block of the integrated file-based media production architecture. Preferably, these different layers are loosely coupled and deployed in a modular way, such that each basic component can be exchanged or replaced by future technological components without compromising the basic architectural model.

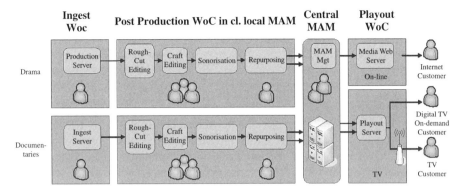

Figure 15.2   Possible production model for project-based media production

15.2.1.7   Use of Standard IT Equipment

The application of generic ICT-based infrastructure is rapidly gaining ground in media production. In fact, a key design parameter for the storage and network environment is to use open technologies and standard IT equipment in order to ensure the necessary throughput and storage capacity. Files for mainstream production quality video tend already to be very large for SDTV (with 25–50 Mbps) and, depending on inter- or intraframe compression, can reach 50–200 Mbps for HDTV. For high-quality HDTV productions compressed video of 440–880 Mbps needs to be considered. The network and storage bandwidth capacities required to allow parallel transfers of hundreds of these very large files are far from trivial, though not impossible with standard IT equipment. Adoption of this technology is particularly important for the central media storage environment (high- and low-resolution video and audio storage), for a number of server-based applications (central media asset management, radio production, online editing and distribution, integration) and for the PC-based user environment. Flexibility, programmability, and higher performance at lower cost are some of the obvious advantages. Specialized equipment is especially found in the work centers (craft editing, ingest, and play-out automation).

15.2.2   Required Functionality and Technical Guidelines

15.2.2.1   Multiple Resolutions

An important requirement in file-based media production is the ability to work with different quality versions of the same video material. In this respect, two main categories can be distinguished: high-resolution and low-resolution media instances. The high-resolution media is encoded during ingest (or generated by means of file-based camera acquisition) and from this material one or more low-resolution proxy versions and key frames are generated on-the-fly.

The high-resolution material is mainly employed in work center-related operations such as craft editing and play-out, and in conform and cutting operations in the central media asset management system. The selection of the video compression scheme typically depends on the type of production, both for SD and HD television. Time-critical productions demanding limited postproduction processing, such as news and sports, usually lend themselves to lower bit-rates than high-quality productions such as drama and documentaries. For file-based compression formats in television production, the bit-rates range from 25 to 50 Mbps for SD and from 50 to 200 Mbps for HD. The most commonly used SD formats are DV- and MPEG-based compression schemes. An overview of available HD compression formats is given in Section 15.4.

A particular requirement is the mixing of different SD or HD compression formats along the production chain. Moreover, as the transition from SD to HD television is gradually taking place, mixing of SD and HD material will become commonplace too, not to mention mixing of progressive and interlaced formats. Finally, even higher resolution formats, such as 2k, 4k, and UHDTV, are expected to enter the broadcast scene as well in the future.

Browse video enables hundreds of people in a broadcasting or media company to access the stored media material for different purposes: programme research, visioning and quality control, viewing, shot selection, rough cut editing, adding voice-overs, etc. These various applications may impose different requirements on the low-resolution format. Consequently, in addition to different high-resolution formats, multiple-browse versions are likely to emerge in the central media repository as well. The most commonly used low-compression formats are MPEG-1, MPEG-2, MPEG-4, H.264 (MPEG-4 AVC), and VC-1 (Windows Media 9).

15.2.2.2 Media Data Model

Keeping track of all these different high- and low-resolution versions and their interrelationship, on the one hand, and ensuring accurate media tracking between finished and original material, on the other hand, is a complex task that is one of the core responsibilities of the central media asset management. This can be accomplished by implementing a sophisticated underlying media object data model. In such a data model, the following concepts are typically defined:

- media object instances, which are merely copies of the same media object;
- media objects, which group media of the same basic type (video, audio, subtitle, key frame), possibly referring to multiple encoding variants;
- media object groups, which group time-code equivalent media objects that semantically correspond to the same content;
- media items and subitems, which are general-purpose containers for media, describing a collection of media objects such as a news story, ingested raw material, etc.; media items are usually the smallest items that carry user-defined, descriptive metadata, while subitems can hold specific parts of an item, such as shots and takes;
- playlists, rundowns, wires, episodes, and programs, which represent high-level entities for storage of media.

Media tracking keeps the relationship between media items up-to-date, aided by unique identification of the material (e.g. UMID = unique material identification).

As the amount of available media material is ever increasing, and correspondingly the complexity of the related media asset management, it is clear that the number of codec versions and media instances should anyhow be limited as much as possible. In future architectures, scalable formats possibly provide one way to deal with these scalability issues by combining multiple versions into one instance.

15.2.2.3 Performance and Quality

In order to turn the file-based paradigm actually into advantage, a per-application processing rate of up to four times real-time is considered an indicative 'rule of thumb' target for multistep workflows with a limited number of head–tail transfer operations. An example of such a workflow is the intake of file-based camera material in the central MAM system and subsequent forwarding to an editing work centre for further processing.

In the case where a nonscalable high-resolution format has been chosen, proxy generation from high-resolution material is required. This is typically a rather process-intensive and time-consuming operation, potentially jeopardizing the faster-than-real-time ingest requirements of time-critical production use cases, such as news. In such cases, the high-resolution material is by preference instantly transferred to the central MAM and directly forwarded to the relevant postproduction work center, even before transcoding to browse media has started. Note that, if the camera format differs from the selected high-resolution production format, transcoding is required as well.

Transcoding operations are generally known as a common bottleneck in media production workflows, for multiple reasons. It usually has a significant impact on performance, quality, and cost. The inevitable explosion of transcoding needs and the corresponding processing power requirements urge for performance-effective solutions. In order to cope with these performance-related demands, hardware and software solutions should be optimized per task and tuned to each other, e.g. by using grid-based processing techniques. The use of royalty-free codecs facilitates the implementation of free and optimized transcoding solutions, which are custom-made for particular use cases.

Repeated reuse of video material in postproduction results in video quality loss, due to multiple generations of decoding and re-encoding. This should be avoided as much as possible by implementing intelligent media tracking techniques whereby derivative program material can be produced as much as possible by referring to the original material.

15.2.2.4 Application of File Formats

To a large extent, the advantages of file-based production – such as the flexibility in integration between different production systems – can be attributed to the adoption of standardized file formats such as the material exchange format (MXF). MXF wraps media (video, audio) and related technical and descriptive metadata. Important technical parameters include time code, aspect ratio, and video/audio parameters. Descriptive metadata relates to user- or application-defined information such as camera type, GPS data, and logging information. In most cases, only the high-resolution material is wrapped in MXF.

The MXF file format ensures compression agnosticism and enables random access to the material, partial file retrieval, and transparent metadata exchange, among other benefits. It largely facilitates integration between production systems, such as central MAM and work centres. Although the complexity of MXF implementations is preferably kept as low as possible, future challenges such as repurposing of the same media to multiple outlets (e.g. linear and on-demand television, Web, handheld devices, etc.) may impel the application of more complex MXF patterns. In addition, the application of descriptive user-defined metadata inside MXF is expected to mature, whenever this can deliver added value in the production processes. Specifically, metadata acquired by media analysis, validation, and feature extraction techniques, such as shots, key frames, region of interest, and so forth, are strong candidates to augment further automated production workflows.

15.3   Codec Requirements for Broadcasting Applications

The broadcast environment has experienced tremendous evolution following the introduction of digital signal processing. This has specifically been the case for video compression technologies. The use of video compression systems (the encoder and decoder system, commonly called codec) has enabled the development of economical devices in complex TV studio environments by reducing the huge baseband bit-rates of uncompressed SDTV, and more recently HDTV, to data rates that can be more easily stored and transmitted via the available interfaces and networks. Developments in video compression have provided the broadcast industry with a significant set of legacy compression systems (see Section 15.4) (AVCHD, XDCAMHD422, HDCAM SR, JPEG2000, DNxHD, ProRes 422, DVCPROHD, AVC-Intra, etc.), all of which have valuable functionalities for specific parts of the broadcast chain but not necessarily enabling an interoperable and cost-efficient platform.

Defining codec requirements depends on the broadcast application in which it is used and the characteristics of the content being processed. In order to define the codec requirements clearly for broadcast applications, a clear overview of the characteristics and constraints of these applications is needed. The next section will describe the different characteristics of a broadcast signal followed by an overview of the different broadcast applications and their respective codec requirements.

15.3.1   Broadcast Content Characteristics

The SDTV and HDTV baseband signals are fully described in ITU-R Rec. 601 and ITU-R Rec. 709 (equivalent SMPTE 274 M). This section offers a short summary of these recommendations.

15.3.1.1   Colorspace and Sampling Scheme

Most broadcast content is produced in YCbCr/YUV color space which corresponds efficiently to the human visual system (HVS) properties. Indeed, the HVS is more sensitive to the luminance than to the chrominance (color) components. This property is the basis for color downsampling, which is an efficient way of reducing the amount of information while preserving the visually most relevant parts. The 4:2:2 (horizontal downsampling of chrominance) and 4:2:0 (horizontal and vertical downsampling of chrominance) sampling schemes are usually applied. The 4:2:2 scheme is preferred for content in production and postproduction since it provides more quality headroom. The 4:2:0 scheme is used in distribution since it further reduces the amount of color information by half.

15.3.1.2 Bit-Depth/Dynamic Range

This parameter represents the number of bits used to represent each color/luminance sample value. The 10-bit and 8-bit per component signals are common in broadcasting. The 10-bit signals are usually used in the studio environment where high quality is required and postprocessing including graphics may occur. A signal with a higher bit-depth provides better precision for postprocessing operations as well as better performance on a plain background, where contouring and banding artifacts can appear due to compression.

Unlike in the computer domain, reference black and white have the value 16 (respectively 64) and 235 (respectively 940) for 8-bit (respectively 10-bit) content. The remaining value range provides footroom and headroom codes that can be used to accommodate filter overshoot and undershoot. For further information on these issues we refer to Poynton (2003).

15.3.1.3 Scanning Format

An image scanning format defines the manner in which a time-varying picture is explored for its luminance and chrominance values. It also defines the number of scanning lines per picture and the number of frames per second. There are two main picture scanning methods used in television broadcasting: the interlaced (noted '') and the progressive (noted 'p') scanning methods (Figure 15.3).

The interlaced method consists of dividing the original image into two separate fields. One field (odd field) contains the odd lines of the original picture while the other field (even field) contains the even lines. On the receiving display, the odd field is first displayed

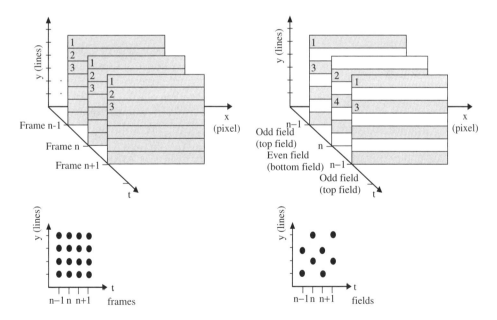

Figure 15.3 Interlaced and progressive scanning methods: (left) progressive format, (ight) interlaced format (Hoffmann, 2007)

followed by the even field 20 ms (1/50 s for a field rate of 50 fields per second) later to complete the picture.

The interlaced scanning method was invented in the 1930s with the primary purpose of reducing the required bandwidth. In fact, at that time a video transmission system at 50 or 60 pictures per second was considered overly expensive. The interlace format helped to reduce the bandwidth needs by a factor of two while reducing the flicker on the display. However, this scanning method generates a serious loss in vertical resolution and motion is not well portrayed. It also requires additional processing operation (deinterlacing) in matrix displays. Interlaced scanning can be defined as a vertical subsampling of the image, which in fact lacks a proper lowpass filter to avoid aliasing. Typical interlace defects are:

- line crawl visible for slow vertical motion;
- sawtooth artifacts;
- interline twitter.

Progressive scanning displays an image one line after the other from left to right and top to bottom. It is not as bandwidth efficient as the interlaced method but it has less visual impairments. Progressive scanning also provides better motion portrayal.

Another mode is the progressive segmented frame (PSF) (noted 'psf'). It is a method used to transmit digitized film content with 25 frames per second as a quasi-interlaced signal. The original frame is subdivided into two fields representing the same time instance, but the fields are displayed successively one line after the other and not one field after the other. PSF is used to transmit film type images (1080psf/24-25) over existing interlaced interfaces or broadcast channels.

15.3.1.4 Spatial Resolution and Frame Rate

These two parameters are coupled since together they define the notion of video resolution. The latter has long been restricted to only include the spatial resolution of the image, which is only valid for still images. For video, the temporal dimension, or more precisely motion, has a tremendous impact on the resolution. Motion portrayal is directly related to the frame rate of a broadcast signal (a higher frame rate allows for better motion portrayal). The actual frame rates used in broadcasting are the following: 60/1.001 Hz, 50 Hz, 30/1.001 Hz, 25 Hz, 24 Hz; 60/1.001 and 30/1.001 Hz frame rates are used in the US and Asia, and 50/25 Hz is used in Europe, Middle East, and Africa.

As far as traditional broadcast is concerned, SDTV and HDTV are the TV signals used worldwide (Figure 15.4). Ultra-high-definition Television (UHDTV) is not yet used

Figure 15.4 Static image formats. Note that this comparison ignores the interlaced and progressive scanning impact and cannot be used to describe the different HDTV formats

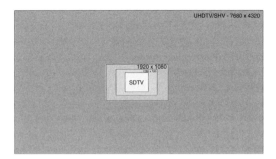

Figure 15.5   Static image format comparison of SDTV, HDTV, and UHDTV

in practical applications even if it is already standardized in ITU-R (ITU-R BT.1296) and SMPTE 2036 (Figure 15.5). SDTV is provided only in interlaced scanning mode and a frame rate of 30/1.001 Hz and 25 Hz. It can be analog or digital. SDTV can be broadcasted in several formats:

- 720 pixels × 480 lines, 60/1.001 Hz field frequency (NTSC);
- 720 pixels × 576 lines, 50 Hz field frequency (PAL);
- 640 pixels × 480 lines (computer format; generally not used for broadcasting).

The concept of high definition was invented by the Japanese in the 1960s. Research conducted by Dr. Fujio at the public service broadcaster NHK defined the basis for today's HDTV formats. HDTV refers to the following image formats and scanning methods:

- 1920 × 1080 (i/p). This is the highest image resolution available in HDTV. The image dimensions are 1920 samples ×1080 lines. This spatial resolution is used in both scanning modes but with different frame rates (1080i/25, 1080psf/24-25, 1080p/50). While 1080i/25 was the first available HDTV standard, today 1080psf/24 content is available in Blu-ray disk and production archives. 1080p/50 is not yet available in live broadcast or retail content but its full rollout is expected in the near future. It is already supported by several display systems and cost-efficient professional inter-faces to cope with its high raw bit-rate (2.98 Gbps) have been standardized (SMPTE 424–425).
- 1280 × 720p. Here the image dimensions are 1280 samples ×720 lines. It is only used in progressive mode. From a numerical point of view it has a lower spatial resolution than the 1080 format, but 720p/50 has a better motion resolution than 1080i/25 since the interlace scanning method reduces the vertical resolution and does not optimally portray transversal movements.

15.3.2   Broadcast Applications

Broadcast applications can be divided into three categories: production, contribution, and distribution. They are described in the following sections.

15.3.2.1 Production (and Archiving)

A production environment is generically composed of four core blocks (Figure 15.6):

- *Ingest/acquisition* represents the process of capturing the content (video and audio).
- *Postprocessing operations* consist of nonlinear editing, metadata insertion and maintenance, color grading and color correction operations, etc.
- *MAM/archive* stores different quality versions of the content and makes it available and easily accessible from other core blocks. With the convergence to file-based production workflows, production activities have become centred in the MAM system, enabling faster processing and transfer of content but also increasing the requirement on this part of the chain.
- *The play-out* centre is the content exit point for distribution to the end-user.

As explained earlier in this chapter (see Section 15.2), the production domain refers to the content origination, creation, and main processing environment. It is the source of all content shared between broadcasters, and ultimately distributed to the end users.

It is a prerequisite that production provides a high technical audiovisual quality. While raw, uncompressed video is the highest quality available, it is not economically feasible to store uncompressed video (e.g. uncompressed HDTV) in all areas of the production chain. Consequently video compression is applied in the production chain, and, to the disadvantage of users, a number of different video compression algorithms have been developed and integrated in the devices along the production chain.

Multiple Studio Compression Technologies
While it has been possible to reach common agreement for a single compression technology in content distribution, first with MPEG-2 and more recently with H.264/AVC, the TV production environment has been the habitat for a number of different and competing video compression technologies (and to a smaller extent for audio compression

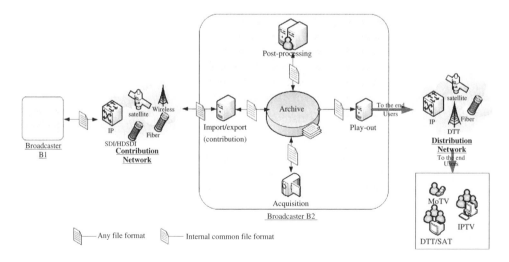

Figure 15.6 Overview of the broadcast chain (see Plate 9)

technologies). This situation puts professional content providers such as broadcasters, telcos, etc., in a difficult position since they must handle different, and not necessarily compatible, video compression formats in their facilities. The development of the digital Betacam video tape recorder (VTR) for SDTV applications by Sony is a prominent example of system incompatibility in the TV production environment.. The VTR used a proprietary algorithm based on the discrete cosine transform (DCT) and was widely used for mainstream applications. Later, still for SDTV, Sony and Panasonic developed two incompatible systems called IMX50 and DVCPRO50. Both used a bit-rate of 50 Mbps but while Panasonic used an algorithm derived from the consumer world (DV-based SMPTE D7), Sony developed a professional version of the MPEG-2 compression standard (IMX). Ultimately, the audiovisual community (broadcasters) faced an economic cost since it needed to handle two incompatible video compression formats in its production facilities. This example illustrates the need for a single standard compression system. In addition, an open standard is also needed to guarantee future support of the compression system, even if the product manufacturer is no longer active.

Need for a Single Archive Codec

For each production core activity listed above (ingest, postproduction, play-out, and archive), different compression systems are proposed to meet their specific requirements (DNxHD and ProRes422 for editing, AVC-I, XDCAMHD422, JPEG2000, etc., for acquisition, and so forth). Various copies of the same content, coded with different compression formats, are archived. Consequently, several transcoding operations, which are time consuming, costly, and lead to a loss of content quality, are necessary. Content quality decreases with each operation due to the accumulation of different types of artifacts. Having a unique compression system in an archive-centric workflow limits the number of necessary transcoding operations. A unique system should optimally be supported by most editing suites.

Generally, compression systems are based on intraframe and/or interframe coding. Cumulating both technologies provides significant compression capability to a compression system since both coding dimensions are exploited (spatial and temporal). For editing functions such as fast reverse, fast forward, or reverse play, interframe coding algorithms result in limited performances while intraframe coding allows direct access to any frame without the need for prior decoding of a particular GOP structure.

Sustainability to Multigeneration Coding (Cascading)

In a typical production environment, the visual content suffers several successive encoding and decoding passes, altering its quality. The number of generations is lower in a file-based workflow since task parallelism can be implemented although a certain number of generations is maintained. The EBU has defined as a quality criterion for production compression algorithms where the perceived quality remains constant, even after seven multigenerations including vertical and horizontal pixel shifts after each generation. This criterion guarantees sufficient quality headroom to maintain quality in the production chain and to reduce the impact on the distribution compression. The criterion requires a production system to have a relatively high sustainability to cascading processes. This property is not intrinsic to interframe coding, which will aggregate the prediction errors from one generation to another, deepening the quality gap between the original and the chain output. In such a high-quality environment as content production, high error resilience is also a desired feature.

Support of Higher Image Resolutions, Higher Frame Rates, Larger Bit Depth, etc

Today's trends show that broadcast image formats are developing toward, and even beyond, digital cinema formats (HD, 2k, 4k). Even a live satellite broadcast of SHV/UHDTV – super high vision (16 times HDTV resolution) – has been demonstrated at IBC 2008 by NHK together with the EBU, RAI, and BBC (BTF – Broadcast Technology Future Group) and has set the expectation for the next broadcast image format. Spatial scalability can be beneficial since several resolutions can be intrinsically supported by the same single stream.

Future broadcast image formats are planned to be fully progressive. Nevertheless, an efficient production coding system should be able to handle both interlaced and progressive scanning format properly since archived content will inevitably include interlaced material (SD and 1080i/25).

It is known that, e.g. Salmon (2008), increasing only the spatial resolution of the image without increasing the frame rate hampers the impression of high detail with the introduction of motion. A higher frame rate allows for better motion portrayal. This principle is well embraced by high frame rate flat panel display manufacturers who try to provide, via an interpolation process, a higher temporal resolution (100/120 Hz) out of a 50/60 Hz input signal. This is less effective than displaying a native signal generated at a higher frame rate but nonetheless provides an improved viewing experience. A future-proof production codec should be able to sustain higher frame rates in conjunction with the higher spatial resolution in order to maintain the 'high-definition experience' envisaged.

A higher bit-depth than the usual 10 bit or full 4:4:4 color sampling are other quality headroom parameters that are also relevant for such a system. The adoption of a higher bit-depth will depend on the improvements of camera sensor technology to allow for acquisition of higher bit-depth content.

Need for Scalability

As shown in Section 15.2, a file-based production workflow includes a central archive, which is the repository for all content issued during the different production activities. These copies include browsing versions, intermediate editing copies, and high-quality master formats, which have different quality levels and/or resolutions. This results in an inefficient use of storage space, which can easily be resolved by a combination of spatial and quality scalability.

Licensing Issues

Not solely related to the production domain, compression technology licensing issues are increasingly becoming a burden to service deployment in broadcasting. Indeed, most of the available compression systems require a license fee that is then redistributed to the technology intellectual property rights (IPR) owners. This is the case for the MPEG system family. The use of efficient, royalty-free compression systems can provide significant savings to the broadcast industry. An example of such a system is DIRAC, an open source royalty-free system developed by the BBC and standardized as SMPTE VC-2.

Codec Support in Reference Exchange File Format

One requirement for a production archive, ideally, is the use of a single file format. This decision is relevant to the management of a production facility since it can have a huge impact on investment and staff training. An archive essence consists of the compressed content, the accompanying metadata, and the wrapper file format. At present, MXF (SMPTE standard) and QuickTime from Apple (not a standard exchange format but well known) are the most used file formats in the professional broadcast community. A production compression format should thus have its mapping in one of these wrapper file formats, clearly defined in order to be adopted for content exchange.

15.3.2.2 Contribution

Contribution is a business-to-business (B2B) application that consists of exchanging audiovisual material between broadcaster production centres. A production centre can be:

- an internal production studio;
- an OB-Van (outside broadcast van) covering an event;
- a regional TV station that feeds the local production to the broadcast headquarters;
- an SNG device (satellite news gathering) handled by a field journalist/correspondent.

The exchanged content can be postprocessed (editing, color correction, etc.) before its final delivery to the end users. Considering the variety of content sources, it is obvious that different network technologies can be involved in the content backhauling. Contribution networks can use:

- *Satellite,* which enables a point-to-multipoint contribution link (used in WANs). Over satellite, the usable channel bandwidth is limited to the 36 MHz bandwidth, which is usually too expensive for allocation to a single HD channel. This bandwidth is usually split into smaller frequency band channels of 18, 9, or 6 MHz to allow for simultaneous contribution of SD and HD signals as well as enabling an efficient and economic use of the available spectrum. Depending on the modulation technique used (DVB-S, DVB-S2), different data rates can be achieved. The maximum data rate currently achieved for a 36 MHz channel using DVB-S2 is around 140 Mbps.
- *Optical fibre network (SONET/SDH),* which is usually used in WANs as a supply to satellite coverage. It offers a point-to-point connectivity with different high bit-rate profiles. Over fiber rings, the bandwidth constraint is less of a problem, but this medium does not offer the multipoint connectivity and wide-area coverage available via satellite. They are usually used for WAN connections. The most widespread optical network connections are OC-3/STM-1 and OC-12/STM-4, delivering respectively a capacity of 155 Mbps and 622 Mbps. The content quality can be maintained close to the uncompressed perceived quality at those bit rates.
- *Coaxial (SDI[1]/HD-SDI) cables,* which are professional product interfaces and thus are mainly used for studio contribution. Over SDI cables, the limitation comes from the cable length. Depending on the type of interface, bit-rates of 270 Mbps (SDI) up to 1.48 Gbps (HD-SDI) can be achieved, allowing the transport of uncompressed SD or

---

[1] Serial digital interface: widely used for standard television transport.

HD (720p/50 or 1080i/25) over a couple of meters. HD-SDI dual link interfaces exist to cope with the 2.98 Gbps raw bit-rate of the 1080p/50 image format, but it is not easy to use. Recently a single-link 3 Gbps capable interface has been standardized at SMPTE (SMPTE424-425), but it is not yet widespread as a professional interfacing technology. Upgrading an SDI (SDTV) network to an HD-SDI network involves significant invest-ments (compatible switches, video routers, etc.). Broadcasters also investigate solutions known as 'Mezzanine' compression systems, which are able to compress in a visually lossless manner an HD uncompressed stream into an SDI bit-rate or a 1080p/50 3 Gbps feed to an HD-SDI bit-rate.

- *IP/Ethernet*. The 100-base-T (gigabit Ethernet) allows transport of video content at up to 1 Gbps. Again, the limitation of this technology resides in cable length.
- *Wireless broadband links* (WiMax) can also be used for the contribution. Technologies such as WiMax offer metropolitan area coverage (up to 50 km range) of up to 70 Mbps in a point-to-point link. Higher bit-rates can be achieved in the vicinity of the base station.

The EBU EUROVISION contribution network is one of the world's largest contribution networks and spans five continents and offers satellite and fiber ring connectivity.

Quality versus Latency Criticality

Contribution networks are used to exchange three quality levels of content: high-quality content corresponding to premium drama productions, medium or mainstream quality, and finally low-quality content exchange equivalent to news gathering. For live events, the quality and the system coding latency are an issue. For offline content exchange, only the quality level needs to be maintained at its maximum. In the case of news production, where the prevalent characteristic is instantaneous search and retrieval, the compression system latency is an issue. For this type of content, the visual quality does not matter as much as the immediate availability of the content. As far as latency is concerned, intraframe systems provide a lower latency than the interframe systems.

Depending on the bandwidth constraints, the system coding efficiency may prevail. It is thus difficult to find a system that matches all network types and application require-ments. At the moment, MPEG-2 is the legacy contribution system for both HD and SD. However, the perceived quality provided by the system is no longer considered satis-factory and, in addition, it is bandwidth greedy. Several alternatives for its replacement are under investigation by the broadcast community. A scalable solution can provide an alternative to the simulcast of SD and HD contribution feeds. In this case, the receiver only selects the packet necessary to reconstruct the format it supports (SD/HD). Contribu-tion feeds are usually transported as MPEG-2 TS. It is thus required for any contribution stream to have its corresponding mapping scheme in a TS stream fully defined and standardized.

15.3.2.3 Distribution

Distribution is the delivery of the content generated and processed in the production facil-ities to multiple end users using various distribution networks. Multiple platforms must be addressed with the same production essence. Mobile TV, IPTV, HDTV, SDTV, and so forth, require different image formats and have different network constraints (variable bit-rate, variable error rates, etc.), which results in enormous equipment investments to be properly handled.

The delivery can use the digital terrestrial network (DVB-T, DVB-T2) for conventional broadcast television, cable networks (DVB-C), satellite (DVB-S), xDSL lines for IPTV, etc. Each of these delivery networks are characterized by limited capacity. Therefore, a distribution/emission codec must enable efficient compression of the actual content at a very low bit-rate. As a result, coding technologies coupling interframe and intraframe compression have always been successfully adopted by the distribution domain. The MPEG compression system suite has been and continues to be the legacy distribution codec. MPEG-2, and more recently H.264/AVC, both perform intra- and interframe compression. Thanks to several additional prediction tools and a high efficiency statistical coder, H.264/AVC provides a 50% coding gain over MPEG-2 for distribution bit-rates (up to 25 Mbps).

In order to address all new media platforms efficiently, scalable video streams that can easily be adapted to the network constraints are needed. However, providing scalability should not be done at the expense of nonbackward compatibility with existing products, such as set-top boxes, or additional cost and time-consuming transcoding operation for the users. The system should also have low complexity in order to run efficiently on power-limited devices such as mobile phones, porTable 15.multimedia players, etc. The scalable video coding (SVC) extension of H.264/AVC was developed in consideration of the above points. SVC is further described in Section 15.7 of this chapter.

15.3.2.4 Summary

A summary of the required codec functionalities can be found in Table 15.1.

15.4 Overview of State-of-the-Art HD Compression Schemes

The main objective in HDTV is to deliver high-quality television programs. Apart from craftsmanship during acquisition and production of the program, a key constraint to reach this goal is to minimize quality loss when encoding the source media and final program material. However, a compromise between the compressed bit-rate and the technology used for storage and sharing (i.e. network bandwidth in IT-based environments) will be necessary.

For obvious reasons, bit-rates in the production backend are considerably higher than in contribution and distribution environments. Consequently, the internal storage and network infrastructure is heavily stressed, both centrally, in the work centres, and in camera equipment. The strategies for reducing file sizes without compromising picture quality typically differ between manufacturers. Camera formats, for instance, are optimized for storage, while postproduction formats are optimized for editing. Contribution, play-out and archive applications each add their own specific requirements as well. As a result, the number of available compression algorithms, different bit-rates, and concatenation of systems give a very large number of possible combinations. This section provides an overview of the legacy and state-of-the-art HD compression algorithms and the most important characteristic parameters that have to be considered.

Table 15.1 Review of codec requirements for broadcast application (M = mandatory)

Features	Production and archive	Contribution	Distribution
Full scalability	M	M	M
High coding efficiency	M	M	M
High bit-depth support ($\geq$10)	M	Desirable	Optional
Colour sampling	All if possible but 4:2:2 and 4:2:0 are mandatory	4:2:2 and 4:2:0	4:2:0
Support for interlaced and progressive scanning	M	M	M
Royalty fee free	Desirable	Desirable	Desirable
Standardized	M	M	M
Low delay	M	M	M
Low complexity	M	M	M
Intraframe	M	M	M
Interframe	Optional	Optional	M
Lossless coding	M	Optional	Optional
Visually lossless (at affordable rates)	M	Optional	Optional
Backward compatibility with other systems	Optional	Optional	M
High error resilience	M	M	M
Support by editing suites	M	Not applicable	Not applicable
Support for higher image formats and frame rates	M	M	Optional
MXF/AAF mapping	M	M	–
MPEG2-TS	–	M	M
RTP packaging	–	M	M

15.4.1 Technical Parameters

15.4.1.1 Quality Level

Different areas of tapeless television production can be identified, each requiring a distinct degree of quality:

- High-end production: premium quality is demanded for landmark and scenic productions.
- Mainstream production: daily soaps, documentaries, sitcoms, magazines, sports.
- News production.
- Video journalism based on consumer-type camera acquisition.

15.4.1.2   Intraframe versus Interframe Compression

In the broadcasting industry, interframe coding is applied mainly in the contribution and distribution domains. It is a powerful technique used drastically to reduce the necessary bit-rate, and is very well suited in cases of limited bandwidth network transmission and viewing-only by the end user. Intraframe-based compression is mainly applied in production formats, such that each frame contains a similar amount of data – in most cases even a constant bit-rate is applied – and can be processed independently of other frames. Intraframe codecs are extremely edit-friendly and the ease of performing basic cutting operations is almost comparable to editing uncompressed video. Interframe-based editing is increasingly becoming a candidate for production environments due to the significant progress in computer power that allows editing of intraframe compressed video with the same picture quality.

15.4.1.3   HDTV Scanning Formats (Interlaced and Progressive)

Interlaced scanning is an old but effective technique for bandwidth reduction of analog TV signals. By alternatively scanning the odd and even lines of a frame, each frame is split into two fields, each containing half of the frame information only. As a consequence, the necessary bandwidth is halved. However, modern compression techniques offer much better performance than the fixed technique of interlacing, by combining several techniques that can be optimized to suit the picture content. Progressive scanning differs from interlaced scanning in that each frame is fully encoded line by line rather than in alternate order, resulting in a more detailed, smoother image. As already mentioned, the most common applied HDTV scanning formats in broadcast production and emission are 720p/50-60 and 1080i/25-30, with respective resolutions of 1280 samples ×720 lines and 1920 samples ×1080 lines. 1080p/50-60 is considered as a scanning format for the future, combining high resolution and frame rate simultaneously.

15.4.1.4   Single-Layer versus Scalable Encoding

Scalable video coding is attractive in broadcast scenarios since it provides a cost-efficient solution for the delivery of different formats of the same content to multiple users. This is particularly interesting in distribution scenarios, e.g. the simultaneous delivery of SD and HD television to the end user. While in the production workflow single-layer encoding techniques currently prevail, scalable video coding is nevertheless attractive due to the capability of reconstructing lower quality versions from partial bit-streams, without the need for a separate browse file. Like interframe coding, scalable video coding adds complexity in such production activities as editing.

15.4.1.5   Subsampling and Quantization

These are common techniques used to further reduce the bit-rate of an encoded video signal. It is generally accepted that in order to obtain high-quality video production, horizontal and vertical subsampling should be avoided as much as possible. The recommended colour sampling scheme is 4:2:2. Most HD compression schemes offer 8-bit and/or 10-bit quantization. The advantages of 10-bit systems generally become obvious in scene-based colour correction, high-quality graphics, and layering. In other cases, picture quality of even multigeneration media turns out to be similar for 8-bit and 10-bit systems, although 10-bit encoding is expected to give more headroom in postproduction.

15.4.1.6   Bit-Rate

Uncompressed HD video requires 1.48 Gbps for 720p/50-60 and 1080i/25-30 and 3 Gbps for 1080p/50-60 scanning formats. Compressed video bit-rates for mainstream applications range between 50 and 200 Mbps for state-of-the-art compression algorithms. A higher bit-rate (440–880 Mpbs as used in HDCAM-SR) is required for high-quality TV productions.

15.4.2   Legacy HD Compression Algorithms for Production

Tape-based HDTV production has been in existence for the past several years using legacy formats such as HDCAM, HDCAM SR, and DVCProHD100. In the lower bit-rate ranges, the most popular legacy formats are XDCAM HD35 or MPEG-2 long GOP at 35 Mbps and HDV.

HDCAM and XDCAM HD35 compression is only available in 1080i/25-30 and 1080p/25-30 image formats. Both HDCAM and HDCAM SR are tape-only formats, while DVCProHD can be used as both tape- and file-based formats. Of all legacy formats, HDCAM SR provides the highest quality for HD production at bit-rates of 440 Mbps, and with an option for 1080p/50-60 or even 880 Mbps. It is often used as a compression format for the high-quality master recording. HDCAM SR is based on the MPEG-4 Part 2 studio profile.

For the higher bit-rate compression algorithms, intraframe coding and (minimally) 4:2:2 color sampling is applied. For lower bit-rate compression schemes, interframe coding and 4:2:0 color sampling are used. In most legacy formats, horizontal subsampling is applied, which results in loss of image details; 10-bit quantization is only used in HDCAM Sr.

Table 15.2 summarizes the main technical parameters of the different legacy compression formats.

15.4.3   Advanced HD Compression Algorithms for Production

As with SDTV, the continuous challenge in HD compression lies in minimizing the bit-rate while maintaining the original quality of the image during encoding as much as possible and throughout the production chain. For production in particular, the strategies for reducing file sizes without compromising picture quality differ between manufacturers. Generally, two main categories can be distinguished: camera-based compression formats (e.g. in an ENG camera), on the one hand, and postproduction formats, on the other hand.

Camera formats are typically optimized to reduce the necessary video data rate to a manageable size for storage on memory or professional BluRay or hard disk. Postproduction compression schemes, on the other hand, tend to be as straightforward as possible, such that the editing process is not unnecessarily complicated. The key question is if and to what extent these two types of compression formats are reconcilable within an integrated

Table 15.2 Technical parameters of legacy HD compression formats

Compression algorithm	Bit-rate (Mbps)	Subsampling and quantization	Intra/interframe coding
HDCAM SR	440 (880 option)	4:2:2 or 4:4:4; 10-bit	Intra
HDCAM	117	4:2:2; 8-bit horizontal subsampling to 1440 samples/line	Intra
DVCProHD	100	4:2:2; 8-bit horizontal subsampling	Intra
XDCAM HD35	35	4:2:0; 8-bit horizontal subsampling	Inter
HDV	20 (720p) or 25 (1080i)	4:2:0; 8-bit horizontal subsampling (interlaced)	Inter

production environment, bearing in mind that unnecessary transcoding scenarios should be avoided as much as possible.

An overview of the prevailing advanced HD compression formats of both camera and postproduction manufacturers is given below. Note that each of the described formats offers compression solutions at multiple-quality levels. The technical parameters of each of the formats are summarized in Table 15.3.

15.4.3.1 Camera Formats

AVC-Intra
Extensions to support high-definition resolutions, higher bit-depths (up to 10) and different color formats (including 4:2:2 and 4:4:4) were subsequently integrated into H.264 under the name of FRExt (fidelity range extension) by means of the addition of a 'high profile.' The Panasonic AVC-I codec is a variant of the professional version of the H.264 standard, which comes in two versions, each based on an intraframe compression method.

They are termed the High 10 Intra profile (50 Mbps mode) and the High 4:2:2 Intra profile (100 Mbps mode), respectively, with a 10-bit bit-depth. Constant bit-rate (CBR) encoding is applied in both cases. SMPTE approved a recommended practice for using these AVC intraframe coding schemes for SSM (solid state machine) card applications.

In order to improve the efficiency of Intra-only compression, the described AVC-Intra codec uses some of the H.264's new compression tools, which were not available in earlier compression schemes such as MPEG-2. Two specific compression methods that provide a major increase in efficiency are:

- Use of the correlation between adjacent pixels within a frame, known as *Intra prediction*.
- Entropy encoding improvement, specifically CABAC (context adaptive binary arithmetic coding), which is considered one of the primary advantages of the H.264/AVC encoding scheme. It requires a considerable amount of processing compared to other similar algorithms such as CAVLC (context adaptive variable length coding). It is especially powerful for lower bit-rates; therefore it is only used in the 50 Mbps mode of AVC-I.

Table 15.3 Technical parameters of state-of-the-art HD compression formats

Codec	Video bit-rate (Mbps)	Subsampling and quantization	Particularities	Standards
AVC-I Class 50	54	I-frame only; 4:2:0; 10-bit horizontal subsampling (1 440 × 1 080 or 960 × 720)	CABAC entropy coding	MPEG-4 AVC/H.264- SMPTE RP 2027
AVC-I Class 100	112	I-frame only; 4:2:2; 10-bit	CAVLC entropy encoding	MPEG-4 AVC/H.264- SMPTE RP 2027
XDCAM HD422	50	Interframe coding; 4:2:2; 8-bit GOP structure: IBBPBBP...	GOP size max. 15 (1920 × 1080) or 12 (1280 × 720)	MPEG-2
JPEG 2000	50/75/100	Intra-frame only; 4:2:2; 10-bit		ISO/IEC 15 444-1
VC-3 (DNxHD)	120/145	Intra-frame only; 4:2:2; 8-bit	720p50/1080i25: 120 Mbps 720p60/1080i30: 145 Mbps	SMPTE 2019
	185/220	4:2:2; 8/10-bit	720p50/1080i25: 185 Mbps 720p60/1080i30: 220 Mbps	SMPTE 2019
ProRes 422	185/220	I-frame only; 4:2:2; 10-bit		–
	120/145	I-frame only; 4:2:2; 10-bit		–

The increased compression efficiency results in higher computational complexity, which should be dealt with, especially in postproduction.

MPEG-2 Long GOP

Sony employs MPEG-2 4:2:2 at HL compression technology, also referred to as XDCAM HD 422, in its XDCAM HD422 product series. By applying interframe coding, the bit-rate can be limited to 50 Mbps only with a bit-depth of 8 bits. Therefore, the required bandwidth is fully equivalent to the standard definition and the network and storage infrastructure is not to be upgraded. This simplifies the implementation and maintenance of a complete, end-to-end workflow in practice.

MPEG-2 is a solid, proven codec and in general less complex than its successor H.264. Different varieties are available and its long GOP version is possibly the most widely used codec in the industry at the moment. A drawback of applying interframe encoding in production workflows is that the editing process becomes more complex and smart rendering techniques are required to resolve this issue. Reuse and rendering of original material also becomes more prone to quality losses in comparison with I-frame only codecs. MPEG-2 codecs offer 8-bit processing only.

JPEG 2000

For example, a practical broadcast implementation of JPEG 2000 is employed by Thomson Grass Valley in its Infinity product series (see Chapter 17). It is available at three different bit rates: 50, 75, and 100 Mbps and with 10-bit bit-depth and 4:2:2 sampling.

The wavelet-based compression technology imparts a number of compelling advantages over the discrete cosine transform (DCT) compression methods used in other state-of-the-art codecs, such as MPEG. One of the most attractive features is scalability, which allows the easy reduction in the number of pixels when required, without having to transcode to a separate low-resolution proxy file. Common visual impairments produced by DCT-based compression, such as blocking artifacts, do not occur.

It also offers low latency, in general less than 1.5 frames for encode or decode, which is – in combination with its inherent robustness to transmission errors – a highly interesting feature in contribution and two-way communication scenarios. By contribution, we mean the backhaul of programmes or content to studios or play-out facilities. Due to the symmetrical complexity of JPEG 2000, the same chips can be used for encoding and decoding. This feature fits well in the architecture of the contribution system too, where the number of transmitters and receivers are relatively equally matched, when compared to distribution scenarios. Up to now, however, MPEG-2 and in the future MPEG-4 AVC are still mostly used for contribution links.

JPEG 2000 is well established in the area of high-end imaging products; in particular it has been chosen as standard format for digital cinema, offering perfect quality at high resolutions. It remains to be seen to which extent its success will trickle down to 'lower' resolutions such as mainstream HDTV.

15.4.3.2 Postproduction Formats

VC-3 (AVID DNxHD)

The DNxHD video codec is employed in Avid non-linear editing systems and has recently been standardized in SMPTE under the name VC-3. It is an intraframe DCT-based CBR compression format that supports both 8-bit and 10-bit sampling depths for 4:2:2 sampling of 1920 × 1080 and 1280 × 720 video rasters. The used techniques in DNxHD are very similar to JPEG. For high-quality encoding, three different flavors are available: a mainstream variant at 120/145 Mbps (the bit-rate depends on the video frame rate) with a bit-depth of 8 bits and two premium quality variants at 185/220 Mbps with bit-depths of, respectively, 8 and 10 bits. Compared to the camera formats, VC-3 is computationally not complex, thus augmenting edit friendliness, especially in the case of multiple-layer editing using sophisticated effects.

Note that Ikegami's camera system is unique in its support for DNxHD, by recording directly to DNxHD encoded video. Thus, material can be made immediately accessible by Avid editing platforms, without the need for transcoding. Nevertheless, it is expected that editing platforms will extend their codec support with certain camera formats (as described above) as well.

ProRes

The ProRes 422 video codec has been developed for use in Apple's nonlinear editing systems. The codec matches many of the features of VC-3, including the available bit-rates, bit-depths, and color sampling scheme. However, it needs to be noted that the details of the codec are not known – it is not standardized and could consequently represent interoperability problems in the chain. It differs in that it uses variable bit-rate (VBR) encoding, allowing complex frames more bandwidth than simple frames. ProRes is primarily meant to be an intermediate codec during video editing, such that higher quality can be retained and multistream real-time editing scenarios can be easily supported. Unlike VC-3, ProRes is also available for advanced resolutions (2k and 4k) and SD.

15.4.4   Summary

Several new, advanced HD codecs have recently emerged in the broadcast industry, making the selection of the appropriate format quite challenging. The world of production codecs is complex and a number of requirements play a role (see previous section). Long GOP codecs result in significant storage and bandwidth savings, but are susceptible to artifacts, typically affecting multiple frames instead of one (e.g. GOP pumping). Advanced intraframe codecs such as wavelet codecs and H.264 deliver better quality compression at lower data rates at the expense of requiring considerably more processing power. Postproduction formats are optimized for editing purposes, but generally require transcoding during the production workflow.

The next section will elaborate the specific advantages when selecting JPEG 2000 as the preferred format in HD production workflows.

15.5   JPEG 2000 Applications

15.5.1   Why is JPEG 2000 Interesting for Broadcasting?

JPEG 2000 was initially intended to address the shortcomings of the JPEG standard. Thus, it is intrinsically an intraframe compression system. In contrast with the MPEG codec family which extensively uses the DCT spatial frequency transform, JPEG 2000 uses wavelet transforms. In many respects, JPEG 2000 is interesting for production and contribution broadcast applications. For distribution applications with drastic bit-rate limitations of the different networks, however, the combination of intraframe and interframe coding is a mandatory requirement that is not fulfilled by JPEG 2000.

The general advantages of JPEG 2000 are listed below.

15.5.1.1   Scalability

This is the most powerful advantage of JPEG 2000. It enables the extraction of a full standalone stream at various spatial, temporal, and quality resolutions. The powerful EBCOT encoding system enables SNR scalability with very fine granularity. Multiresolution scalability is provided by the wavelet transform. A JPEG 2000 compressed image is also component-wise scalable since each color space component is treated independently.

15.5.1.2   Error Resilient Algorithms

In addition to its error resilience methods (resynchronization markers inserted by the encoder in the code-stream, error correction tools, etc.), JPEG 2000 Part 3 (Motion JPEG 2000) is purely intraframe, thus avoiding error propagation in a group of pictures and improving editing possibilities of the video stream.

15.5.1.3   Artifacts

Code-block construction is a major difference between JPEG 2000 and DCT-based codecs. With JPEG 2000, the code-blocks are defined after the transform for each subband while in the DCT the blocks are predefined (fixed size) before the transform is applied on each block independently. The former technique avoids blocking artifacts. Nevertheless, blocking artifacts can still appear in a JPEG 2000 image if tiling is heavily used. Artifacts in JPEG 2000 are more visual friendly than DCT artifacts. Indeed, a JPEG 2000 artifact occurs by blurring the zone concerned. It corresponds to a lack of information (missing high frequencies) in the specific regions. In uniform or gradient regions, the visual effect of those artifacts will be quasi-lossless. Contouring and banding experienced in DCT-based systems are not observable with JPEG 2000.

15.5.1.4   Compression

JPEG 2000 increases its coding efficiency by the use of wavelet technology together with an arithmetic coder (MQ-coder). The use of arithmetic coding instead of variable-length coding enhances the coding efficiency of the system. It has been shown in several comparative studies that JPEG 2000 (Motion JPEG 2000) performs better than DCT-based codecs such as JPEG and H264/AVC-Intra at high bit-rates. Furthermore, as an essentially intraframe compression system, the JPEG 2000 stream is more robust to prediction error propagation compared with GOP (group of picture)-based systems such as MPEG-2 and H.264/AVC. An Intra codec is easier to handle in editing suites since all frames can be accessed on a single basis while a GOP-based system will require additional decoding, which inevitably introduces latency.

15.5.1.5   Compressed Domain Image Editing

This feature enables the processing of images in the compressed domain, thus allowing the extraction of the relevant bytes contributing to a particular quality level or resolution without decompress and compress operations. This feature can save several cascades in a production environment.

15.5.1.6   Symmetric Coding Technology

As outlined in Section 15.4.3.1, JPEG 2000 is a symmetrical system enabling coding and decoding at the same latency with the same chip.

15.5.1.7   Lossless Coding (Mathematically and Visually)

Using the 5/3 LeGall wavelet filter, mathematically lossless compression can be achieved, reducing the input bit-rate by half approximately (2:1). Using the 9/7 Daubechies lossy filter, visually lossless compression can be achieved at even lower bit-rates, as further described in Section 15.6.2. The input bit-rate can be decreased by a ratio of 8 to 10:1.

15.5.1.8 License- and Royalty-Free (to a Certain Extent) and Open Codec

Currently, several companies with patents in the image processing domain have allowed the use of their technology without demanding any royalty fee payments.

15.5.2 JPEG 2000 for Production and Archiving

As described in Section 15.3, today's production environment needs a standard compression system that can ensure support for future large image sizes as well as higher frame rates. The wavelet transform being a simple filtering process, the treatment of higher image resolutions results in an additional filtering process or decomposition level. Tiling of the image can be used for performance reasons. An example of the easy upgrade to higher resolutions is a migration between the 2k and 4k digital cinema image formats. It is recognized that only an additional decomposition level or the independent processing of two 2k tiles is necessary to be 4k compliant.

The intraframe characteristic of JPEG 2000 allows for any possible frame rate since each image is coded independently. It also provides low latency for editing and fast access purposes.

The high-quality requirement of the production environment combined with the high-quality expectation of HDTV are fully met (as far as high bit-rates are used, i.e. higher than 100 Mbps) by JPEG 2000. Visually lossless compression can be achieved to maintain the highest perceived quality of the content at affordable bit-rates (roughly 10:1 of the input rate).

As will be demonstrated in Section 15.6.2, JPEG 2000 is highly robust to multigeneration processes, one of the major requirements for broadcast applications.

JPEG 2000 supports higher bit-depths up to 16 bits as well as all color sampling schemes. As it is scalable componentwise, the 4:2:2 and 4:2:0 sampling schemes can be easily obtained by partially dropping the chrominance component subbands. Scalability enables the extraction of several quality versions from the same file. JPEG 2000 will thus enable a smooth transition to HD for those broadcasters that are not yet offering an HD service and provide a future proof production facility for larger image formats.

Low-resolution images used for fast content browsing in postproduction are easily extracted from the same JPEG 2000 stream by identifying stream packets that correspond to a low resolution. Those images are usually extracted and encoded using a different coding system than the mainstream content. With JPEG 2000 in production, a unique compression system can be used for multiple-production activities avoiding multiple coding and decoding passes and maintaining a high quality.

For content exchange, JPEG 2000 mapping into MXF has been standardized by the SMPTE (2006) for digital cinema applications.

15.5.3 JPEG 2000 for Contribution

As explained in previous sections, contribution applications require high quality and/or low coding latency, depending on the transmitted content and the bit-rate capacity of the network. Video coding latency can cause a serious delay between audio and video transmission. GOP-based systems usually have a longer coding delay due to the complex prediction operation involved.

The error resilience as well as error concealment features allow for robust transmission with regard to packet losses, which is highly valuable in error-prone media such as wireless links.

JPEG 2000's scalability feature and the excellent code-stream structure allow for bandwidth adaptation of the transmitted stream. The progressive transmission of different quality and resolution levels helps to improve the video quality gradually as more data are received. This particularly improves the time to display, i.e. the user does not need to wait for the full quality download to view the content.

JPEG 2000 is a symmetric compression system in that the complexity is shared equally in the encoder and the decoder. Therefore, it is ideally suited for contribution applications where a comparable number of systems is required at both ends of the network.

The full range of bit-rates addressed by contribution networks can be handled by JPEG 2000, delivering (visually) lossless-to-lossy compression quality.

15.5.4 Issues with JPEG 2000

Despite the increasing availability of JPEG 2000 products (cameras, real-time encoder/decoder chipsets, and powerful FPGA implementations), JPEG 2000 support by legacy editing systems was still under development. In addition, since JPEG 2000 is part of a new family of codecs based on wavelet technology, it is not backward compatible with most of the legacy broadcast compression technologies, which are DCT based. While the standardized mapping of JPEG 2000 into the MXF file format has been taken care of one of the most widespread transport stream structures, MPEG-2 TS, does not yet have a JPEG 2000 definition. These issues will need to be addressed in order to facilitate the adoption of JPEG 2000 by the broadcast industry.

15.6 Multigeneration Production Processes

As described in the previous section, a typical production environment is characterized by a cascade of legacy compressions systems, each having an impact on the quality of the final output of the chain. In order to adequately assess the performance of the different production CODECs proposed by the industry, the EBU has designed a process to simulate a typical production environment. This process is described in the following section. The JPEG 2000 compression system was tested in this simulation in order to evaluate its intrinsic performance in a typical production environment and to determine the adequate settings to obtain visual transparency for a variety of test materials. The results of this test are described in Section 15.6.2.

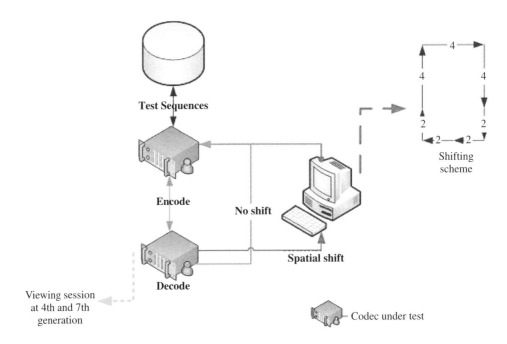

Figure 15.7   Test setup showing on the right the shifting scheme applied to the sequences

15.6.1   Test Setup

The codec test environment is designed to be closely representative of an actual production environment. The production chain is simulated by cascaded encodings up to the 7th generation, with and without a pixel shift, as can be seen in Figure 15.7.

Spatial or temporal shifting is a very common method used in broadcasting to simulate the shifting that can be introduced by some devices during transcoding or processing of the visual content. The pixel shift can be applied by software or hardware and reflects the nonlinearity of production processes. It is assumed that the CODEC tested does not add any spatial shift during compression.

As shown in Figure 15.7, the original uncompressed material is fed to the CODEC and the encoded material becomes the source for the next encoding. Each frame is shifted horizontally and vertically according to the following scheme (see also Figure 15.7):

- *1st generation*: shift of 4 pixels horizontally and 4 vertically;
- *2nd generation*: shift of 2 pixels from vertically;
- *3rd generation*: shift of 2 pixels horizontally in the opposite direction to the first horizontal shift;
- *4th generation*: same as generation 3;
- *5th generation*: shift of 2 pixels vertically;
- *6th generation*: shift of 4 pixels vertically.

Subjective image quality assessment is a very difficult and time-consuming task (some experts consider it an 'art'). The most quoted procedure of assessing visual quality has been normalized in the ITU-R recommendation BT.500-11 (ITU-R, 2003). In this recommendation, several criteria, such as the viewing conditions, assessment procedures, criteria for viewer selections, and so on, are defined.

An alternative, which is less time consuming, is to use a model of the HVS. These evaluations are objective in that they can be repeated at any time and always provide the same results. The scientific community uses so-called 'objective metrics,' which use various models of the HVS. Simple metrics measure the pixel-to-pixel error for faster quality assessment. More complex metrics such as the SSIM (structural similarity) are also used. It is also common to validate these objective measures such as PSNR (peak signal-to-noise ratio) and the SSIM against subjective expert viewings. For the EBU JPEG 2000 test, an expert viewing session with members from the broadcast industry was held during the 4th and the 7th generations to assess the visual quality of the output sequence.

The test sequences used are carefully selected to cover different levels of spatial and temporal criticality. All test sequences (see the overview in Figure 15.8), are compliant with the relevant recommendations (ITU-R 709 and ITU-R 1543 and SMPTE 296 M, respectively EBU 3299).

Each sequence is 10 seconds long so that consistent subjective quality assessment can be properly performed. All sequences of the same format are concatenated to form a single clip. At the beginning of the latter, a second clip of the HD universal Essert test chart (zone plates) is added.

15.6.2 Results

15.6.2.1 Sustainability to Multigeneration Environments

The JPEG 2000 codec has been used without the contrast sensitivity weights (CSF weights) at 50 Mbps and 100 Mbps to align with the other intra codec (Panasonic AVC Intra HD) operational rates. All encodings have been made using the Kakadu

Figure 15.8 Overview of a set of EBU test sequences. Some sequences are freely available. More information can be found on the EBU technical website http://tech.ebu.ch

software Version 5.2. The test sequences were provided in uncompressed 10-bit 1080i/25, 1080psf/25, and 720p/50 format. The yuv10 files were first converted into a. VIX file, which is basically a. YUV wrapper file with a content description header, containing the YUV components of each frame one after the other, in planar mode. This file was fed to the kdu_v_compress.exe tool with the appropriate settings (code-block 64 × 64, no weights, five decomposition levels). As illustrated in the following PSNR graphs (Figures 15.9 and 15.10), JPEG 2000 has a great sustainability to cascaded encodings (without shift). For all HDTV formats, at both 100 Mbps and 50 Mbps, the PSNR level is nearly the same from one generation to another up to the 7th generation. In fact the losses are quasi-null. However, JPEG 2000 experiences losses when the pixel shift is applied, with up to 2 dB lost in the shift-cascading process (Figures 15.9 and 15.10).

During the viewing session very disturbing flickering artifacts at 50 Mbps and 100 Mbps were noticed at the 1st generation for very critical sequences, such as the 'CrowdRun'

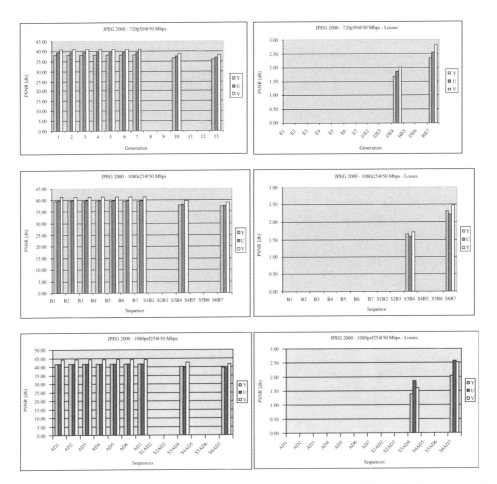

Figure 15.9 Sustainability at 50 Mbps for 1080i/25, 720p/50, and 1080psf/25. Note that only the PSNR values for the 4th and the 7th generations with shift were computed for the shifted generations

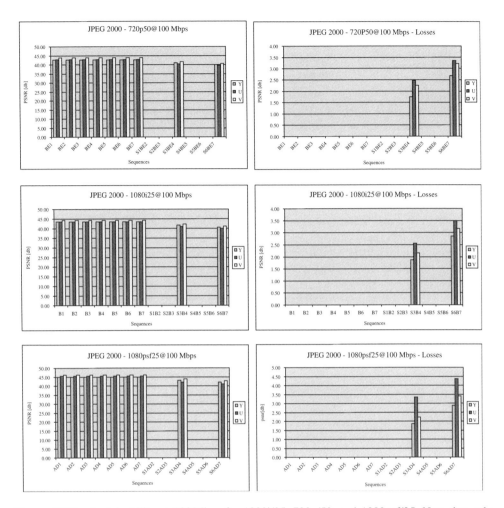

Figure 15.10 Sustainability at 100 Mbps for 1080i/25, 720p/50, and 1080psf/25. Note that only the PSNR values for the 4th and the 7th generations with shift were computed for the shifted generations

sequence – VID 09' in Figure 15.8. The artifacts are much more disturbing at 50 Mbps than at 100 Mbps.

15.6.2.2 Finding the Optimal JPEG 2000 Settings for the HDTV Production Environment

In order to define the optimal parameters for the application of JPEG 2000 for TV production, two factors have to be considered:

- the encoder parameters;
- the content type (scanning format, criticality of the sequence, HD/SD format (number of pixels), uncompressed/compressed).

For this part of the test, a description of the influence of each parameter on the visual quality as well as how they can be tuned to improve the visual quality will be provided. The test content consists of uncompressed 10-bit, 4:2:2, YUV sequences.

Scanning Modes

In the standard (ISO 15444 Part 3), there is no particular handling for the interlaced scanning format as opposed to other standards (e.g. PAFF, MBAFF in H.264/AVC). In fact, the same core encoding is applied either on the full frame (frame coding) or on the separate single fields (field coding).

This part shows which encoding method maintains the highest level of quality. For this experiment, two different sequences were chosen: a fast panning sport sequence (Crowdrun – VID09, see Figure 15.8) and an almost still scene (e.g. 'Vegetables' – VID06, see Figure 15.8). Each frame is split into two fields using software that simulates interlaced camera behaviour.

In theory, it is obvious that splitting and encoding the two fields separately will seriously impair the overall quality since the correlation between adjacent pixels is broken. This

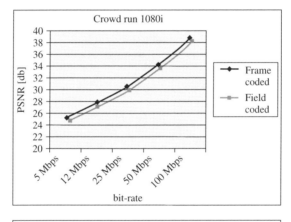

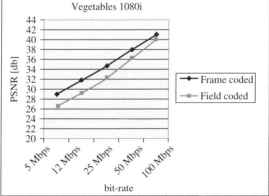

Figure 15.11 Field coding versus frame coding using the Crowdrun and the Vegetables test sequences

Table 15.4 PSNR values and differences for field coded and frame coded sequences

Sequences	Bit-rate (Mbps)	PSNR (dB)		
		Frame coded	Field coded	Difference
	5	25.34	24.72	0.62
	12	27.67	26.92	0.75
Crowdrun	25	30.53	29.77	0.76
	50	34.24	33.62	0.62
	100	38.88	38.38	0.5
Vegetables	5	29.41	27.03	2.38
	12	32.23	29.57	2.66
	25	35.38	32.87	2.51
	50 M	38.74	37.03	1.71
	100 M	42.03	40.88	1.15

fact is confirmed by the PSNR values shown in Figure 15.11. An average loss of 0.5 to 2 dB is reached depending on the criticality of the sequences. However, this is only an objective measure. The additional blurring effect of the interlaced format makes the visual quality in the field coded sequences even worse.

Using JPEG 2000 in the frame coding mode requires twice the memory than for field coding. However, it performs better in terms of visual quality and has better coding performance. Based on the PSNR curves, it can be determined that field coding needs 3/2 to twice its actual bit-rate to achieve the same quality as the frame coding method (see Table 15.4).

Code-Block Size

For an image with 8-bit precision per component, the maximum sample size of a JPEG 2000 code-block is 4096. This constraint thus limits the dimensions of the code-blocks.

As shown in Figures 15.12 and 15.13, and in the corresponding tables (Tables 15.5 and 15.6), from a PSNR point of view, the image quality improves along with the code-block size. Regardless of the bit-rate, the optimal block size is between 64×64, 8×512, 16×256, and 32×128, although 64×64 and 32×128 are slightly above the others (they all contain the maximum authorized number of samples).

Looking at the visual effect of another code-block dimension, a slightly better quality for the 32×32 block size can be noticed. Furthermore, the PSNR value of the latter code-block size is very close to the maximum achievable with the maximum number of samples. The use of smaller code-blocks can yield better visual quality, as is experienced in this case. A code block size of 32×32 is thus advisable.

Image Partitioning

Partitioning in JPEG 2000 can happen at several points in the encoding process. At the start of the encoding, the picture can be tiled into several subimages in order to reduce the amount of memory needed to store the image sample and simultaneously to enable parallel

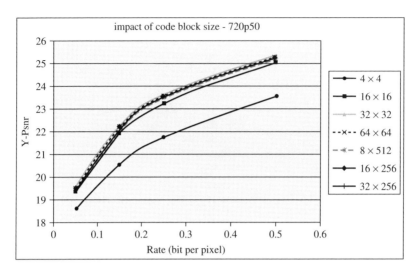

Figure 15.12 Impact of cod- block size on video quality for 720p/50

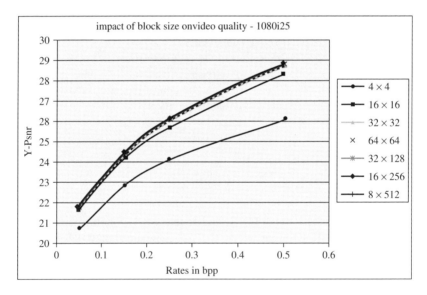

Figure 15.13 Impact of code-blocks on video quality for 1080i/25

encoding of the different parts. While this technique is useful for memory management, it has a negative effect on the visual quality of the content. Indeed, tiling (blocking) artifacts can appear at the border of each tile. After the wavelet transform, the wavelet coefficients can be divided into a set of code-blocks, known as precincts (optional), which obey the same restrictions (dimensions should be a power of 2). Although partitioning (precincts, etc.) helps for a better arrangement of the code-stream, it has no real influence on the

Table 15.5 Y-PSNR code-block comparison for 720p/50

720p50	dim\rate(bpp)	0.05	0.15	0.25	0.5
Code-block size	4 × 4	18.57	20.59	21.75	23.56
	16 × 16	19.34	21.87	23.28	25.04
	32 × 32	19.44	22.09	23.54	25.25
	64 × 64	19.47	22.2	23.58	25.3
	8 × 512	19.37	22.1	23.55	25.23
	16 × 256	19.42	22.16	23.57	25.27
	32 × 128	19.43	22.19	23.59	25.3

Table 15.6 Y-PSNR code-block comparison for 1080i/25

1080i25	Dim\rate(bpp)	0.05	0.15	0.25	0.5
Code-block size	4 × 4	20.68	22.8	24.07	26.04
	16 × 16	21.61	24.19	25.71	28.32
	32 × 32	21.71	24.42	26.01	28.7
	64 × 64	21.73	24.5	26.14	28.79
	32 × 128	21.71	24.49	26.12	28.78
	16 × 256	21.67	24.46	26.07	28.74
	8 × 512	21.6	24.4	26	28.77

coding and, consequently, on the visual quality. Neither of these two methods is required for a better image quality.

Filters

As shown in the JPEG 2000 description earlier in the document, two wavelet filters are available. Their use also depends on the application. Since the production operations are more likely to be lossy (cascaded encoding) it is obvious that one would prefer the 9/7 filter, which provides better performance against lossy coding. Even if the short 5/3 integer filter is more attractive and is supported by cheaper hardware implementation, the 9/7 is more appropriate for such applications.

Decomposition Levels

The question at this point is how many wavelet decomposition levels are necessary in order to maintain quality. The decomposition levels are related to the different available resolution levels, e.g. five decompositions implies six image resolutions.

CSF (Contrast Sensitivity Function) Weights

Contrast sensitivity weights are used to reduce the flickering artifacts that appear in a JPEG 2000 encoded video but are indistinguishable on a single still image. They are used in the rate control algorithm in order to adapt the quantization to the human visual system, which is less sensitive to image changes in higher spatial frequencies. Thus, all image compression systems try to make use of this property in order to remove as much visually irrelevant data as possible.

In Taubman and Marcellin (2000), several weighting tables have been proposed in order to reduce the flickering artifacts that might appear after the encoding. Each table was tested using the 'CrowdRun' (VID09) test sequence at 100 Mbps with the 720p format. Only the progressive format was used, given the advantages of this scanning mode as detailed in Section 15.3.

After several expert viewing sessions, it was assessed that the generic weights recommended by the standard gave the best results in terms of visual quality. Henceforth, in the following experiments this set of CSF weights will be used. Optimal values for broadcast applications should be investigated in future studies.

Bit-Rate

The bit-rate is the 'expensive' parameter in broadcasting, which has resulted in much research on video compression. In production, the bit-rate is kept as low as possible with the requirement that the quality after multiple generations (seven generations is the maximum) remains almost transparent to the original material.

The quality depends on the quantization, but in the case of JPEG 2000, the rate-control algorithm will drop packets (blocks in the bit-stream), which do not contribute to a particular quality level below the requested encoding rate. Thus there is a limit to the influence of quantization on quality for JPEG 2000.

Summary

Since JPEG 2000 is essentially an intraframe codec it would not be appropriate to expect superb visual quality at a low bitrate (50 Mbps) for HDTV content. In order to have a better understanding of the quality progression with respect to the bit-rate, the selected test sequences were gradually encoded from 50 Mbps up to 150 Mbps (for 1080i and 720p). In the first part, any optimization effect (weights, etc.) was switched off. During the expert viewing sessions, the gap in quality was larger between 50 and 100 Mbps than between 100 and 150 Mbps. Obviously, the appropriate bit-rate for 720p/50 and 1080i/25 is in the area of 100 to 125 Mbps.

Using the CSF weights, the quality at 100 Mbps becomes acceptable but the flickering artifacts were still too visible, especially for the 720p/50 format, which is upscaled on the 1920×1080 pixel panel. For 1080i/25, the quality was comparable to the original[2] but a higher bit-rate is preferred to provide sufficient quality headroom.

For 1080p/50, a different bit-rate range (100 to 250 Mbps) was tested. After viewing the same 'Crowd run' sequence encoded at the different bit-rates, it was noted that at 200 Mbps the 7th generation sequence was very similar to the original. At 150 Mbps, the visual quality was considered sufficient but the amount of noise was somewhat disturbing. For 175 Mbps, the limit of the barely noticeable difference was already reached at the 1st generation. A bit-rate of 200 Mbps is safer since it offers more quality headroom.

---

[2] Frame coded; otherwise we would need 10 to 15 Mbps more to achieve the same quality in the field coded mode.

After several viewing sessions (at 3 H viewing distance), the following bit-rates can be suggested:

- 1080i/25: 115 Mbps (sufficient headroom for critical sequences);
- 720p/50: 115 Mbps (just perceptible impairments for very critical sequences);
- 1080p/50: 200 Mbps (sufficient enough headroom for critical sequence).

15.7 JPEG 2000 Comparison with SVC

The functionalities and coding performances outlined in the previous sections demonstrate that JPEG 2000 is an adequate solution for today's broadcast applications. However, another standard may challenge JPEG 2000 since it not only provides scalability but also a high-efficiency baseline coding system, namely scalable video coding (SVC).

The next sections will provide an overview of the MPEG-4 AVC/H.264 SVC technology and compare the key features of SVC with those of JPEG 2000.

15.7.1 SVC Overview[3]

SVC is not the first attempt by the ISO/IEC's MPEG committee and the ITU SG16 committee to provide scalability for one of their single-layer standards. For instance, a scalable profile is available in the MPEG-2 standard but the added complexity and the coding efficiency losses were too high. SVC was designed to address those requirements.

SVC scalability is structured into base and enhancement layers, as represented in Figure 15.14. An enhancement layer is a collection of H.264/AVC stream NAL (network adaption layer) units coupled with an additional 3 bytes SVC NAL header. The SVC NAL header includes layer identifiers as well as other information useful for the decoding. Each layer represents a scalable layer carefully identified by a specific tag in the SVC NAL header.

Each scalability layer is attached to a single *dependency layer* and references the layer from which it is predicted in the SVC NAL header. A dependency layer corresponds to one spatial resolution.

Temporal scalability is provided by using the hierarchical B and P prediction structures already used in H.264/AVC. SVC only adds the necessary tagging information (temporal layer identifier) to identify frames that are part of the same frame rate layer. This has the advantage of not decreasing the coding efficiency of the system. The number of temporal layers thus depends on the size of the decoded picture buffer since it stores the reference pictures used for the hierarchical prediction.

Spatial scalability is provided via intralayer prediction and motion compensation processes within the layer images coupled with interlayer predictions. The efficiency and low complexity of the SVC approach to spatial scalability is derived from two innovative interlayer prediction features:

- *Interlayer macroblock prediction* (intralayer and motion vectors) mechanism. A process by which macroblocks of an upper layer are predicted depending on the prediction

---

[3] The reader is expected to be familiar with H.264/AVC technology. An overview of H.264/AVC technology can be found in Wiegand *et al.* (2003).

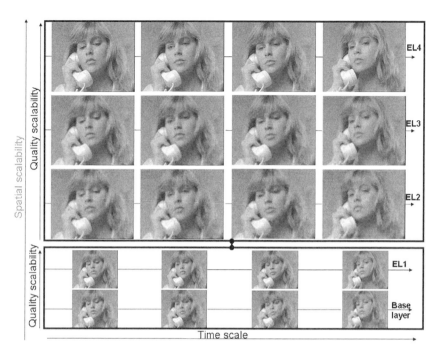

Figure 15.14   SVC layers overview (Kouadio *et al.*, 2008)

scheme of the co-located macroblock (intra- and/or interlayer predicted) in the reference layer.

- *Interlayer macroblock residual signal prediction* mechanism, which helps to reduce significantly the amount of information needed to transmit interlayer coded blocks of residual information; i.e. only the difference between the upsampled reference layer residual information and the predicted enhancement layer residual information is coded. Any picture resolution ratio can be considered between two successive enhancement layers, the only restriction being that the next enhancement should have a higher or equal spatial resolution than the reference layer.

Quality scalability is only considered within pictures of the same spatial resolution (dependency layer). It uses the same prediction mechanism as the spatial scalability features, which is further enhanced by a clever drift control mechanism (*key picture concept*) and provides an efficient tradeoff between the coding efficiency and drift control.

The key picture concept consists in enabling fast resynchronization of the encoder and decoder during the motion compensation loop, by referencing frames with the coarsest temporal resolution of the stream as the key reference picture. The drift is then limited to the closest coarse temporal picture.

For those particular frames, motion vectors do not change from the base to the enhancement layer, avoiding additional prediction processes. Furthermore, the base quality reconstruction of those frames is inserted in the DPB (decoded picture buffer) allowing no drift at this particular temporal level. Other temporal layers will use the highest available quality as a reference.

The SVC quality scalability feature allows for two levels of granularity:

- *coarse grain scalability (CGS)*, which allows a quality difference of 25% between successive quality layers;
- *medium grain scalability (MGS)*, which allows a quality difference of 10% between successive quality layers.

The fine grain scalability (FGS) mode was abandoned during the standardization process due to its high level of complexity. In terms of performance, a subjective quality test run by the MPEG verification committee showed that, depending on the sequence complexity, a 10% bit-rate increase is required to achieve the same visual quality as the single-layer H.264/AVC counterpart.

Further details on SVC can be found in Kouadio *et al.* (2008), ISO/IEC (2008), and Wiegand, Marpe, and Schwarz (2007).

15.7.2 JPEG 2000 versus SVC

Being based on H.264/AVC provides a significant advantage to SVC. Indeed, H.264/AVC is used in production as one of the acquisition formats (AVC-I) is also the default distribution compression system and is a strong candidate for the replacement of MPEG-2 for HD contribution applications (although its coding gain over MPEG-2 has still to be verified for contribution bit-rates). SVC feeds will thus be backward compatible with a range of products in diverse broadcast areas. SVC inherits all of the benefits of H.264/AVC, which include high coding efficiency at a low bit-rate, and is highly valuable in distribution environments. In a production or contribution environment, SVC uses interlayer prediction, even in intralayer mode. This results in a higher error potential than the filtering process of JPEG 2000. SVC also inherits AVC performance for large high-quality images. A higher quality performance on large image format can be expected from JPEG 2000 since the potential for prediction errors is relatively higher for SVC (due to interlayer predictions) than for AVC. Temporal scalability is easily provided by both standards without increasing the system complexity. However, SVC's temporal scalability relies on a set of predicted frames, where the quality can be lower or equal to an intralayer coded frame. Considering quality scalability, JPEG 2000 can achieve finer granularity by means of its difference layers than SVC, which is limited to the two types previously cited (CGS, MGS).

The asymmetry in the SVC standard together with the issues enumerated above makes it more suitable for distribution while JPEG 2000 remains the best choice for production and high-quality contribution.

15.8 Conclusion

This chapter has shown that the broadcast industry is experiencing fundamental changes:

- migration from traditional sequential production chains to file-based workflows centred also on the archive;
- migration from SDTV to HDTV with substantial higher quality requirements;
- multiple-delivery platforms (Mobile TV, IPTV, etc.) to be addressed with the same master format.

Table 15.7 Proposed settings for the use of JPEG 2000 in HDTV production

Parameters\HDTV formats	720p/50 – 1080i/25	1080p/50
Tiling	No	No
DWT	9/7	9/7
Coding scheme	Frame coded	Frame coded
Code-block size	32 × 32	32 × 32
CSF (weights)	Yes[a]	Yes[a]
(Minimal) bit-rate (Mbps)	115	200

[a]Optimal values should be investigated.

All of these changes bring new compression system requirements, which can mostly be addressed by JPEG 2000 (real full scalability, high error resilience, high quality for higher image format, future proof, open standard, graceful degradation, etc.) as far as production and contribution applications are concerned. The EBU evaluation highlighted JPEG 2000's high sustainability to the cascading process as well as optimal settings that allow for adequate production quality (see Table 15.7).

The use of CSF weights can improve the content visual quality; it is thus worth investigating optimal CSF values for broadcast application. SVC, a challenger system also providing temporal, spatial and, to some extent, quality scalability, appears to be more valuable on the distribution than in high-quality demanding environments such as production and contribution. Further practical comparisons of both systems need to be conducted.

With regards to all of its functionalities, JPEG 2000 has recently generated much interest from the broadcast industry. Digital cinema has paved the way by adopting JPEG 2000 as a compression system. The broadcast industry is slowly following with the appearance of acquisition products as well as real-time encoder/decoders for contribution as well as for archival applications. However, missing components such as JPEG 2000 PES mapping into the MPEG-2 TS stream hampers its full application by the broadcast industry. Activities are underway in the respective standardization committees to solve these issues. At the moment, a broadcast profile for JPEG 2000 is being defined by the JPEG committee to define adequate settings of JPEG 2000 for broadcast applications.

References

EBU (2008) Current status of high definition delivery technology, *European Broadcast Union (EBU)*, Geneva Tech 3328. Available at: tech.ebu.ch.

EICTA (European Industry Association) (2007) HD Ready 1080p. Conditions for high definition labelling of display devices, EICTA.

Gilmer, B. (2004) *File Interchange Handbook for Images, Audio, and Metadata*, Elsevier.

Hoffmann, H. (2007) Image quality considerations for HDTV formats in the flat panel display environment, PhD Thesis, Brunel University, UK.

ISO/IEC (2004) JPEG 2000 Image Coding System – Part 1: Core Coding System, ISO/IEC International Standard 15444-1:2004.

ISO/IEC (2007) JPEG 2000 Image Coding System – Part 3: Motion JPEG 2000, ISO/IEC International Standard 15444-3:2007.

ISO/IEC (2008) Scalable Video Coding – Part 10, ISO/IEC International Standard 14496-10:2008.

ITU (International Telecommunication Union) (2003) Methodology for the subjective assessment of the quality of television pictures, International Telecommunication Union, Geneva, ITU-R BT.500-11.

Kouadio, A., Bottreau, V., Noblet, L. and Clare, M. (2008) SVC a highly scalable codec, EBU, Technical Review Q2 2008.

Poynton, C. (2003) *Digital Video and HDTV: Algorithms and Interfaces*, Morgan Kaufmann.

Salmon, R. (2008) High frame rate television, in *Conference Proceedings IBC 2008*.

SMPTE (2004) Material exchange format (MXF), in *Society of Motion Pictures and Television Engineers*, SMPTE 377M.

SMPTE (2006) Material exchange format (MXF) – Mapping JPEG 2000 codestreams into the MXF generic container, in *Society of Motion Pictures and Television Engineers*, SMPTE 422M-2006.

Taubman, D. S. and Marcellin, M. W. (2000) JPEG 2000 *Image Compression Fundamentals, Standards and Practices*, Springer.

Wells, N. (2006) *The MXF Book: An Introduction to the Material eXchange Format*, Elsevier.

Wiegand, T., Marpe, D. and Schwarz, H. (2007) Overview of the scalable video coding extension of the H.264/AVC Standard, *IEEE Transaction on Circuits and Systems for Video Technology*, **17**(9).

Wiegand, T., Sullivan, G. J., Bjontegaard, G. and Luthra, A. (2003) Overview of the H.264/AVC Video Coding Standard, *IEEE Transactions on Circuits and Systems for Video Technology*, **13**(7), 560–576.

16

JPEG 2000 in 3-D Graphics Terrain Rendering

Gauthier Lafruit, Wolfgang Van Raemdonck, Klaas Tack, and Eric Delfosse

16.1 Introduction

Three-dimensional (3-D) content is an essential component in 3-D graphics applications (Figure 16.1) ranging from visually appealing geographic information systems (GIS) (Figure 16.2; see also Section 12.2.5) to 3-D games (Figure 16.3) with walkthrough/flyover animations (Figure 16.4). 3-D content is typically modeled as a combination of geometry (i.e. the object's 3-D shape) and appearance (i.e. textures and color rendering). A popular way for modeling the geometry is to tile the surface of the 3-D object with triangles (see Figure 16.1, left). The shape guides the 3-D rendering engine in calculating the screen pixels covered by the 3-D object. The color of these pixels is defined with a combination of a mathematical model of materials and 2-D textures (see Figure 16.1, right). 2-D textures add important visual information to graphical applications requiring a high level of realism and immersion (Figures 16.2 and 16.3).

GIS and 3-D games applications – and especially their terrains – represent vast amounts of 2-D texture data (Figure 16.5). For example, terrains covering a horizon of 1 km width, at a three-color pixel density of 16×16 pixels per m^2 and a flyover speed of 20 m/s, trigger a texture transfer speed of 123 Mbit/s, hardly achievable even with performant LAN connections. Not all network connections and device processors are capable of providing such high bandwidth and associated processing demands, and hence special software and hardware precautions have to be taken into account to guarantee cross-network and cross-platform interoperability of 3-D content. A texture compression scheme with high compression capabilities as JPEG 2000 is mandatory, but, as will be shown in the present chapter, much more technology is needed.

The JPEG 2000 Suite Edited by Peter Schelkens, Athanassios Skodras and Touradj Ebrahimi
© 2009 John Wiley & Sons, Ltd

Figure 16.1 3-D scene with 3-D meshes wrapped up with 2-D textures (reproduced by permission of GeoID) (see Plate 10)

Figure 16.2 Toward more realism in GIS applications using mega-textures (reproduced by permission of GeoID)

The following sections address the technological solutions in a tutorial-oriented bottom-up approach with simple examples and figures of merit. The reader is gradually guided through the anatomy of 3-D content transmission, focusing on the JPEG 2000 texture streaming of large terrains in high-quality 3-D rendering. Step by step we move forward toward integrating all ingredients into a 3-D online terrain flyover where textures are JPEG 2000 compressed and adapted according to the network and platform resources. The reader is referred to the JPIP protocol (Chapter 6) as an instantiation of the JPEG 2000 packet transmission approach used in this context.

Figure 16.3   3-D gaming with mega-texture terrain rendering (reproduced by permission of Larian Studios)

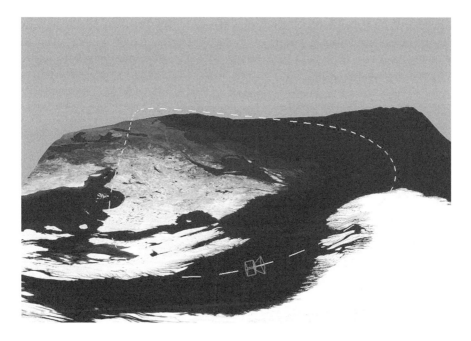

Figure 16.4   3-D walkthrough path (dashed) with user specified viewing directions (red camera view representation) over the 3-D scene (reproduced by permission of NASA) (see Plate 11)

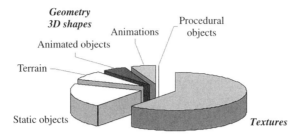

Figure 16.5 The relative bandwidth requirements for the transmission of 3-D content information (3-D shapes with procedural/mathematically defined objects, 3-D meshes and their animation, augmented with 2-D textures) for the application in Figure 16.3

16.2 Tiling: The Straightforward Solution to Texture Streaming

The straightforward solution to texture streaming of large textures/terrains consists in subdividing the texture into tiles that are transmitted independently from each other upon the 3-D player client's request. In this way the texture loading is spread over time, hence considerably reducing the *instantaneous* bit-rate through the network. Additionally, the software framework might choose to transmit a low-resolution version of some specific tiles to overcome network latency temporarily: the higher resolution version of these tiles will be transmitted only under ample network bandwidth availability conditions (Taubman, 2000; Taubman and Marcellin, 2001). Figures 16.6 to 16.8 show the distracting visual discontinuities that occur in such approaches with low-bandwidth/high-latency network connections (e.g. GPRS-EDGE provides only 200 kbps): an annoying continuous tile-based texture update is visible over time (over a couple of seconds). The quality discontinuity between the tiles is even so severe that feature misalignments sometimes become visible (see, for example, the arrow labeled with '*' in Figure 16.8). When time

Figure 16.6 Tiling effect (arrows) with low-bandwidth/high-latency network 3-D transmission and rendering (reproduced by permission of GeoID)

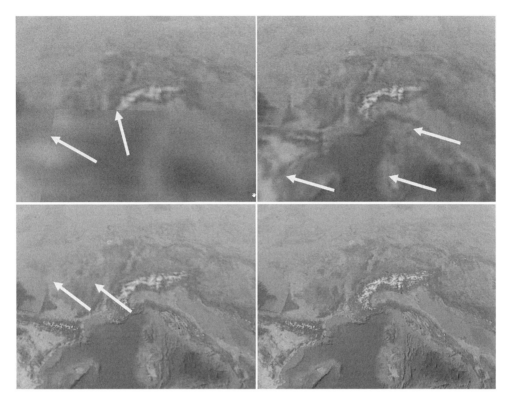

Figure 16.7 Gradual texture improvement with vanishing tiling effects (arrows) over texture download time. When time passes by, more texture information is transmitted, enhancing the visual quality of the rendering over low-bandwidth/high-latency networks (reproduced by permission of NASA) (see Plate 12)

passes by, the amount of downloaded data is sufficient to make the tile boundaries (almost) imperceptible. The quality of the terrain then improves gracefully in accordance with a legitimate requirement of the average user; more and more details are added without abrupt visual changes, reaching the high-quality rendering of Figure 16.7 (bottom-right) and Figure 16.8 (bottom).

As shown in Figure 16.9, the introduction of JPEG 2000 achieves these results without tiling, drastically reducing visual artifacts anytime during the rendering (even at start-up) – using the techniques explained in detail in following sections – albeit at the cost of a more complex texture transmission bookkeeping protocol. A simpler (non-JPEG 2000) alternative would consist in sending small tiles (e.g. 16×16 pixels), taking care to allow only gradual quality changes over adjacent tiles. Unfortunately, the tiles being then considerably small, any texture compression scheme (and in particular JPEG 2000) will fail in conciliating quality with high compression (see also Section 1.5). It hence becomes mandatory to keep large tiles (e.g. larger than 256×256 pixels), bringing back the original problem statement of visual discontinuities to be resolved.

To solve this dilemma of '*tiles or no tiles,*' JPEG 2000 technology provides the capability to use tiles in the wavelet domain – the so-called code-blocks and precincts – which

Figure 16.8 Gradual texture improvement with vanishing tiling effects (arrow '*' indicates feature misalignments) over texture download time. When time passes by, more texture information is transmitted, enhancing the visual quality of the rendering over low-bandwidth/high-latency networks (reproduced by permission of Agentschap voor Geografische Informatie Vlaanderen, Belgium) (see Plate 13)

are blended into one another through the filtering operations of JPEG 2000's wavelet transform (WT). The code-blocks/precincts hence semantically correspond to the afore-mentioned 'small tiles,' while the full texture is logically equivalent to 'one large tile,' which does not impede JPEG 2000's high compression performance. Summarized, tiles are present in the wavelet domain (the precincts), while tiles are absent in the spatial domain (the one single texture). Of course, in reconstructing a texture region correctly, all data dependencies (between all involved code-blocks/precincts) in the successive filtering operations of the inverse wavelet transform (IWT) must be respected to recreate the spatial domain texture information correctly. This issue is closely related to mipmap and clipmap in high-quality 3-D rendering, as will be explained in Section 16.3. The JPIP protocol is then used to transmit the precincts involved in constructing a texture region after decoding and I(L)WT transformation. Section 16.3.4 gives a simple JPIP example to clarify the relation between JPIP caching, the wavelet transform and mipmaps/clipmaps.

In principle, the same texture region packet selection approach might be used over all textured 3-D objects (i.e. not only terrains), but in practice the associated texture region selection bookkeeping for objects with limited sized textures (low to moderate spatial resolution) results in an important workload overhead that is not justified in view of the moderate instantaneous bit-rate savings. In this case, the only viable option for further bit-rate reductions is spatially to fully decode the object texture, but at a lower

Figure 16.9 JPEG 2000 reduced tiling effect, i.e. even over very short download times a smooth texture transition is obtained (top), using a prioritized texture packet transmission protocol. Zoom-in (bottom) shows very smooth texture transitions without any abrupt tiling effect (reproduced by permission of Agentschap voor Geografische Informatie Vlaanderen, Belgium) (see Plate 14)

resolution using simple bit-stream cutting with JPEG 2000's resolution level–position–component–layer (RPCL) bit-stream ordering (see Section 1.3.6).

Finally, to be complete, for a more pleasant user experience, the adaptation process should effectively find optimal system operating points based on the appropriate characterization of the expected workload as a function of content and device parameters. A precise description hereof is beyond the scope of the present chapter: the reader is referred to Andreopoulos *et al.* (2001, 2002, 2003), Aouadi (2005), and Tack *et al.*

(2004, 2006) for 3-D graphics workload modeling of respectively JPEG 2000 textures and MPEG-4 3D meshes, and the associated performance tradeoff control leading to optimal visual rendering (Delfosse, Lafruit, and Bormans, 2002; Lafruit *et al.*, 2003, 2004; Tack *et al.*, 2004). We will restrict ourselves to giving some performance hints in Section 16.4.

16.3   View-Dependent JPEG 2000 Texture Streaming and Mipmapping

16.3.1   View-Dependent Streaming

An interesting observation is that to obtain the visual appealing results of Figure 16.9 not all texture regions have to be transmitted at the highest resolution, hence drastically reducing (1) the JPEG 2000 code-blocks/precincts to be transmitted and (2) the associated decoding execution time.

Figure 16.10 indeed suggests that the visible portion of the object's texture resolution to be displayed is adjusted to the *viewing angle* α and the viewing-distance-dependent *texture texels/pixel ratio* R; i.e. large low-resolution portions of the texture are used for further away, tangentially viewed areas (red – D′), while small high-resolution texture portions are used for the nearby scenery (yellow – A), with:

```
cos(α) = polygon_normal * (-camera_position),

R = terrain_width × cos(α) × 2tan(field_of_view/2) ×
distance_to_polygon,
```

where '*' and '×' represent a vectorial dot product and a scalar multiplication, respectively.

Figure 16.11 shows the gradual change in local texture pixel-density/texture-resolution *ratio* R that is needed to be transmitted in order to obtain a visually qualitatively 3-D scene: for a given viewing distance/angle, red regions correspond to almost invisible texture regions that can be transmitted at very low resolution/quality, green and yellow regions should be transmitted at moderate quality (moderate viewing angle between the viewing direction and the normal vector on the 3-D object's triangle), and purple/blue/turquoise regions should be transmitted at the highest quality (where the 3-D object's triangles are looked at straightly). The subtle relationship between this phenomenon and the visual quality of Figure 16.9 is further explained in the next subsections.

16.3.2   Mipmaps and Spatial Scalability

The visual quality phenomenon hinted in Figures 16.9 to 16.11 is actually closely related to what is commonly known in 3-D graphics as 'mipmapping' for avoiding aliasing effects in the rendered images (Foley, van Dam, and Feiner, 1997). Figure 16.12 shows its basic principle: instead of directly rendering texels from the high-resolution texture toward the pixel screen (Figure 16.12, left), several resolutions of the same texture image are blended together to create the agreeable effect of the antialiased texture of Figure 16.12 (right). The blending operation consists of a linear interpolation/filtering between two adjacent texture resolutions for each screen pixel that is rendered. This creates – together with the

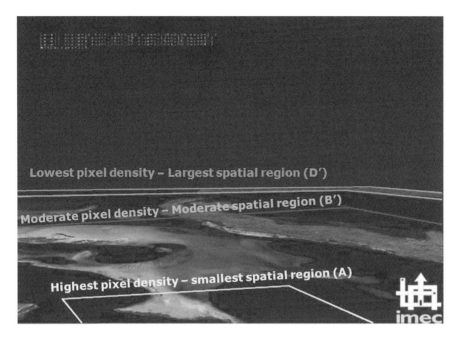

Figure 16.10 Texture texel/pixel densities as a function of viewing distance in the 3-D scene (reproduced by permission of NASA) (see Plate 15)

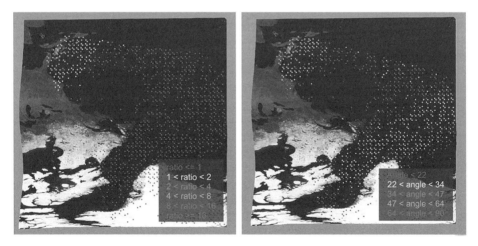

Figure 16.11 3-D terrain scene with distribution of the viewing distance and accompanying ratio between visualized pixels and texture texels resolution (left) and viewing angle (right) (reproduced by permission of NASA) (see Plate 16)

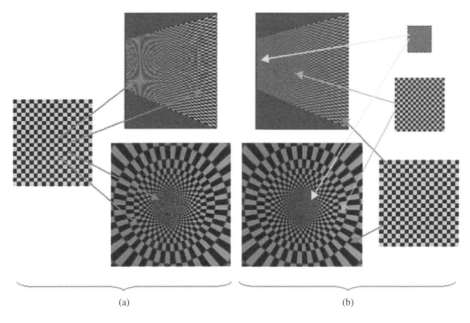

(a)   (b)

Figure 16.12 Rendering of a flat wall (top) and interior of a tube (bottom) without mipmapping: (a) one texture resolution is used for any viewing conditions and with mipmapping; (b) different, antialiased texture resolutions are used for different viewing conditions (see Plate 17)

2-D image filtering operations for creating intermediate texture resolutions – a so-called trilinear filtering operation for each pixel to be rendered on to the screen.

Similarly, in a 3-D graphics rendering application, because of the mipmapping, a user request for visualizing the 3-D scene along the viewing cone frustum F in the viewing direction V of Figure 16.13 will need to have the terrain texture filtered and downsampled at successive resolutions A (highest resolution), B, C, and D (lowest resolution) of Figure 16.13. Interestingly, only a very small portion of the texture is needed at the highest resolution (A), while a big portion is actually only required at moderate-to-low resolution (B, C, and D), effectively reducing the transmission bandwidth and decoding time.

In classical 3-D rendering engines, the creation of these successive texture resolutions for mipmapping starts at the highest full texture resolution that is gradually filtered and downsampled toward lower resolution replicas (from bottom to top in Figure 16.12, right). This solution would require downloading the full terrain texture at the highest resolution first (with all the latency and tiling problems discussed in Section 16.2) and then performing all the trilinear filtering operations afterwards. It would, however, be more appropriate to transmit the different mipmap textures regions of Figure 16.13 directly at their appropriate (mostly low) resolutions. This is what is made possible with the multiresolution representation of JPEG 2000's wavelet transform. Figure 16.14 (bottom) shows how JPEG 2000's inverse wavelet transform (IWT) creates, through IWT filtering operations, intermediate, antialiased downsampled spatial resolutions of the original texture for a large terrain texture, e.g. the 16k × 16k island terrain of Figure 16.14 (top-left). For its distinctive and easily recognizable shapes at any resolution, this will be further used as a tutorial example throughout this chapter.

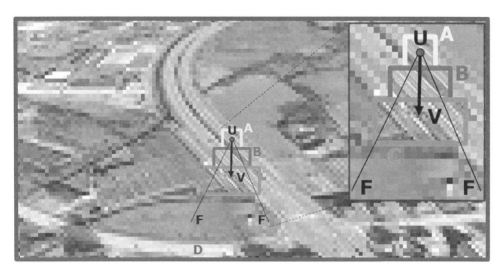

Figure 16.13 Mipmap texture resolutions (A, B, C, D) required for the user standing in position U and looking in the viewing direction V along the viewing cone frustum F (see Plate 18)

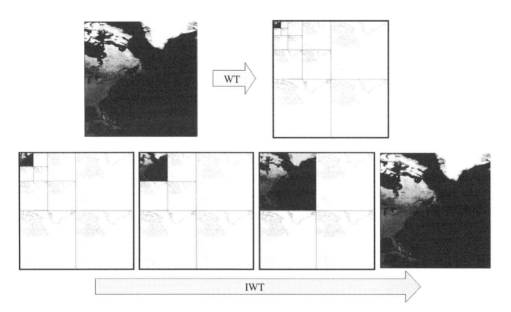

Figure 16.14 The wavelet transformed image (top) creates intermediate spatial resolutions during the inverse wavelet transform process (bottom)

The close relationship between JPEG 2000's wavelet transform and mipmapping is further discussed in following subsection.

16.3.3   Data Dependencies in JPEG 2000 and Mipmaps

In practice, the implementation of the mipmapping technique described in the previous subsection is a bit more complex since an arbitrarily large texture cannot be stored into a finite amount of physical memory on to the 3-D graphics processor. Fortunately, however, at any moment in time, only a portion of the large texture is visible, and hence only the corresponding information of the mipmap has to be transmitted, decoded, and stored into the processor's texture memory. A clipmap serves this purpose exactly: it is a dynamic mipmap where the largest texture layers (those with the highest level of spatial resolution) are *clipped*/restricted to only the visible viewport/frustum (cf. Figure 16.13).

Figure 16.15 shows an example of a clipmap and its relation to mipmaps and the wavelet transform. The original texture 5-V is once wavelet transformed according to the representation of Figure 16.15 (top-right). The lowest spatial resolution image E corresponds to the first WT resolution level 1. By injecting detail information (the LH, HL, and HH code-blocks of the WT in Figure 16.15, top-right) and iteratively applying IWT filtering operations, levels 2, 3, 4, and 5 are successively reconstructed. Level 5 of this multiresolution mipmap texture corresponds to the original resolution of the texture. Through memory constraints, only portions of this mipmap can be stored. If each level of the so-obtained clipmap has, for example, only enough memory capacity to store the texture size of level 2 (region D in Figure 16.15, top), then levels 1 and 2 can be stored fully in the clipmap memory, but levels 3, 4, and 5 are clipped to contain, respectively,

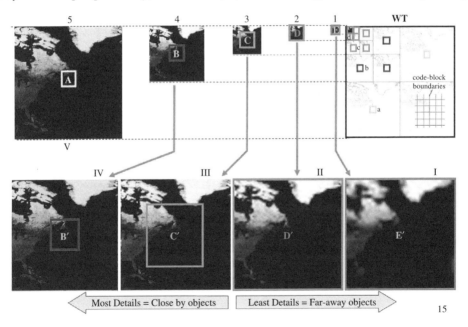

Figure 16.15   The relation between the WT (top-right), the mipmaps (top-left), and the clipmaps (bottom) for the terrain texture (see Plate 19)

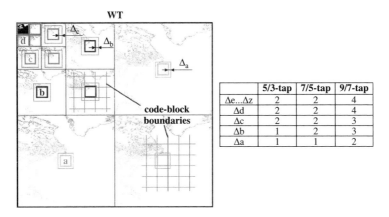

	5/3-tap	7/5-tap	9/7-tap
$\Delta e...\Delta z$	2	2	4
Δd	2	2	4
Δc	2	2	3
Δb	1	2	3
Δa	1	1	2

Figure 16.16 Code-blocks in JPEG 2000 used for the regions A, B, C, etc., in Figure 16.15 (left) and the additional amount of wavelet data for proper IWT filtering (right) (see Plate 20)

only regions C, B, and A, which are constructed out of the detail information/code-blocks c, b, and a of the WT data, through successive one-step upsampling operations (hence a, b, c, etc., are one quarter of the size of A, B, C, etc.). The bottom row of Figure 16.15 shows how this A, B, C, D, and E clipmap data will actually appear on the display through mipmap filtering and 3-D rendering, to obtain the regions B', C', D', and E', which are blended into the visual result of Figure 16.10.

To perform the IWT filtering operations involved in the reconstruction of regions A, B, C, etc., of Figure 16.15 correctly, not only do the wavelet data regions a, b, c, etc., have to be transmitted/cached but also the surrounding data, as shown in Figure 16.16. This means that if regions a, b, c, etc., correspond exactly to one code-block each, also some data of the surrounding code-blocks have to be transmitted. The reader can verify that the additional number of surrounding wavelet coefficients to transmit corresponds to the values of Figure 16.16 (right), which is at most four in the default wavelet filter configurations (5/3-tap and 9/7-tap) in JPEG 2000. Since code-blocks are minimally 4×4 wavelet coefficients large, each transmitted code-block a, b, c, etc., should be accompanied by its eight surrounding code-blocks to perform proper IWT filtering operations. This observation highly simplifies the data flow control in JPEG 2000 texture transmission, decoding, and clipmap processing in the 3-D graphics pipeline renderer, for achieving high-performance JPEG 2000 texture streaming 3-D graphics implementations. The reader is referred to the associated block-based 'local wavelet transform' (LWT and its inverse, the ILWT) technology first proposed in Lafruit *et al.* (1999), which achieves the target of only performing the strictly necessary (I)WT filtering operations in a block-by-block processing order for better cache data locality exploitation and associated performance increase (Andreopoulos *et al.*, 2001; see also Chapter 17).

16.3.4   Time-Dependent Mipmap/Clipmap Caching and JPEG 2000-JPIP Streaming

As slightly touched on in the previous subsection, the appropriate code-blocks/precincts of the JPEG 2000 texture have to be cached in time for a proper 3-D rendering experience.

Though one might follow an explicit simplified caching mechanism, such as the one proposed in Figure 16.16, it is from a software reuse point of view that the JPIP protocol and its caching mechanism is recommended for use, as explained in Chapter 6.

Indeed, for the examples of Figures 16.10, 16.13, and 16.15, with four different required texture resolutions distributed over different clipmap levels, the code-block data that are common between the different levels will automatically only be transmitted once, thanks to JPIP's caching mechanism. The individual clipmap layers requests can hence safely be transformed into the following JPIP requests without redundant data transmission and decoding:

```
GET /world.jp2?stream=0&len=2000&fsiz=16384,16384,closest&roff=
6007,10786&rsiz=273,682&cid=JPH_A8136D31668832F3 HTTP/1.1
GET /world.jp2?stream=0&len=2000&fsiz=8192,8192,closest&roff=
2867,5256&rsiz=136,614&cid=JPH_A8136D31668832F3 HTTP/1.1
GET /world.jp2?stream=0&len=2000&fsiz=4096,4096,closest&roff=
1297,2560&rsiz=102,443&cid=JPH_A8136D31668832F3 HTTP/1.1
GET /world.jp2?stream=0&len=2000&fsiz=1024,1024,closest&roff=
196,571&rsiz=51,255&cid=JPH_A8136D31668832F3 HTTP/1.1
```

Each clipmap layer generates its own JPIP requests, which cover the corresponding clipmap area. The yellow clipmap layer A of Figures 16.10 and 16.15 requests the JPIP server for a small region (rsiz = 273, 682) of the highest available resolution (fsiz = 16384, 16384). On the contrary, the red clipmap layer D/D' requests a relatively big region (rsiz = 51, 255) of the best fitting resolution (fsiz = 1024, 1024). The corresponding 2-D texture clipmaps are shown in Figure 16.17 (left).

Each 3-D walkthrough action will result in a corresponding 2-D translational movement over the 2-D texture. When the user moves upwards over the 2-D texture, as shown in

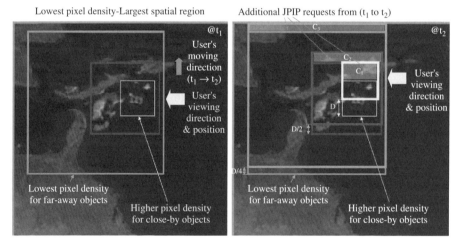

Figure 16.17 Clipmap data with different pixel densities at two successive time stamps t_1 and t_2, when the user viewpoint moves toward the top of the texture over distance D. Clipmap data have to be updated over respectively distances D, $D/2$, and $D/4$ with data additions C_1, C_2, and C_3 (reproduced by permission of NASA) (see Plate 21)

Figure 16.17 (left), the 3-D terminal asks for new texture regions at the quality levels that are needed according to the new viewpoint of the 3-D scene. In the example of Figure 16.17 (right), a translation over a distance D in the highest resolution clipmap layer will require new data to be cached according to the region C_1. Subsequent clipmap layers will also update their data over displacement heights of $D/2$, $D/4$, etc. The JPIP caching mechanism will take care of only transmitting the data of these newly visible regions, and will hence not retransmit old data in the overlapping regions of Figure 16.17 (right).

If the user's movement is sufficiently slow, all newly required data will reach the terminal in time to be properly rendered. However, if the data transmission is relatively slow compared to the user's movement, newly visible regions might not have gathered all the information required for highest quality rendering, creating smoothing effects as shown in Figure 16.9. In contrast to the visually disturbing situation depicted in Figures 16.6 to 16.8, no tiling boundary effects are visible, since block boundaries in the wavelet domain are smoothed by the IWT filtering operations of Section 16.3.3 to obtain the spatial domain texture. The end result is hence visually more appealing in Figure 16.9 than in Figures 16.6 to 16.8.

16.4   JPEG 2000 Quality and Decoding Time Scalability for Optimal Quality–Workload Tradeoff

Section 16.3 has shown that for achieving high-quality JPEG 2000 streaming results (e.g. Figure 16.9), the JPEG 2000 decoding should be closely linked to the mipmapping rendering techniques of the graphical processor. Counterintuitively, this technique provides the means to reduce the decoding and rendering execution time by exploiting the close relationship between mipmaps and JPEG 2000's spatial scalability, as suggested in Figure 16.13.

Furthermore, as will be shown in the remainder of this section, one can take as a rule-of-thumb that the number of decoded spatial wavelet levels is probably the most important parameter in JPEG 2000 decoding execution time, largely justifying the spatial-resolution-driven mipmapping technique of the previous section. Nevertheless, though the spatial layer parameter has a predominant impact on JPEG 2000's decoding time, the quality layer parameter has still its word to say in decoding conditions where only a limited number of decoded wavelet levels are used. Hence this section surveys some general rules-of-thumb for efficient and performant JPEG 2000 streaming in a variety of conditions, allowing some quality workload tradeoffs.

Figures 16.18 to 16.21 show the root mean square (RMS) error between the encoded and reference textures, respectively the execution time for decoding the JPEG 2000 texture at different wavelet and quality levels/layers. The execution time has been measured on an Intel Centrino 1.86 GHz laptop. It can clearly be seen that the number of decoded spatial wavelet levels has a tremendous quadratic impact on the execution time and a linear impact on the decoded texture quality (RMS error). The number of quality layers also contributes – but to a lesser extent than the number of wavelet levels to the decoded texture quality, while the impact on the execution time is negligible, except when decoding a limited number of wavelet levels.

Every JPEG 2000 image can be used as a source for view-dependent transmission, but the bit-rate and execution time overhead might become large with nonoptimally chosen

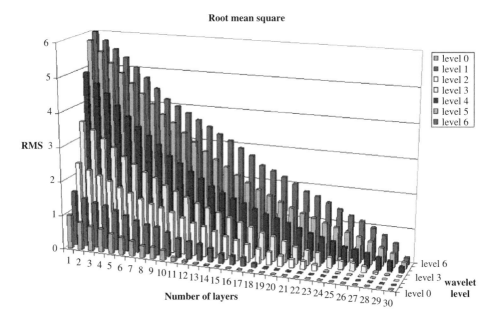

Figure 16.18   RMS error as a function of the wavelet level and number of layers

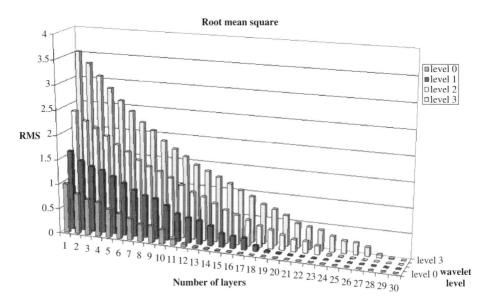

Figure 16.19   RMS value as a function of the wavelet level and number of layers for the lowest resolution wavelet levels

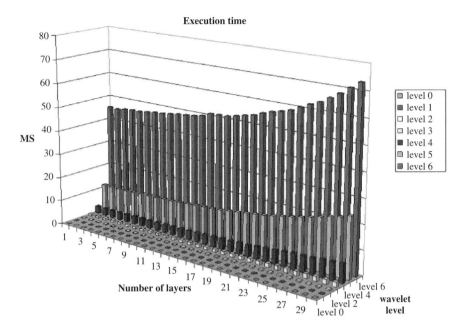

Figure 16.20 Execution time (ms) as a function of the wavelet level and number of layers

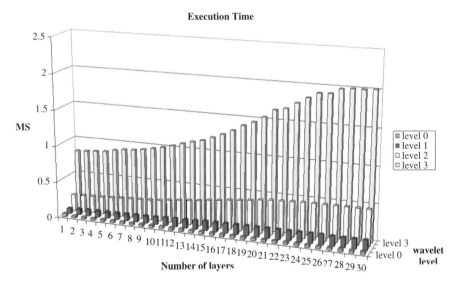

Figure 16.21 Execution time as a function of the wavelet level and number of layers for the lowest resolution wavelet levels

encoding parameters. Furthermore, the set of compression parameters used to compress an image can make a significant difference in decompression performance, while at the same time a wide range of different compression parameters have a relatively small effect on the coding efficiency. For low-complexity operational modes, we hence recommend the following settings:

- Tiles can drastically reduce the memory consumption for the JPEG 2000 encoder, albeit at a possibly high visual quality penalty. An efficient implementation of the wavelet transform, as in Lafruit *et al.*(1999) and Andreopoulos *et al.* (2001) can overcome tiling and guarantee high performances.
- The number of wavelet decomposition levels determines the lowest available resolution and therefore the commonly used default of five levels should be increased for large images.
- Two default wavelet kernels are available in JPEG 2000. The 5/3-tap wavelet filter intended for lossless compression has been chosen for its low computational complexity, but has a slightly worse image quality compared to the 9/7-tap wavelet filter.
- The overhead introduced by layers is very small for JPEG 2000 textures. Since layers are the key to quality and spatial scalability, it is generally recommended to have many layers to achieve fine-grained scalability and decode only the really required information for any given viewpoint.
- Consequently, when small regions of interest are frequent, we recommend using small code-block sizes; e.g. 32×32 and 64×64 code-blocks provide a good tradeoff between coding efficiency and fine-grained texture region selection at the appropriate quality and spatial resolution.
- A good progression order is one that reduces disk/cache thrashing for large textures. For a texture with many layers, the resolution level–position–component–layer or position–component–resolution level–layer order (see Section 1.3.6) should be preferred to keep all layers of a precinct at one position on the hard disk and/or cache. It is worth noting that the progression order influences the client only if it is accessing the JPEG 2000 texture directly. When using JPIP, the progression order is irrelevant for the decoder, but still important for the JPIP server's performance.
- The precinct size can also be used to reduce the code-block size. By using different precinct sizes per resolution, it is possible to use smaller code-block sizes only for the lower resolutions. The Kakadu manual[1] suggests such a setting when encoding large untiled textures, because it allows incremental flushing and therefore reduces the memory usage of the encoder. Of course this has the side effect of decreasing the coding efficiency for the low-resolution code-blocks. Luckily, there are not many low-resolution code-blocks; therefore the overall compression performance is often hardly altered. However, in a view-dependent texture streaming 3-D application, it might not be so uncommon that only low-resolution data are required for certain clients. In this case, the worsened coding efficiency matters. That is why such 'encoding tricks' should be used with care, preferably only when strictly necessary; e.g. the encoder runs out of memory.
- The use of PLT (packet length information for tile parts) marker segments for allowing the stream to be accessed in random order is recommended. Moreover, experiments

---

[1] www.kakadusoftware.com.

show that, without this, view-dependent decoding of a large image without tiles slows down drastically, with up to 50% longer execution times.

16.5 Conclusion

Transmitting, decoding, and rendering 3-D content requires a significant amount of resources both in the network and on the terminal. In the case of scarce/insufficient resources, the content richness is scaled down for obtaining real-time operational conditions. In this context, this chapter has been devoted to the study of one part of 3-D content data: the 2-D textures coded in JPEG 2000 format. We have shown how bit-rate and workload can be spread over time by (1) view-dependent streaming of JPEG 2000 textures, taking care to avoid visually disturbing artifacts, and (2) model-driven scaling down the processing in the visually least important texture regions, using JPEG 2000 scalability properties.

References

Andreopoulos, Y., Zervas, N., Lafruit, G., Schelkens, P., Stouraitis, T., Goutis, C. and Cornelis, J. (2001) A local wavelet transform implementation versus an optimal row–column algorithm for the 2D multilevel decomposition, in *Proceedings of the IEEE International Conference on Image Processing (ICIP)*, pp. 330–333.

Andreopoulos, Y., Masselos, K., Schelkens, P., Lafruit, G. and Cornelis, J. (2002) Cache misses and energy-dissipation results for JPEG 2000 filtering, *International Conference on Digital Signal Processing (DSP)*, CD-ROM T2C.2, pp. 1–9.

Andreopoulos, Y., Schelkens, P., Lafruit, G., Masselos, K. and Cornelis, J. (2003) High-level cache modelling for 2D-discrete wavelet transform implementations, *Kluwer Journal of VLSI Signal Processing Systems*, **34**(3).

Aouadi, I. (2005) Optimisation de JPEG 2000 sur système sur puce programmable, techniques avancées, PhD Thesis, ENSTA. Available at: [run on] http://pastel.paristech.org/bib/archive/00001658/01/THESE_ENSTA_I.AOUADI.pdf.

Delfosse, E., Lafruit, G. and Bormans, J. (2002) Streaming MPEG-4 textures: a 3D view-dependent approach, in *International Conference on Acoustics, Speech and Signal Processing*, 2002.

Foley, J. D., van Dam, A. and Feiner, S. K. (1997) *Computer Graphics: Principles and Practice*, Addison-Wesley Systems Programming Series, ISBN 0-201-84840-6, Addison-Wesley, Reading, MA.

Lafruit, G., Nachtergaele, L., Bormans, J., Engels, M. and Bolsens, I. (1999) Optimal memory organization for scalable texture codecs in MPEG-4, *IEEE Transactions on Circuits and Systems for Video Technology*, **9**(2), 218–243.

Lafruit, G., Pham Ngoc, N., van Raemdonck, W., Tack, N. and Bormans, J. (2003) Terminal QoS for real-time 3D visualization using scalable MPEG-4 coding, *IEEE Transactions on Circuits and Systems for Video Technology*, **13**(11), 1136–1143.

Lafruit, G., Delfosse, E., Osorio, R., Van Raemdonck, W., Ferentinos, A., Bormans, J. (2004) View-dependent, scalable texture streaming in 3-D QoS with MPEG-4 visual texture coding, *IEEE Transactions on Circuits and Systems for Video Technology*, **14**(7), 1021–1031.

K. Tack, K., F. Moran, F., G. Lafruit, G. and R. Lauwereins, R. (2004) 3D graphics rendering time modeling and control for mobile terminals, 3D technologies for the World Wide Web, in *Proceedings of the 9th International Conference on 3D Web Technology*, ACM, pp. 107–119.

Tack, K., Lafruit, G., Catthoor, F. and Lauwereins, R. (2006) Platform independent optimization of multi-resolution 3D content to enable universal media access, *The Visual Computer*, **22**(8), 577–590.

Taubman, D. S. (2000) High performance scalable image compression with EBCOT, *IEEE Transactions on Image Processing*, **9**(7), 1158–1170.

Taubman, D. S. and Marcellin, M. W. (2001) JPEG 2000: *Image Compression Fundamentals, Standards and Practice*, ISBN 0-7923-7519-X, Kluwer Academic Publishers, Boston, MA.

17

Conformance Testing, Reference Software, and Implementations

Peter Schelkens, Yiannis Andreopoulos, and Joeri Barbarien

17.1   Introduction

The complexity of JPEG 2000 is often perceived to be high compared to that of the older JPEG standard. Notwithstanding the more advanced nature of the JPEG 2000 coding technology, this perception is merely due to the lack of hands-on knowledge concerning the efficient implementation of wavelet-based technologies. Together with a fear of claims based on submarine patents, which are faced by royalty-free standards, the perception of JPEG 2000's high complexity has considerably inhibited the broad adoption of the standard on the consumer market. Of course, employing a global transform and entropy coding based on sequential bit-plane scanning does not encourage labeling a compression scheme as inexpensive in terms of computational and memory complexity. Nonetheless, industry and academia have shown that efficiently implementing JPEG 2000 is feasible. Hence, at the end of this chapter we will give some insight into a number of currently available JPEG 2000 products and their performance. It was not our intention to be complete, neither to give a list of the best products available. The overview is solely meant to provide examples of design options taken for solutions currently available on the market.

It is clear that implementing a complex standard such as JPEG 2000 also requires a verification of the functionality of the produced software. In order to facilitate this process, the JPEG committee has defined JPEG 2000 Part 4, to formalize the conformance testing procedure. In addition, Part 5 of JPEG 2000 offers reference software. Until now, we have been very silent in this book about Part 4 and Part 5 of the JPEG 2000 suite of standards and we will cover both parts in subsequent sections.

The JPEG 2000 Suite   Edited by Peter Schelkens, Athanassios Skodras and Touradj Ebrahimi
© 2009 John Wiley & Sons, Ltd

17.2 Part 4 – Conformance Testing

To realize a fast proliferation of standards, implementations need to be tested for conformance to the specification in order to ensure interoperability. JPEG 2000 provides a framework to enable compliance testing. This includes, for example, functionality, quality, and speed performance testing. Since JPEG 2000 incorporates a significant amount of features, which can be used to support a wide variety of applications, different *compliance profiles* with several attached *compliance classes* have been defined.

17.2.1 Definition of Profiles

Each compliance profile retains a subset of technology features from the JPEG 2000 Part 1 standard. JPEG 2000 specifies three generic profiles, which allow for successively increasing functionality sets, i.e. *Profile-0* is a subset of *Profile-1*, which is in turn a subset of *Profile-2*. Since it embeds all JPEG 2000 Part 1 functionality, the latter profile is also known as the *No Restrictions Profile*. The first two are specified in ITU-T Rec. T.800 (08/2002) | ISO/IEC 15444-1:2004. For compliance testing, only *Profile-0* and *Profile-1* are considered. *Profile-1* compliant codecs will also pass *Profile-0* compliance testing. The specific parameter ranges and functional options activated are listed in Table 17.1. As mentioned earlier in this book, additional profiles have been or are being specified for specific application domains, e.g. digital cinema (ITU-T Rec. T.800 (2002)/Amd.1 (09/2005) | ISO/IEC 15444-1:2004/Amd.1:2006 and ITU-T Rec. T.800 (08/2002)/Amd.2| ISO/IEC 15444-1:2004/FPDAM 2) and broadcasting (ITU-T Rec. T.800 (08/2002)/Amd.4| ISO/IEC WD 15444-1:2004/Amd.4).

17.2.2 Definition of Compliance Classes

Whereas profiles limit the code-stream syntax parameters, compliance classes define the limitations to which JPEG 2000 implementations are subjected. Hence, an implementation of a specific compliance class, indicated as Cclass, should support the corresponding required parameter spectrum. Table 17.2 summarizes the parameter ranges for each of the three distinct Cclasses.

17.2.3 Encoder and Decoder Compliance Testing

During decoder compliance testing, a normalized procedure determines the compliance class (Cclass) a decoder belongs to, given a specific profile. The procedure is designed based on the abstract test suite (ATS) specification, which is listed as an informative section of JPEG 2000 Part 4 and which describes the JPEG 2000 functionality and syntax to be tested. The practical embodiment of the ATS is the executable test suite (ETS), which is constructed from code-streams, reference decoded images, and tolerance values for MSE and peak error. The tolerance range will determine the Cclass attributed to a profile-X decoder. Exceeding the tolerance boundaries will lead to compliance failure of the decoder under test.

Similarly, an encoder can be tested for compliance by having a reference decoder parsing and reconstructing the images. Compliance is in this case verified by sweeping all

Table 17.1   Main qualitative specification of the restricted profiles: Profile-0 and Profile-1. A detailed specification is contained by ITU-T Rec. T.800 (08/2002) | ISO/IEC 15444-1:2004

Restrictions	Profile-0	Profile-1
SIZ marker		
Profile indication	Rsiz = 1	Rsiz = 2
Image size	Vertical and horizontal image sizes should be smaller than 2^{31}	Vertical and horizontal image sizes should be smaller than 2^{31}
Tiles	Tile dimension smaller or equal to 128 × 128 or one tile for the whole image	Tile dimension smaller or equal to 1024 × 1024 samples or one tile for the whole image
Image and tile origin	Image and tile origin is (0, 0)	Image and tile origin are located with the image size range
RGN marker segment	Maximum ROI shift is 37	Maximum ROI shift is 37
Subsampling	Possible horizontal and vertical subsampling factors are 1, 2, and 4	No restriction
Code-blocks		
Code-block size	32 × 32 or 64 × 64 pixels	Smaller or equal to 64 × 64 pixels (any dyadic size is possible, e.g. also 16 × 64)
Code-block style	Only the following options are selectable: termination on each coding pass, predictive termination, and segmentation symbols	No restriction
Marker locations		
Packed headers (PPM, PPT)	Disallowed	No restriction
COD/COC/QCD/QCC	Main header only	No restriction
Subset requirements		

Additional subset requirements relate to LL resolution, parsability, tile parts, and precinct sizes.

parameters (e.g. image sizes, number of components, decomposition levels, etc.) over the complete ranges for which the encoder has been designed and checking the decodability of the produced code-streams by a reference decoder.

17.3   Part 5 – Reference Software

For the purpose of testing JPEG 2000 encoder and decoder implementations and to help implementers fully comprehend the algorithmic structure and functionality of the JPEG 2000 Part 1 standard, the JPEG committee has also standardized ITU-T rec. T.804| ISO/IEC 15444–5, containing two independently designed JPEG 2000 implementations. The first implementation was delivered by the University of British Columbia, and is

Table 17.2 Definitions of compliance classes (Cclass). A detailed spec is contained by ITU-T Rec. T.803 (11/2002) | ISO/IEC 15444-4:2004

Parameter	Cclass 0	Cclass 1	Cclass 2
Image size ($W \times H$)	128×128	2048×2048	16384×16384
Number of components (C)	1	4	256
Code-block parsing guarantee (N_{cb}): minimum number of code-blocks to be parsable	48	66 048	268 468 224
Component parsing guarantee (N_{comp}): the minimum number of components that needs to be parsable (or the decoder is required to buffer information) without necessarily decoding the information	64	256	16 384
Coded data buffering guarantee (L_{body}): the parameter defines the minimum amount of packet bytes that should be storable before the decoder effectively starts decoding. It concerns those packet bytes whose precincts are relevant for the dimensions and components for which compliance is being claimed	8192 bytes	2^{23} bytes	2^{30} bytes
Decoded bit-plane guarantee (M): the decoder must decode correctly the M most significant bit-planes per code-block	11	15	30
9-7I precision guarantee (P): transform precision needed to meet the decoding quality bounds associated with each compliance class	Low enough to allow 5-3I decoding of 9-7I[a]	16-bit fixed-point implementation	3jpeg 2000-bit single precision floats
5-3R precision guarantee (B): component bit-depths for which the decoder should guarantee lossless reconstrunction	8	12	16

(*continued overleaf*)

Table 17.2   (*continued*)

Parameter	Cclass 0	Cclass 1	Cclass 2
Transform level guarantee (T_L): minimum amount of levels the inverse DWT is able to synthesize	3	7	12
Layer guarantee (L): the minimum amount of layers the decoder is able to decode	15	255	65 535
Progressions	All basic progressions in COD. Only need to decode the first progression per tile	Limited only by the number of levels, layers, and components	Limited only by the number of levels, layers, and components
Tile parts	Decode only first tile part per tile	Decode all tile parts up to N_{cb} or L_{body} limits	Decode all tile parts up to N_{cb} or L_{body} limits
Precincts	Decode first precinct per subband	Decode all precincts up to N_{cb} or L_{body} limits	Decode all precincts up to N_{cb} or L_{body} limits

[a] In order to meet the decoder resource constraints it is possible to replace a 9-7I inverse transform with a 5-3I inverse transform, hence reducing the decoding complexity. This type of replacement will, however, give rise to signal-dependent noise, manifesting itself especially in the high spatial frequency regions of the image.

known as Jasper,[1] while the second one was developed by Canon, EPFL, and Ericsson, and is known as JJ2000.[2] Jasper is implemented in C, while JJ2000 is a Java™ implementation. Both implementations were designed to ensure a fast proliferation of the JPEG 2000 Part 1 standard. These reference encoders can be utilized in the compliance testing procedures, as outlined in the previous section.

17.4   Implementation of the Discrete Wavelet Transform as Suggested by the JPEG 2000 Standard

Before addressing additional – commercially available – JPEG 2000 solutions, this section will address implementation problems typically encountered when aiming for an efficient implementation of JPEG 2000. The section will especially focus on the wavelet transform since this component is, due to its multirate character, particularly suited for parallelization, both in software (e.g. GPU processing) and hardware.

---

[1] http://www.ece.uvic.ca/~mdadams/jasper/.

[2] http://jj2000.epfl.ch/.

In his seminal paper, Mallat (1989) introduced a discrete pyramidal filterbank decomposition with two important properties:

1. It is nonredundant, i.e. C signal samples are transformed to C coefficients.
2. It is a multirate decomposition, i.e. downsampling is applied for each level of the pyramidal decomposition.

In order to provide an artifact-free decomposition and perfect reconstruction, the analysis filters are designed following certain conditions, which ensure that, when the filters are iterated for a number of times over an impulse, a scaling or wavelet function is formed (Daubechies, 1992; Mallat, 1998) that has certain regularity and smoothness properties.

This decomposition immediately became very popular and came to be known as the discrete wavelet transform (DWT). Since then, there has been a flurry of research papers on how to use this new signal processing tool for a variety of applications. A good summary of such applications is presented in Mallat's book (1989). An excellent introduction to the DWT from an engineering perspective is given in the book of Strang and Nguyen (1996).

The most prominent application of the DWT in electrical engineering is in the field of digital image and video compression and coding (Antonini *et al.*, 1992; Ohm, 2005). Eventually, these research efforts led to the realization of new, wavelet-based image compression schemes with excellent compression performance in comparison to the JPEG standard (Shapiro, 1993; Said and Pearlman, 1996). This fact led to the inclusion of the DWT in the JPEG 2000 standard (Taubman and Marcellin, 2002).

In the following subsection we review the DWT. Subsection 17.4.2 surveys the main elements of typical hardware designs, while Subsection 17.4.3 provides some pointers for further study.

17.4.1 Introduction to the Discrete Wavelet Transform

The one-dimensional fast discrete wavelet transform was initially proposed as a critically sampled time-frequency representation realized by filtering the C-sample input signal $\mathbf{x} = [x[0] \;\cdots\; x[C - 1]]^T$ with a T_{low}-tap lowpass filter $\mathbf{h} = [h[T_{\text{low}} - 1] \;\cdots\; h[0]]$ and a T_{high}-tap highpass filter $\mathbf{g} = [g[T_{\text{high}} - 1] \;\cdots\; h[0]]$ followed by dyadic downsampling (Mallat, 1989). Based on the first Noble identity (Strang and Nguyen, 1996), the signal analysis can be expressed in the time domain as $\mathbf{s} = \mathbf{E} \cdot \mathbf{x}$, where $\mathbf{E}$ is the $C \times C$ analysis matrix. Assuming (for illustration purposes) that $\mathbf{h}$ and $\mathbf{g}$ have equal maximum degrees in the Z domain, the analysis matrix is (in block Toeplitz form):[3]

$$
\mathbf{E} = \begin{bmatrix}
\ddots & & \ddots & & & \ddots & & & \ddots \\
\cdots & h[0] & h[1] & h[2] & \cdots & h[T_{\text{low}} - 2] & h[T_{\text{low}} - 1] & 0 & 0 & \cdots \\
\cdots & g[0] & \cdots & g[T_{\text{high}} - 1] & 0 & \cdots & 0 & 0 & 0 & \cdots \\
\cdots & 0 & 0 & h[0] & h[1] & h[2] & \cdots & h[T_{\text{low}} - 2] & h[T_{\text{low}} - 1] & \cdots \\
\cdots & 0 & 0 & g[0] & \cdots & g[T_{\text{high}} - 1] & 0 & \cdots & 0 & \cdots \\
\ddots & & \ddots & & & \ddots & & & \ddots
\end{bmatrix}.
$$

$$(17.1)$$

---

[3] In order to treat the signal borders, the border coefficients of $\mathbf{E}$ are modified to perform symmetric extension, as appropriate.

The output of the decomposition, s, is

$$\mathbf{s} = \begin{bmatrix} x_{\text{low}}[0] & x_{\text{high}}[0] & x_{\text{low}}[1] & x_{\text{high}}[1] & \cdots & x_{\text{low}}[C/2-1] & x_{\text{high}}[C/2-1] \end{bmatrix}, \tag{17.2}$$

consisting of alternating low-frequency and high-frequency coefficients $x_{\text{low}}[i]$ and $x_{\text{high}}[i], 0 \le i < C/2$. The decomposition can iterate on low-frequency coefficients $x_{\text{low}}[i]$ of s for L decomposition levels, thereby forming a multiresolution transform decomposition of the input signal $\mathbf{x}$ (Mallat, 1989).

Daubechies and Sweldens (1998) demonstrate in their tutorial paper that any analysis matrix $\mathbf{E}$ allowing for perfect reconstruction of $\mathbf{x}$ from $\mathbf{s}$ (i.e. $\mathbf{E}$ has to be invertible) can be factorized in a succession of liftings (updates) and dual-liftings (predictions), up to shifting and multiplicative constants:

$$\mathbf{E} = \Upsilon \cdot \prod_{k=1}^{K} \mathbf{U}_k \mathbf{P}_k, \tag{17.3}$$

where matrices $\mathbf{P}_k$ and $\mathbf{U}_k$ are breaking the realization of the highpass and lowpass filters (respectively) into a series of K pairs of lifting steps. The factorization of Equation (17.3) is scaled by the diagonal matrix $\Upsilon = \text{diag}\,(\Upsilon\,{}^1/\gamma \cdots \Upsilon\,{}^1/\gamma)$ with alternating scaling factors Υ and ${}^1/\gamma$. The conceptual explanation of the factoring algorithm is based on the notion that a highpass filter 'predicts' its current input signal value from its neighbouring context, while a lowpass filter 'updates' its current input signal value based on the prediction error (Daubechies and Sweldens, 1998). The respective lifting steps are expressed as the progressively shifted row of filter coefficients alternating with one row consisting of zeros and the unity sample, which propagates the output of the previous step:

$\mathbf{P}_k =$

$$\begin{bmatrix}
\ddots & & & & & \ddots & & & & & \ddots & & \\
\cdots & 0 & 0 & \cdots & & 0 & 0 & 1 & 0 & 0 & 0 \\
\cdots & a_{p(k)}[0] & 0 & \cdots & a_{p(k)}\left[\frac{P_{p(k)}}{2}-1\right] & 0 & a_{p(k)}\left[\frac{P_{p(k)}}{2}\right] & 1 & a_{p(k)}\left[\frac{P_{p(k)}}{2}+1\right] & 0 \\
\cdots & 0 & 0 & 0 & 0 & \cdots & 0 & 0 & 1 & 0 \\
\cdots & 0 & 0 & a_{p(k)}[0] & 0 & \cdots & a_{p(k)}\left[\frac{P_{p(k)}}{2}-1\right] & 0 & a_{p(k)}\left[\frac{P_{p(k)}}{2}\right] & 1 \\
\ddots & & & & & \ddots & & & & & \ddots &
\end{bmatrix}$$

$$\begin{bmatrix}
\ddots & & & \ddots & & & & \ddots \\
0 & & \cdots & 0 & 0 & 0 & 0 & \cdots \\
a_{p(k)}\left[\frac{P_{p(k)}}{2}+2\right] & \cdots & 0 & a_{p(k)}\left[T_{p(k)}-1\right] & 0 & 0 & \cdots \\
0 & & 0 & 0 & \cdots & 0 & 0 & \cdots \\
a_{p(k)}\left[\frac{P_{p(k)}}{2}+1\right] & 0 & a_{p(k)}\left[\frac{P_{p(k)}}{2}+2\right] & \cdots & 0 & a_{p(k)}\left[T_{p(k)}-1\right] & \cdots \\
\ddots & & & & \ddots & & & \ddots
\end{bmatrix}$$

$$\tag{17.4}$$

$\mathbf{U}_k =$

$$
\begin{bmatrix}
\ddots & & & & \ddots & & & & & & & & \ddots & \\
\cdots & a_{u(k)}[0] & 0 & \cdots & a_{u(k)}\left[\dfrac{P_{u(k)}+1}{2}-2\right] & 0 & a_{u(k)}\left[\dfrac{P_{u(k)}+1}{2}-1\right] & 1 & a_{u(k)}\left[\dfrac{P_{u(k)}+1}{2}\right] & 0 \\
\cdots & 0 & 0 & 0 & \cdots & 0 & 0 & 0 & 1 & 0 \\
\cdots & 0 & 0 & a_{u(k)}[0] & 0 & \cdots & a_{u(k)}\left[\dfrac{P_{u(k)}+1}{2}-2\right] & 0 & a_{u(k)}\left[\dfrac{P_{u(k)}+1}{2}-1\right] & 1 \\
\cdots & 0 & 0 & 0 & 0 & \cdots & 0 & 0 & 0 & 0 \\
\ddots & & & & \ddots & & & & & & \ddots
\end{bmatrix}
$$

$$
\begin{bmatrix}
& a_{u(k)}\left[\dfrac{P_{u(k)}+1}{2}+1\right] & \cdots & 0 & & a_{u(k)}[T_{u(k)}-1] & 0 & 0 & \cdots \\
& 0 & \cdots & 0 & & 0 & 0 & 0 & \cdots \\
& a_{u(k)}\left[\dfrac{P_{u(k)}+1}{2}\right] & 0 & a_{u(k)}\left[\dfrac{P_{u(k)}+1}{2}+1\right] & \cdots & 0 & a_{u(k)}\left[T_{u(k)}-1\right] & \cdots \\
& 1 & 0 & 0 & \cdots & 0 & 0 & \cdots \\
& & & \ddots & & & & \ddots
\end{bmatrix}
$$

(17.5)

with $a_{p(k)}[t]$ and $a_{u(k)}[t]$ the tth nonzero (and nonunity) predict and update filter coefficients, respectively, out of $T_{p(k)}+1$ and $T_{u(k)}+1$ nonzero taps, and $P_{p(k)}$ and $P_{u(k)}$ the respective maximum degrees in the Z-domain for the predict and update filters. Since the predict and update filters are applied to the even and odd polyphase components of the input, $P_{p(k)}$ is an even number (or zero) and $P_{u(k)}$ is an odd number. A unity tap is always placed at the position of the 'current' input sample[4] in order to enable 'prediction' or 'update' of the sample based on its context. The corresponding signal flow graph is given in Figure 17.1. The notation $\mathbf{P}_{i,\text{even}}$ and $\mathbf{U}_{i,\text{even}}$ indicates the even rows of the ith step prediction or update matrix (correspondingly for the odd rows), where we assume that rows in all matrices are indexed starting from zero.

As an example, for the 9/7 filter-pair we have: $K = 2$, $T_{p(k)} = T_{u(k)} = 2$, and $P_{p(k)} = 0$, $P_{u(k)} = 1$, $a_{p(k)}[0] = a_{p(k)}[1]$, $a_{u(k)}[0] = a_{u(k)}[1]$ for each $k = \{1, 2\}$. For the 5/3 filter-pair we have the same settings but with $K = 1$ (one predict and one update

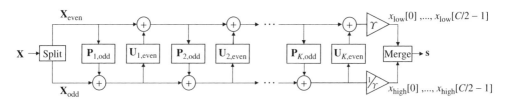

Figure 17.1 Signal flow graph for the lifting decomposition using K pairs of predict and update steps

---

[4] In our analysis, the 'current' input sample is defined as the filter position corresponding to zero degrees in the Z-domain representation of the update filter and to -1 degrees in the Z-domain representation of the predict filter (because we assume that the highpass filter is applied to the odd polyphase components of the input signal).

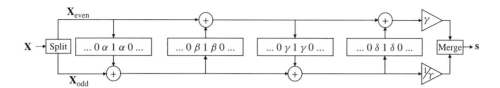

Figure 17.2   Signal flow graph for the lifting decomposition

step). The signal flow graph of the analysis (decomposition) using the 9/7 filter-pair is given in Figure 17.2. Reconstruction is performed symmetrically by switching the split and merge steps and the scaling factors, traversing the graph from right to left (reversed signal flow), and flipping the signs of the predict and update coefficients.

A recently proposed modification of the lifting signal flow graph is the so-called flipping structure (Huang, Tseng, and Chen, 2004; Xiong, Tian, and Liu, 2006). An example of the flipping structure for the 9/7 filter-pair is given in Figure 17.3. Removing the multiplication from the third accumulation point of Figure 17.3 leads to a reduced critical path in comparison with the conventional design. Identical results with the conventional lifting-based or convolution-based implementation are achieved by the system of Figure 17.3 by including the factor $\gamma\delta$ in the fourth multiply-accumulated point. The coefficients of the lifting decomposition of the 9/7 filter-pair are given in Table 17.3. We also provide the corresponding values for the 5/3 filter-pair in this table, which only involve the first two stages of Figure 17.2 (coefficients α and β) since $K = 1$ for this case.

Since the factorization algorithm is based on polynomial division (Daubechies and Sweldens, 1998), there are many possible factorizations that allow for the realization of a given filter-bank with the lifting scheme. In addition, even though the factorization

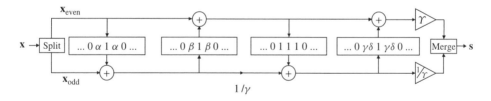

Figure 17.3   Signal flow graph for the lifting decomposition using the flipping structure

Table 17.3   Coefficients of the lifting-based decomposition of the 9/7 and 5/3 DWTs; the detailed analysis equations for both filters are given later in Equations (17.6) to (17.13)

Coefficient	9/7 DWT, approximate value	5/3 DWT, value
α	−1.586134342	−0.5
β	−0.052980118	0.25
γ	0.882911075	N/A
δ	0.443506852	N/A
K	1.149604398	N/A

presented in Figure 17.1 starts with a prediction matrix, the order of steps ($\mathbf{P}_k$ and $\mathbf{U}_k$) can be interchanged. The lifting scheme combines many advantages, such as low implementation complexity, by reducing the required number of multiplications and additions, support for integer-to-integer transforms, and complete reversibility even if nonlinear or adaptive filters are used in the prediction and update steps. Finally, the lifting scheme allows for in-place mapping, i.e. the results of applying each step are stored in place, overwriting the results of the previous step or the input signal.

Since the DWT is a separable transform, it is conventionally implemented in two dimensions via rowwise filtering followed by columnwise filtering of the intermediate results. This is shown in Figure 17.4, where the original image is filtered rowwise (with in-place storage) followed by columnwise filtering. The coefficients of LL1 are assumed to be a new input for the higher decomposition levels l, $1 \leq l \leq L$. Figure 17.4(a) shows the dyadic decomposition, while Figure 17.4(b) shows the same decomposition albeit with in-place mapping.

For each row or column, the lifting equations of the 9/7 filter-pair are (ISO/IEC, 2004)

$$Y_{2n+1}^1 = X_{2n+1} + \alpha \times (X_{2n} + X_{2n+2}), \tag{17.6}$$

$$Y_{2n}^1 = X_{2n} + \beta \times (Y_{2n-1}^1 + Y_{2n+1}^1), \tag{17.7}$$

$$Y_{2n+1}^2 = Y_{2n+1}^1 + \gamma \times (Y_{2n}^1 + Y_{2n+2}^1), \tag{17.8}$$

$$Y_{2n}^2 = Y_{2n}^1 + \delta \times (Y_{2n-1}^2 + Y_{2n+1}^2), \tag{17.9}$$

$$Y_{2n+1}^3 = K \times Y_{2n+1}^2, \tag{17.10}$$

$$Y_{2n}^3 = Y_{2n}^2 / K. \tag{17.11}$$

The one-dimensional application of the transform occurs as follows: first, Equations (17.6) and (17.7) are applied to the input samples of each decomposition level l ($0 \leq n < N/2^l$):

$$\{\dots X_{2n-1}, X_{2n}, X_{2n+1}, X_{2n+2}, \dots\}.$$

Point-symmetric mirroring is used at the input signal borders. The output coefficients

$$\{\dots, Y_{2n-1}^1, Y_{2n}^1, Y_{2n+1}^1, Y_{2n+2}^1, \dots\}$$

are further modified (lifted) by Equations (17.8) and (17.9) into coefficients

$$\{\dots, Y_{2n-1}^2, Y_{2n}^2, Y_{2n+1}^2, Y_{2n+2}^2, \dots\}.$$

Finally, a scaling operation formulates the final values (Equations (17.10) and (17.11)). Each pair of equations formulates a predict-and-update step. In practice, the final scaling step of Equations (17.10) and (17.11) is either incorporated within Equations (17.8) and (17.9) or performed during the entropy quantization stage, leading to $K = 2$ passes for the application of the lifting transform.

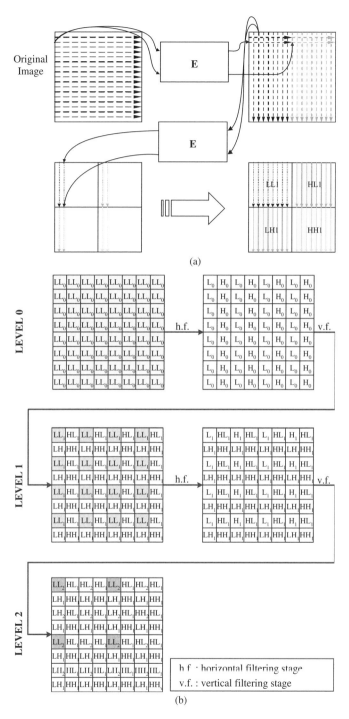

Figure 17.4 Single-level row–column decomposition of an input image. The output consists of low-frequency subband LL1 and high-frequency subbands HL1, LH1, HH1: (a) dyadic decomposition ('Mallat'); (b) dyadic decomposition with in-place mapping (further decomposition to two levels is also shown)

Similarly, for the 5/3 filter-pair, the lifting equations are (ISO/IEC, 2004)

$$Y_{2n+1}^{1} = X_{2n+1} - \left\lfloor \frac{X_{2n} + X_{2n+2}}{2} \right\rfloor, \tag{17.12}$$

$$Y_{2n}^{1} = X_{2n} + \left\lfloor \frac{Y_{2n-1}^{1} + Y_{2n+1}^{1} + 2}{4} \right\rfloor. \tag{17.13}$$

Notice that the 5/3 filter pair is applied in $K = 1$ passes. The floor operator $\lfloor \bullet \rfloor$ provides an integer-to-integer (reversible) transform (ISO/IEC, 2004). The $+2$ factor in Equation (17.13) is included to provide the equivalent of the rounding operation.

Apart from the 1-D decomposition, a proposal for direct 2-D computation of the 9/7 filter-pair using the lifting scheme can be found in the literature (Meng and Wang, 2000; Andreopoulos and van der Schaar, 2008), and is called the spatial combinatorial lifting algorithm (SCLA). There, the single-level decomposition is applied directly to the input $R \times C$ image array $\mathbf{X}$, by refactoring the two-dimensional application of Equation (17.3):

$$\mathbf{S} = \mathbf{E} \cdot \mathbf{X} \cdot \mathbf{E}^{T} \Leftrightarrow \tag{17.14}$$

$$\mathbf{S} = \mathbf{\Upsilon}(\mathbf{U}_{K}(\mathbf{P}_{K}(\cdots(\mathbf{U}_{1}(\mathbf{P}_{1} \cdot \mathbf{X} \cdot \mathbf{P}_{1}^{T})\mathbf{U}_{1}^{T})\cdots)\mathbf{P}_{K}^{T})\mathbf{U}_{K}^{T})\mathbf{\Upsilon}^{T}, \tag{17.15}$$

where $\mathbf{P}_k$, $\mathbf{U}_k$, and $\mathbf{\Upsilon}$ were defined previously, and $\mathbf{S}$ is the 2-D array of output wavelet coefficients.

The last equation is computing the 2-D DWT as a series of basic computation steps in two dimensions, starting from the inner part of Equation (17.15), i.e. $\mathbf{P}_1 \cdot \mathbf{X} \cdot \mathbf{P}_1^{T}$, and working outwards to the final result S. The computation is performed in squares of 2×2 input samples (two-dimensional polyphase components). The computation order of predict and update steps of Equations (17.4) and (17.5) can be seen in Figure 17.5. The sequence of steps shown in the figure originates from the matrix row containing only the unity tap (placed at the position of the 'current' input column sample) because this requires a simple assignment operation. It then proceeds clockwise to complete the computation for each square of 2×2 input samples.

Based on this calculation and the predict matrix defined in Equation (17.4), the prediction part $\mathbf{M}_k^{P} = \mathbf{P}_k \cdot \mathbf{M}_{k-1}^{u} \cdot \mathbf{P}_k^{T}$ of the kth lifting step of Equation (17.15), where $\mathbf{M}_{k-1}^{u}$

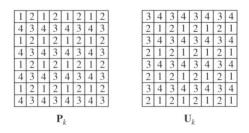

Figure 17.5 Sequence of steps in the 2-D polyphase components of the input of each predict and update step (Meng and Wang, 2000; Andreopoulos and van der Schaar, 2008)

is the previously computed result (with $\mathbf{M}_0^u \equiv \mathbf{X}$), can be written for all output coefficients $p_k[2i + \{0, 1\}, 2j + \{0, 1\}]$ $(0 \le i < R/2, 0 \le j < C/2)$ of $\mathbf{M}_k^p$ as (Andreopoulos and van der Schaar, 2008)

$$p_k[2i, 2j] = u_{k-1}[2i, 2j], \tag{17.16}$$

$$p_k[2i, 2j + 1] = u_{k-1}[2i, 2j + 1] + \sum_{l=0}^{T_{p(k)}-1} a_{p(k)}[l] \cdot u_{k-1}[2i, 2(j + l) - P_{p(k)}], \tag{17.17}$$

$$p_k[2i + 1, 2j + 1] = u_{k-1}[2i + 1, 2j + 1]$$
$$+ \sum_{l=0}^{T_{p(k)}-1} a_{p(k)}[l](u_{k-1}[2i + 1, 2(j + l) - P_{p(k)}]$$
$$+ p_k[2(i + l) - P_{p(k)}, 2j + 1]) \tag{17.18}$$

$$p_k[2i + 1, 2j] = u_{k-1}[2i + 1, 2j] + \sum_{l=0}^{T_{p(k)}-1} a_{p(k)}[l] \cdot u_{k-1}[2(i + l) - P_{p(k)}, 2j], \tag{17.19}$$

where $u_{k-1}[r, c]$ are the input coefficients of matrix $\mathbf{M}_{k-1}^u$ $(0 \le r < R, 0 \le c < C)$, with $u_0[r, c] \equiv x[r, c]$. Each of the above calculations in Equations (17.16) to (17.19) is performed in the order presented (and illustrated in the left side of Figure 17.5) and for all coefficients of the output, before initiating the calculation of the subsequent step. For the kth update step, i.e. $\mathbf{M}_k^u = \mathbf{U}_k \cdot \mathbf{M}_k^p \cdot \mathbf{U}_k^T$, we have (Andreopoulos and van der Schaar, 2008)

$$u_k[2i + 1, 2j + 1] = p_k[2i + 1, 2j + 1], \tag{17.20}$$

$$u_k[2i + 1, 2j] = p_k[2i + 1, 2j] + \sum_{l=0}^{T_{u(k)}-1} a_{u(k)}[l]p_k[2i + 1, 2(j + l) - P_{u(k)}], \tag{17.21}$$

$$u_k[2i, 2j] = p_k[2i, 2j] + \sum_{l=0}^{T_{u(k)}-1} a_{u(k)}[l \left(p_k[2i, 2(j + l) - P_{u(k)}] \right.$$
$$+ u_k[2(i + l) - P_{u(k)}, 2j]), \tag{17.22}$$

$$u_k[2i, 2j + 1] = p_k[2i, 2j + 1] + \sum_{l=0}^{T_{u(k)}-1} a_{u(k)}[l]p_k[?(i + l) - P_{u(k)}, 2j + 1]. \tag{17.23}$$

Each of the calculations of Equations (17.20) to (17.23) is performed in the order illustrated on the right side of Figure 17.5.

It is important to remark that to ensure complete reversibility, the inverse equations are applied in a completely symmetric manner, i.e. the row–column decomposition is inverted by a column–row reconstruction, and the equations of the SCLA are also inverted in the reverse order.

17.4.2 Hardware Designs for the DWT

The discussion of the previous subsection hinted that there are several parameters one can tune in order to derive an efficient hardware design for the DWT. Starting with the lifting decomposition itself, there are many potential candidates to choose from, depending on how the polynomial division is performed (Daubechies and Sweldens, 1998). Assuming the standard lifting decomposition of the 9/7 and 5/3 filter-pairs recommended in the JPEG 2000 standard (ISO/IEC, 2004), the first major concern for hardware designs is the fixed-point quantization of the lifting coefficients, since the vast majority of FPGA-based or custom-hardware-based designs use fixed-point arithmetic.

Initial proposals (Spiliotopoulos *et al.*, 2001; Andra, Chakrabarti, and Acharya, 2002; Grangetto *et al.*, 2002) follow a direct approach to converting the coefficients of the 9/7 DWT (Table 17.4) to integers. The aim is to minimize the complexity of the complement-two representation of the fixed-point coefficients such that the multipliers can potentially be implemented with shift-add operations. This leads to small number of logic gates (small circuit area) per multiplier. However, the selection of which representation to use has an impact on the reconstruction quality, which needs to be carefully studied. More recently, this problem has been studied in more detail for the 9/7 filter-pair (Kotteri, Bell, and Carletta, 2004; Reza and Zhu, 2005) and derived the representation found in Table 17.4. This representation is using rational coefficients and was derived by moving two of the four analysis lowpass filter zeros away from $z = -1$. It has been reported (Kotteri, Bell, and Carletta, 2004) that a nearly identical performance to the floating-point representation can be achieved with this representation. The multiplications with all coefficients except γ and K can be implemented with (up to) two additions and shifts in a fixed-point representation.

If we exclude the scaling by K and $1/K$ from the implementation of the 9/7 DWT, the flipping structure presented in Figure 17.3 can provide a multiplier-free implementation under the parameters of Table 17.4. This is due to the fact that all multiplications with γ are replaced by multiplications with $1/\gamma$ and $\gamma\delta$, both of which are realizable with additions and shifts.

Table 17.4 Approximation coefficients (Kotteri, Bell, and Carletta, 2004) for the 9/7 lifting-based DWT

Coefficient	Approximation
α	$-3/2$
β	$-1/16$
γ	$4/5$
δ	$15/16$
K	$4\sqrt{2}/5$

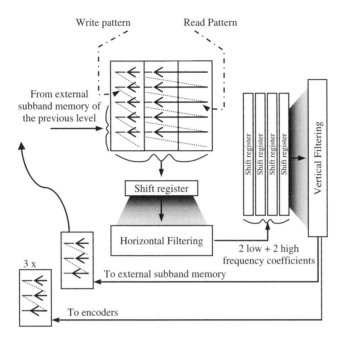

Figure 17.6   A schematic for the line-based architecture

Once the choice of coefficients has been fixed for a particular hardware design (convolution, lifting, flipping structure, SCLA), the traversal schedule becomes important. For all presented designs, several traversal schedules have been proposed in the literature, starting from the conventional row–column decomposition. After careful consideration of the EBCOT requirements of JPEG 2000, the line-based separable architecture is recommended for implementation (Chrysafis and Ortega, 2000; ISO/IEC, 2000). A schematic of such an architecture is given in Figure 17.6. For L decomposition levels, this architecture can be repeated L times. Apart from the components depicted in Figure 17.6, an external subband memory (ESM) is used to buffer the low-frequency output lines of the previous level or the input samples (for the first level). The central element of Figure 17.6 is a memory bank with three components that are written and read as depicted. Their size depends on the filter pair and the code-block height of EBCOT. Each part of the memory bank is written independently from the ESM of the previous level and the contents of all three banks are pushed via a shift register to the horizontal filtering unit, the results of which (low- and high-frequency coefficients of decomposition level l) are pushed into the vertical filtering unit. The high-frequency outputs from this unit are sent to the encoders while the low-frequency outputs are stored in the ESM of the current level, so as to be used by the next level. This is depicted in Figure 17.7. Regarding details of single-port or dual-port memories that can be used for such a design in custom hardware, the reader is referred to Andra, Chakrabarti, and Acharya (2002) and Angelopoulou et al. (2008).

The design of the LBWT as presented in this section is obviously suitable for a multiprocessor architecture where a processing element of Figure 17.6 and a number of encoders exist for every decomposition level. Of course, with the appropriate scheduling

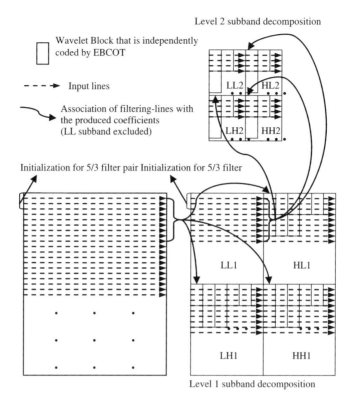

Figure 17.7   An example of the line-based scanning with the 5/3 filter-pair

of the processing (folding), one unit for the transform and the coding can be used. This is the case with the implementation of the VM of JPEG 2000 (ISO/IEC, 2000).

Once the image traversal schedule is decided, the main filtering unit for the particular DWT of choice can be designed, along with the overall architecture. We present in Figure 17.8 such a design for the 5/3 lifting-based filtering (Angelopoulou *et al*., 2008), corresponding to Equations (17.12) and (17.13). Figure 17.9 presents an example of the overall architecture designed for the line-based 5/3 filter-pair under the lifting implementation corresponding to Equations (17.12) and (17.13). Further details can be found in Angelopoulou *et al*. (2008).

Concerning the SCLA, a VLSI design has been proposed in Liu *et al*. (2004). A block-based computation of this transform (9/7 filter-pair) has been proposed in Hu, Yan, and Chung (2004). However, since SCLA requires a slightly modified edge treatment in comparison to the conventional separable implementation, designs based on SCLA forward (inverse) may not be completely reversible near the subband borders if the separable decomposition of Equations (17.6) to (17.11) is used for the inverse (forward).

17.4.3   Pointers for Further Study on DWT Implementations

Apart from the references already discussed, we conclude this section by providing additional literature pointers for the interested reader. Prominent hardware designs have

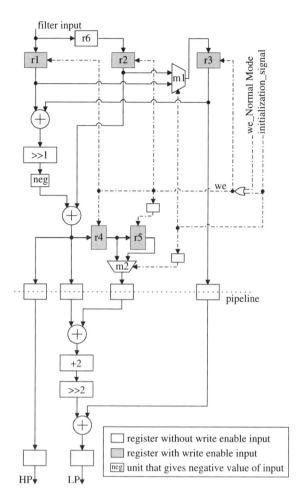

Figure 17.8 Hardware implementation of the 5/3 filter-pair (Angelopoulou *et al.*, 2008), designed to perform the 1-D DWT. The write enable signal (we) determines if the registers with write enable inputs will be written

been proposed in Cheng and Parhi (2003), Bartholoma *et al.* (2008), Angelopoulou *et al.* (2008), Liao, Mandal, and Cockburn (2004), Cheng *et al.* (2007), and Lafruit *et al.* (1999), addressing various aspects such as critical path latency, memory organization and memory hierarchy, on-chip register minimization, etc. Theoretical comparisons of convolution-based DWT using different traversal schedules and memory hierarchies have been presented in Zervas *et al.* (2001). Lifting-based DWT architectures have been designed and compared in Jiang and Ortega (2001). Additional details on comparisons between lifting-based designs and convolution-based designs have been presented in Kotteri *et al.* (2005). The SCLA-based design has the interesting property that it reduces the arithmetic operations for the DWT for the case of the 9/7 filter-pair (Meng and Wang, 2000). The corresponding hardware design of Liu *et al.* (2004) obtains quite high performance figures. Finally, the issue of transform production and consumption by codecs has

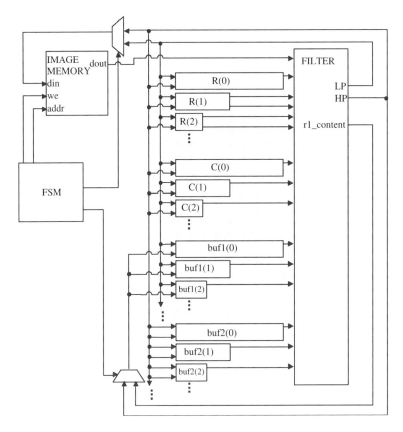

Figure 17.9 Block diagram of the line-based architecture for the 2-D DWT. The shaded area includes the on-chip memory buffers used in the architecture (Angelopoulou *et al.*, 2008)

been theoretically studied early on (Vishwanath, 1994). Overall, the degrees of freedom of the lifting-based designs allow room for further research on issues such as throughput, approximation of coefficients of the decomposition, and scheduling for efficient hardware designs. This means that future advances in this area should be anticipated.

17.5 JPEG 2000 Hardware and Software Implementations[5]

17.5.1 Kakadu

Kakadu is a complete implementation of the JPEG 2000 standard Part 1, i.e. ISO/IEC 15444-1 and a significant amount of Part 2 and Part 3. The software was developed by David Taubman of the University of New South Wales (UNSW), Australia, who is also

---

[5] Disclaimer by the authors. A plethora of JPEG 2000 implementations has currently reached the marketplace. Since it would be beyond the scope and capacity of this book, we list in the following sections only a limited number of implementations, illustrating different design options and solutions for different processing platforms, ranging from software to hardware realizations. Hence, this overview does not have the intention to be complete and neither to give a quality judgment on the solutions listed or not listed.

known as the designer of EBCOT, the core coding component of JPEG 2000. Kakadu is named after the 'Kakadu National Park,' located in Australia's Northern Territory.

The Kakadu software framework, which is available under research or commercial licensing schemes, has been adopted by a large range of JPEG 2000 products (e.g. Apple's Quicktime v6 for MAC, Yahoo's Messenger, which utilizes JPEG 2000 for live video, and MicroImages TNT products for geospatial imagery).

Currently, Kakadu is used in compression/decompression of JPEG 2000 images and video, medical imaging applications, geospatial imaging applications, interactive image rendering applications, remote browsing of large images and collections, and digital cinema applications. Kakaku can be considered as a comprehensive, heavily optimized, fully compliant software toolkit for JPEG 2000 developers. It supports, for instance, multithreaded processing to utilize fully parallel processing resources (multiple CPUs, multicore CPUs, or hyperthreading). Moreover, Kakadu provides a carefully engineered thread scheduler, so once you have created a multithreaded environment and populated it with one thread for each physical/virtual processor on your system, close to 100% utilization of all computational resources is typically achieved.

Kakadu additionally supports Part 2 features such as general multicomponent transforms and arbitrary wavelet transform kernels. The toolkit offers extensive support for interactive client–server applications, implementing most features of the JPIP (JPEG 2000 Internet Protocols) standard (see Chapter 6). More information can be found on the Kakadu website: www.kakadusoftware.com.

17.5.2 OpenJPEG

The OpenJPEG library is an open-source JPEG 2000 library developed to promote the use of JPEG 2000. The main part of the project consists of a JPEG 2000 codec compliant with Part 1 of the standard (Class-1 Profile-1 compliance). Besides this main codec, OpenJPEG integrates several other modules:

- JP2 (JPEG 2000 standard Part 2, on handling of JP2 boxes and extended multiple component transforms for multispectral and hyperspectral imagery);
- MJ2 (JPEG 2000 standard Part 3);
- JPWL (JPEG 2000 standard Part 11);
- OPJViewer, a GUI viewer for J2K, JP2, JPWL, and MJ2 files;
- OPJIndexer, a code-stream indexer to view information about the headers and packets localization, and the rate-distortion contribution of each packet to the image.

The OpenJPEG library is written in C language, released under the BSD license and targets Win32, Unix, and Mac OS platforms. The library is developed by the Communications and Remote Sensing Lab (TELE) of the Université Catholique de Louvain (UCL), with the support of the CS company and CNES. The JPWL and OPJViewer modules are developed and maintained by the Digital Signal Processing Lab (DSPLab) of the University of Perugia (UNIPG).

Thanks to the constant contributions of many developers from the open source community, OpenJPEG has gained throughout the years in flexibility and performance. It is integrated in many open source applications, such as Second Life and Gimp. More information can be found on the website: http://www.openjpeg.org.

17.5.3   IRIS Codec

IRIS-JP3D is an implementation of Part 1 and Part 10 of the JPEG 2000 standard. It is developed by the Department of Electronics and Informatics (ETRO) of the Vrije Universiteit Brussel (VUB) and the Interdisciplinary Institute for Broadband Technology (IBBT). The codec is written in ANSI/C, in order to allow compilation on a large set of different platforms. IRIS-JP3D was initially created to serve as the reference software for JPEG 2000 Part 10. Thus, it provides a functional, complete, high-quality, and extremely comprehensive implementation of both JPEG 2000 Part 1 and Part 10 that is open to scientific research. The codec is written such that the applied algorithms are easily recognizable in the source code. By doing so, it tries to match the source code and the JPEG 2000 specifications as closely as possible, without jeopardizing the readability of the source code. Thus, the code forms a great starting point for researchers, students, and private companies that need to learn more about the JPEG 2000 compression technologies, both 2-D (Part 1) and 3-D (Part 10). The IRIS-JP3D codec is designed to conform to Profile-1 of the JPEG 2000 Part 1 specification. Moreover, by delivering high-quality code-streams that have highly optimized rate-distortion characteristics, the IRIS-JP3D can, in that respect, easily compete with other commercial implementations. The codec is available under two licenses from the website (http://www.irissoftware.eu). Educational and research institutes can receive the software package for free under a noncommercial license. Companies, on the other hand, are allowed to use the software package for commercial purposes when buying a commercial license.

17.5.4   Wavescale™ Monolithic IC Implementations from Analog Devices, Inc.

In 1997, Analog Devices introduced an industry-first, monolithic, wavelet-based video codec. The ADV601 would become the first of a multigenerational family of products. Analog Devices' involvement in the JPEG 2000 standardization activities resulted in the introduction of the ADV-JP2000 in 2001, which was the first application-specific integrated circuit (ASIC) implementation compliant to the JPEG 2000 (ISO/IEC 15444-1) standard.

Thereafter, the ADV202 Wavescale™ codec and the ADV212 Wavescale™ codec were introduced respectively in 2003 and 2007 (ADI, 2007). The Wavescale codecs are symmetric video codecs capable of compressing or decompressing images or video up to 65 Mpixels/s. Multiple Wavescale devices can be configured to support applications that require higher pixel rates. All Wavescale devices are fully compliant with the JPEG 2000 standard and employ the patented SURF™ (spatial ultra-efficient recursive filtering) technology (Greene, Rossman, and Hallmark, 1999). The SURF technology executes the wavelet transform efficiently and reduces the memory requirements making external memory unnecessary, resulting in lower system costs (see Section 14.4.2).

Due to the high performance of Wavescale codecs they have become a widely used hardware implementation of the JPEG 2000 standard. As the only ASIC solution that can scale to resolutions up to 4k DCI (See Chapter 9), Wavescale codecs have been successfully deployed in a broad range of applications, such as digital cinema projection servers, high-definition (HD) broadcast and movie cameras, broadcast storage servers, nonlinear video editing systems, surveillance digital video recorders, digital signage, and

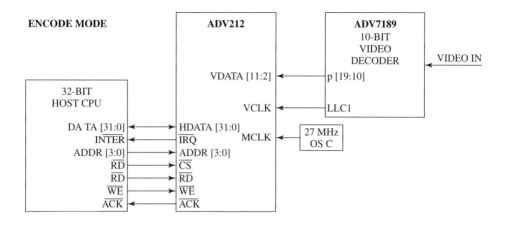

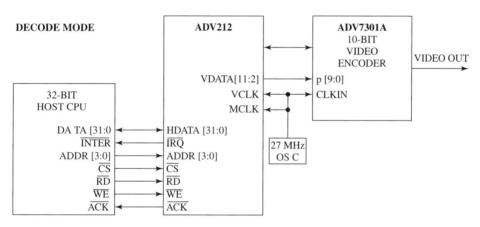

Figure 17.10   Encode and decode mode of the ADV212 (reprinted with permission from Analog Devices, Inc.)

many more professional and industrial applications. Benefits inherent to the JPEG 2000 code-stream have resulted also in the use of Wavescale codecs in a new class of consumer applications that convey high-definition video over wireless network technology.

A single ADV212 can be used to either encode or decode with appropriate firmware. Figure 17.10 illustrates a generic configuration of the ADV212 in a television application. In the encode mode of operation the ADV7189 converts an analog television signal into a digital video signal compatible with the input of the ADV212. Similarly, in the decode mode of operation the ADV212 outputs a digital signal, which is converted by the ADV7301A into an analog video signal.

This section primarily describes the ADV212 as configured for encoding purposes. When configured for decoding, the opposite process and flow applies. Figure 17.11 depicts the internal architecture of the ADV212. Video or pixel data enter the ADV212's pixel interface, which formats the data for encoding. The wavelet engine decomposes each tile or frame into subbands. The resulting wavelet coefficients are written to internal memory for

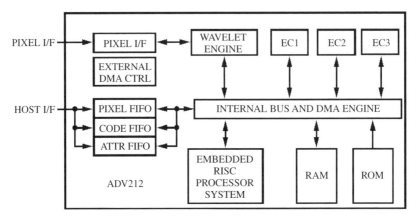

Figure 17.11 The ADV212 core architecture (reprinted with permission from Analog Devices, Inc)

subsequent operations such as entropy coding by the entropy codecs (EC1, EC2, EC3). A direct memory access (DMA) engine provides high-bandwidth memory-to-memory transfers as well as high-performance transfers between functional blocks and memory.

The pixel interface provides flexible input support for ITU.R-BT656, SMPTE125M PAL/NTSC, SMPTE274M, SMPTE293M (525p), ITU.R-BT1358 (625p), or any video format with a maximum input rate of the active image resolution at 65 Mpixels/s for the irreversible mode or 40 Mpixels/s for the reversible mode. Two or more ADV212s can be combined to support HDTV or Blu-ray standards such as SMPTE274M/SMPTE292M (1080i, 1080p) and SMPTE296M (720p). The pixel interface supports input and output video data or still-image data in 8/10/12-bit monochrome or YCbCr 4:2:2 format. Multiple chips can be configured to process 4:4:4 YCbCr or RGB formats.

The host interface typically interfaces to a host/system processor or system level ASIC and operates in the style of an asynchronous static random access memory (SRAM). The ADV212 can connect to 16- or 32-bit host processors and ASICs using an asynchronous SRAM-style interface. The ADV212 host interface can be configured as 16 or 32 bits for control and as 8, 16, or 32 bits for data transfers. The control and data channel bus widths can be specified independently, allowing the ADV212 to support applications that require control and databuses of different widths. The host interface is used for configuration, control, and status functions as well as for transferring compressed data-streams. In certain modes of operation, raw (without video timing) uncompressed video or still-image data can be input/output on the host interface in 8/10/12/14/16-bit raw pixel formats.

The ADV212 implements a dedicated wavelet transform engine. The wavelet processor supports the 9/7 irreversible wavelet transform and the 5/3 wavelet transform in reversible and irreversible modes with up to six levels of decomposition. The 5/3 wavelet has a programmable tile or image size with widths up to 2048 pixels in the pixel interface's three-component 4:2:2 mode, and up to 4096 pixels in the single-component mode. For the 9/7 wavelet, the programmable tile or image size widths are 1024 pixels in the pixel interface's three-component 4:2:2 mode, and up to 2048 pixels in the single-component mode.

The recursive nature of the wavelet transform combined with the need to process large frames or tiles requires a large amount of memory for the temporary storage of intermediate data. The SURF technology implements a Mallat wavelet transform (Mallat, 1989) and methods of performing filtering, synthesis, and analysis over portions of the frame/tile using a common set of filter banks, reducing the need for temporary storage of large amounts of data. This reduction in memory has allowed the complete memory system to be integrated on-chip, hence reducing system costs and eliminating the performance bottleneck that is incurred when going off-chip.

Entropy coding is the most computationally intensive operation in the JPEG 2000 compression process; the ADV212 implements three dedicated hardware entropy codecs. Each entropy codec performs context modelling and arithmetic coding on a code-block of wavelet coefficients. Additionally, these blocks also perform the distortion metric calculations during compression required for optimal rate-distortion performance. Traditional JPEG 2000 entropy coding processes each bit in the code-block individually. A key patented invention employed by the Wavescale codecs breaks the code-block into smaller arrays on which context modelling can be performed in parallel, significantly improving performance and latency. This approach allows entropy coding performance of one clock cycle per bit, where traditional methods require approximately 3–6 cycles per bit. Several improved parallel processing schemes have been proposed since the filing of the Analog Devices patent, but the best performance that any have been able to achieve is 1.25–1.3 cycles per bit (Zhang and Xu, 2005; Dyer, Nooshabadi, and Taubman, 2007). The ADV212 incorporates an embedded 32-bit RISC processor that configures and manages the dedicated hardware functions and performs portions of the JPEG 2000 process such as generating and parsing the JPEG 2000 code-stream. Other than basic bus and I/O configurations that must be set up through the host interface, on-chip firmware configures, controls, and schedules ADV212 functions.

Flexibility, performance, scalability, and overall cost effectiveness of the Wavescale codecs have resulted in a broad range of applications. The earliest applications of the ADV6xx wavelet processors and the ADV2xx Wavescale codecs were primarily in security applications. More recently, the Wavescale codecs have been used in a wide spectrum of professional and industrial video applications, where JPEG 2000 is selected because compression is necessary for channel bandwidth or storage capacity reasons and video quality must be preserved.

Although each implementation tends to be unique in its primary function as well as the specific design, the video processing core comprised of ADV212 codecs tends to fall into a set of typical configurations. Key factors that help define the configuration include resolution (SD, 720p, 1080i, 1080p, 2K DCI, 4K DCI), video format (monochrome, 4:2:2, 4:4:4), tile size, pixel rate (which is a function of the previous factors), and the bit-rate of the compressed stream.

Although there are many other uses, the majority of new applications tends to fall into the broad categories of HD: DCI, consumer video applications, and broadcast applications including cameras (see also the next section), storage recorders, and channel compression. Some example resolutions and typical codec hardware are:

- HD 1080i 4:2:2 applications would include two ADV212 devices with one processing luma and the other handling chroma.
- HD 1080i 4:4:4 applications can be accomplished with three ADV212 devices, each one processing a single component.

- HD 1080p60 applications require an additional two (4:2:2) or three (4:4:4) ADV212 devices over that required for 1080i, with each 'set' operating on one half of the frame. This is because 1080i presents a pixel rate near the limits of the 'set' of ADV212 devices and 1080p doubles the pixel rate.
- DCI 2k applications typically use four or six ADV212 devices depending on the frame rate.

17.5.5 Thomson JPEG 2000 Products

The Infinity Product Range of Thomson Grass Valley aims at field acquisition applications. It comprises a camcorder called the DMC1000, a digital media recorder or deck called the DMR1000, and a file-based media and drive solution, respectively the Thomson's REV PRO and DMD.

Field use places important constraints such as low power consumption, portability, and ruggedness on these products. Despite these constraints, increasingly more production work is pushed out into the field or 'grassland.' As a result, nomadic journalism requires an increasing amount of file-based content processing, allowing for real-time editing of high-quality HD material on a normal laptop computer, and the ability to interoperate with other field production devices.

17.5.5.1 Design Decisions and Tradeoffs

For this set of motivations, Thomson chose JPEG 2000 as the main HD codec for the Infinity Product Range. A close collaboration with Analog Devices allowed all the above requirements to be met. These requirements led to the following criteria set for the format support:

- full HD raster (either 1080i or 720p);
- full 10 bits per pixel encoding, as defined by the SMPTE 292M standard for HD-SDI;
- full 4:2:2 color sampling.

Thomson balanced these requirements with the need to permit their customers to make recordings of a reasonable length onto affordable media and they used video tape as their guide for interpreting the use of the concepts 'reasonable' and 'affordable.' Hence, recording about 45 minutes of HD on to a piece of media in the same price range as a digibeta tape became their goal and led to the selection of REV PRO as the main media type used by the camcorder and deck.

After exhaustive tests, it was determined that, in general, bit-rates above 1 bpp delivered acceptable image quality, whereas below this rate compression artifacts become more noticeable. This experimental result, and the desire to allow the user to make the picture quality versus record time tradeoff according to their own particular needs, led to the design of three profiles for JPEG 2000 encoding within the Infinity Product Range:

- 100 Mb/s, which is equivalent to1.58 bits/pixel for 1920 × 1080i at 30 fps. This mode is intended for electronic fiction production (EFP) type work.
- 75 Mb/s, which is equivalent to 1.2 bits/pixel and suitable for electronic news gathering (ENG) work.

- 50 Mb/s, which is below the threshold at 0.8 bits/pixel but provided so that the user has the option of working in a 'long play' mode if necessary.

17.5.5.2 Real-Time on a Laptop

Being able to play back and make rough cut edits in the field requires a code-stream that can be decoded in real-time on a regular laptop. At the time of writing, it was not possible to decode full HD JPEG 2000 at 100 Mb/s on a normal laptop, but by exploiting some of the unique scalability and flexibility features of the JPEG 2000 code-stream, it is possible to provide functionality that enables decoding up to an acceptable quality level for editing purposes.

Figure 17.12 depicts the layering arrangement within the JPEG 2000 bit-stream for the Infinity Product Range and illustrates how these layers are mapped into the wavelet space for an HD bit-stream (a slightly different mapping is used for SD). Basically, resolution scalability is deployed in order to support the different reconstruction layers and bit-rates. Actually, the camera realizes support for three quality levels by exploiting the resolution scalability mode of JPEG 2000. This design option was chosen because of complexity constraints.

Hence, a quarter-by-quarter resolution version of the original image can be extracted from the 6 Mbps layer. The number of bits per pixel for this arrangement is 1.54 bpp, offering a good visual quality and an acceptable resolution of 480×270 pixels for field editing. On a mid-range workstation, pictures can be decoded from the intermediate layer operating at 25 Mb/s. The third layer is used to decode to the full resolution, regardless of which of the three profiles are being used. Figure 17.13 summarizes the main differences and correspondences between the digital cinema profiles and the infinity configuration.

17.5.5.3 Thomson HD Wireless Camera

In TV production, studios, or outside broadcasting, there is a market demand for wireless cameras that can be used everywhere in the field. Especially nowadays with the

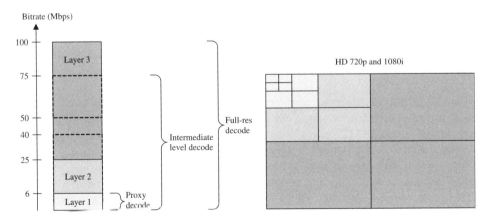

Figure 17.12 The resolution layers in the Infinity JPEG 2000 code-stream (left) can be easily mapped to wavelet subbands (right)

Parameter	2k Digital Cinema	4k Digital Cinema	Infinity
Wavelet Transform	9/7	9/7	9/7
Number of Wavelet Transform Levels	5	6	5
Codeblock Size	32 × 32	32 × 32	128 × 32
Resolution Layers	1	1	3
Progression Order	CPRL	CPRL, with data necessary to decode 2k images separated from the balance required for 4k	LRCP
Colour Space	R'G'B'	R'G'B'	YCrCb
Bit depth	12	12	10
Colour Sampling	4:4:4	4:4:4	4:2:2

Figure 17.13 Comparison between the infinity configuration and the digital cinema profiles

transition to HDTV, this application is very demanding with respect to rate-distortion performance because of the very low available bit-rate for robust wireless transmission. Current state-of-the-art products cannot exceed a few tens of Mb/s (<100 Mb/s) on the wireless links. Other very important constraints, like latency, weight, power consumption, and ergonomics are imposed on the HD wireless camera.

All those constraints led Thomson Grass Valley to adopt JPEG 2000 compression for their HD wireless (HDW) product. The complete HDW system uses real-time HDTV compression and decompression and can handle 1080i and 720p HDTV standards. The JPEG 2000 engine starts from 1200 Mb/s of useful image signal (1080–60i at 10 bits 4:2:2) and goes down to about 30 Mb/s for small bandwidth channels (10 MHz) and to 75 Mb/s for larger bandwidth channels (20 MHz).

The hardware solution is based on Analog Devices integrated circuits (see the previous section), together with a local programmable logic device. The Thomson architecture is symmetrical, i.e. the same hardware can in turn perform either compression or decompression.

The complete hardware resources necessary for the implementation of full JPEG 2000 encoding (or decoding) are based on a pair of ADV212 ASICs and a single FPGA from ALTERA (STRATIX series), the latter being used for control and interfacing. Less than 15k cells are necessary to implement the real-time control, including an embedded microcontroller. Specific software is running to actually manage, in real-time, interfacing and all aspects of ADV chips. The printed circuit board (PCB) is about 3×3 cm^2 in size excluding the FPGA used mainly for other purposes, and power consumption is less than 4 W. This is a very small, efficient low-power solution which was – according to the company – impossible to achieve with MPEG technology at design time.

Picture quality and low latency were the leading ideas of the HDW concept. JPEG 2000 ensures constant picture quality along frames as it is a pure 'intra' mode compression. The very low latency is mandatory for using wireless and wired cameras together. Depending on the supported TV standard, 40 or 33 ms represent the maximum delays introduced by compressing or decompressing two fields (1080i), in the case of interlaced recording, or one frame (720p), when progressive scanning is used. Hence, it leads to extremely low latency of the complete system. Moreover, it includes support for the studio synchronization (i.e. genlocking), and contributes to a natural use and seamless mix of wireless cameras in a wired cameras environment for 'live' production.

17.5.6   Barco Silex IP Cores for FPGA and ASIC

17.5.6.1   General Description

With the emergence of high-definition formats, Barco Silex has – based on its JPEG 2000 IP cores dedicated to FPGA and ASIC – released encoder and decoder engines optimized to meet the real-time requirements of digital cinema and video broadcast applications. The architecture of both cores offers a flexible and high-speed solution to the performance challenges of cinema, broadcast, and postproduction applications. It is able to sustain the high encoding and decoding requirements of the large DCI frame formats, including 4096×2160 resolution and frame rates up to 60 frames per second. They perform JPEG 2000 processing on untiled large color images with 4:4:4 or 4:2:2 color resolutions and are able to support lossy and lossless compression. The cores also offer multichannel capabilities.

The BA110 encoder core performs the following video compression operations of the normalized encoding process: color transform (ICT/RCT), discrete wavelet transform (DWT) with both 5/3 and 9/7 filters, quantization, entropy encoding, and rate allocation. It expects pixel data in 4:4:4 or 4:2:2 color formats at its input interface, with a bit-depth of up to 12-bit per sample. It generates a j2c JPEG 2000 stream at its output interface. In the case of 4:4:4 operation this stream is DCI compliant.

The BA109 decoder core performs the complete set of video decompression operations of the normalized decoding process: the stream parsing and header decoding, the entropy decoding, inverse quantization, inverse discrete wavelet transform (IDWT) with both 5/3 and 9/7 filters, and inverse color transform (ICT/RCT). It supports a JPEG 2000 j2c file at its input interface and generates decoded samples at its output interface under the following formats: 4:4:4 or 4:2:2 color resolutions with up to 12-bit samples.

Both cores are optimized for speed and are able to deal with the demanding DCI and HD processing speeds: they provide a single-chip FPGA solution for all 2k at 24/48/60 fps, 4k at 24/30 fps, 720p30/60, 1080i, and 1080p30/60 distributions. The flexible FPGA architecture allows the user to build a secure codec by integrating Barco Silex cryptography engines (DCI AES).

Both IP cores can be used in a wide variety of applications: digital cinema (compliant to DCI recommendation), digital video broadcasting, postproduction, and the wireless home/office. The core can also be integrated in other applications with similar high-demanding processing requirements. The underlying architecture of the core enables a broad range of features and performance options, as well as specific customizations.

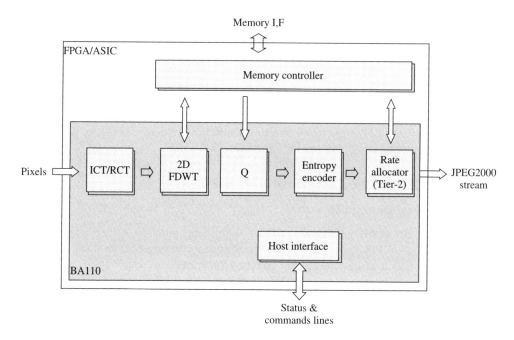

Figure 17.14   The Barco Silex BA110 encoder block diagram

17.5.6.2   Technical Description

Figure 17.14 illustrates a simplified block diagram of the Barco BA110 JPEG 2000 encoder showing the internal modules and the interfaces. The IP core features four main interfaces: a pixel interface, a stream interface, a control interface, and a memory interface for intermediate large data storage. The encoder control interface allows the user to parameterize the core, trigger encoding, and monitor the status of the encoding process.

The encoder requires a single external memory containing intermediate buffered data. The core implements these memory channels as a generic multiport interface to a memory controller. This will typically be a DDR/DDR2 SDRAM controller. The following modules are integrated in the BA110 encoder core:

- *ICT/RCT.* If enabled, this unit performs the irreversible or reversible color transform on the input samples. This optional operation converts samples from RGB to YUV space. It is used to improve the compression efficiency. The ICT/RCT module accepts samples of any size up to 12 bits. The operation is enabled by the user.
- *2-D FDWT.* This module performs the forward 2-D wavelet transform. It is able to process frames without tiling (with a size up to 4096×2160). It performs wavelet decomposition with a programmable number of decomposition levels up to 6, as specified by the user. Both 9/7 and 5/3 filters are supported. The wavelet engine stores its result in external memory.
- *Quantizer.* The quantizer applies quantization steps as specified by the user. A different quantization step is programmable for each subband, resulting in differently weighted frequency subbands. This unit also divides the decomposed subbands into code-blocks

and dispatches the various code-blocks to the entropy encoding channels. This is done by reading the code-blocks from the external memory at the right location inside the subband to which they belong.

- *Entropy encoder (bit modeller and arithmetic encoder).* This module generates compressed coding passes from the quantized code-blocks received from the quantizer. The bit modeller decomposes the code-block bit-plane by bit-plane from the most significant to the least significant plane and identifies significant bits according to a stripewise scan in each bit-plane. Moreover, it computes the context information needed by the arithmetic encoder. The contexts and binarized code-blocks are processed by the arithmetic encoder, which generates the bit-stream ready to be encapsulated in a JPEG 2000 file container.
- *Rate allocator.* This module performs optimal rate allocation based on the rate/distortion information produced by the entropy encoder, for the purpose of constraining the generated bit-rate. This module also embeds the generated code-stream in a JPEG 2000 file container ready for decoding by any third party JPEG 2000 decoder. The rate allocator is configured by the user to target one of the three following modes: constant quality, constant bit-rate, and constant R-D slope. The user sets the mode parameter for each picture.
- *Host interface.* This interface provides configuration, status, and command lines to the user. These are used to parameterize the JPEG 2000 process, trigger encoding of a JPEG 2000 picture, and follow the status of the encoding process. The user intervention is minimized.

The block diagram of the BA109 JPEG 2000 decoder is very similar to that of the encoder. The decoder core features a stream parser for analysis of headers and unpacking of compressed data, an entropy decoder, an inverse quantizer, a 2-D IDWT engine, an ICT/RCT module, and a host interface. External memory is required for intermediate large data storage and stream parsing.

The BA110 and BA109 IP cores target high-performance devices from the two major FPGA manufacturers, including the Virtex4 and Virtex5 from Xilinx and the StratixII and StratixIII from Altera. Also available on Xilinx Spartan3 or Altera CycloneIII for lower resolutions, these cores offer the advantages of a programmable solution at low cost. Finally, the BA110 and BA109 are suitable for ASIC integration utilizing process technologies of 130 nm and below.

17.5.6.3 Barco Silex Acceleration Boards

Barco Silex also provides the DCPB-2000, a family of PCI-Express acceleration boards, which perform JPEG 2000 compression and decompression. Plugged into a PC, a member of the DCPB-2000 product family allows for real-time encoding and decoding performance for very high resolution pictures, such as 2k digital cinema or 1080p HD.

The DCPB-2000 family is able to process untiled images at resolutions up to DCI 4k. Based on FPGA technology, members of the DCPB-2000 family integrate the IP cores from Barco Silex, offering the advantages of high performance, high quality, flexibility, and upgradeability.

The PCI-Express connectivity is ideal for file-to-file processing and represents an alternative to software solutions running on PC clusters. The DCPB-2000 boards can also be equipped with video interfaces in order to perform combined acquisition and encoding or decoding and display.

For the digital cinema market, the DCPB-2100 and DCPB-2140 offer compression and decompression solutions compliant with the Digital Cinema Initiative. They support the full JPEG 2000 coding chain in hardware. The DCPB-2100's rate allocation allows DCI code-streams to be generated at a constant bit-rate or constant quality. The DCPB-2100 and DCPB-2140 products can be used at different stages in the digital cinema workflow, including acquisition or projection, production, archiving, and mastering.

For the broadcast market, the DCPB-2120 and DCPB-2160 perform compression and decompression of one 1080p60 HD stream or multiple streams at lower resolutions, in real-time. Those products are suitable among others for acquisition, storage, archiving, production, contribution, and distribution.

As illustrated in Figure 17.15, the PCI-Express board comes with a Software Development Kit supporting Windows and Linux operating systems. The Application Programming Interface enables simple parameterization of the compression and decompression, and provides an efficient handshake between the input and output video streams. Users of the DCPB-2000 boards can build a large amount of files stored on a hard drive, on top of various applications of the SDK like compressing or uncompressing, or on the fly coding of streaming video.

17.5.7 Alma Technologies JPEG 2k-E IP Core

Alma Technologies JPEG 2k-E is an encoder IP core, launched in 2002, suitable for FPGA and ASIC implementations. Originally Alma's core supported only tier-1 coding, and the tile size was limited by the fact that it had to store two tiles in local memory (on-chip). Today (2008) Alma Technologies offers a second generation core, with a broader support for the JPEG 2000 features and having double the processing speed of its predecessor. The standard functionalities supported by the JPEG 2k-E are summarized in Table 17.5, while supported input formats are provided in Table 17.6. The JPEG 2k-E can encode grayscale or color images with up to four components, having 8 up to 16 bits per sample in all widely used formats, including, but not limited to, 4:4:4, 4:2:2, 4:2:0, 4:1:1, and 1:1:1:1.

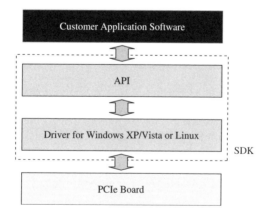

Figure 17.15 DCPB-2000 system overview

Table 17.5 Standard support

Feature	JPEG 2k-E
Tile size	Up to 4096 × 4096
2-D DWT	5/3 and 9/7
Quantization	Programmable tables
Coding switches	RESTART, RESET, SEGMARK ERRTERM
Code-block size	16 × 16, 16 × 32, 16 × 64, 32 × 16, 32 × 32, 32 × 64, 64 × 16, 64 × 32, 64 × 64
Precinct size	Maximal (no spatial segmentation of subbands)
Rate control	Yes
Output format	Stream format (.jpc), file format (jp2), and proprietary format
Progression order	CPRL, PCRL, LRCP (only for grayscale images)
Quality layers	1–20

Table 17.6 Supported input formats

Feature	JPEG 2k-E
Maximum image size	65 535 × 65 535
Number of image components	1 up to 4
Number of bits per sample	8–16
Subsampling factors	1, 2, or 4 on either dimension

Being a hardware module, the JPEG 2k-E is targeted for applications requiring high processing rates. Video surveillance and defence imaging, as well as digital cinema, are the main application domains being addressed. The recent releases of JPEG 2k-E support both the basic image interchange format (BIIF) (ISO/IEC, 1998) and DCI profiles.

17.5.7.1 Hardware Architecture and Processing Speed

The block diagram of JPEG 2k-E is illustrated in Figure 17.16. The encoder employs a pipelined three-staged architecture, with the 2-D DWT being the first stage, quantization and entropy coding being the second stage, and rate control and stream syntax being the third stage. Each pipeline stage is capable of processing a tile. Data between pipeline stages is exchanged via an external memory to which the accesses are arbitrated by a module within the JPEG 2000 core. The core's external memory interface was designed to be independent of the memory type – supporting SRAM, SDRAM, or DDR1/2/3 – and tolerant to the large delays and latencies typically present on shared bus architectures.

In any such architecture, the throughput of the entire architecture depends on the throughput of the entropy coding stage. Hence, parallelization of entropy coding is required and supported on a code-block basis in the JPEG 2k-E core. The throughput of the JPEG 2000 encoder obviously depends on the number of instantiated entropy coders (which is a synthesis time parameter). In its maximum speed configuration, the JPEG 2k-E core is able to provide a throughput of 1.4 clock cycles/sample. The clock frequency depends on the implementation technology (ASIC, FPGA, etc.). Table 17.7

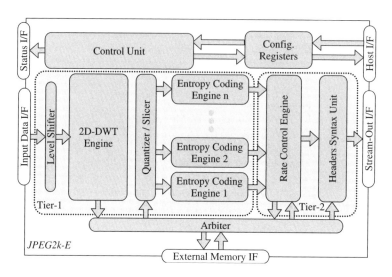

Figure 17.16   Block diagram of JPEG 2k-E

Table 17.7   Sample implementation data

Technology	Clock frequency (MHz)	Throughput (Msamples/s)
Altera Cyclone-III	100	71
Xilinx, Virtex4 (speed grade-1)	133	95
Altera Stratix-III (speed grade-2)	140	100
TSMC 90 nm	210	150

provides indicative performance figures for different example implementations, in terms of frequency at which the core can be clocked and throughput in terms of Msamples/s. From these data it is evident that the JPEG 2k-E core can meet the processing speed requirements of even the most demanding applications like HDTV. JPEG 2k-E embeds a sophisticated rate control hardware module that practically matches the rate-distortion performance of the Kakadu software implementations. Apart from its rate-distortion performance, the rate control module allows for accurate bit-rate control, as the maximum deviation in terms of compressed stream size is less than 3% when the target compression ratio is less or equal to 50:1 and less than 5% for higher compression ratios.

17.5.8   Sesame and Falcon IP Cores by MegaChips

Last but not least it is worthwhile mentioning the JPEG 2000 series of LSIs (large-scale integrations) developed by MegaChips Corporation and Kyoto University. Sesame (MA05137) is the JPEG 2000 encoder/decoder LSI developed in 2004. It supports, utilizing single-tile processing, maximum image input sizes of 8192 × 8192 pixels, corresponding to 64 megapixels. Falcon, its successor and developed in 2007, is also a JPEG 2000 encoder/decoder RTL IP, and it even allows for a maximum input size of 65 536 × 65 536 pixels (4 gigapixels), hence supporting extremely large image sizes. This section covers the Falcon JPEG 2000 RTL IP core. The advantage of this implementation is its potential to handle extremely large tile sizes, hence avoiding distortions typically occurring at the tile boundaries. Supporting single-tile processing requires large amounts of memory and bus bandwidth. The Falcon implementation reduces memory complexity by exploiting overlapping stripe regions.

17.5.8.1   The Internal Architecture

The input–output controller (IOC) of the Falcon core (Figure 17.17) controls all processing data flow and enables internal/external DMA transfer. The discrete wavelet transform unit (DWTU) incorporates both the forward and inverse discrete wavelet transform. During encoding the output of the DWTU is temporarily stored in a group of buffer memories contained by the code-block unit (CBU). In the decoding mode, the CBU stores the output of the inverse wavelet transform (DWTU) and entropy codec unit (ECU). The latter unit consists of multiple entropy codecs (ECs), each of which consists of a coefficient bit modelling (CBM) unit and an arithmetic codec (AC). The ECU also controls the synchronized reset start signals when needed. Finally, the packet header unit (PHU) generates and analyzes the packet headers. It supports precinct sizes from 32×32 to 256×256.

The Falcon codec (for specifications see Table 17.8) is able to choose the optimal circuit scale (Table 17.9) depending on the transaction speed of a target product. The logic core scalability is realized by scaling the entropy coder (EC) between 1 and 16 units.

To overcome the memory complexity drawbacks of a single-tile system architecture the Falcon integrates three principles to reduce the memory footprint and memory transfer:

1. Utilization of a line-based DWT (see Section 17.4.2).
2. Division of the input data in vertical stripes.
3. Limiting the number of decompositions.

Unfortunately, the line-based DWT encounters the problem that the line buffer size becomes larger when the horizontal width of tile is large. Hence, a tile is divided into

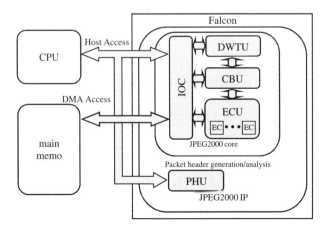

Figure 17.17   JPEG 2000 IP core Falcon block diagram

Table 17.8 JPEG 2000 IP core Falcon specification

Items	Content
Input	YUV4:2:0,YUV4:2:2,YUV4:4:4
Colour conversion	Not supported (processed by external block)
Image size	Max. H65 536 × V65 536, 4 gigapixels
Tiling	Max. H65 536 × V65 536, processed in 1 tile
Bit precision	Support from 8 to 12 bits
Code-block size	32 × 32
Filter	9/7(lossy), 5/3(lossless)
Layer	Single layer only
Packet header	Precinct size from 32 × 32 to 256 × 256
Entropy coder(EC)	Support from 1 to 16
Rate control	Custom solution

Table 17.9 JPEG 2000 IP core Falcon gate scale[a]

	EC × 1	EC × 6	EC × 16
Logic (gate)	270 k	727 k	1640 k
RAM (bit)	419 k	585 k	918 k

[a]Synthesis example by 65 nm Generic Library. Gate, SRAM bit, clock speed, and performance depend on the process. Pure random logic is 35%. Tiny memories synthesized as logic dominate this size.

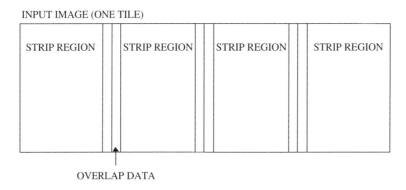

INPUT IMAGE (ONE TILE)

STRIP REGION   STRIP REGION   STRIP REGION   STRIP REGION

OVERLAP DATA

Figure 17.18 Dividing a input data into strip regions and overlap data

vertical stripe regions (Figure 17.18) and the data from the individual stripes is processed sequentially by the line-based DWT on a stripe-by-stripe basis, consequently reducing the line buffer size (Mizuno and Sasaki, 2003; Tsutsui, 2005).

However, because of the support width of the wavelet basis functions, coefficient values near the stripe edge in neighbouring stripe regions need to be accounted for. Consequently, these data have to be buffered in an overlap memory to guarantee a correct calculation of the forward or backward wavelet transform (Figure 17.18). In this context, it needs to be noted that the size of overlap depends on the DWT level. The higher the level of the DWT, the larger the overlap will be. The overlapped pixels are added to the normal input of line-based DWT, hence increasing the memory bandwidth. The size of the overlap can be calculated by accounting for the decomposition level of the DWT m, having the stripes starting at even sample positions, and having even lengths when employing the JPEG 2000 9/7 filter:

$$Right\ side : R_m = 3 \times 2^m - 3, \tag{17.24}$$

$$Left\ side : L_m = 2^{m+2} - 4. \tag{17.25}$$

When the DWT level m equals 2, the overlap buffer spans an area of 9 samples on the right side (R_m) and 12 samples (L_m) on the left side. For a horizontal size l stripe region, Figure 17.19 indicates the increasing memory transfer rate attributed to the overlap buffer. As can be observed, this excess rate can be reduced by limiting the amount of wavelet decomposition levels m and maximizing the stripe width l. The Falcon LSI will therefore also limit the amount of wavelet decomposition levels in order to minimize the memory transfer overhead.

The Falcon aims at Profile-0 compliance; the CBU will process code-blocks of $2^{xcb} \times 2^{ycb}$ samples, where $2^{xcb} = 2^{ycb} = n$. Hence, the CBU cannot start processing until the DWTU outputs its first decomposition level. The latter can be effectuated when the first $2n$ lines of the input image are available to the DWTU. Intermediate results are therefore recursively stored in the main memory.

The CBU core can handle units of 128×128 coefficients, containing multiple code-blocks. Figure 17.20 depicts such a unit, containing 32×32 code-blocks labelled according to covered subbands. The stream grouping for CBU processing is described in Table 17.10.

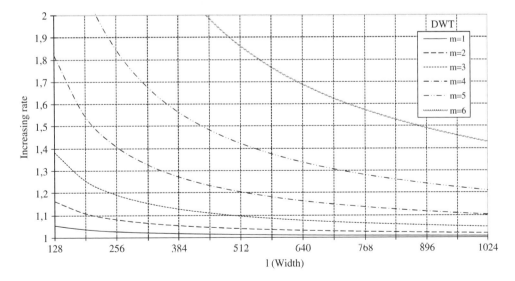

Figure 17.19   Influence of the number of wavelet decomposition levels m and the stripe region width l on the memory transfer rate

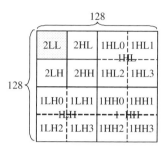

Figure 17.20   A CBU data unit containing 32×32 code-blocks, corresponding to the subband resulting from a two-level DWT on an 128×128 image input matrix

Table 17.10   Streaming groups

Group	Content
A0	1HL0, 1HL1, 1LH0, 1LH1, 1HH0, 1HH1
A1	1HL2, 1HL3, 1LH2, 1LH3, 1HH2, 1HH3
B	2HL, 2LH, 2HH
LL	2LL (code-stream or raw data)
A0	1HL0, 1HL1, 1LH0, 1LH1, 1HH0, 1HH1
A1	1HL2, 1HL3, 1LH2, 1LH3, 1HH2, 1HH3

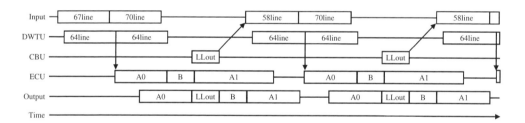

Figure 17.21   Encoding flow diagram

For the encoding procedure, the DMA transfer operates on input units corresponding to one line and output units corresponding to a code-block. Since the LL data are only encoded for the highest decomposition level (i.e. 2LL), it is recursively stored by the CBU and transferred to the DWTU for the next wavelet decomposition with DMA and a unit size equal to the LL size. Figure 17.21 shows the timing diagram for the full encoding process.

17.6   Conclusions

A broad knowledge base currently exists supporting the implementation of JPEG 2000 technology. Via Part 4 (compliance testing) and Part 5 (reference software), the standard itself is offering reference software implementations and guidelines to evaluate the performance and quality of proprietary JPEG 2000 implementations. Moreover, as

Section 17.5 illustrates, a plethora of JPEG 2000 implementations, both commercial and noncommercial, are currently widely available both in software and in hardware. The latter are even addressing high-end applications such as digital cinema and beyond, putting significant stress on the underlying computing architecture since very high resolution images and video need to be processed in real-time.

Acknowledgments

The authors would like to thank David Taubman (University of New South Wales, Australia), François-Olivier Devaux and Benoit Macq (Université Catholique de Louvain-la-Neuve, Belgium), Tim Bruylants (Vrije Universiteit Brussel, Belgium), Ambarish Natu, Andrew Lambrecht, and Dale Stolitzka (Analog Devices, Australia/USA), Guillaume Boisson (Thomson, France), Laurent Petit and Frederic Devisch (Barco Silex, Belgium), Nikos D. Zervas (Alma Technologies), and Yusuke Mizuno and Gen Sasaki (Megachips, Japan) for their contributions and advice with respect to the available market solutions section.

References

ADI (2007) JPEG 2000 Video Codec, ADV212, Analog Devices, Inc., Norwood, MA. Available at: http://www.analog.com/en/audiovideo-products/video-compression/ADV212/products/product.html.

Andra, K., Chakrabarti, C. and Acharya, T. (2002) A VLSI architecture for lifting-based forward and inverse wavelet transform, *IEEE Transactions on Signal Processing*, **50**(4), April, 966–977.

Andreopoulos, Y. and van der Schaar, M. (2008) Incremental refinement of computation for the discrete wavelet transform, *IEEE Transactions on Signal Processing*, **56**(1), January, 140–157.

Angelopoulou, M., K. Masselos, K., Cheung, P. Y. K. and Andreopoulos, Y. (2008) Implementation and comparison of the 5/3 lifting 2-D discrete wavelet transform computation schedules on FPGAs, *Journal of Signal Processing Systems for Signal, Image and Video Technology*, **51**(1), April, 3–21.

Antonini, M., Barlaud, M., Mathieu, P. and Daubechies, I. (1992) Image coding using wavelet transform, *IEEE Transactions on Image Processing*, **1**(2), April, 205–220.

Bartholoma, R. R., Greiner, T. T., Kesel, F. F. and Rosenstiel, W. W. (2008) A systematic approach for synthesizing VLSI architectures of lifting-based filter banks and transforms, *IEEE Transactions on Circuits and Systems – I: Regular Papers*, **55**(7), August, 1939–1952.

Cheng, C. and Parhi, K. K. (2003) High-speed VLSI implementation of 2-D discrete wavelet transform, *IEEE Transactions on Signal Processing*, **56**(1), January, 393–403.

Cheng, C., Huang, C.-T., Chen, C.-Y., Lian, C.-J. and Chen, L.-G. (2007) On-chip memory optimization scheme for VLSI implementation of line-based two-dimensional discrete wavelet transform, *IEEE Transactions on Circuits and Systems for Video Technology*, **17**(7), July, 814–822.

Chrysafis, C. and Ortega, A. (2000) Line-based, reduced memory, wavelet image compression, *IEEE Transactions on Image Processing*, **9**(3), March, 378–389.

Daubechies, I. (1992) *Ten Lectures on Wavelets*, SIAM CBMS-NSF Regular Conference Series in Applied Mathematics, vol. 61.

Daubechies, I. and Sweldens, W. (1998) Factorization wavelet transforms into lifting steps, *Journal of Fourier Analysis and Applications*, **4**, 247–269.

Dyer, M., Nooshabadi, S. and Taubman, D. (2007) Analysis of multiple parallel block coding in JPEG 2000, in *IEEE International Conference on Image Processing (ICIP)*, **5**, October, 173–176.

Grangetto, M., Magli, E., Martina, M. and Olmo, G. (2002) Optimization and implementation of the integer wavelet transform for image coding, *IEEE Transactions on Image Processing*, **11**(6), June, 596–604.

Greene, R., Rossman, M. and Hallmark, P. (1999) Method and apparatus for using minimal and optimal amount of SRAM delay line storage in the calculation of an XY separable Mallat wavelet transform, US Patent 5 984 514, November 16, 1999.

Hu, C.-K., Yan, W.-M. and Chung, K.-L. (2004) Efficient cache-based spatial combinative lifting algorithm for wavelet transform, *Signal Processing*, **84**(9), September, 1689–1699.

Huang, C.-T., Tseng, P.-C. and Chen, L.-G. (2004) Flipping structure: an efficient VLSI architecture for lifting-based discrete wavelet transform, *IEEE Transactions on Signal Processing*, **52**(4), April, 1080–1089.

ISO/IEC (1998) Information Technology – Computer Graphics and Image Processing – Image Processing and Interchange – Functional Specification – Part 5: Basic Image Interchange Format, ISO/IEC International Standard 12087-5, December 1, 1998.

ISO/IEC (2000) JPEG 2000 Verification Model 8.6, ISO/IEC JTC1/SC29/WG1N1894, 2000.

ISO/IEC (2004) ITU-T Recommendation T.800 (08/2002) | ISO/IEC 15444-1:2004, JPEG 2000 Part I, 2004.

Jiang, W. and Ortega, A. (2001) Lifting factorization based discrete wavelet transform architecture design, *IEEE Transactions on Circuits and Systems for Video Technology*, **11**(5), May, 651–657.

Kotteri, K. A., Bell, A. E. and Carletta, J. E. (2004) Design of multiplierless, high-performance, wavelet filter banks with image compression applications, *IEEE Transactions on Circuits and Systems – I: Regular Papers*, **51**(3), March, 483–494.

Kotteri, K. A., Barua, S., Bell, A. E. and Carletta, J. E. (2005) A comparison of hardware implementation of the biorthogonal 9/7 DWT: convolution versus lifting, *IEEE Transactions on Circuits and Systems – II: Express Briefs*, **52**(5), May, 256–260.

Lafruit, G., Nachtergaele, L., Bormans, J., Engels, M. and Bolsens, I. (1999) Optimal memory organization for scalable texture codecs in MPEG-4, *IEEE Transactions on Circuits and Systems for Video Technology*, **9**(2), March, 218–243.

Liao, H., Mandal, M. and Cockburn, B. F. (2004) Efficient architectures for 1-D and 2-D lifting-based wavelet transforms, *IEEE Transactions on Signal Processing*, **52**(5), May, 1315–1326.

Liu, L., Chen, N., Meng, H., Zhang, L., Wang, Z. and Chen, H. (2004) A VLSI architecture of JPEG 2000 encoder, *IEEE Journal of Solid-State Circuits*, **39**(11), November, 2032–2031.

Mallat, S. G. (1989) A theory for multiresolution signal decomposition: the wavelet representation, *IEEE Transactions on Pattern Analysis and Machine Intelligence*, **11**(7), July, 674–693.

Mallat, S. G. (1998) *A Wavelet Tour of Signal Processing*, Academic Press, San Diego, CA.

Meng, H. and Wang, Z. (2000) Fast spatial combinative lifting algorithm of wavelet transform using the 9/7 filter for image block compression, *IEE Electronics Letters*, **36**(21), October, 1766–1767.

Mizuno, Y. and Sasaki, G. (2003) Wavelet Processing Apparatus and Wavelet Processing Method, US Patent 7 184 604.

Ohm, J.-R. (2005) Advances in scalable video coding, *Proceedings of the IEEE*, **93**, January, 42–56.

Reza, A. M. and Zhu, L. (2005) Analysis of error in the fixed-point implementation of two-dimensional discrete wavelet transforms, *IEEE Transactions of Circuits and Systems – I: Regular Papers*, **52**(3), March, 641–655.

Said, A. and Pearlman, W. A. (1996) A new fast and efficient image codec based on set partitioning in hierarchical trees, *IEEE Transactions on Circuits and Systems for Video Technology*, **6**, June, 243–250.

Shapiro, J. M. (1993) Embedded image coding using zero trees of wavelet coefficients, *IEEE Transactions on Signal Processing*, **41**(12), December, 3445–3462.

Spiliotopoulos, V., Zervas, N. D., Andreopoulos, Y., Anagnostopoulos, G. and Goutis, C. E. (2001) Quantization effect on VLSI implementations for the 9/7 DWT filters, in *Proceedings of IEEE International Conference on Accoustics, Speech, and Signal Processing (ICASSP)*, vol. 2, May 2001, pp. 1197–1200.

Strang, G. and Nguyen, T. (1996) *Wavelets and Filter Banks*, Wellesley-Cambridge Press.

Taubman, D., and Marcellin, M. (2002) *JPEG 2000: Image Compression Fundamentals, Standards and Practice*, Kluwer Acadademic Publishers, Boston, MA.

Tsutsui, H. (2005) Implementation of JPEG 2000 Codec for Embedded Systems, PhD Thesis, Kyoto University (in Japanese).

Vishwanath, M. (1994) The recursive pyramid algorithm for the discrete wavelet transform, *IEEE Transactions on Signal Processing*, **42**, March, 673–676.

Xiong, C.-Y., Tian, J.-W. and Liu, J. (2006) A note on 'Flipping structure: an efficient VLSI architecture for lifting-based discrete wavelet transform', *IEEE Transactions on Signal Processing*, **54**(5), May, 1910–1916.

Zervas, N. D., Anagnostopoulos, G. P., Spiliotopoulos, V., Andreopoulos, Y. and Goutis, C. E. (2001) Evaluation of design alternatives for the 2-D discrete wavelet transform, *IEEE Transactions on Circuits and Systems for Video Technology*, **11**(12), December, 1246–1262..

Zhang, Y.-Z. and Xu, C. (2005) *Analysis and High Performance Parallel Architecture Design for EBCOT in JPEG 2000*, ISBN 0-7803-9134-9, IEEE.

18

Ongoing Standardization Efforts

Touradj Ebrahimi, Athanassios Skodras, and Peter Schelkens

18.1 Introduction

The JPEG committee has been continuing its activities on the definition of international standards related to still image coding systems in order to prepare the ground for inter-operable solutions, products, and services that make use of digital images. Currently, the JPEG committee is pursuing three such efforts, namely JPSearch, which deals with annotation, search, and retrieval of digital images and image sequences; JPEG XR, which seeks a solution for compression of high dynamic range images, in environments such as digital photography, where there is a need for solutions with strong complexity requirements; and AIC, which aims at preparing the ground for advanced image compression systems beyond the JPEG 2000 family of standards.

In the following section, each of these activities is briefly described.

18.2 JPSearch

18.2.1 Motivations

Powerful image compression and coding standards, along with rapid evolution in computer technologies, image sensors, storage devices, and networking, have made digital imaging a tremendous success for consumers and businesses. As a result, an impressive growth in personal, professional, and shared digital image collections has been observed in recent years. However, searching and managing these large distributed collections represents a considerable challenge. With the evolution from film to digital camera, it has become easy, fast, and free to take lots of photographs. Yet, to organize the resulting digital photo shoebox, a modern variation of the traditional photo shoebox, remains cumbersome. Photo organizer applications allow for a simple classification of images to generate albums

The JPEG 2000 Suite   Edited by Peter Schelkens, Athanassios Skodras and Touradj Ebrahimi
© 2009 John Wiley & Sons, Ltd

or slideshows, but often require manually annotating the whole collection, a very tedious and time-consuming process. Furthermore, because the portability of data and metadata is not guaranteed, the consumer is de facto locked into one system.

Stock photography companies, such as Getty Images or Corbis, hoast tens of millions of digital photographs. These photographs are labeled and organized based on professionally developed taxonomies. The tremendous cost of generating the annotation is justified and recouped by the high commercial value of the content. More recently, photo sharing websites, such as Flickr and Picasa have gained in popularity, fueled by innovative online community tools that allow users to categorize images by use of tags, in order to allow searchers to find images concerning a certain topic easily. The resulting folksonomy consists of collaboratively generated, open-ended labels authored by the users, and sometimes submitters, of the content. However, again these systems are closed and lock up the user. Recognizing the need for an interoperable and open standard in image search and retrieval systems, JPEG has recently launched a new activity, JPSearch, also known more formally as ISO/IEC 24800 (ISO/IEC, 2006, 2007). The aim of this standard is to allow different image management systems to interoperate. It is foreseen that JPSearch will enable more complete solutions and give consumers and businesses confidence in the longevity of their annotations and collection maintenance effort.

18.2.2   JPSearch Overview

JPSearch aims at interoperability between devices and systems by defining the interfaces and protocols for data exchange between them, while restricting as little as possible how those devices, systems, or components perform their task (Dufaux, Ansorge, and Ebrahimi, 2007). Existing search systems are implemented in a way that tightly couples many elements of the search process. JPSearch provides an abstract framework as well as a modular and flexible search architecture that allows an alignment of system design to a standard framework. In this framework, interoperability can be defined in many ways, e.g. between self-contained vertical image search systems that interact to provide federated search, between the different layers of image search so that these components could be supplied by different best-of-breed vendors, or at the metadata level such that different systems may add, update, or query metadata for images and image collections.

In particular, JPSearch facilitates the use and reuse of metadata. A user makes a heavy investment when annotating a collection of images. With JPSearch, the portability of the metadata is guaranteed, hence allowing a user subsequently to migrate to applications or systems that best suit his/her needs. In community-based image sharing systems, this portability enables the owner of an image collection to merge community metadata back into his/her own management system, hence helping to overcome the manual annotation bottleneck.

Similarly, JPSearch makes possible the use and reuse of ontologies to provide a common language for contexts. Indeed, searching for images always takes place in a context, either implicit or explicit. A common format for handling context allows a user to carry his/her context with him/her to different search engines. It also allows the context to be owned by the user and not by the system, hence protecting the user's privacy.

JPSearch also provides a common query language, giving search providers a reference standard to remove ambiguity in the formulation of a query and to make searching over shared repositories consistent. The common query language also defines a query management process such as relevance feedback.

Finally, by providing a solution for the carriage of image collections and associated metadata between compliant devices and systems, JPSearch enables image search and retrieval functionality across multiple repositories. Therefore, it allows leveraging the generally high cost of creating metadata.

18.2.3 JPSearch Architecture and Elements

A JPSearch architecture has been defined, as illustrated in Figure 18.1. This architecture is generic in the sense that most existing image search systems can be straightforwardly mapped to it. The architecture is divided into four layers, namely:

- *User layer.* The user layer aims at the personalization of the search service.
- *Query layer.* The query layer deals with the formulation of the query and presentation of the results.
- *Management layer.* The management layer handles the distribution of a search task over multiple-image repositories.
- *Content layer.* The content layer includes the image repositories and the associated metadata, as well as the corresponding schema and ontology.

Note that while the query, management, and content layers are in the scope of JPSearch, the user layer is outside the scope. In turn, four independent functional modules are involved in the architecture, as detailed below:

- *Query process.* The query process is part of the query layer. It aims at the efficient execution of search tasks. More specifically, it forms a machine-understandable query from the user's search task, which is then conveyed to the subsequent management and content layers. Conversely, it validates the search results and ranks them according to the user's criteria. Finally, the query process may also generate a new query based on relevance feedback mechanisms.
- *Repository management process.* The repository management process belongs to the management layer. Its purpose is to allow users simultaneously to search multiple distributed image repositories with the same query. Another purpose is comprehensively to aggregate the results returned from the image repositories.
- *Image repository process.* The image repository process is located in the content layer and supplies basic search functionalities. More specifically, this includes receiving a set of queries, executing the matching of these queries with the stored metadata, and forming a result.
- *Metadata creation and update process.* The metadata creation and update process is also part of the content layer. On the one hand, this process enables metadata to be built using a proper schema and ontology definition. On the other hand, this process also provides with the functionality to update the metadata, e.g. by adding, replacing, and removing all or part of the metadata. In particular, whenever the image content is modified, it verifies that the metadata are suitably brought up to date.

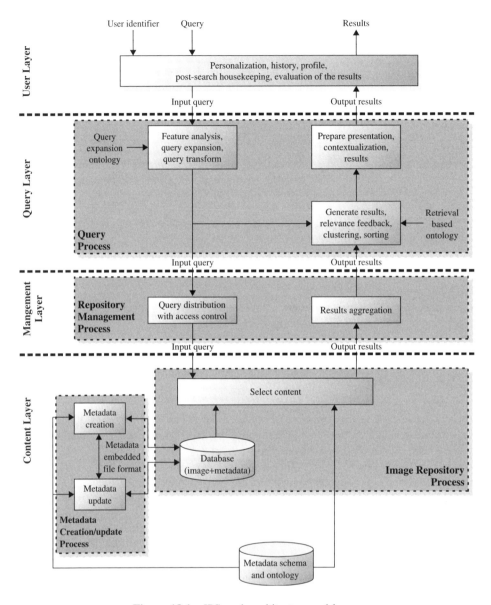

Figure 18.1   JPSearch architecture and layers

18.2.4   JPSearch Specification

JPSearch is a multipart specification. Five parts are currently envisioned, as detailed hereafter:

- *Part 1: Framework and system components.* This part is a type-3 technical report that provides a global view of JPSearch. In particular, the technical report reviews the traditional approaches to image search and motivates the importance of the user in the search process. It describes a number of use cases in order to identify user needs and requirements. It then explains the overall search and management process. This leads to the introduction of a JPSearch architecture composed of four layers: user layer, query layer, management layer, and content layer. Finally, the technical report outlines the organization of the JPSearch specification.

- *Part 2: Schema and ontology registration and identification.* This part standardizes a platform-independent format for the import, export, and exchange of ontologies. It also defines a registry of ontologies that can be imported into a JPSearch compliant system. Finally, it standardizes basic functions to query and manipulate one or more ontologies in a repository.
- *Part 3: JPSearch query format.* This part provides three standardized functionalities between users and image repositories. First, it allows users to express their search criteria. Second, it allows users to describe the aggregated return result sets for user presentation or machine consumption. Third, it defines query management processes such as relevance feedback.
- *Part 4: Metadata embedded in image data (JPEG-1 and JPEG-2000) file format.* This part specifies the image data exchange format with associated metadata to accelerate the reuse of metadata. It supports two functionalities, namely the mobility of metadata and the persistent association of metadata with image data.
- *Part 5: Data interchange format between image repositories.* This part standardizes a format for the exchange of image collections and respective metadata between JPSearch-compliant repositories. The data interchange format enables the synchronization of repositories in order to facilitate simple and fully interoperable exchange across different devices and platforms.

18.3 JPEG XR

JPEG XR – ISO/IEC 29199 – is a recent standardization effort to put forward an image compression algorithm, which provides high compression efficiency while requiring a lower complexity when compared to the JPEG 2000 compression algorithm. The JPEG XR basic approach is similar to that of the HD Photo image compression system proposed by Microsoft Corporation.

JPEG XR is based on a block transform. Its coding structure, which shares many similarities with other traditional image coding techniques, is composed of the following steps: color conversion, reversible lapped biorthogonal transform (LBT), flexible quantization, interblock prediction, adaptive coefficient scanning, and entropy coding of transform coefficients. The distinguishing features are the LBT (Malvar and Staelin, 1989; Malvar, 1998) and advanced coding of coefficients (see Figure 18.2).

JPEG XR supports a wide range of pixel formats, including high dynamic range (HDR) and wide gamut formats (Srinivasan *et al.*, 2008), as well as a number of color formats including RGB and CMYK. To convert spatial domain image data to the frequency domain, HD Photo uses an integer hierarchical two-stage LBT, which is based on a flexible concatenation of two operators: the photo core transform (PCT) and the photo overlap transform (POT). PCT is similar to the widely used DCT and exploits spatial correlation within the block. However, it fails to exploit redundancy across block boundaries and may introduce blocking artifacts at low bit-rates. To alleviate these drawbacks, POT

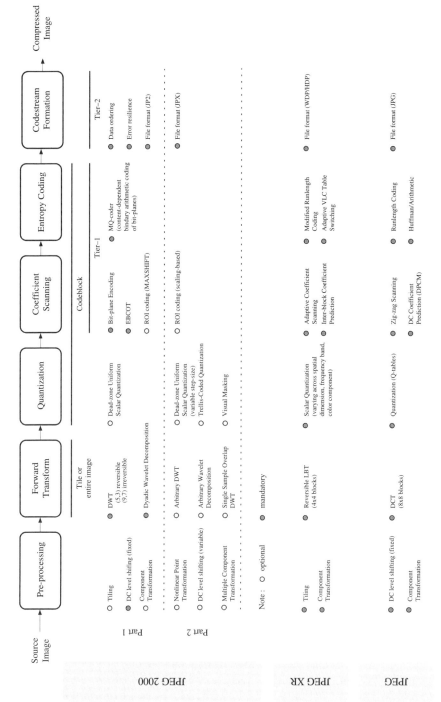

Figure 18.2 Overview of the technologies included in the different components of a JPEG, JPEG XR, and JPEG 2000 encoder

is designed to exploit the correlation across block boundaries. Hence, POT improves compression efficiency, while at the same time reducing blocking artifacts.

The transform is performed in a two-stage hierarchical structure. For the sake of simplicity, we consider the case of the luminance channel. At the first stage, a 4×4 POT is optionally applied, followed by a compulsory 4×4 PCT. The resulting 16 DC coefficients of all 4×4 blocks within a 16×16 macroblock are grouped into a single 4×4 DC block. The remaining 240 AC coefficients are referred to as the highpass (HP) coefficients. At the second stage, the DC blocks are further processed. Another optional 4×4 POT is first performed on the DC blocks, followed by the application of a compulsory 4×4 PCT. This yields 16 new coefficients: one second-stage DC coefficient and 15 second-stage AC coefficients, referred to as the DC and lowpass (LP) coefficients, respectively. The DC, LP, and HP bands are then quantized and coded independently. All transforms are implemented by lifting steps. The chrominance channels are processed in a similar way. Whenever POT and PCT are concatenated, the transform becomes equivalent to LBT. In order to enable the optimization of the quantization parameters (QP) based on the sensitivity of the human visual system and the coefficient statistics, JPEG XR uses a flexible coefficient quantization approach; namely QP can be varied based on the spatial location within the image (e.g. at frame, tile, or macroblock level), the frequency band, and the color channel.

To improve compression efficiency further, interblock coefficient prediction is then used to remove interblock redundancy in the quantized transform coefficients. More specifically, three levels of interblock prediction are supported: prediction of DC subband coefficients, prediction of LP subband coefficients, and prediction of HP subband coefficients. This prediction step can be compared to the intra prediction used in H.264/AVC. Adaptive coefficient scanning is then used to convert the 2-D transform coefficients within a block into a 1-D vector to be encoded. Scan patterns are adapted dynamically based on the local statistics of coded coefficients. Coefficients with a higher probability of nonzero values are scanned earlier. Finally, the transform coefficients are entropy coded. For this purpose, advanced entropy coding using adaptive VLC table switching is used. A small set of fixed VLC tables are defined for each syntax element. The best table is then selected based on local coefficient statistics. The choice of table is computed based on previously coded values of the corresponding syntax element. Furthermore, HP coefficients are layer coded. More precisely, the coefficients are partitioned into two components by adaptive coefficient normalization. After the partition, the significant information is entropy coded and the remainder is signaled using fixed-length codes.

18.4   Advanced Image Coding and Evaluation Methodologies (AIC)

The state-of-the-art in rate-distortion performance assessment has been so far the measurement of the PSNR, MSE, and/or maximum absolute difference (MAXAD) between the original and the decoded images for a given bit-rate in order to obtain objective measures for the reconstruction quality. To estimate the effective visual quality, these measurements are typically combined with subjective visual tests under controlled environments (a well-controlled set of test persons, well-defined viewing distances, and illumination conditions, etc.). Since the latter procedure is time consuming and thus a costly activity, and, moreover, the objective measures are not guaranteeing a reasonable correlation with human visual perception, the image quality research community has massively subscribed itself to the quest for more reliable objective perceptual metrics. In this context, the JPEG committee has preexamined several metrics that potentially contribute to an improved objective quality assessment (De Simone *et al.*, 2008):

- The monochannel *mean structural similarity index* (MSSIM) (Wang *et al.*, 2004), which builds upon the hypothesis that the human visual system is adapted to extract structural information from the scene.
- The monochannel *visual information fidelity measure in the pixel domain* (VIF-P) (Sheikh and Bovik, 2006), which is based on the assumption that images of the human visual environment are all natural scenes with similar statistical properties. Gaussian scale mixture (GSM) models in the wavelet or spatial domain are used to represent the statistical properties by natural scene statistics (NSS) models.
- The monochannel *PSNR human visual system masking metric* (PSNR-HVS-M) is a modified version of the PSNR, incorporating a model of the human visual system contrast sensitivity function (CSF) and contrast masking property (Ponomarenko *et al.*, 2007).
- The multichannel *DCTune metric* (Watson, 1993) is based on the computation of an error map as the arithmetic difference between DCT coefficients of the test and reference images, whereas the error DCT coefficients are weighted by their absolute visibility as a function of the DCT frequency, the display mean luminance, and resolution. Contrast masking further corrects the error sensitivity.
- The multichannel *weighted PSNR metric* weighs separately the contributions from the different component planes (RGB, YcbCr, etc.).
- The multichannel *weighted MSSIM metric* weighs in a similar way the contributions from the different components.

Despite these metrics, investigations by the JPEG committee have shown that further efforts appear to be necessary in order to come forward with reliable and well-characterized perceptual visual metrics.

Advanced Image Coding and Evaluation Methodologies (AIC) is a new work item in JPEG 2000 with the aim to explore those new metrics and test procedures, and moreover to design the next-generation image compression systems beyond those offered by the JPEG 2000 family of standards. The initial goal of this activity is to assess the functionalities of current JPEG, JPEG 2000, and JPEG XR algorithms such as their compression efficiency and other offered features, when compared to state-of-the-art procedures. To this end, a call for evidence for advanced image coding technologies and evaluation methodologies has been issued, with responses expected to include a comparison between various func-tionalities (including compression efficiency) of the JPEG, JPEG 2000, and JPEG XR algorithms compared to alternative technologies and potential candidates for AIC.

The findings of this exploratory phase will be summarized in a technical report, which will include an evaluation process and conditions, as well as the content, evaluation metrics, and results of such assessments. If one or more technologies exist that provide significantly better existing functionalities or new functionalities useful for a large enough class of applications, a call for proposals will be issued in order to create a new image coding standard known as AIC. The objectives of AIC could be either to offer significantly better compression efficiency (e.g. twice or more better) when compared to current JPEG, JPEG 2000, and JPEG XR standards, and/or to offer advanced functionalities such as better, more efficient, or new ways to process and represent new types of images. The standardization timeframe is planned to take place in 2010 or beyond.

References

De Simone, F., Ticca, D., Dufaux, F., Ansorge, M. and Ebrahimi, T. (2008) A comparative study of color image compression standards using perceptually driven quality metrics, in *Proceedings of SPIE on Optics and Photonics 2008, Applications of Digital Image Processing*, vol. 7073, San Diego, CA, August 2008.

Dufaux, F., Ansorge, M. and Ebrahimi, T. (2007) Overview of JPSearch: a standard for image search and retrieval, in *Content-Based Multimedia Indexing, CBMI'07*, London, June 2007.

ISO/IEC (2006) JPSearch Executive White Paper, ISO/IEC JTC1/SC29/WG1N4108, December 2006.

ISO/IEC (2007) ISO/IEC PDTR 24800-1: Framework and System Components, ISO/IEC JTC1/SC29/ WG1, 2007.

Malvar, H. S. (1998) Biorthogonal and nonuniform lapped transforms for transform coding with reduced blocking and ringing artifacts, *IEEE Transactions on Signal Processing*, **46**(4), 1043–1053.

Malvar, H. S. and Staelin, D. H. (1989) The LOT: transform coding without blocking effects, *IEEE Transactions on Acoustics, Speech, and Signal Processing*, **37**(4), 553–559.

Ponomarenko, N., Silvestri, F., Egiazarian, K., Carli, M., Astola, J. and Lukin, V. (2007) On between-coefficient contrast masking of DCT basis functions, in *Proceedings of the Third International Workshop on Video Processing and Quality Metrics for Consumer Electronics (VPQM-07)*, Scottsdale, AZ, January, 25–26, 2007.

Sheikh, H. R. and Bovik, A. C. (2006) Image information and visual quality, *IEEE Transactions on Image Processing*, **15**(2), February, 430–444.

Srinivasan, S., Zhou, Z., Sullivan, G. J., Rossi, R., Regunathan, S., Tu, C. and Roy, A. (2008) Coding of high dynamic range images in JPEG XR/HD Photo, in *Proceedings of SPIE on Optics and Photonics 2008, Applications of Digital Image Processing*, vol. 7073, San Diego, CA, August 2008.

Wang, Z., Bovik, A. C., Sheikh, H. R. and Simoncelli, E. P. (2004) Image quality assessment: from error visibility to structural similarity, *IEEE Transactions on Image Processing*, **13**(4), April, 600–612.

Watson, A. B. (1993) DCTune: a technique for visual optimization of DCT quantization matrices for individual images, in *Society for Information Display Digest of Technical Papers XXIV*, pp. 946–949.

Index

The JPEG 2000 Suite   Edited by Peter Schelkens, Athanassios Skodras and Touradj Ebrahimi
© 2009 John Wiley & Sons, Ltd